GEOGRAPHICAL
INFORMATION SYSTEMS

GEOGRAPHICAL INFORMATION SYSTEMS

Principles, Techniques, Management, and Applications

Second Edition, Abridged

Edited by

PAUL A. LONGLEY,

MICHAEL F. GOODCHILD,

DAVID J. MAGUIRE,

DAVID W. RHIND

John Wiley & Sons, Inc.

Published by John Wiley & Sons, Inc., Hoboken, New Jersey
Published simultaneously in Canada

For general information about our other products and services, please contact our Customer Care Department within the United States at (800) 762-2974, outside the United States at (317) 572-3993 or fax (317) 572-4002.

Wiley also publishes its books in a variety of electronic formats. Some content that appears in print may not be available in electronic books. For more information about Wiley products, visit our web site at www.wiley.com.

ISBN-13: 978-0-471-73545-8
ISBN-10: 0-471-73545-0

Printed in the United States of America

10 9 8 7 6 5 4 3 2 1

Contents

Part 2: Technical Issues

(a) GIS architecture issues

(b) Spatial databases

(c) Technical aspects of GIS data collection

(d) Data transformation and linkage

Part 3: Management Issues

NOTE: Chapters marked with a CD-ROM icon in this Table of Contents are only available as full-text PDFs on the CD-ROM included with this book. Color images referred to in some of the chapters (denoted as "Plates" in the text), are only available in a digital image gallery on the included CD-ROM.

New Developments in Geographical Information Systems: Principles, Techniques, Management and Applications

P A LONGLEY, M F GOODCHILD, D J MAGUIRE, D W RHIND[1]

1 INTRODUCTION

We are delighted that the publisher of **Geographical Information Systems: Principles, Techniques, Management and Applications** (the 'Big Book' of GIS) has decided to publish this abridged edition of the original two-volume set. Since its publication in 1999, we believe that the title has become established as the premier reference compendium in its field, continuing a tradition begun with **Geographical Information Systems: Principles and Applications** in 1991. Both titles, we hope, are part of the history of GIS: as encyclopaedic reference works they have provided reliable reference points during the evolution of the subject; and as thoroughly cross-referenced essays, commissioned from leading authors, we think they still bear comparison with the best of all the encyclopaedias and handbooks that are now appearing in an increasingly crowded market place.

Our motivations for producing this abridged version are straightforward and simple. Following publication of this major work in 1999, we produced what we believe was an equally successful textbook, designed as an advanced introductory guide to GIS – the marketplace seemed to think this too, for that title, **Geographic Information Systems and Science,** sold over 25,000 copies in the three years following its publication. In the meantime, the 'Big Book' experienced a sustained level of sales to libraries and research organisations, and in 2004 a successful Chinese language version was published. We thought of the textbook (Longley, Goodchild, Maguire and Rhind 2001) as providing a primer to the fast-developing field of GIS, while the 'Big Book' would remain a key reference point that would reinforce its key concepts, techniques and management practices. Fundamental to our view of GIS is that principles and best practices are enduring, while the more ephemeral issues that pervade more techno-centric guides are not. Thus when the success of the textbook led the publisher to ask us for an expanded Second Edition (Longley, Goodchild, Maguire and Rhind 2005), we were keen to maintain a link with the research compendium that has been formative to our own understanding of GIS.

We were also keen that the book reach the widest possible audience – for example, individuals who would use the material to provide background information while undertaking courses in GIS education; those working in institutions that are new to GIS and might not have access to a copy of the original boxed set in their libraries; and small organisations for whom the pricing of the original work was prohibitive. We are very pleased that the publisher, John Wiley and Sons, Inc., was amenable to the view that their substantial initial investment in the second edition of 'Big Book' had been largely amortised, and that a low-cost abridged version could be published. We should add that we are grateful that similar magnanimity underlay their agreement to post those chapters of **Geographical Information Systems: Principles and Applications** (the first edition of the work: Maguire, Goodchild and Rhind 1991) on the Website of the textbook (www.wiley.com/go/longley), some years before free downloads became widespread. The economics of this project have nevertheless dictated that we reduce the number of chapters that are available in print form – although all of the contents of the original work are retained on the CD that is provided with the book.

---

[1] Incorporating written suggestions of many of the contributors to the second 'Big Book' of GIS.

GIS is, without doubt, a fast-moving field, albeit one that is centred upon principles, techniques and management practices that are less transient than its software or its technology. We were guided by three criteria in selecting the chapters that would be retained in print form. Our first criterion was the relevance of the material that each retains. Much has happened since this work was first published in 1999, and it would be foolish to suggest that the practices of actually *doing* GIS have also not developed and changed in recent years. Because the use of GIS is now so pervasive, it is also the case that GIS is rarely thought of as a novel solution. For this reason, we have retained the original *Applications* part of the book in the CD-ROM only. Interested readers are invited to consult the Applications boxes in our textbook (Longley, Goodchild, Maguire and Rhind 2005) for recent case studies that we think exemplify particularly new or novel characteristics of real-world applications. Our second criterion concerned the balance of this abridged edition: we wished to retain material that was representative of all of the parts and sections in all of the remaining three parts (Principles, Techniques and Management) of the original. The third criterion was the degree to which the material directly reinforces that discussed in the second edition of *Geographic Information Systems and Science* – in order to reinforce our own view that the two works can be viewed as complementary. Thus the reader will find that nearly all of the chapters included in this abridged edition are cited in the *Guides to Further Reading* that are at the chapter ends in the Longley, Goodchild, Maguire and Rhind (2005) textbook.

In most all instances, we believe that the textbook enables the full benefits of this reference work to be unlocked. In the remainder of this extended update, however, we provide some additional observations that help to update the chapters in the original. These have been provided with the very significant assistance of the authors of many of the original chapters. We are very grateful, therefore, to the following for their inputs: Prue Adler, Luc Anselin, Richard Aspinall, Kate Beard, Yvan Bédard, Tor Bernhardsen, Antonio Câmara, Heather Campbell, David Coleman, Michael Curry, Peter Dale, Bill Davenhall, Ian Dowman, Sue Elshaw Thrall, Robert Fincham, Manfred Fischer, Peter Fisher, Pip Forer, Greg Forsyth, Art Getis, Steve Guptill, Gerard Heuvelink, Ron Johnston, Russ Johnson, Menno-Jan Kraak, Werner Kuhn, Art Lange, Mary Larsgaard, Robin McLaren, Dave MacDevette, Jeff

Meyers, Helena Mitasova, Nancy Obermeyer, Peter van Oosterom, John Pickles, Mark Sondheim, David Swann, Grant Thrall, David Unwin, Nigel Walters, Rob Weibel and Mike Worboys.

2 PRINCIPLES (PART ONE)

(a) Space and time in GIS

A significant proportion of the opening part of the work focuses upon the disciplinary setting of GIS. It is interesting that most of the recent handbooks and encyclopaedias on GIS pay rather less attention to issues of spatial representation and the apparent central relevance of geography to them. Indeed, whilst some (e.g., Bossler et al 2002) are undoubtedly *about* GIS as an area of activity, they prefer to use the term 'geospatial' (implying a subset of the adjective 'spatial' applied specifically to the Earth's surface and near-surface) rather than 'geographical' – particularly when referring to technologies or generic techniques.

On balance, we think this is a pity – the term 'GIS' is distinctive, widely recognised in government and business, and suggests the clear and demonstrable link to *geographical information science* (GIScience) that underlies our own view of the development of the field. We suggest that the newer term, by contrast, which fails to confer any unequivocal advantages, is likely to confuse the uninitiated and seemingly presents applications as devoid of real-world scientific context. For us, the term 'geospatial' seems to imply a preoccupation with technology and low-order data concepts, rather than the higher levels of the chain of understanding (evidence, knowledge and wisdom) that we have discussed elsewhere (Longley, Goodchild, Maguire and Rhind 2005: Section 1.2). A new term may nevertheless be attractive to some, such as those developing instruments and technologies (e.g. GPS and remote sensing) associated with academic disciplines other than geography, and new business entrants to the field that may not have track records in GIS.

The rise of the term 'geospatial' also seems to convey to some a greater sense of 'hard science' and has sometimes gone un-noticed in the discipline of Geography. In some parts of the world (but notably not the United States), it seems to us that the discipline of Geography has developed a very complacent attitude to investment in its flagship specialism: GIS is widely recognised in high schools, is central to transferable geographical skills, is key to outreach to potential new recruits to the discipline and is crucial to justification of education funding as a part laboratory-based discipline. We observe that a decreasing real share of new student recruitment, a need to recruit skilled researchers from other disciplines, and a downturn in real share of research awards are seeming consequences of such neglect.

These are concerns with GIS as an applied problem-solving technology, and are thus not the central preoccupations of Helen Couclelis (Chapter 2: *Space, time, geography*) or Ron Johnston (Chapter 3: *Geography and GIS*). But these concerns nevertheless have ramifications for the practice of Geography and the vitality of the discipline. Today, the Geography taught in many quite respectable universities largely fails systematically to address the core organizing principles that are inherent to any widely recognized definition of the subject (see also the contribution of Pip Forer and David Unwin, Chapter 54: *Enabling progress in GIS and education*). Ron Johnston observes the periodic disciplinary rumblings about the inevitable ethical inconsistency of anything that can be measured, allied with assertions that quantitative measurement is either passé, no longer practiced, or relevant only to 'thin' preliminary description of geographical problems (e.g., Cloke et al 2004). Some traditional quantitative geographers have suggested that numbers are relevant to the vitality of the discipline, in ways that scarcely begin to capitalize on the potential and achievements of GIScience (Johnston et al 2003), while some external observers remain bemused at the failure of the discipline to agree on issues of content, coherence and relevance. Meanwhile, GIS practitioners continue to respond to interest within other disciplines and remain part of the GIS success story.

Six years on, it seems odd to recall the passions that were sparked by the early debates between social theorists and GIScientists, and that were so accurately recorded in the chapter by John Pickles (Chapter 4: *Arguments, debates and dialogues: the GIS–social theory debate and the concern for alternatives*). The dialog that began with the Friday Harbor meeting in 1993, and continued through Research Initiative 19 of the National Center for Geographic Information and Analysis (NCGIA) has continued, and today it would be inconceivable that a text in GIS written by a geographer would not address the social context and social impacts of the technology. As the author writes, referring to the well-known arguments presented by C.P. Snow in the 1950s about deep divisions in society between the technological orientation of the sciences on the one hand and the more reflective orientation of the humanities on the other (Snow 1961), 'the best of our students and young faculty members working in GIS and social theory are no longer...wrapped up in the 'Two Cultures' arguments and ways of thinking.'

Recently several significant new directions have emerged in the dialog. Most conspicuous of these, perhaps, is the growing interest in actor-network theory and the work of Bruno Latour, led by such GIScientists as Chrisman, Harvey, and Schuurman. It asks, in essence, how advances in GIScience can be understood in the context of the individuals who made them, their networks of interaction, and other aspects of their social setting. What, for example, can we learn from the particular circumstances surrounding Roger Tomlinson's proposal for a Canada Geographic Information System in the 1960s (Foresman 1998), or Ian McHarg's (1995) interest in modeling planning decisions through the overlay of transparent representations of different variables? Who were they talking with in that period, and what was the nature of their relationships with others?

Pickles also notes a third recent trend: '...the convergence of discussions about information and communication technologies, public-participation GIS (PPGIS), and broader writings about representational technologies. One branch of PPGIS has to do with issues of cost, access, open source, technical assistance, user-interface design, the Web, etc. These have moved along quickly, as [we] know. Another part of this, however, is the question of agency and voice – crudely put, who acts, who decides, who designs, and who inputs?' This second area seems to Pickles to be moving much more slowly, but the questions it asks seem more important, and he anticipates increasing interest in the years to come. Issues of public participation in GIS are addressed in detail by Craig et al (2002).

Several real breakthroughs have been achieved in the past six years in the area of map generalisation. Notable among them are the work of Bader and Barrault on the application of energy-minimisation techniques from engineering physics, and the careful analysis of feature shape employed in techniques developed for road generalisation at IGN. Progress has also been made in algorithms to control the generalisation process, by exploiting optimal selections and combinations from among the host of specialized techniques that have been developed by researchers over the past two decades. Ware et al (2003) used combinatorial optimisation techniques including simulated annealing and genetic algorithms, integrating several algorithms into a comprehensive process. In the AGENT project (http://agent.ign.fr), a multi-agent system was built that is even more flexible and basically allows integrating any generalisation algorithm into a comprehensive process using a constraint-based approach (Barrault et al 2001).

Author Robert Weibel writes: 'there is an increasing demand by map and data producers, in particular public mapping agencies, to be able to automate the production process of maps and databases more thoroughly, owing to new product requirements such as data or maps for navigation and for location-based services (LBS). In response to these requirements, the leading software manufacturers have improved their products considerably in recent years, both in terms of functionality and quality. For example, Laser-Scan, which has a tradition of concentrating on the mapping market, was a partner in the AGENT project. The agent-based prototype system of AGENT has since been turned into a commercial product called Clarity. Intergraph, another software vendor with a strong presence in the mapping market, offers its own generalisation software under the name DynaGen. Even ESRI, where mapping is but one of many markets, has added more generalisation-related functions into its system in recent years. Additionally, there are smaller vendors that offer their own solutions, such as Axes Systems with the Axpand software.'

Generalisation research is no longer restricted to automation of the production of paper maps, but is moving in new directions in response to new products, technologies, and demands. The extremely constrained screen area and resolution of hand-held devices, such as cellphones, is one such area, and others include automated generalisation of three-dimensional objects such as buildings, and the creation of the multi-resolution databases that are needed in vehicle guidance systems.

By 1999 it was abundantly clear that the Internet and Web were having a profound impact on many aspects of human activity. Over the past six years since the original publication of this book, several

new directions have emerged in cartographic research and practice, driven in large part by the potential of these new communication technologies. As we also comment elsewhere in this review and update, dramatic reductions in the cost of software, and simultaneous increases in power, have enabled large numbers of non-experts to engage in map-making, and to make their products freely available to others. A specific design approach has emerged, motivated by the limited bandwidth of the Internet on the one hand, and by easy access to techniques of transparency, blinking and shading on the other. The Scalable Vector Graphics (SVG) format has been adopted widely in the cartographic community, leading to increasing facility of interaction and interoperability.

'In the GIScience community', writes author Menno-Jan Kraak (Chapter 11: *Visualising spatial distributions*), 'the abundance of geospatial data has created a new role for maps in exploratory environments where they are used to stimulate (visual) thinking about geospatial patterns, relationships and trends. The context where maps like this operate is the world of *geovisualisation* (Dykes et al 2004) which can be described as a loosely bounded domain that addresses the visual exploration, analysis, synthesis and presentation of geospatial data by integrating approaches from disciplines including cartography with those from scientific visualisation, image analysis, information visualisation, exploratory data analysis and GIScience.'

(b) Data quality

Interest in representation and ontologies is also relevant to the issues of data quality and uncertainty that are inherent in them. Shi et al (2002) provide perhaps the starkest statement of this problem when they claim that 'for many types of geographical data, there is no clear concept of *truth*, so that models that address the differences between measurement and truth are clearly inappropriate' (cited by Heuvelink 2003, 817). This invites consideration of semantics (termed *discord* in Peter Fisher's Chapter 13: *Models of uncertainty in spatial data*) and how discord arises through the social construction of information (also alluded to by John Pickles in his comments, above). Peter Fisher's contribution remains a very useful framework for thinking about uncertainty in GIS. However, the Shi et al (2002) volume now provides an additional valuable and wide-ranging update on spatial data quality issues of theory, method and application, and provides evidence that we now have a much fuller and more rounded view of the sources, operation and consequences of uncertainty in GIS. This volume also demonstrates the broadening of interest that has occurred in the topic over time, and illustrates how we have moved well beyond preoccupations with uncertainty propagated in map overlay operations and statistical models for representing positional uncertainty (discussed by Howard Veregin in Chapter 12: *Data quality parameters*). Today, there is a much more established interest in visualisation and communication of uncertainty, the pitfalls of decision-making under uncertainty, and the quest to develop error-sensitive GIS.

Within the statistical perspective, Gerard Heuvelink's treatment of uncertainty (Chapter 14: *Propagation of error in spatial modelling with GIS*) remains a holistic overview of the sources and operation of quantitative attribute errors that is of enduring importance. He identifies three respects in which the methodological discussion of Chapter 14 might be updated. First, he observes that Monte Carlo simulation is very much taking over from the Taylor series approximation method and other more analytically based methods for uncertainty propagation analyses. Algorithms for stochastic simulation of spatial phenomena, he argues, have become much richer over recent years – particularly with regard to problems involving categorical spatial data and the stochastic simulation of objects. Second, and related to this, he notes progress in problems of uncertainty involving categorical spatial attributes (e.g. Kyriakidis and Duncan 2001). He observes that problems of uncertainty with categorical attributes are more difficult than their continuous-attribute counterparts. Third, he sees progress in handling error propagation arising out of positional uncertainty, as discussed by Shi et al (2002).

More generally, the recent literature provides some evidence that spatial uncertainty analysis is enjoying greater real-world professional application. However, it is the case today that uncertainty analysis remains the preserve of the GIS specialist, and that it remains a labour-intensive task. Time will tell whether uncertainty analysis and risk analysis will become more standard elements in spatial analysis.

Kate Beard and Barbara Buttenfield's contribution (Chapter 15: *Detecting and evaluating errors by graphical methods*) remains an important early graphical evaluation of uncertainty (or more specifically, errors). In 1999 it seemed that further advances in the detection and evaluation of errors by graphical methods depended on further refinement of error models. Since then, new methods and models have been developed, and a general trend is evident away from summary-level measurements, such as root-mean-square error (RMSE), toward local statistics that reflect non-stationarity. They concur with Gerard Heuvelink that greater quality assessment is now supported through use of Monte Carlo simulations to generate multiple realisations of error processes. Author Kate Beard writes 'These improvements have raised new challenges for graphic display. Finer detail in spatial variation can be more difficult to communicate and the side-by-side and sequenced graphic displays originally proposed for quality depiction are less able to support user perception and association of fine spatial variations in quality with the data distribution. With simulations the visualisation challenge is to convey effectively the uncertainty represented by large sets of possible realisations.'

Progress has been made in the development of techniques linked to specific error types and assessment contexts, and such work has become more central to the visual analysis of spatial distributions discussed in Section 1(a) above. For example, Menno-Jan Kraak (Chapter 11: *Visualising spatial distributions*) has recently described a visualisation tool for fuzzy attribute classification that uses a collection of multiple and dynamically linked visual displays including images, parallel coordinate plots, and a 3D feature space plot that users can interact with to explore the classification process. Rather than simply viewing static displays of error, this approach creates an environment for interactive exploration that potentially leads to a better understanding of uncertainty.

Gerard Heuvelink, Kate Beard and Barbara Buttenfield share a frustration that little of this research has found its way into the commercial GIS packages and into mainstream use. The need for quality assessment has if anything grown in the past few years with the explosion of Web applications and the growing availability of spatial data. While the majority of the spatial data distributed over the Web now have metadata associated with them, quality descriptions are not as complete or as useful as they might be for prospective users. Work by

Devillers et al (2002) and Bédard et al (2004) makes some progress on this front. Their work targets data-quality exploration in an interactive environment associated with a SOLAP (spatial on-line analytic processing) architecture that allows users to isolate and explore individual data-quality variables or to view data-quality metadata at several different aggregation levels from an individual data value, to object classes, to an entire data set.

(c) Spatial analysis

Spatial analysis is very much the engine that drives research applications of GIS. In technological terms, the chapters in this section assimilate many of the implications of the ongoing improvements in computer memory and processing speed, and these developments have continued to impact profoundly upon the field of GIS. The chapter by Stan Openshaw and Seraphim Alvanides (Chapter 18: *Applying geocomputation to the analysis of spatial distributions*) exemplifies an approach to spatial analysis that continues to develop using the increased power of computing: although widely applicable analytical solutions to the modifiable areal unit problem remain as elusive as ever, the geocomputational approach to zone design provides an example of the use of research techniques in spatial analysis in applied problem solving – for example, the 2001 UK Census zones were designed around some of the principles set out in Chapter 18.

The contributions to this section of the book also variously flag a number of longstanding issues about the ways in which we carry out our scientific investigations and seek to improve the kind of findings that we are able to generate. The linkage of these issues to those of spatial, temporal and cognitive representation, and to visualisation (all discussed in Section 1(a) of this book) has strengthened in recent years. This is largely because improvements in computation have facilitated greater disaggregation and a greater focus upon individuals and micro-scale events and occurrences, alongside improved representation of temporal dynamics and, in socio-economic applications, more realistic simulation of spatial behaviour.

Developments in digital data infrastructures have also fuelled interest in spatial analysis techniques that perform well using large numbers of georeferenced observations. Data-mining techniques (Miller and Han 2001) in the geocomputational paradigm

provide the most obvious examples. Tremendous progress has been made in the past six years in the development and dissemination of readily accessible tools for advanced forms of spatial analysis. Luc Anselin's (Chapter 17: *Interactive techniques and exploratory spatial data analysis*) GeoDa consolidates and extends the available set of tools for exploratory spatial data analysis, local statistics and spatial regression. It has been developed through the Center for Spatially Integrated Social Science (www.csiss.org) and has been downloaded over 4,000 times. At the same time, ESRI's ArcGIS 9.0 has greatly expanded the set of tools available either as extensions or as parts of the core of this popular GIS; GeoVISTA Studio from Pennsylvania State University has become a powerful collection of open-source tools with an emphasis on visualisation (www.geovistastudio.psu.edu); and Serge Rey's STARS (Space-Time Analysis of Regional Systems; stars-py.sourceforge.net) adds another open-source package focused on statistical techniques to this rapidly growing collection.

Historically, GIS has provided a medium within which spatial analysis techniques, that often predate the innovation of GIS by decades, could be rehabilitated for real-world problem solving in data-rich environments. The pace of development is evident in the increasing recognition of the importance of space in academic disciplines outside Geography. As this rehabilitation nears its completion, the techniques that are most used are those that deliver the most in real-world applications. Thus while some of the techniques reviewed by Art Getis (Chapter 16: *Spatial statistics*) and Manfred Fischer (Chapter 19: *Spatial analysis: retrospect and prospect*) have withered in usage, others – notably Bayesian and non-parametric statistics, have become much more widely used. It is also important to note that the software environment of GIS is also leading to the development of new spatial analysis methods – geographically weighted regression (Fotheringham et al 2002) being perhaps the best example. This is an example of a technique that allows the unique characteristics of localities to be examined within what is ultimately a global statistical generalisation – this kind of statistical sensitivity to context is also illustrated by locally sensitive measures of spatial autocorrelation, such as the Ord and Getis O statistic (Ord and Getis 2001).

Finally, it is also clear that practical problems of

extending theory in applied contexts and accommodating ethical concerns are becoming increasingly important: problems of ownership of data, copyright, and the limits posed by changing conceptions of ownership are changing the way we are able to share and communicate spatial information, and require ever more of spatial analysts in relation to privacy issues (see Michael Curry, Chapter 55: *Rethinking privacy in a geocoded world*).

3 TECHNIQUES (PART TWO)

(a) GIS architecture issues

As suggested above, the original edition of this book postdated the innovation of the Internet, but far-reaching changes in peer-to-peer networking (David Coleman, Chapter 22: *GIS in networked environments*) have occurred in recent years. The notion of using peer-to-peer technology and standards to access distributed spatial data holdings began attracting the attention of the GIS community in late 2000. In suggesting the scale of change that has occurred since this book was first published, David Coleman cites OECD statistics that suggest the number of Internet subscribers worldwide has more than tripled to over 300 million since 1999, and the number of those subscribers with broadband connections has increased to over 100 million over the same period. He points out that the sharing of digital data (whether as downloaded music, movies, images, games, software or geographical data) through peer-to-peer networks continues to increase at an unprecedented rate: the number of people logged on simultaneously to popular file-sharing networks approached 10 million in April 2004, a rise of 30% from the same period a year earlier.

GIS interoperability (Mark Sondheim, Kenn Gardels and Kurt Buehler, Chapter 24: *GIS interoperability*) is key to driving such data exchange. Using Open Geospatial Consortium (OGC; www.opengeospatial.org), U.S. Federal Geographic Data Committee (FGDC; www.fdgc.gov) and ISO specifications along with IT industry standards (e.g. SOAP/XML, .Net and Java), emerging Web mapping services based upon common standards are enabling users not only to access geospatial data from different servers around the world, but also to determine their fitness for a particular application,

conflate (not just register) them to a common base, and then communicate the results to others. Geoportals (Maguire and Longley 2005) such as Geospatial One Stop (www.geodata.govv: see also Section 4(a) below) have become important ways of facilitating access to data (see also David Rhind, Chapter 56: *National and international geospatial data policies*). Much of this progress has been driven by developments in the wider information technology arena, and their adaptation to the special needs of GIS. For example, the success of XML (eXtensible Markup Language) has led to the development of GML (Geography Markup Language) as a GIS data transfer format; and several of the leading database management systems have added spatial capabilities, in the form of IBM DB2 Spatial Extender and Oracle Spatial, for example.

Author Mark Sondheim writes 'The Open Geospatial Consortium and related ISO (International Standards Organisation) activities have been increasingly successful in developing specifications highly relevant to GIS interoperability. The most popular of these is certainly the Web Map Service; however, Web Feature Service, Web Coverage Service, Location Based Services, Web Coordinate Transformation Service, Web Registry Service (as a profile of the OGC Catalogue Service) and others are beginning to be implemented as well. Of course GML is a key development of the OGC and integral to some of these services. It is of note that both commercial and open-source versions of some of these services are now available. This bodes well for their broader acceptance.'

An important area of GIS that we did not anticipate in 1999 is the convergence of networked GIS with communications and positioning technologies in *location-based services*. Today, GIS data and software are increasingly accessed remotely, allowing the user to move away from the desktop and hence to apply GIS anywhere – as 'distributed GIS'. Limited GIS services are already available in common mobile devices such as cell phones, and are increasingly being installed in vehicles. In the future it is clear that GIS will become both more mobile and ubiquitous, and will be based around distributed data, distributed users and distributed software (Longley, Goodchild, Maguire and Rhind 2005: 241-59). Author David Coleman notes that the number of cell phone users worldwide has almost quadrupled since this book was published in 1999;

and that many of these cell phones contain tiny Global Positioning System (GPS) receivers that allow service providers to know the location of the caller in the event of an emergency or (with user consent) to monitor his or her activity patterns. In a similar way, sensor webs are being created, composed of intelligent digital devices that exchange many kinds of information with people and other machines around them. These are creating new types of networks which will ultimately provide important sources of input to GIS and Web-mapping systems.

Distributed GIS offers enormous advantages, in reducing duplication of effort, allowing users to take advantage of remotely located data and services through simple devices, and providing ways of combining information gathered through the senses with information provided from digital sources. However, progress on a number of fronts is difficult because of complications resulting from the difficulties of interacting with devices in field settings, limitations placed on communication bandwidth and reliability, and limitations inherent in battery technology. Thus for the foreseeable future, we are likely to continue to associate GIS with the desktop, where rapid developments in software (Elshaw Thrall and Thrall, Chapter 23: *Desktop GIS software*) remain central to the development of the field.

Sue Elshaw Thrall and Grant Thrall reflect that the compounded effect of developments in desktop software has led to a bifurcation in GIS. On the one hand, specialised GIS software continues to add spatial analysis applications while, on the other, many core functions are no longer recognised as GIS *per se*. With respect to the latter, they point out that end mass-market GIS users are only rarely aware that they are using high-technology geography. Thus interactive street displays and route finders are standard fixtures in up-market automobiles; camping stores sell GPS-enabled wristwatch-like devices that use GPS and digital terrain maps for off-road route finding; joggers use similar devices as high-tech alternatives to the pedometer; and Internet mapping has become ubiquitous, through services such as mapquest.com and yell.com. Perhaps most impressive of all, Microsoft's MapPoint finds routes, calculates drive times, and interfaces with both GPS devices and standard Excel software, thereby becoming much more than geographically enabled spreadsheets. MapPoint is not yet presented as a serious contender to challenge mainstream GIS functionality: but if and when the general public becomes familiar with and expects GIS functionality, MapPoint will be there to seamlessly deliver to and interface with Microsoft Office applications.

An update of core GIS software offerings and the ways in which they may be customised (David Maguire, Chapter 25: *GIS customisation*) is provided by Longley, Goodchild, Maguire and Rhind (2005: 157-75). Proprietary geographically enabled programming languages have been replaced with integration of GIS functionality within standard programming languages such as Microsoft's Visual Basic, Visual C++, C# and Java. These development tools have assisted deployment of GIS via wireless interfaces to a range of hand-held devices, and as such have contributed to the development of location-based services. A major new development in this arena is the adoption of distributed architectures (such the .Net and Java frameworks) for implementing enterprise systems. Web services and service-oriented architectures (SOA) are the underlying technologies that sew together legacy and new systems.

A final point is that GIS software and data products are sensitive to the end uses to which they are put. Recent years have seen increasing usage of GIS in the areas of business and business geography (Thrall and Campins 2004). It is sometimes the case that such users use different spatial terminology (e.g. 'trade areas') and are reluctant to become acquainted with the established terminology of GIS. Thus GIS market development has entailed repackaging of standard offerings into business-specific application packages, such as ESRI's Business Analyst, that use novel user interfaces to guide the user through solving a business problem, rather than requiring mastery of a list of GIS functional logic.

(b) Spatial databases

Relational database technology (Mike Worboys, Chapter 26: *Relational databases and beyond*) remains and looks set to continue to be at the heart of geographical data management and analysis for the foreseeable future. Author Mike Worboys reflects that object-oriented database management systems (DBMS) have not turned out to be replacements for relational systems. Instead, he sees that the most popular model, and that implemented by the large database players, has been to incorporate object-oriented concepts in extended relational database systems. The challenges set out in Mike Worboys' chapter still remain: constraint databases remain an elusive route towards incorporating more expressive power; and the dynamic nature of the world is still not addressed by the current round of database technology (e.g. see Worboys 2005). Elsewhere more progress has been in evidence: geosensor data management suggests clear and important goals for the future of database science in the context of real-time data management; the linkage between agent-based computational models and database technology provides interesting possibilities for the management and analysis of dynamic spatial data; and event-oriented models are becoming important for the development of our ability to represent and reason about dynamic geographical phenomena.

Most of the current generation of DBMSs in commercial systems (e.g. Oracle, Informix, DB2, etc.) and open source software (e.g. PostgreSQL, MySQL) support geographical data to some degree. This represents a significant improvement on the situation when Peter van Oosterom (Chapter 27: *Spatial access methods*) prepared his contribution, although the spatial indexing methods that are supported remain limited to variants of simple grids, R-trees and quadtrees. He notes that there remain challenges to using these methods to manage a greater range of topological structures, such as linear networks and TINs, which have been responded to in part by Oracle's, LaserScan's and ESRI's commercial systems. He sees a growing areas of interest in creating methods to support 3D data within GIS and geo-DBMSs: this entails implementation of appropriate data types (e.g. polyhedra), operators (e.g. for computing volume and intersection), 3D spatial access methods and complex 3D structures (polyhedral partitions of 3D space, TINs, etc.). He also notes that progress towards the development of multi-scale spatial storage and access methods has been very limited.

The past six years have seen rapid development and adoption of the database design environments described by Yvan Bédard (Chapter 29: *Principles of spatial database analysis and design*). Modelling techniques are now generally accepted in the development of geographic databases for GIS, LBS and other position-aware applications, and several GIS products now include modelling tools. Over the past six years, UML has clearly established itself as a standard and is nowadays used on a regular basis

by the GIS industry, academia and standardisation bodies like the ISO/TC-211 and OGC. A good example of this evolution is the thousands of downloads of the spatially extended UML CASE tool 'Perceptory' and its use in over 40 countries – a major increase in usage rates since the late 1990s. In the meantime, very flexible approaches to system design have emerged in reaction to the highly disciplined approaches typically represented by RUP (Rational Unified Process). These light but robust approaches, stemming from the philosophy of Extreme Programming, are described by the umbrella terms 'Agile Development' and 'Agile Modelling'. Yvan Bédard observes that numerous books describing these new approaches have appeared in the past six years, including books specifically focused upon agile database techniques. Books comparing 'heavy methods' with 'light methods' have also appeared in order to appraise users of their relative merits.

New sources and types of data may generate the need for new access methods. In particular, Peter van Oosterom identifies the need to support a basic 'point cloud' data structure, in order to manage the massive amounts of point cloud data that are now collected by laser scanning (or multi-beam sonar), and for which traditional (raster and vector) data structure access methods are not very effective.

Human interaction with GIS (Max Egenhofer and Werner Kuhn, Chapter 28: *Interacting with GIS*) is also changing in detail, if not in its fundamental nature. Author Werner Kuhn identifies a number of developments in current thinking about user interface issues. First, he observes that the user interface of many applications is now a Web browser, or at least is being accessed through one. This has generally improved the possibility of transferring knowledge from one application to another (and thereby the usability of both). Second, Web interfaces have also led to an increased emphasis on the use of remote services in preference to locally installed functions. This has had the effect of vastly simplifying the functionality of GIS. Third, the range of devices and their user interfaces has broadened: some of the most challenging user-interface problems are now those of small, mobile devices such as cellphones (see the discussion of location-based services in Section 2(a) above). Fourth (and also echoing David Coleman's comments in Section 2(a) above), user interaction is increasingly becoming implicit, as more and more information about users becomes available through sensors, including the determination of user location. Fifth, 'clickable' maps and other forms of dynamic user interaction are now commonplace. And sixth, interactive visualisation of 3D and 4D data remains generally awkward and confined to research prototypes.

(c) Technical aspects of GIS data collection

There have been significant developments in the image data available for use in GIS, subsequent to the publication of Ian Dowman's chapter (Chapter 31: *Encoding and validating data from maps and images*). He observes that not only has photogrammetry moved into the digital age, but a number of new and important satellite sensors have come into service (e.g. see Donnay et al 2001). Digital photogrammetric workstations were available at the time that this book was originally published, but these have now become standard for photogrammetric map production in most developed countries. In very recent years digital airborne cameras have reached the market and are already making an impact by cutting out film processing and extending the range of conditions under which photography can be obtained. Ian Dowman notes that airborne LiDAR and interferometric synthetic aperture radar (SAR) are now used for directly generating digital elevation models (DEMs) and have allowed dense networks of accurate elevation points to be obtained for applications such as 3D city models and flood prediction and management. High-resolution optical sensors on satellites, with pixel sizes of 0.6m, are now available, thus extending the accuracy of images available from space (cf. Mike Barnsley, Chapter 32: *Digital remotely-sensed data and their characteristics* and Chapter 48: Jack Estes and Tom Loveland: *Characteristics, sources, and management of remotely-sensed data*). More DEMs are now available using data from space sensors, the most notable being a near-global DEM at 90m spacing and 10m vertical accuracy from the Shuttle Radar Topography Mission (SRTM); these data are freely available over the Internet.

While there have been a number of technical improvements in Global Positioning System equipment, the basic technology fundamentally remains the same (Art Lange and Chuck Gilbert, Chapter 33: *Using GPS for GIS data capture*). Art

Lange observes that there are now more and better sources of differential correction such as the Federal Aviation Administration's Wide Area Augmentation System (WAAS) and its counterpart in Europe and Japan. Virtual Reference Station (VRS) networks are another new source of differential correction being implemented across Europe and more recently in select areas within the United States. The appeal of VRS technology is easy to understand: a mobile GIS user equipped with a standard mapping-grade GPS unit connected to a cell phone with data transfer capability can routinely collect location data with sub-meter accuracy – without assistance from a dedicated receiver used as a base station. More importantly, differential correction is constant within the VRS network area. There is no degradation of correction accuracy as the user moves away from the differential signal source.

Art Lange anticipates that within the next few years, a new generation of GPS satellites will be launched with improved access to the L2 Civilian (L2C) signal (the 'second frequency'). This will lead to improved measurement accuracy by removing some of the errors associated with atmospheric GPS signal distortion. The general trend of GPS receivers for GIS applications has been to provide greater accuracy and more user-friendly features at a generally lower cost. The widespread availability of consumer-grade GPS receivers with built-in large map data bases has changed the way many non-GIS professionals depend upon, and are affected by, the accuracy of the coordinates in GIS databases. GIS software for processing field-collected data has continued to evolve – for example, the GPS extensions to ArcGIS and ArcPad, which make the process of transferring field-collected GPS/GIS data to the user's database a seamless operation.

While GPS-enabled devices have opened the door for location-based technologies, the emergence of Wi-Fi-based systems promises to bring the same location-tracking services indoors to hospitals, warehouses and large industrial complexes. Creating what has been sometimes called 'the Internet of Things' or 'indoor GIS', these systems are being used to track moveable assets, such as machines, merchandise, animals, and even people carrying or wearing a new generation of tiny RFID (Radio Frequency Identification) tags which will store important attribute information about the object or wearer of the tag.

Here once again, the potential of these systems also highlights the important growing debate over what constitutes 'appropriate' access to personal data and a person's current location (see Chapter 55, Michael Curry: *Rethinking privacy in a geocoded world*).

(d) Data transformation and linkage

Spatial interpolation continues to be a vitally important part of GIS functionality, and the set of tools has matured significantly over the past six years. Increasingly, spatial interpolation services are being offered remotely over the Web, as the field of scientific computing moves to a more integrated view of how standard operations are packaged and made available.

The most dramatic recent development has been the increase in the size of the datasets that require spatial interpolation or approximation. New mapping technologies, such as LiDAR, real-time kinematic GPS, and automated sensors have made data acquisition orders of magnitude more efficient. However, data processing and analysis often lags behind data acquisition – for example, millions of georeferenced points can be measured by LiDAR within a single hour but it takes much longer to produce a bare-Earth digital elevation model. The properties of data produced by automated technologies also require smoothing of noise and, as a result, spatial approximation rather than exact interpolation has become increasingly important.

While the principles of the most commonly used methods (Kriging, splines, inverse distance weighting, nearest neighbour, triangulated irregular network) remain the same, their implementation has become more robust and better adapted to data heterogeneity, noise and large datasets. Research into model-based interpolation continues and is being used for specialised applications (meteorology, topography, groundwater). A fully automated methodology that would select the most appropriate method and optimise its parameters for a given dataset and application is still not available. However, the tools that help to select a suitable method (geostatistical analysis, visualisation, etc.) have improved significantly. Nevertheless, writes author Helena Mitasova, 'spatial interpolation and approximation remain a challenging task for many GIS users that requires a solid understanding of available methods and modelled phenomena.'

Author Antonio Câmara sees four main trends underlying the integration of GIS and virtual environments over the past six years. First, the promise of immersive systems largely failed to deliver, and the field remains dominated by non-immersive desktop solutions. These benefit from the development of an extension to VRML (Virtual Reality Markup Language) known as GeoVRML, which provides the capability to browse multi-resolution, tiled data that are streamed over the Web (Reddy et al 1999, 2001). Second, interoperability efforts, such as those promoted by the Open GeoSpatial Consortium (www.opengeospatial.org), are also facilitating the integration of GIS with virtual environments. SRI International is in the implementation phase of an OGC Web Map Service (WMS) capable of generating GeoVRML output.

Third, improved representations of virtual terrains have become possible thanks to new techniques of detail management and multi-resolution modeling of geographical data. The work of Losasso and Hoppe (2004) and Cignoni et al (2003) on interactive rendering of large-sized textured terrain surfaces is worth mentioning. Finally, mobility is a new trend in GIS due to the emergence of the wireless Internet and the availability of communication-enabled mobile devices. Romao et al (2003) discuss the promise of location-based augmented reality services, where three-dimensional synthetic images from databases are superimposed on real images in mobile devices.

Many of the issues of emerging data infrastructures, provision and access that are raised by Mike Goodchild and Paul Longley (Chapter 40: *The future of GIS and spatial analysis*) are becoming still more significant with the gradual maturation of advanced information economies. They observe that this is particularly apparent in the socio-economic realm, where small-area geodemographic measures of social and economic conditions are becoming key to GIS-based models of resource allocation. Current research in geodemographics illustrates the interdependences between classification method, variable selection and data source when devising classifications that work in the real world. They reiterate the point already made several times above that, in measurement terms, more data are collected about more aspects of our individual lifestyles than at any point in the past, through routine interactions between humans and machines. They see enlightened approaches to public and academic data access (e.g. through geoportals: Maguire and Longley 2005) as key to making wide dissemination of socio-economic data a reality, and making possible an open debate about the remit and potential of social measurement at neighbourhood scales. Geodemographic systems based on framework socio-economic data can be successfully 'fused' to census sources to provide richer depictions of lifestyles – yet lifestyles data sources are usually not scientific in collection and require reconsideration of the practices of science in the ways identified in Chapter 40. If this can be successfully undertaken, the toolkit of spatial analysis in GIS now makes it easier than ever before to match diverse data sources and accommodate the uncertainties created by scale and aggregation effects.

4 MANAGEMENT (PART THREE)

As in most other things, the GIS world of 2005 is a rather different place to that of 1999 when the original version of this book was published. Before we discuss what has changed in the world of GIS, we first rehearse (very briefly) the external factors which impact on our subject matter and which differ from those of 1999.

(a) The changing context

One factor is common to the drivers for change noted in the Principles and Techniques section. The continuing evolution of information technology has been rapid. This has had some direct effects and many indirect ones. The growth of storage capacity – typically around 20GB in 1999 and around 100GB now – exemplifies the more/smaller/faster trend which enables software functionality and applications that were hitherto impossible to become commonplace. The most obvious change is in regard to location-based services (LBS) which have penetrated both consumer cellphone and industrial and military tracking markets. The dramatic growth in broadband uptake and Internet bandwidth has also fuelled the interest in such geographic Web services.

Aside from the technology, the most obvious change of GIS drivers is that governments and their leaders have changed. Nowhere has this been more important than in the USA, the largest individual

'engine' of GIS. The replacement of Clinton and Gore by Bush and Cheney resulted in a loss of overt support for GIS based on the former administration's seemingly altruistic approach – this was the underpinning of the 1994 Presidential Executive Order which triggered the US National Spatial Data Infrastructure. Yet federal government support in the USA has not died: rather it has been transformed into something driven by the two imperatives of getting more efficient government and Homeland Security issues (see below). Like many other governments, the US federal one has launched an e-government initiative to enhance efficiency and service to the citizens, with one of the initial batch of projects being a GIS metadata service: the Geospatial One Stop initiative (see Section 3(a) above and Box 20.6 in Longley, Goodchild, Maguire and Rhind 2005) builds upon much of the work described by Guptill (Chapter 49: *Metadata and data catalogues*). Underpinning the present situation is that metadata standards have continued to evolve as they have made their way through the various national and international standards bodies. While this has helped in the syntax and coding of metadata, the underlying content has remained pretty much intact. As more diverse user groups have reviewed the metadata standards they have added elements of value to their specialty. As a general result, the number of possible data elements has grown quite considerably. The major difficulty was, and continues to be, that very few parties are willing to fully populate both the 'mandatory' and 'optional' fields. Mandatory fields are usually reduced to the equivalent of 'title, author, date' and are usually populated. Since many other fields are optional and left unfilled, any third party data user is left unaware as to the value of the data and its fitness for particular uses. The only way out of this problem is if GIS software becomes a lot smarter and automatically encodes more fields of metadata.

The number of catalogues has continued to grow but there is as yet no good catalogue of catalogues. Thus much surfing is still required to discover what is needed (or if it exists). The user interfaces for the catalogues are also different, creating a learning curve for novice users. Private sector sites have emerged which point to their products and the products of their partners. Some of the products are repackaged government data sets. This can cause some user confusion.

More generally, whilst the formal policies of

national governments described in Rhind (Chapter 56: *National and international geospatial data policies*) have not changed dramatically, some blurring has occurred at the edges, some potentially important international agreements have been launched (see Longley, Goodchild, Maguire and Rhind 2005, Chapter 20) and the advent of commercial suppliers has rendered the practical import of some policies much more limited: government matters less in GIS than it did. Perhaps because of this, the nature of the National Map is evolving rather differently in many countries (see, for instance, Longley, Goodchild, Maguire and Rhind 2005, Section 19.3) – most dramatically between the USA and the UK.

The 'new world order' has transformed many aspects of our world since 1999 and impacted on GIS. The growth of terrorism, notably manifested in the events of 9/11 in the USA but replicated in many ways in other countries as far afield as Colombia, India and Russia, has triggered a renewed focus on 'homeland security' and military campaigns such as those in Afghanistan and Iraq. Immediately after 9/11 there was a flurry of activity in removing GI from Websites so as to minimise aid to terrorists; after various studies it was concluded that this was not really necessary (see Baker et al 2004). But, more generally, GIS is clearly relevant to homeland security issues. We can think of five stages in any major disaster, natural or man-made. They are:

- Risk assessment
- Preparedness
- Mitigation
- Response
- Recovery

It is now widely accepted that each of these inherently involves use of geographic information and GIS. Equally, some parts of all of these stages involve human judgement, understanding of the characteristics of other organisations as well as of data and a clear understanding of what needs to be done for the greater good – and a strong code of ethics. The consequences for those in GIS of terrorism acts and of natural disasters alike (like the South East Asia tsunami disaster of December 2004) has been a further impetus to the development of hardware and software, including sensing platforms ranging from aircraft a few centimetres across to commercial satellite imaging (supplied to the military, the 'anchor customers') with resolutions

as great as 60cm (and getting still finer – 40cm is predicted by 2007).

But the consequences of the changed geopolitical situation are not all those of Homeland Security. The addition of a further 10 countries to the European Union, now an entity with 450 million people, creates a formidably large trading bloc and one which is determined to become a major player in high technology and its uses; GIS is part of this, the European Union proposing far-reaching plans for a new GI directive.

In one sense a 'counterbalance' to terrorism and homeland security is the use of GIS to support humanitarian efforts arising both from natural and human causes. Both in-field hand-held mobile GIS and office-based systems have been used to great effect during such crises – the former to collect data about safety, infrastructure damage, disease, for example, and the latter to plan missions such as food drops, and temporary hospital locations, as well as to brief executives and members of the press. Rapid access to current data can help save lives and reduce human suffering.

It is not only military concerns and homeland security issues that have become global. Global science has come of age. Concerns with 'big issues', such as climate change and global warming and dimming, have become widespread. Models of past and projected changes to our habitat have typically used GIS: the geographical manifestations of any changes have become highly charged political issues. And natural disasters – such as the South East Asia tsunami but also many others – have forced a need for much-improved monitoring and early warning tools in which GIS has a key role to play.

(b) The results

Over the past few years we have seen a maturing recognition of the potential of GIS in its original heartland (North America, some parts of Western Europe and Australasia) but also in the emerging economies of Asia, Europe and Latin America. We have estimated (Longley, Goodchild, Maguire and Rhind 2005) that the likely number of active GIS professional users must be well over two million people world-wide. At least double that number of individuals will have had some direct experience of GIS and perhaps an order of magnitude more people (i.e. well over 10 million) will have heard about it (e.g. through such events as GIS Day). Yet more will – sometimes unknowingly – have used elementary GIS capabilities in passive Web services such as local mapping. We estimated that, in 2004, the total global expenditure caused by the use of GIS could not be less that $20 to $25 billion annually; the sum is not precisely knowable and is partly dependent on how we define GIS – but it is large and continuing to grow every year. One indicator of this growth is the numbers attending software conferences: that hosted by ESRI has seen numbers grow from a start of 23 attendees in 1982 to 12,000+ in 2004. Even if we take our highest estimate (10 million globally) of those having had some experience of GIS, this still means that only one person in 600 on Earth falls in this category. Clearly there is still a long way to go! Part of this growth has come about by a spread of the same type of applications worldwide and part from innovative applications. Section 5 describes how these have expanded and diversified.

The GIS industry has also become more mature. This is manifested in various ways. There is now widespread international agreement on the use of standards of various kinds, on mainstream functionality and even the terminology used – with one exception (see below). The growth of the commercial elements in GIS – software, system integration, consultancy, data provision, etc. – has become global and has led to some consolidation. The US hegemony is being challenged: for example, the data and services company Tele-Atlas of the Netherlands purchased its equivalent, GDT of the USA.

In becoming more mature, the industry has become more commercial. We now know much more about the business of selling data and services based upon sound economic principles (see for instance, Box 19.10 in Longley, Goodchild, Maguire and Rhind 2005 or Shapiro and Varian 1999). We have already identified the ease with which global monitoring by commercial satellite and data-serving organisations is carried out. To an extent, GIS has become democratised: mapping is now produced more frequently by individuals via the Web than by national mapping organisations and the underlying service is funded by advertising – a dramatic change to the situation of a few years ago. The early work described by Shiffer (Chapter 52: *Managing public discourse: towards the augmentation of GIS with multimedia*) has become much more commonplace and public participation in GIS (PPGIS) has

blossomed with the fall of GIS software costs and the spread of GIS skills (see Box 20.2 in Longley, Goodchild, Maguire and Rhind 2005).

Although business applications of GIS have not developed at the rate anticipated when this book was first published (see the comments of Sue Elshaw Thrall and Grant Thrall in Section 3 above), there have nevertheless been three sustained developments in the area described by Mark Birkin, Graham Clarke and Martin Clarke in the book (Chapter 51: *GIS for business and service planning*). First, businesses have continued to create ever-increasing volumes of data about their operations, customer behaviour, and competitive environment (e.g. through loyalty cards, lifestyle data, stock control systems, etc). Many of these data are spatially referenced with respect to the points of delivery and of consumption. However, Mark Birkin, Graham Clarke and Martin Clarke suggest that there 'is as yet little evidence that retail and service businesses are able to determine key actions based on this information, for example in the optimisation of product mix or network configuration.' In their view, off-the-shelf GIS packages offer appropriate components for operations such as like spatial interaction modelling, but still lack the flexibility and sophistication required to support business decision-making, especially at the strategic level. Second, some of the increasing interest in spatial decision support systems (e.g. Geertman and Stillwell, 2002) has been directed at business applications, though this area of activity remains small relative to environment and physical planning applications. Third, there has been interest in new modelling techniques such as intelligent agents and cellular automata (see Longley and Batty 2003). Here again, however, the emphasis has not been upon the tactical or strategic needs of business, or upon activity patterns and spatial behaviour, but rather has addressed more general problems of urban form and structure. There has, however, been an upsurge in interest in the use of geodemographics (a tool traditionally focused upon business applications) to issues of efficiency and effectiveness of *public* service delivery, and these are addressed in the update on Tony Yeh's contribution (Chapter 62: *Urban planning and GIS*) in Section 5(a) below.

In general, the capabilities of GIS to profile entire populations of a country (or beyond), grouping those millions of individuals on the basis of their inferred (e.g. purchasing) characteristics, has created commercially valuable knowledge. But this has a downside. The black forebodings about the capacity of the technology to destroy privacy expressed by Michael Curry (Chapter 55: *Rethinking privacy in a geocoded world*) have more than come to pass. Knowing where people are at any moment clearly has some benefits – e.g. if someone is being attacked and calls for help, the emergency can be targeted effectively in real time. But combining this with a detailed or inferred knowledge of a variety of the victim's characteristics – and potentially producing a cradle-to-grave narrative of an individual's life poses real challenges for the sort of society we wish to be. Curry has gone so far as to suggest that the impact of homeland security and the developments in locational technology mean that we have returned in some senses to the 1960s and 1970s. He argues that whilst in the 1980s and 1990s it was widely believed that business was the most significant threat to individuals' rights to privacy and autonomy, the more significant threat now appears to arise from government.

If the technology and practical experience of designing, selecting and using GIS has come some way since 1999, the legal aspects remain complex. The law touches everything. During a career in GIS, we may have to deal with several manifestations of the law. These could include copyright and other intellectual property rights (IPR), data protection laws, public access issues enabled (e.g.) through Freedom of Information Acts (FOIA), and legal liability issues. But since laws of various sorts have several roles – to regulate and incentivise the behavior of citizens and to help resolve disputes and protect the individual citizen – almost all aspects of the operations of organizations and individuals are steered or constrained by them. One complication in areas such as GIS is that the law is always doomed to trail behind the development of new technology; laws only get enacted after (sometimes long after) a technology appears. All those using GIS need to be aware that, whilst commerce is global, law – for the most part – is not. In essence, there is a geography of the law, the legal framework varying from country to country. The creation, maintenance and dissemination of 'official' (government-produced) geographic information are strongly influenced by national laws and practice. The best recent overall summary of this is Cho (2005).

The management of GIS has also come some way: chapters by Bernhardsen (Chapter 41: *Choosing*

a GIS), by Obermeyer (Chapter 42: *Measuring the benefits and costs of a GIS*) and Sugarbaker (Chapter 43: *Managing an operational GIS*) still contain much of value. But since 1999 much focus has been put on 'interoperability' and the concept of spatial data infrastructures (SDI). These two topics have given a new dimension to the process of choosing a GIS. They are technically based on the development of standards created by ISO and OGC, such as the geography markup language (GML) and the Web map service (WMS). These technical developments are discussed in Section 3(a) above. The driving force is the user's requirements for access to data, and thus to increase the social benefit of data collected with only minor increase in the costs. Any organisation producing georeferenced data, and which is in the process of choosing a GIS, should now choose a system not only on the basis of their internal user requirements, but should seek to integrate their own new system in the regional or national spatial data infrastructure. Our understanding of how to assess costs and projected benefits of systems has also improved considerably (see Thomas and Ospina 2004 and Tomlinson 2003). There are, for instance, far more examples of real-world benefit-cost analyses for GIS implementations than there were in 1999; good examples include the 'best-GIS' ESPRIT-ESSI Project n.21.580 at http://www.gisig.it/best-gis/Guides/chapter9/nine.htm or that written by Darlene Wilcox at http://www.geoplace.com/gw/2000/0200/0200wlcx.asp.

Given the much-enhanced commercial interest in GIS and GI, it is no great surprise to find that university departments outside the traditional ones of geography and surveying are now beginning to teach and research in our subject area. The highest-ranked business school world-wide on many occasions in the past decade has been the Wharton School of the University of Pennsylvania. It now has a senior member of staff working in GIS and this trend is being followed elsewhere.

All this begs the question of what is needed and what is changing in GIS education. In truth, we know relatively little about the backgrounds of many people who are active GIS practitioners since some at least seem to have become 'accidental geographers', drawn into GIS by the need to carry out spatial operations in their own, original, field of endeavour. What they felt they needed is therefore less than crystal-clear. Traditionally, however, much education in GIS has really been training, especially in how to use a particular set of software tools (see Forer and Unwin's Chapter 54: *Enabling progress in GIS and education*). Forer has argued that we 'may be producing a range of people who have a simple tool for simple problems, and aren't impatient with the tool or aware of its multiple limitations'. Unwin, on the other hand, has claimed that four developments have occurred since 1999 in GIS and education:

- The development of several Web-based distance learning courses. This, he argues, says more about universities' needs to expand their market in general than it does about GIS;
- A near universal concentration on GIScience as opposed to GISystems, though training remains a thriving industry;
- Greatly increased teaching of GIScience in contexts other than academic geography. Mainly this seems to be in applications areas such as archaeology and environmental science and management rather than computer science.
- A divergence in trajectories between the US and UK. In the US, geography has done well out of embracing GIScience, which has been the spearhead of a very substantial revival in the discipline's fortunes throughout education. This is not so in the UK where, despite the best efforts of many people, it has not been greatly used in an 'embedded' mode, to teach about other parts of the discipline (see also the comments on Ron Johnston's Chapter 3: *Geography and GIS*, in Section 2(a) above).

Given all this and since the world of applications has spread ever-wider, surely the old approaches are no longer appropriate and new core competences are needed? To maximise the utility of GIS, we see the need for education offerings in GIS and GI-related areas to extend beyond the ambit of commercial firms and universities. We see the need for some GIS education to be much more sophisticated in relation to critical theory and its application to understand the downsides as well as the upsides of GIS and to help understand how it is changing some aspects of society. We therefore conjecture the need for GIS education now to include:

- Entrepreneurial skill development and leadership
- The principles of geographical science
- Understanding of and familiarity with GIS technologies

- Understanding of organisations
- Finance, investment criteria and risk management
- Human resources policies and practice and ethics relating to use of GIS
- Legal constraints to local operations
- Cultural differences between disciplines
- Awareness of international differences in culture, legal practice and policy priorities
- Formal management training, including staff development, team working using GIS, and presentational and analytic skills
- Attempts to embed GIS and GIScience in the mainstream of the academic discipline of geography as well as other disciplines

Clearly not all courses and learning needs to include all of this material and some of it is not best learned from courses. Some elements will be particularly relevant to those engaged (as all professionals should be) in continuing professional development. The introduction of local legal, cultural and application-related elements to GIS courses – as well as buttressing the global technical, business and management issues – will be to the benefit of GIS practitioners, the discipline and business (used in a wide sense) alike.

One characteristic which has become ever more commonplace is the role of GIS partnerships. These now operate at all levels, ranging from the very local (e.g. where lobby groups share resources and pool skills to produce maps to illustrate threats), to the continental (like the Permanent GIS Committee comprising 55 countries in Asia) to the global. They also range from the informal through to those defined via international treaty obligations, with commercial partnerships being somewhere in the middle. The most common manifestation of these partnerships is a national spatial data infrastructure: some 39 countries are now said to have one though what really is happening on the ground is somewhat variable. How partnerships are best made to work in an era of Homeland Security, where safeguarding access to information is seen by some as crucial, is not immediately clear.

5 APPLICATIONS (PART FOUR)

In the Introduction to the Applications section in the book we classified GIS applications as traditional, developing and new. Traditional applications include military, government, education, utilities and also natural resources. In the last six years these traditional areas have continued to prosper and still constitute the lion's share of GIS application activity and revenue. The US National Geospatial-Intelligence Agency (NGA) continues to be the largest single spender on GIS in the world. The developing applications of the mid-1990s involved general business uses such as banking and financial services, transportation logistics, real estate and market analysis. Consistent with Sue Elshaw Thrall and Grant Thrall's comments in Section 3 above, we observe that in the ensuing years these applications have not made the progress we had expected. In part this is a manifestation of the downturn in the US economy in the early 2000s (its stock market has yet to regain its all time high of early 2000), but other contributory factors include commercial company inertia, an inability for GIS vendors and consultants to simplify the technology, and a poor presentation of the business case. Nonetheless, there are notable showcase examples of GIS in business, such as home delivery of electrical goods at Sears in the USA, reinsurance risk modelling by PartnerRe in Europe and, at a more local scale, direct marketing to banking customers by Arrowhead Credit Union, California (Thomas and Ospina 2004). Back in 1999 we hypothesized that small office/home office (SOHO) and personal or consumer applications would represent the next wave of new GIS applications. While we were incorrect in our SOHO prediction we were on target with our suggestion about personal/consumer applications. There has been rapid growth in interest in location-based services. Some things that we did not foresee in 1999 included growth in interest in homeland security, in geoportals and in spatial data infrastructures (SDIs: Maguire and Longley 2005). It was the events of September 11, 2001 that put the term 'homeland security' into common parlance. The interest and funding for homeland security geographic information infrastructure projects has helped propel the rather stagnant SDI community into the modern era (see Section 4).

Looking forward we envision a bright future for hand-held and mobile GIS applications, as well as development of a new field that may be characterized as 'indoor GIS'. The latter is concerned with the location and movement of resources within buildings. Using RFID (radio frequency identification) tags (see Section 3(c) above) and other technologies it is possible to monitor the movement of people, and other inanimate resources, around buildings. This has many implications for security and for facility and resource management.

(a) Operational applications

In the new millennium GIS continues to be used to even greater effect in operational application areas. There can scarcely be a government in the world that does not use GIS either directly or indirectly in one way, shape or form for managing its assets, be they land and property parcels and easements, road/street/highway networks, or information about citizens (Tony Yeh, Chapter 62: *Urban planning and GIS*). GIS has played a central role in many of the big digital/electronic government initiatives of the past few years (Curtain et al 2004; Greene 2001; Song 2003). A similar picture can be painted for utilities (Jeff Meyers, Chapter 57: *GIS in the utilities;* Caroline Fry, Chapter 58: *GIS in telecommunications*) where GIS is extending from core network maintenance and mapping applications to a plurality of applications such as tree trimming, customer recruitment and retention, environmental management, pipeline routing, mobile workforce management, and transportation logistics. There is also a trend toward integrating GIS with other existing enterprise applications such as network optimization, SCADA (Supervisory Control and Data Acquisition), CRM (Customer Relationship Management) and ERP (Enterprise Resource Planning) systems. After a period of rapid expansion, the telecommunications industry is now in a period of stability, and consolidation is the order of the day in most geographies.

In Jeff Meyers' opinion, two major and related changes have driven the utility GIS industry in the past five years. First, GIS has become mainstream Information Technology (IT). This change began with the ability to store the geometry (spatial characteristics) of features in garden-variety, open relational database management systems. Partly enabled by improved server and network performance, the open storage of GIS features brought GIS into the fold of IT within utilities, initially encouraging and then demanding that IT professionals review platform requirements in the context of corporate standards. The standard technology trend continued with the development of core GIS tools written with standard programming languages. As programming standards emerged within the IT industry, GIS vendors (for the most part) adopted those standards, making GIS implementation and extension using standard programming languages possible.

The implications of standard technology are many. Among the most important, open GIS technology has reduced the barriers to IT acceptance within the utility organisation, and provided a means for more standard IT support. Software engineers and computer scientists with standard education can now be utilized as resources in GIS. Through data storage in non-proprietary formats, management can confidently invest in GIS to manage assets more effectively, without the fear that data might become obsolete or be costly to migrate from one proprietary format to another. Standard technology has led to integration of GIS within the backbone IT of the enterprise, sharing data and even function between core IT systems through standard interface techniques. In this respect, open GIS technology has drastically reduced the life-cycle cost of spatially enabled systems. And, perhaps most significantly, the availability of GIS technology based on computing industry standards has changed the debate about vendor selection from a preference-based choice between proprietary platforms to a focused decision based on business benefit and feature function.

The adoption of open, standard GIS technology has been a key driver in the second key change in utility GIS. As predicted and planned by industry experts and observers, utilities have moved GIS from a departmental tool to an enterprise solution. Enabled by standard technology, and driven by competitive factors within the energy industry, many of the best performing companies in the utility industry today use GIS as an everyday tool for asset management, work design, and outage management. Through the integration of disparate data sources about network assets, utilities use GIS today as an everyday resource for managing assets and communicating change across enterprises that serve large geographic areas. Additionally, standard computing technology has enabled the development of focused end-user GIS applications that allow business people to interact with data about customers and assets in a spatial context without becoming 'GIS experts'. End user GIS tools that solve business problems lead to the demand for more

and more functionality. And since they can easily be supported through standard development environments, GIS core teams are happy to oblige with more development, which in turn leads to demands for more data and function, and the enterprise cycle sustains itself.

In terms of new GIS developments in transportation, Nigel Waters (Chapter 59: *Transportation GIS: GIS-T*) draws attention to Simon Lewis' lists of GIS-T 10 accomplishments and 10 challenges for the future (http://www.gis-t.org/yr2003/gist2003sessions/gist2003session5.htm). The accomplishments are largely technological while the challenges for the future relate to concepts and frameworks, people and institutions. Others besides Lewis have their own lists (e.g. Fletcher 2003; http://www.gis-t.org/yr2003/gist2003sessions/gist2003session4.htm). Looking forward Waters believes that there is an ongoing need for GIS-T Science integration, temporal modelling of transportation data, integration of web-based modelling, a new data model, markup language and data standards, and greater public participation in GIS-T (for more discussion see: http://esri.com/industries/federal/gis-business/transportation2.html). Waters' chapter on GIS-T discussed some of the pioneering attempts at integrating GIS-T and Intelligent Transportation Systems. This discussion concentrated on work being carried out on the ROMANSE project at Southampton, UK, and other parts of the European Community (http://www.romanse.org.uk). The field has moved forward since then but perhaps more slowly than might have been expected. This is especially true in Canada where a report by Transport Canada and Intergraph has detailed the problems and challenges of integrating ITS and GIS-T (http://www.its-sti.gc.ca/en/downloads/execsum/tp13224e.htm). The three main difficulties outlined in this report are the cost-recovery pricing policies of the Canadian Government and Statistics Canada, the lack of public-private partnerships (one of the great strengths of the ROMANSE project) and the lack of government involvement (another strength of ROMANSE). The UK government, like its Canadian counterpart, also implements punitive cost recovery policies, while the US remains one of the few countries where this is not the case. Thus it is in the US that GIS-T and ITS integration is likely to take off in the coming years. Key new works in this field include Lang's review of applications (Lang

1999) and Miller and Shaw (2001) that describes GIS-T principles and applications.

GIS provides the core organising framework for emergency management (Tom Cova, Chapter 60: *GIS in emergency management*). Today emergency management activities are focused on three primary objectives: protecting life, property and the environment. All phases of emergency management depend on geographic information from a variety of sources. The use of GIS ranges from displaying the effects of events, to managing the actual incidents themselves at command posts or emergency operations centres. GIS is the only tool capable of providing a common operating picture for emergency management, planning, response and recovery. State of the art emergency management centres allow managers to fuse real time GIS data (pertaining to the area affected or threatened by an event) with fixed or static GIS infrastructure data. This provides users with dynamic 'actionable information' in near real time. This information can be further enhanced within a GIS with event modelling tools and by importing real time weather data. With these tools, emergency managers can close roads, order evacuations, and route public safety responders efficiently and accurately. Radke et al (2000) review the application challenges for GIScience and their implications for research, education, and policy for risk assessment, emergency preparedness and response.

The awareness of the benefits of good land administration has grown over the last five years according to Peter Dale and Robin McLaren (*GIS in Land Administration,* Chapter 61), in part because of the better understanding of the role of land in poverty reduction. As a World Bank Policy Research Report states (World Bank 2003) 'Land Policies are of fundamental importance to sustainable growth, good governance, and the well-being of and economic opportunities open to rural and urban dwellers – particularly poor people'. De Soto (2000) has argued that open, efficient and enforceable property rights could release trillions of dollars of 'dead capital' in poor countries.

The recognition of the need for greater openness with regard to property rights has led to increasing opportunities to exploit land-related data. This has become possible as a result of technological developments both with regard to networking and also through the use of GIS as a data analysis tool – in addition to its established roles in creating and maintaining cadastral maps. Most countries are introducing computerized systems for handling their land administration data and some are now making these data available across networks in moves towards full electronic conveyancing, supported by e-signatures. In Europe, for example, all countries have computerized systems and the development emphasis has shifted from 'design and build' to 'sustain and maintain'. There are also moves through the European Union Land Information Service (EULIS) to exploit data across national boundaries so that all European countries might participate in a pan-European land market. Conveyancing is increasingly being perceived as an integral part of e-government initiatives and this service is being delivered by governments alongside other 'life events', such as the registration of births, companies, marriages and deaths, using 'citizen accounts'.

The money to pay for the upgrade of hardware and software systems has to come either from government or from charges for products and services. To achieve the latter, the agencies delivering land administration services must operate on a business basis and although they may not be run for profit, they are increasingly being asked to recover all their operating costs. While this is relatively easy for those handling text data, such as computerized versions of old deeds and paper documents, it still remains a problem for many mapping agencies whose costs are not so easy to recover.

What started as a question of how to fund the purchase of new equipment has led to a fundamental change in the culture of land administration agencies, making them much more customer focused and conscious of the costs of their operations. In addition it is leading to a reappraisal of the role of the private sector in land administration and the creation of various forms of public-private partnerships. Ultimately, in many countries the risk remains with the central government, but the delivery of various services now involves a wider range of stakeholders and is based on greater exploitation of the data that are now available. Many countries now provide information services to a wide range of stakeholders, including financial services, taxation agencies, economic development agencies, police, estate agents and the citizen, within the constraints of their legislative frameworks for access to information. Open access to Land Administration information is sowing the seeds for the introduction of participatory democracy.

These various developments are relevant to the use of GIS in planning (Tony Yeh, Chapter 62: *Urban planning and GIS*), where it has established roles in development control and general administration. The more strategic use of GIS in plan making has been explored by Benenson and Torrens (2004), particularly with regard to the development of applications of cellular automata. Yeh's discussion makes little mention of the use of GIS to plan expenditure on services that is incurred at the local level: recent years have seen increasing interest in the use of *geodemographics* (Harris et al 2005; Longley 2005) to understand and prescribe local spatial patterns of resource expenditure in policing, health and education. The greatly increased use of GIS in policing is discussed in detail by Chainey and Ratcliffe (2005).

David Swann observes that the last five years have seen an acceptance that GIS provides critical infrastructure for defence and intelligence (Chapter 63: *Military applications of GIS*). There has been a very swift deployment of commercial-off-the-shelf (COTS) GIS throughout the defence sector, representing a considerable advance from previous niche usage. The rapid uptake of GIS is part of a broader recognition that IT adoption creates and requires a transformation of defence activities. This revolution in military affairs manifests itself as 'Network Centric Operations' (NCO) – the use of the network to connect decision making across multiple defence domains. The horizontal nature of NCO demands a move to a modern services-oriented architecture that provides transparent interoperability between domains.

Modern COTS GIS platforms are capable of providing enterprise-class IT infrastructure for military applications. The resulting spatial infrastructure couples the sensors that monitor the battlespace to distributed geodatabases that contain knowledge of the battlespace (data, data models, process models, maps and metadata). These distributed geodatabases are coupled to distributed geoprocesses that provide powerful analysis and information filtering tools. The resulting information can be served across the network for fusion into embedded geovisualisation components. At every stage, interoperability is required with other enterprise class technologies such as supply chain management (SCM) and enterprise resource planning (ERP).

As a tool providing defence-wide infrastructure, GIS is simultaneously able to support war fighting missions such as command and control, business missions such as installation management, and a variety of strategic intelligence missions. Given the enormous expense of spatial data collection and production, this re-use across a common spatial information infrastructure offers immediate cost savings in addition to important new capabilities.

The massive increase in electronic information is not without its challenges. Prue Adler and Mary Larsgaard (Chapter 64: *Applying GIS in libraries*) highlight an explosive growth in the acquisition and use of electronic resources. These include resources that are available locally and via the Web. Research resources may include digital information, resources that are neither owned nor licensed by the library, resources digitized by the library or in partnership with others, and electronic publishing projects. As a result, there has been a dramatic rise in the digital services provided by research libraries. At the same time the continued evolution of information technologies including network services has greatly increased the ability of the research user to manipulate, analyse, and integrate data and information. There remain continuing challenges for long-term preservation and access to GIS and related datasets such as curation, meeting the needs of data users, data authors, and data managers, and the degree of centralisation of the support that is provided.

Today many in the academic and research communities are embracing open access – the sharing of information and data without restriction. In support of these efforts, many research libraries are establishing institutional repositories to capture, index and preserve the intellectual content of the researchers, faculty and staff in digital format. MIT Library's DSpace is illustrative of this digital library initiative and is available worldwide as open source software. As a consequence, a large and growing body of GIS information and datasets is available without restriction. At the same time as we are witnessing a move towards open access publishing, so there is also a rise in the use of licensing by many GIS vendors that may restrict how a user may utilize data, such as restrictions on how data may be shared or used. Today, nearly all major map libraries have a GIS facility of some kind. As a general rule, this does not include teaching classes, but may well include brief introductory sessions for individuals and groups.

Standard cataloging of Web resources is now commonplace. Libraries are now also deeply involved in the creation of metadata records for digital geospatial data. It is both understood and accepted that metadata records may be loaded not into the library's online catalogue but rather into some alternative database, with some form of Web interface to facilitate query.

(b) Social and environmental applications

David MacDevette, Robert Fincham and Greg Forsyth's account of the role of GIS in nation-building (Chapter 65: *The rebuilding of a country: the role of GIS in South Africa*) was written during a time of major upheaval. They offer some observations that are pertinent to contextualising GIS in today's developing African sub-continent where the situation regarding GIS has continued to alter, in many respects. Most notably, it is evident that GIS must be irrevocably tied into a strategic context of poverty alleviation and development. These are now major policy concerns of government and they require pursuance of service delivery on a scale unprecedented in the past.

They also observe that there is a need for a change in mind set of those who are purveyors of the GIS story. The technology has it roots outside of the continent – a European- and American-centric capacity that has been transferred to the developing world. 'Proudly South African' is a new motto in business that signals a new pride in indigenous capacities. For those working in the GIS arena it means an African context with a real commitment to the ideals and aspirations of the country as a whole. From an educational point of view there has to be a strong commitment to the notion of African scholarship. Such an ideal offers great prospects for contributing to better standards of living for more of the country's people than was the case in the past.

MacDevette et al's original contribution reminds us how GIS can contribute to major world events and can effect social and political change. In some respects GIS's role in South Africa is comparable to the events surrounding 9/11 and the December 2004 South East Asian tsunami referred to throughout this chapter. Since 1999 GIS in South Africa has followed a course more similar to that in many developed countries in America, Europe and Asia

Pacific. One recent exceptional GIS project in South Africa has been Willem van Riet's Peace Parks Foundation of South Africa. Under his leadership the Foundation has established transfrontier conservation areas (TFCAs), also known as peace parks, which are large tracts of land that cross international boundaries. The purpose of these parks is to employ conservation as a land use option to benefit local people. Their pioneering spirit is forging international cooperation and is resulting in areas where wildlife can roam freely across borders. GIS has been used not only to help spark the adoption of the original peace parks concept, but to maintain existing parks and promote the concept in other countries.

The use of GIS in *Health and health care applications* (Tony Gatrell and Martin Senior, Chapter 66) has experienced significant recent growth, both in numbers of GIS practitioners and in organizations using GIS. The outbreaks of SARS in Asia, West Nile Virus in the US and foot-and-mouth in the UK, and the threat of global bio-terrorism all illustrate why GIS activities are essential. Among the ranks of the newest users are public health program managers responsible for health policy impact analysis, public health preparedness and disease surveillance. For example, in England and Wales there is now a network of regionally based Public Health Observatories in which GIS figures quite prominently. Increasingly, academic health science centres offer GIS as part of their health informatics and healthcare training programs. For example, US National Institutes of Health have recently funded a biomedical imaging research program that is GIS-centric, as well as a Geospatial Medicine program at Duke University Medical Center. Large public and private hospitals are establishing GIS departments to foster the wider use of GIS to meet the challenges of improving performance and lowering costs. On a global scale, the use of GIS to geographically enable national public health surveillance systems is one of the fastest growing markets for GIS technology and services. Perhaps one of the greatest unrealized values of GIS to health organizations lies in creating new knowledge about the convergence of environmental factors on human health status and healthcare delivery outcomes. Cromley and McLafferty (2002) provide a useful review of progress in the field.

Recent years have certainly seen very many profound political changes. Mark Horn's discussion of *GIS and the geography of politics* (Chapter 67) described the principles of gerrymandering and malapportionment and showed how they have been violated. While the effects of geographical jurisdictions were writ large on the outcome of the 2000 US Presidential Election (when the winning candidate failed to win the popular vote), it is argued that understanding of geodemographic detail is crucial to the US out-turn in 2004 and that in the UK in 2005. It seems that the availability of new technology through GIS can serve the interests of those that seek to manipulate the geography of voter turn out for political gain. It will be interesting to see if broadening of participation and engagement through PPGIS can counter such effects. On a brighter note GIS has recently been used successfully to communicate election results. CBS and other television stations used GIS in the 2004 US Presidential Elections: novel visualization techniques such as 3D perspective views and cartograms were used in near real time, perhaps for the first time on primetime television. In planning for future elections GIS-based redistricting remains as popular as ever.

The use of GIS in *Monitoring land cover and land-use for urban and regional planning* (Peter Bibby and John Shepherd, Chapter 68) dates back to the very origins of the GIS field. Recent advances in satellite geo-imaging have allowed the extents of land cover to be estimated ever more accurately and cheaply. Precise ascription of land-use, however, remains a difficult task, not least because of new and challenging policy requirements to differentiate between greater numbers of land use classes. Bibby and Shepherd concentrated on how GIS has been used to generate information for policy purposes, but, today, there remains a remarkable lack of awareness and willingness for senior politicians and policy makers to embrace science and technology.

GIS and landscape conservation (Richard Aspinall, Chapter 69) is a field closely related to land cover and land-use estimation, in that both involve wildlife and scenic resources, and have significant environmental policy implications. Richard Aspinall suggests that in the past few years there has been wider use of more flexible regression methods (generalized linear models and generalized additive models) for predictive modelling of species distribution (e.g. Guisan et al 2002) and development of rendering of landscapes for landscape assessment (e.g. Bishop et al 2001).

The business of agriculture is also an inherently geographic practice from the local to the global scale. In Chapter 70, *Local, national and global applications of GIS in agriculture*, John Wilson illustrated this with a number of examples. Improvements in global satellite-based sensors have provided many new opportunities for monitoring and modelling global changes. At a more local scale some of the advances in geoprocessing and visual modelling (for example, in Clark Labs IDRISI and ESRI ArcGIS) have helped progress simulation analysis and modelling efforts.

The final chapter in this section examined *GIS in environmental monitoring and assessment* (Lars Larsen, Chapter 71). A feature of this chapter was the interest in real-time monitoring, data quality and mathematical modelling. Increasing concern with the impact of pollution on the environment and the exploitation of natural resources have been major concerns for the environmental lobby and have led to interesting GIS applications in the last few years. The proceedings of the 4th International Conference on Integrating Geographic Information Systems and Environmental Modeling (Parks et al 2003) provide a useful update on progress in the field.

6 CONCLUSIONS

So, is everything progressing well, with the continuing advent of new technology driving ever-greater benefits to mankind? That is an unsustainably simplistic view of the GIS world. Yes, many of the current GIS tools are being used for real benefits to society and helping to generate wealth, employment, safety and improvements of the quality of life of some peoples at least. But, as ever with technology, we are exposed to risks through over-enthusiastic use of databases or the pursuit of wealth or power and influence irrespective of wider considerations. The GIS community exhibits all the usual characteristics of any group of human beings: it is sometimes fractious (even disputing the best name for the field — GIS or geospatial systems/engineering, as discussed in Section 2(a) above) or finding it difficult to establish consensus (e.g. on the best way forward for national spatial data infrastructures). It often lacks professionalism compared to some older professions or disciplines (such as engineering) so greater focus on this, on ethics and on regulation will probably be appropriate. But what we have now is a global movement which recognises that geography and the processes and actions which are manifested geographically can influence the lives of all humans. The next decade or so will show us whether this optimistic scenario really is what happens.

References

Baker J, Lachman B, Frelinger D, O'Connell K, Hou A, Tseng M, Orletsky D, Yost C 2004 *Mapping the risks: assessing the homeland security implications of publicly available geospatial information.* Rand Corporation Monograph (see http://www.rand.org/publications/MG/MG142)

Barrault M, Regnauld N, Duchene C, Haire K, Baeijs C, Demazeau Y, Hardy P, Mackaness W, Ruas A, Weibel R 2001 Integrating multi-agent, object-oriented, and algorithmic techniques for improved automated map generalization. *Proceedings of the 20th International Cartographic Conference,* Beijing China, 3: 2110-6

Bédard Y, Devillers R, Gervais M, Jeansoulin R 2004 Toward multidimensional user manuals for geospatial datasets: legal issues and their considerations into the design of a technical solution *Third International Symposium on Spatial Data Quality,* Austria

Benenson I, Torrens P 2004 *Geosimulation: Automata-based modeling of urban phenomena.* Chichester, John Wiley & Sons

Bishop I D, Wherrett J R, Miller D R 2001 *Landscape and Urban Planning* 52: 227-39

Bossler J D, Jensen J R, McMaster R B, Rizos C 2001 *Manual of geospatial science and technology.* London, Taylor and Francis

Chainey S, Ratcliffe J 2005 *GIS and crime mapping.* Chichester, John Wiley & Sons

Cho G 2005 *Geographic information science: mastering the legal issues.* Chichester, John Wiley & Sons

Cignoni P, Ganovelli F, Gobbetti E, Marton F, Ponchio F, Scopigno R 2003 Planet-sized batched dynamic adaptive meshes P-BDAM *Proceedings IEEE Visualization* IEEE Computer Society Press: 147-55

Cloke P J, Cook I, Crang P, Goodwin M, Painter J, Pluto C 2004 *Practising Human Geography.* London, Sage

Craig W J, Harris T M, Weiner D (eds) 2002 *Community participation and geographic information systems.* London, Taylor and Francis

Cromley E, McLafferty S 2002 *GIS and public health.* New York, Guilford Press

Curtain G G, Sommer M H, Vis-Sommer V (eds) 2004 *The world of e-government* Haworth, Haworth Press

De Soto H 2000 *The mystery of capital: Why capitalism triumphs in the West and fails everywhere else.* New York, Basic Books

Devillers R, Gervais M, Bédard Y, Jeansoulin R 2002 Spatial data quality: from metadata to quality indicators and contextual end user manual *OEEPE/ISPRS Joint Workshop on Spatial Data Quality Management,* Istanbul

Donnay J-P, Barnsley M J, Longley P A 2001 *Remote sensing and urban analysis.* London, Taylor and Francis

Dykes J, MacEachren A M, Kraak M J editors 2004 *Exploring geovisualization* Amsterdam, Elsevier

Foresman T W editor 1998 *The history of geographic information systems: perspectives from the pioneers.* Upper Saddle River, NJ, Prentice Hall

Fotheringham A S, Brunsdon C, Charlton M 2002 Chichester, John Wiley & Sons

Geertman S, Stillwell J C H (eds) 2002 *Planning support systems in practice.* Heidelberg, Springer

Greene R W 2002 *Open access: GIS in e-government.* Redlands CA, ESRI Press

Guisan A, Edwards T J, Hastie T 2002 Generalized linear and generalized additive models in studies of species distributions: setting the scene. *Ecological Modelling* 157: 89-100

Harris R, Sleight P, Webber R 2005 *Geodemographics, GIS and neighbourhood targeting.* Chichester, John Wiley & Sons

Heuvelink G 2003 Review of 'Spatial data quality' (Eds Shi W, Fisher P F, Goodchild M F). *International Journal of Geographical Information Science* 17: 816-8

Johnston R J, Hepple L W, Hoare A G, Jones K, Plummer P 2003 Contemporary fiddling in human geography while Rome burns: has quantitative analysis been largely abandoned – and should it? *Geoforum* 34: 157-61

Kyriakidis P C, Dungan J L 2001 A geostatistical approach for mapping thematic classification accuracy and evaluating the impact of inaccurate spatial data on ecological model predictions. *Environmental and Ecological Statistics* 8: 311-30.

Lang L 1999 *Transportation GIS.* Redlands CA, ESRI Press

Longley P A 2005 A renaissance of geodemographics for public service delivery. *Progress in Human Geography* 29: in press

Longley P A, Goodchild M F, Maguire D J, Rhind D W 2001 *Geographic information systems and science.* Chichester, John Wiley & Sons

Longley P A, Goodchild M F, Maguire D J, Rhind D W 2005 *Geographic information systems and science (Second edition).* Chichester, John Wiley & Sons

Losasso F, Hoppe H 2004 Geometry clipmaps: terrain rendering using nested regular grids. *Proceedings, ACM SIGGRAPH 2004* ACM Press: 769-76

Maguire D J, Goodchild M F, Rhind D W 1991 *Geographical information systems: principles and applications.* Harlow, Longman

Maguire D J, Longley P A 2005 The emergence of geoportals and their role in spatial data infrastructures. *Computers, Environment and Urban Systems* 29: 3-14

McHarg I L 1995 *Design with nature*. New York, John Wiley

Miller H J, Han J, 2001 *Geographic data mining and knowledge discovery*. London, Taylor and Francis

Miller H J, Shaw S-L 2001 *Geographic information systems for transportation: principles and applications*. New York, Oxford University Press

Ord J K, Getis A 2001 Testing for local spatial autocorrelation in the presence of global autocorrelation. *Journal of Regional Science* 41: 411-32

Parks B O, Clarke K M, Crane M P (eds) 2003 *Proceedings of the 4th International Conference on Integrating Geographic Information Systems and Environmental Modeling: Problems Prospectus and Needs for Research* Boulder, University of Colorado Press

Photogrammetric Engineering and Remote Sensing 2003 Special issue on 'the National Map'. Volume 69.

Radke J, Cova T, Sheridan M, Troy A, Lan M, Johnson R 2000 Application challenges for GIScience: implications for research, education, and policy for risk assessment, emergency preparedness and response. *URISA Journal* 12:15-30

Reddy M, Iverson L, Leclerc Y, Heller A 2001 GeoVRML: open Web-based 3D cartography. *Proceedings of the International Cartographic Conference ICC2001*, Beijing, 6-10 August

Reddy M, Leclerc Y, Iverson L, Bletter N 1999 TerraVision II: visualizing massive terrain databases in VRML *IEEE Computer Graphics and Applications* Special Issue on VRML 192: 30-8

Romao T, Dias A E, Danado J, Correia N, Trabuco A, Santos C, Nobre E, Câmara A 2002 Augmented reality with geo-referenced information for environmental management *ACM GIS 2002* ACM Press: 175-180

Shapiro C, Varian H R 1999 *Information rules*. Cambridge, Mass., Harvard Business School Press

Shi W, Fisher P F, Goodchild M F, eds 2002 *Spatial data quality*. London, Taylor and Francis

Snow C P 1961 *The two cultures and the Scientific Revolution*. New York, Cambridge University Press

Song H-J 2003 E-government: lessons learnt and challenges ahead. In *Proceedings, 8th International Seminar on GIS: envisioning cyber-geospace and spatially enabled e-government*: 1-113 Seoul, Korea Research Institute for Human Settlements

Thomas C, Ospina M 2004 *Measuring up: the business case for GIS*. Redlands, CA, ESRI Press

Thrall G I, Campins M 2004 Mapping the geospatial community: parts 1-4. *GeoSpatial Solutions* (July, September, November) 14 (numbers 7, 9, 11); also 2005 15 (number 1)

Tomlinson R 2003 *Thinking about GIS: geographic information system planning for managers*. Redlands, CA, ESRI Press

Ware M, Jones C, Thomas N 2003 Automated map generalization with multiple operators: a simulated annealing approach. *International Journal of Geographical Information Science* 178: 743-69

Worboys M F 2005 Event-oriented approaches to geographic phenomena. *International Journal of Geographic Information Science*, in press

World Bank 2003 *Land Policies for Growth and Poverty Reduction*. Washington DC, World Bank

Addenda

Lubos Mitas and Helena Mitasova, Chapter 34: *Spatial interpolation*

Equation 3 should read:

$$\gamma(\mathbf{h}) = \frac{1}{2} Var\left[\left\{z(\mathbf{r}+\mathbf{h}) - z(\mathbf{r})\right\}\right] \approx \frac{1}{2N_h}\sum_{(ij)}^{N_h}\left[z(\mathbf{r}_i) - z(\mathbf{r}_j)\right]^2$$

Barry Boots, Chapter 36: *Spatial tessellations*.

Unfortunately some of the illustrations in this chapter were corrupted in the production process, and it has not been possible to replace them in this abridged edition. Professor Boots has made correct versions available for interested readers to consult at the following URL: info.wlu.ca/~wwwgeog/facstaff/bootscorrect.htm

Additional Acknowledgment to the Abridged Edition

Paul Longley's contribution to the Abridged Edition was completed during funding for ESRC AIM Fellowship RES-331-25-0001.

List of contributors

THE EDITORS

Paul A Longley

Professor of Geographic Information Science at University College London. Research interests: geographical information systems and science; spatial analysis; geodemographics; data integration, especially involving remote sensing and socioeconomic sources; social survey research practice; management of geographical information; fractal geometry.

Centre for Advanced Spatial Analysis (CASA) and Department of Geography, University College London, Gower Street, London WC1E 7HB, UK
Tel: +44-20-7679-1782; Fax.: +44-20-7813-2843;
E-mail: plongley@geog.ucl.ac.uk

Michael F Goodchild

Chair of the Executive Committee of the National Center for Geographic Information and Analysis (NCGIA), and Professor of Geography at the University of California, Santa Barbara. Research interests: GIS; environmental modelling; geographical data modelling; spatial analysis; location theory; accuracy of spatial databases; statistical geometry.

Department of Geography, University of California, Santa Barbara, CA 93106-4060, USA
Tel: +1-805-893-8049; Fax: +1-805-893-7095;
E-mail: good@ncgia.ucsb.edu

Dr David J Maguire

Director of Products and International at the Environmental Systems Research Institute (ESRI) in California. Research interests: spatial databases; GIS customisation; GIS implementation and object-oriented systems.

Environmental Systems Research Institute Inc., 380 New York Street, Redlands, California 92373, USA
Tel: +1-909-793-2853; Fax: +1-909-793-5953;
E-mail: dmaguire@esri.com

David W Rhind

Vice-Chancellor and Professor of International Science in City University, London; formerly Director General and Chief Executive of the Ordnance Survey of Great Britain. Current research interests include information and data policy issues nationally and internationally; the workings of government; public/private sector interactions and how to make organisations succeed.

City University, Northampton Square, London EC1V 0HB, UK
Tel: +44-20-7040-8002; Fax: +44-20-7040-8596
E-mail: d.rhind@city.ac.uk

THE OTHER CONTRIBUTORS

Prudence S Adler

Associate Executive Director of the Association of Research Libraries in Washington, DC. Research interests: information policies; telecommunications; copyright and intellectual property issues.

Association of Research Libraries, 21 Dupont Circle, N.W., Suite 800, Washington, DC 20036 USA
Tel: +1-202-296-2296 x104; Fax: +1-202-872-0884;
E-mail: prue@arl.org

Seraphim Alvanides

Lecturer in Geography in the School of Geography, Politics and Sociology at the University of Newcastle upon Tyne. Research interests: spatial analysis and modelling in GIS environments; scale and aggregation issues; systematic aggregation of areal data.

School of Geography, Politics and Sociology, University of Newcastle-upon-Tyne, Newcastle-upon-Tyne NE1 7RU, UK
Tel: +44-191-222-6358; Fax: +44-191-222-5421;
E-mail: S.Alvanides@newcastle.ac.uk

Luc Anselin

Professor and Director of the Spatial Analysis Laboratory in the Department of Geography, University of Illinois. Research interests: regional economic and demographic analysis; spatial econometrics and spatial statistics; statistical computing; GIS and spatial analysis.

Department of Geography, University of Illinois, 330 Davenport Hall (MC-150), 607 South Mathews Avenue, Urbana IL 61801-3671, USA
E-mail: anselin@uiuc.edu

Richard J Aspinall

Professor and Chair of the Department of Geography, Arizona State University. Research interests: environmental applications of GIS and spatial analysis; data quality issues; integrating socioeconomic and environmental modelling for land-use applications.

Department of Geography, Arizona State University, P.O. Box 870104, Tempe, AZ 85287-0104, USA.
Tel: +1-480-965-7533; Fax: +1-480-965-8313;
E-mail: Richard.Aspinall@asu.edu

Lawrence E Band

Voit Gilmore Distinguished Professor and Chair at the Department of Geography, University of North Carolina. Research interests include hydrology; geomorphology; GIS and environmental modelling.

Department of Geography, University of North Carolina, Chapel Hill, NC 27599-3220, USA
E-mail: lband@email.unc.edu

Mike Barnsley

Research Professor of Remote Sensing and GIS at the University of Wales Swansea. Research interests: estimation of land-surface biophysical properties by remote sensing; mapping, monitoring, and analysis of urban areas using very high resolution satellite images, including the development of graph-based spatial analytical tools; scaling and generalisation in remote sensing and GIS.

Department of Geography, University of Wales Swansea, Singleton Park, Swansea SA2 8PP, Wales
Tel: +44-1792-295647; Fax: +44-1792-295955;
E-mail: m.barnsley@swansea.ac.uk

Michael Batty

Professor of Spatial Analysis and Planning at the Centre for Advanced Spatial Analysis, University College London. Research interests include visualisation; urban systems modelling; urban morphology; planning and design processes.

CASA, University College London, 1–19 Torrington Place, London WC1E 7HB, UK
Tel: +44-20-7679-1781; Fax.: +44-20-7813-2843;
E-mail: m.batty@ucl.ac.uk

M Kate Beard-Tisdale

Professor and Chair, Department of Spatial Information Science and Engineering, University of Maine. Research interests: spatial data quality; metadata; automated generalisation.

Department of Spatial Information Science and Engineering, University of Maine, 5711 Boardman Hall, Room 348, Orono, Maine 04469-5711, USA
Tel: +1-207-581-2147; Fax: +1-207-581-2206;
E-mail: beard@spatial.maine.edu

Yvan Bédard

Professeur Titulaire at the Département Sciences géomatiques, Université Laval. Research interests: spatial database analysis; spatial data warehousing; spatio-temporal reasoning; organisational issues related to the implementation of geomatics technologies.

Dép Sciences Géomatiques, Centre de recherche en géomatique, Local 1355, Pavillon Casault, Université Laval, Québec, QC G1K 7P4, Canada
Tel: +1-418-656-2131 ext. 3694; Fax: +1-418-656-7411;
E-mail: yvan.bedard@scg.ulaval.ca

Tor Bernhardsen

Senior Consultant at Asplan Viak in Norway. Research interests include implementing GIS and cost/benefit analysis.

Asplan Viak, PO Box 1699, N-4801, Arendal, Norway
Tel: +47-37-035560; Fax: +47-37-023280;
E-mail: tor.bernhardsen@asplanviak.no

Peter R Bibby

Lecturer in Town and Regional Planning at the University of Sheffield. Main research interest is representational systems in urban planning.

Department of Town and Regional Planning, Geography and Planning Building, University of Sheffield, Winter Street, Sheffield, S3 7ND, UK
Tel: +44-114-222-6181; Fax. +44-114-272-2199;
E-mail: p.r.bibby@sheffield.ac.uk

Mark Birkin

Lecturer at the School of Geography and Director of the Informatics Institute, at the University of Leeds, UK. Main research interests: generation of market intelligence from spatial data; application of geographical models to commercial markets; the use of GIS to improve decision-making within business.

School of Geography, University of Leeds, Woodhouse Lane, Leeds, LS2 9JT, UK
Tel.: +44-113-34-36838; Fax: +44-113-34-33308;
E-mail: M.Birkin@geography.leeds.ac.uk

Barry Boots

Professor of Geography and Environmental Studies at Wilfrid Laurier University, Ontario. Research interests: modelling spatial processes; spatial patterns; spatial statistics.

Department of Geography and Environmental Studies, Wilfrid Laurier University, Waterloo, Ontario N2L 3C5, Canada
Tel: +1-519-884-1970; Fax: +1-519-725-1342;
E-mail: bboots@mach1.wlu.ca

Kurt Buehler

President of Image Matters LLC. Research interests include open systems; geographical data models, standards; open GIS; object-oriented databases; and interoperability.

Image Matters LLC, 1580 Whisnand Road, Bloomington, IN 47408, USA
E-mail: kbuehler@opengis.org

Barbara P Buttenfield

Associate Professor in the Department of Geography at the University of Colorado. Research interests: visualisation; spatial data delivery on the Internet; interface design and evaluation.

Department of Geography, University of Colorado, Boulder, CO 80309, USA
Tel: +1-303-492-3618; Fax: +1-303-492-7501;
E-mail: babs@colorado.edu

Antonio Câmara

Associate Professor in Environmental Systems Analysis at the New University of Lisbon. Research interests: environmental simulation; geographical information systems and multimedia; virtual reality; ecological modelling; water quality modelling.

Department of Environmental Science and Engineering, New University of Lisbon, Monte de Caparica, 2875, Portugal
Tel: +351-1-295-4464 ext. 0104; Fax: +351-1-294-2441;
E-mail: asc@mail.fct.unl.pt

Heather J Campbell

Professor and Head of Department at the Department of Town and Regional Planning, University of Sheffield. Research interests: technological innovation; GIS implementation; planning theory.

Department of Town and Regional Planning, University of Sheffield, Western Bank, Sheffield S10 2TN, UK
Tel: +44-114-222-6306; Fax: +44-114-272-2199;
E-mail: h.j.campbell@sheffield.ac.uk

Richard L Church

Professor of Geography at the Department of Geography and the NCGIA, University of California at Santa Barbara. Research interests: location model development; transportation planning and logistics; land management and ecosystems planning; and GIS.

Department of Geography and NCGIA, University of California at Santa Barbara, Santa Barbara, CA 93106 4060, USA
Tel: +1-805-893-4217; Fax: +1-805-839-3146;
E-mail: church@geog.ucsb.edu

Graham P Clarke

Professor of Retail and Marketing Geography at the School of Geography, University of Leeds. Research interests: urban geography; retail and marketing geography; GIS and spatial modelling.

School of Geography, University of Leeds, Woodhouse Lane, Leeds, LS2 9JT, UK
Tel.: +44-113-34-33323; Fax: +44-113-34-33308;
E-mail: g.p.clarke@leeds.ac.uk

Martin Clarke

Professor of Geographical Modelling at the University of Leeds and currently seconded to GMAP Consulting. Research interests: the development and application of spatial modelling methods including spatial interaction and microsimulation; the development of decision support systems that integrate these methods with GIS.

GMAP Consulting, One Park Lane, Leeds LS3 1EP, UK
Tel: +44-113-242-4334; Fax: +44-113-242-4554;
E-mail: martin.c.clarke@btinternet.com

David J Coleman

Dean of the Faculty of Engineering and Professor of Geodesy and Geomatics Engineering at the University of New Brunswick. Research interests: system performance determination in network environments and application of computer-supported cooperative work (CSCW) concepts to geomatics production workflow design.

Department of Geodesy and Geomatics Engineering, University of New Brunswick, Head Hall, Rm. C-28, 15 Dineen Drive, Fredericton, New Brunswick E3B 5A3, Canada
Tel. +1-506-453-4570; Fax +1-506-453-4569;
E-mail: dcoleman@unb.ca

Helen Couclelis

Professor in the Department of Geography at the University of California. Research interests: planning; geographical methodology; theories of space and time; cellular automata; geographical data modelling.

Department of Geography, University of California, Santa Barbara, CA 93106, USA
Tel: +1-805-893-2196; Fax: +1-805-893-4146;
E-mail: cook@geog.ucsb.edu

Thomas J Cova

Assistant Professor at the Center for Natural & Technological Hazards, University of Utah. Research interests: transportation; hazards; GIS and operations research.

Department of Geography, University of Utah, 260 S. Central Campus Dr., Rm. 270, Salt Lake City, UT 84112-9155, USA
Tel: +1-801-581-7930; Fax: +1-801-581-8219;
E-mail: cova@geog.utah.edu

Michael R Curry

Professor in the Department of Geography, University of California, Los Angeles. Research interests: geographical aspects of technological change; cultural and ethical aspects of geographical technologies; history of geographical ideas.

Department of Geography, University of California, Los Angeles, CA 90095, USA
Tel: +1-310-825-3122; Fax: +1-310-206-5976;
E-mail: curry@geog.ucla.edu

Peter F Dale

Retired. Research interests: land, land information and land management; cadastral systems and land registration; professional practice.

5 The Avenue, Bar, Girvan, South Ayrshire KA26 9TX, UK
Tel: +44-1465-861227
E-mail: peter.f.dale@btinternet.com

Ian J Dowman

Professor of Geomatic Engineering at University College London. Research interests: digital photogrammetry; automation of mapping processes; and mapping from satellite data.

Department of Geomatic Engineering, University College London, Gower Street, London WC1E 6BT, UK
Tel: +44-20-7679-7226; Fax: +44-20-7380-0453;
E-mail: idowman@ge.ucl.ac.uk

Geoffrey Dutton

Research interests include spatial data modelling; the generalisation of spatial data; and geospatial metadata.

20 Payson Road, Belmont, MA 02478, USA
Tel: +41-1-635-52-55; Fax: +41-1-635-68-48;
E-mail: Geoff.Dutton@mathworks.com

J Ronald Eastman

Research interests: the development of decision support routines and methodologies using GIS; incorporation of error and uncertainty in GIS analysis; sustainable implementation of information systems technology; change and time series analysis; and community-based mapping.

The Clark Labs for Cartographic Technology and Geographic Analysis, Clark University, Worcester, MA 01610, USA
Tel: +1-608-793-7526; Fax: +1-508-793-8842;
E-mail: REastman@clarku.edu

Max J Egenhofer

Associate Professor in Spatial Information Science and Engineering and Director of NCGIA at the University of Maine. Research interests: geographical database systems; spatial reasoning; GIS user interface design.

NCGIA, University of Maine, Orono, ME 04469-5711, USA
Tel: +1-207-581-2114; Fax: +1-207-581-2206;
E-mail: max@spatial.maine.edu

Susan Elshaw Thrall

Professor of Computer Science at Lake City Community College in Florida. Research interests: GIS application programming and programming languages; desktop GIS; geographically enabling data; GIS education.

Business/Industrial Division, Lake City Community College, RT 19, Box 1030, Lake City, Florida 32025, USA
Tel: +1-904-752-1822 ext. 1366; Fax: +1-352-335-7268;
E-mail: sthrall@cox.net

John E Estes

Deceased, formerly Professor of Geography at the University of California and Director of the Remote Sensing Research Unit.

Robert J Fincham

Professor and Director at the Centre for Environment & Development (CEAD), University of KwaZulu-Natal, RSA. Research interests: nutrition surveillance and nutrition information systems development; GIS applications in health and development.

Centre for Environment & Development (CEAD), University of KwaZulu-Natal, Private Bag X01, Scottsville 3209, Republic of South Africa
Tel: +27-33-260-5193/6223; Fax: +27-33-260-6118;
E-mail: fincham@ukzn.ac.za

Manfred M Fischer

Professor of Economic Geography at the Vienna University of Economics and Business Administration. Research interests: GIS and spatial analysis; spatial neurocomputing; spatial behaviour and processes, regional labour and housing markets; transportation, communication, and mobility; technological change and regional economic development.

Department of Economic Geography & Geoinformatics, Vienna University of Economics and Business Administration, Nordbergstraße 15/4/A, A-1090 Vienna, Austria
Tel: +43-1-31336-4836; Fax: +43-1-31336-703;
E-mail: Manfred.Fischer@wu-wien.ac.at

Peter F Fisher

Professor of Information Science at City University, London. Special research interests include uncertainty modelling and visualisation, and visible area analysis.

Department of Information Science, City University, Northampton Square, London EC1V 0HB, UK
Tel: +44 020 7040 8381; Fax: +44 020 7040 8584;
E-mail: pff1@soi.city.ac.uk

Leila De Floriani

Department of Computer and Information Sciences at the University of Genova. Research interests: geometric models and algorithms for GIS; terrain models; structures and algorithms for visibility computations; visualisation.

Department of Computer and Information Sciences, University of Genova, Via Dodecaneso 35, 16146 Genova, Italy
Tel: +39-10-353-6704; Fax: +39-10-353-6699;
E-mail: deflo@disi.unige.it

Pip Forer

Professor of Geography and Geographic Information Studies at the University of Auckland. Pip Forer has symbiotic interests in GIS, individual human geographies, space-time modelling and educational technology. He is currently engaged in applying GIS to urban structural analysis, tourism planning, Maori economic development, and implementing enhanced learning environments.

School of Geography and Environmental Science, The University of Auckland, Private Bag 92019, Auckland, New Zealand
Tel: +64-9-373-7599 ext. 85183; Fax: +64-9-3737-434;
E-mail: p.forer@auckland.ac.nz

Greg G Forsyth

GIS Project Manager at CSIR in Stellenbosch, South Africa. Research interests: environmental information systems; GIS in natural resource management; GIS in integrated rural development.

CSIR, PO Box 320, Stellenbosch 7599, South Africa
Tel: +27-21-887-5101; Fax: +27-12-886-4659;
E-mail: gforsyth@csir.co.za

Carolyn Fry

Freelance science writer and former editor of *Geographical* Magazine. Reseach interests: science and technology journalism.

41 Alexandra Drive, London SE19 1AW, UK.
Tel: +44-208-670-1075;
E-mail: carolyn@fry.mistral.co.uk

John C Gallant

Postdoctoral Fellow at the Centre for Resource and Environmental Studies at the Australian National University. Research interests: terrain analysis; wavelet and spectral analysis; scale issues in land-surface; and hydrological modelling.

Centre for Resource and Environmental Studies, Australian National University, Canberra, ACT 0200, Australia
Tel: +61-6-249-0666, Fax: +61-6-249-0757;
E-mail: johng@cres.anu.edu.au

Kenn Gardels

Deceased

Anthony C Gatrell

Professor and Dean, Faculty of Social Sciences, University of Lancaster. Research interests: geography of health; spatial data analysis; socioeconomic applications of GIS.

Institute of Health Research, Lancaster University, Lancaster LA1 4YB, UK
Tel: +44-1524-593754; Fax: +44-1524-592401;
E-mail: a.gatrell@lancaster.ac.uk

Arthur Getis

Distinguished Professor of Geography at the Department of Geography, San Diego State University, California. Research interests include the development of spatial statistics, particularly with regard to analysis using large datasets, and the spatial study of disease distributions and urban land-use change.

Department of Geography, San Diego State University, San Diego, CA 92182, USA
Tel: 1-619-594-6639; Fax: 1-619-594-4938;
E-mail: arthur.getis@sdsu.edu

Chuck Gilbert

Technical Services Manager with Trimble Navigation Ltd in California. Special research interest: global positioning systems.

Trimble Navigation Ltd, 645 North Mary Avenue, Sunnyvale, CA 94086, USA
Tel: +1-408-481-2812; Fax: +1-408-481-8699;
E-mail: chuck_gilbert@trimble.com

Stephen C Guptill

Scientific Advisor, US Geological Survey. Research interests: data quality; data structures; GIS design; federated geospatial data systems; GIS policy issues.

US Geological Survey, Reston, VA 20192, USA
Tel: +1-703-648-4520; Fax: +1-703-648-5542;
E-mail: sguptill@usgs.gov

Gerard B M Heuvelink

Associate Professor in Pedometrics at the Wageningen University and Research Centre. Research interests are geostatistics and error propagation in GIS.

Wageningen University and Research Centre, P.O. Box 37, 6700 AA Wageningen, The Netherlands
Tel: +31-317-474628; Fax: +31-317-482420;
E-mail: gerard.heuvelink@wur.nl

Mark E T Horn

Research Scientist at CSIRO Mathematical and Information Sciences in Australia. Research interests: decision support systems; locational analysis; transport planning.

CSIRO Mathematical and Information Sciences, GPO Box 664, Canberra ACT 2601, Australia
Tel: +61-2-6216-7054; Fax: +61-2-6216-7111;
E-mail: mark.horn@csiro.au

Gary J Hunter

Senior Lecturer in the Department of Geomatics and Deputy Director of the Centre for GIS and Modelling at the University of Melbourne. Research interests: data quality and uncertainty in GIS; spatial data algorithms.

Department of Geomatics, University of Melbourne, Parkville, Victoria 3010, Australia
Tel: +61-3-8344-6806; Fax: +61-3-9347-2916;
E-mail: garyh@unimelb.edu.au

Michael F Hutchinson

Senior Fellow at the Centre for Resource and Environmental Studies at the Australian National University. Research interests: spatial interpolation; digital elevation modelling; spatial and temporal analysis of climate; scale issues in ecological and hydrological modelling.

Centre for Resource and Environmental Studies, Australian National University, Canberra, ACT 2601, Australia
Tel: +61-6-249-4783; Fax: +61-6-249-0757;
E-mail: hutch@cres.anu.edu.au

Ron J Johnston

Professor of Geography at the University of Bristol. Research interests: political geography; electoral geography; political economy of the environment.

Department of Geography, University of Bristol, University Road, Bristol BS8 1SS, UK
Tel: +44-117-9289116; Fax: +44-117-9287878;
E-mail: r.johnston@bristol.ac.uk

Menno-Jan Kraak

Professor of Cartography at ITC - the International Institute of Geo-Information Science and Earth Observation. Research interests: 3-dimensional, temporal, and dynamic visualisation of spatial data.

ITC, Department of Geo-Information Processing, PO Box 6, 7500 AA Enschede, The Netherlands
Tel: +31-534-784463; Fax: +31-534-874335;
E-mail: kraak@itc.nl

Werner Kuhn

Associate Professor of Geoinformatics and Digital Cartography at the University of Münster, Germany. Research interests: semantics of spatial information; human-computer interaction; workflows with GIS.

Department of Geoinformatics, University of Münster, Robert-Koch-Strasse 26–28, D-48151, Münster, Germany
Tel: +49-251-8334707; Fax: +49-251-8339763;
E-mail: kuhn@ifgi.uni-muenster.de

Art Lange

Product Manager with Trimble Navigation Ltd in California. Special research interest: global positioning systems.

Trimble Navigation Ltd, 645 North Mary Avenue, Sunnyvale, CA 94086, USA
Tel: +1-408-481-2994; Fax: +1-408-481-6074;
E-mail: art_lange@trimble.com

Lars C Larsen

Chief Engineer at the Hydro Informatics Centre in Denmark. Research interests include environmental modelling and information systems.

Hydro Informatics Centre, Danish Hydraulic Institute, Agern Alle 5, DK 2970, Hørsholm, Denmark
Tel: +45-45769555, Fax: +45-45762567;
E-mail: lcl@dhi.dk

Mary L Larsgaard

Assistant Head of the Map and Imagery Laboratory at the University of California, Santa Barbara. Research interests: metadata for georeferenced information and 20th-century topographic and geologic maps.

Map and Imagery Laboratory, Davidson Library, University of California, Santa Barbara, CA 93106, USA
Tel: +1-805-893-4049; Fax: +1-805-893-8799;
E-mail: mary@library.ucsb.edu

Thomas R Loveland

US Geological Survey in South Dakota. Research interests: large-area land cover mapping; remote sensing applications; role of geographic data in image classification.

US Geological Survey, EROS Data Center, Sioux Center, SD 57198, USA
Tel: +1-605-594-6066; Fax: +1-605-594-6529;
E-mail: Loveland@edcmail.cr.usgs.gov

David R MacDevette

Director of Empowerment for African Sustainable Development (EASD). Research interests: environmental information systems; decision support systems; GIS for education and public information; information for African development.

PO Box 165, Green Point 8051, Cape Town, South Africa
Tel: +27-21-434-6012; Fax: +27-21-434-6134;
E-mail: dave@easd.org.za

Paola Magillo

Department of Computer and Information Sciences at the University of Genova. Research interests: geometric models and algorithms for GIS; terrain models; structures and algorithms for visibility computations; visualisation.

Department of Computer and Information Sciences, University of Genova, Via Dodecaneso 35, 16146 Genova, Italy
Tel: +39-10-353-6704; Fax: +39-10-353-6699;
E-mail: magillo@didi.unige.it

David M Mark

National Center for Geographic Information and Analysis, State University of New York at Buffalo. Research interests: cognitive science; cognitive models of geographical phenomena; critical social history of GIS; languages of spatial relations; qualitative spatial reasoning.

NCGIA/Department of Geography, University of Buffalo, Buffalo, NY 14261, USA
Tel: +1-716-645-2545 ext. 48; Fax: +1-726-645-5957;
E-mail: dmark@geog.buffalo.edu

David Martin

Professor of Geography at the University of Southampton. Research interests: socioeconomic GIS applications; census analysis; medical geography.

Department of Geography, University of Southampton, Southampton SO17 1BJ, UK
Tel: +44-23-8059-3808; Fax: +44-23-8059-3295;
E-mail: d.j.martin@soton.ac.uk

Robin A McLaren

Director of Know Edge Ltd in Edinburgh, Scotland. Research interests: business modelling in NLIS; management of change; visualisation; Web-based information services.

Know Edge Ltd, 33 Lockharton Avenue, Edinburgh EH14 1AY, UK
Tel: +44-131-443-1872; Fax: +44-131-443-1872;
E-mail: robin.mclaren@knowedge.com

Jeffery R Meyers

President of Miner & Miner, Consulting Engineers Inc. in Colorado. Research interests: electrical and gas utilities; GIS design and implementation.

Miner & Miner, Consulting Engineers, Inc., 4701 Royal Vista Circle, Fort Collins, CO 80528, USA
Tel: +1-970-223-1888, x110; Fax: +1-970-223-5577
E-mail: jeff.meyers@miner.com

Lubos Mitas

Associate Professor at North Carolina State University. Research interests: computational and quantum physics; Monte Carlo methods; spatial interpolations; and modelling of landscape processes.

Department of Physics, Center for High Performance Simulation, North Carolina State University, 110 Cox Hall, 2700 Stinson Rd, Raleigh, NC 27695-8202, USA
E-mail: lmitas@unity.ncsu.edu

Helena Mitasova

Works at North Carolina State University. Research interests: surface modelling and analysis; multidimensional dynamic cartography; modelling of landscape processes; and visualisation.

Department of Marine, Earth and Atmospheric Sciences, North Carolina State University, 1125 Jordan Hall, NCSU Box 8208, Raleigh, NC 27695-8208, USA
Tel: +1-217-333-4735; Fax: +1-217-244-1785;
E-mail: hmitaso@unity.ncsu.edu

Jorge Nelson Neves

PhD student in the Environmental Systems Analysis Group at the New University of Lisbon. Research interests: virtual environments; geographical information systems.

Department of Environmental Science and Engineering, New University of Lisbon, Monte de Caparica, 2875, Portugal
Tel: +351-1-2954464 ext. 0104; Fax: +351-1-2942441;
E-mail: jnn@mail.fct.unl.pt

Nancy J Obermeyer

Associate Professor of Geography at the Indiana State University. Research interests: institutional and societal issues related to the implementation of GIS; political/administrative geography.

Indiana State University, Terre Haute, Indiana 47809, USA
Tel: +1-812-237-4351; Fax: +1-812-237-2567;
E-mail: genancyo@isugw.indstate.edu

Harlan J Onsrud

Associate Professor at the Department of Spatial Information Science and Engineering at the University of Maine. Research interests: legal, policy, and institutional issues surrounding geographic information.

Department of Spatial Information Science and Engineering, University of Maine, Orono, ME 04469-5711, USA
Tel: +1-207-581-2175; Fax: +1-207-581-2206;
E-mail: onsrud@spatial.maine.edu

Peter van Oosterom

Professor of GIS technology at the Delft University of Technology. Research interests: Spatial databases (DBMS), GIS architectures, spatial analysis, generalisation, querying and presentation; Internet/interoperable GIS and cadastral applications.

Delft University of Technology, OTB, Section GIS-technology, Jaffalaan 9, 2628 BX Delft, The Netherlands
Tel: +31-15-2786950; Fax. +31-15-2782745;
E-mail oosterom@geo.tudelft.nl

Stan Openshaw

Retired because of ill health; previously Professor of Geography at the University of Newcastle upon Tyne and the University of Leeds. Research interests: computer modelling and spatial analysis; artificial intelligence, high-performance computing; GIS.

Donna J Peuquet

Professor at the Department of Geography at Pennsylvania State University. Research interests: spatial and spatio-temporal representations; spatial cognition; spatial languages; GIS design methodologies; and general issues relating to GIS.

Department of Geography, 302 Walker Building, The Pennsylvania State University, University Park, PA 16802, USA
Tel: +1-814-863-0390; Fax: +1-814-863-7943;
E-mail: peuquet@geog.psu.edu

John Pickles

Earl N. Phillips Distinguished Chair of International Studies at the University of North Carolina. Research interests: social theory; philosophy of science; political economy of technology and socio-spatial change; and regional development in South Africa and Eastern Europe.

Department of Geography and the Curriculum in International and Area Studies, University of North Carolina, Chapel Hill, NC 27599-3220, USA
Tel: +1-606-257-1362; Fax: +1-606-258-1969;
E-mail: jpickles@unc.edu

Jonathan F Raper

Professor of Information Science at City University, London. Research interests: philosophy of spatial and temporal representation; 3-dimensional geometric modelling; spatio-temporal modelling; location-based services; coastal geomorphology.

Department of Information Science, City University, Northampton Square, London EC1V 0HB, UK
Tel: +44-20-7040-8415; Fax: +44-20-7040-8584;
E-mail: raper@soi.city.ac.uk

Francois Salgé

Directeur Exécutif of the Groupe MEGRIN at the Conseil National de l'Information Géographique (CNIG). Research interests: geographic information in all its aspects — production, data management, use and application, economy of the GI sector, quality and quality management, legal and institutional aspects.

Conseil National de l'Information Géographique (CNIG), Paris, France
E-mail: francois.salge@cnig.gouv.fr

Hermann Seeger

Retired

E-mail: Hermann.Seeger@t-online.de

Martyn L Senior

Senior Lecturer in the Department of City and Regional Planning at University of Wales Cardiff. Research interests: geography of health; resource allocation; transport planning.

Cardiff School of City and Regional Planning, Cardiff University, Glamorgan Building, King Edward VII Avenue, Cardiff CF10 3WA, UK
Tel: +44-29-2087-6008; Fax: +44-29-2087-4845;
E-mail: SENIORML@cardiff.ac.uk

John W Shepherd

Professor of Geography at Birkbeck College, University of London. Research interests: urban landuse change; urban and regional planning applications of GIS.

Department of Geography, Birkbeck College, University of London, 7–15 Gresse Street, London W1P 2LL, UK
Tel: +44-207-631-6483; Fax: +44-207-631-6498;
E-mail: j.shepherd@geog.bbk.ac.uk

Michael J Shiffer

Part-time Associate Professor of Urban Planning and Policy at the University of Illinois at Chicago and Vice President responsible for Planning and Development at the Chicago Transit Authority (CTA). Research interests: public planning processes; planning support systems; spatial multimedia; urban rail transit.

Chicago Transit Authority, P.O. Box 7602, Chicago, IL 60680-7602, USA
Tel.: +1-312-733-7000 x6760; Fax: +1-312-413-2134;
E-mail: mshiffer@uic.edu

Karen Siderelis

Associate Director for Geospatial Information at the US Geological Survey. Research interests: managerial and institutional factors in GIS; national and global spatial data infrastructures.

USGS National Center, Mail Stop 108, USGS National Center, 12201 Sunrise Valley Drive, Reston, VA 20192, USA
Tel.: +1-703-648-4000
E-mail: ksiderelis@usgs.gov

Neil Smith

Formerly Senior Consultant at Ordnance Survey, now retired but acting as independent consultant.

E-mail: NSmith8757@aol.com

Jane Smith Patterson

Executive Director, North Carolina Rural Internet Access Authority; Chair, North Carolina Healthcare Information and Communications Alliance; Senior Advisor to the Governor of North Carolina for Science and Technology; Former Vice President, ITT Corp.

E-mail: jpatterson@e-nc.org

Mark Sondheim

Head of the Strategic Developments Unit in Geographic Data British Columbia, an agency of the government of British Columbia. Research interests: interoperability; geographic object modelling; large geographic databases; open GIS, object-oriented databases; terrain analysis.

Geographic Data BC, 1802 Douglas Street, Victoria, British Columbia V8T 4K6, Canada
Tel: +1-250-387-9352; Fax: +1-2501-356-7831;
E-mail: Mark.Sondheim@gems3.gov.bc.ca

Larry J Sugarbaker

GIS Manager at the Department of Natural Resources in Washington. Research interests: GIS management; geographical data integration; natural resource applications of GIS.

Department of Natural Resources, PO Box 47020, Olympia, WA 98504-7020, USA
Tel: +1-360-902-1546; Fax: +1-360-902-1790;
E-mail: larry.sugarbaker@wadnr.gov

David Swann

Defense Marketing Manager, ESRI Inc. Main research interest: military applications of GIS, especially communication information.

ESRI Inc., 380 New York Street, Redlands, CA 92373, USA
Tel: +1-909-793-2853; Fax: +1-909-793-5953;
E-mail: dswann@esri.com

Grant I Thrall

Professor of Geography at the University of Florida. Research interests: spatial analysis of urban commercial and residential land markets; commentaries on geographic technology as an emerging business sector.

Department of Geography, 3121 Turlington, University of Florida, Gainesville, FL 32611, USA
Tel: +1-352-392-0494; Fax: +1-352-392-8855;
E-mail: thrall@geog.ufl.edu

David Unwin

Retired and part time consultant. Research interests: visualisation methods; the development and application of local statistics; and the construction of virtual enhancements to field courses, geographical and GISc education.

Rectory Cottages, Draughton Road, Maidwell, Northamptonshire NN6 9JF, UK
Tel: +44-1604-686526
E-mail: 106464.743@compuserve.com

Howard Veregin

Works at Rand McNally Corporate Headquarters. Research interests: geospatial data quality assessment; simulation modelling of error; error propagation; metadata analysis; classification accuracy; and the effects of scale and resolution on data quality.

Rand McNally Corporate Headquarters, 8255 Central Park Ave., Skokie, IL 60076, USA

Nigel M Waters

Full Professor of Geography at the University of Calgary. Research interests: GIS in transportation; and spatial analytical methods.

Department of Geography, University of Calgary, 2500 University Drive, NW Calgary, Alberta T2N 1N4, Canada
Tel: +1-403-220-5367; Fax: +1-403-282-6561;
E-mail: nwaters@acs.ucalgary.ca

Robert Weibel

Assistant Professor of Spatial Data Handling at the University of Zurich's Department of Geography. Research interests: generalisation of spatial data, digital terrain modelling, and cartographic visualisation.

Department of Geography, University of Zurich, Winterthurerstrasse 190, 8057 Zurich, Switzerland
Tel: +41-1-636-52-55; Fax: +41-1-635-68-48;
E-mail: weibel@geo.unizh.ch

John P Wilson

Professor of Geography; Director, USC GIS Research Laboratory at the University of Southern California. Research interests: terrain analysis; environmental modelling; and environmental applications of GIS.

Department of Geography, University of Southern California, Los Angeles, CA 90089-0255, USA
Tel: +1-213-740-1908; Fax: +1-213-740-9687;
E-mail: jpwilson@rcf.usc.edu

Michael F Worboys

Professor at the NCGIA and Department of Spatial Information Science and Engineering at the University of Maine. Research interests include computational foundations of GIS; geospatial database technology.

NCGIA and Department of Spatial Information Science and Engineering, Room 327, 5711 Boardman Hall, University of Maine, Orono, ME 04469-5711, USA
Tel: +1-207-581-3679; Fax: +1-207-581-2206
E-mail: worboys@spatial.maine.edu

Anthony Gar-On Yeh

Professor and Director at the GIS Research Centre in Hong Kong and Assistant Director of the Centre of Urban Planning and Environmental Management. Research interests: urban and regional applications of GIS; urban development; and planning in China and Asia.

University of Hong Kong, Pokfulam Road, Hong Kong
Tel: +852-2859-2721; Fax: +852-2559-0468;
E-mail: hdxugoy@hkucc.hku.hk

Acknowledgements

Luc Anselin wishes to acknowledge the support of the US National Science Foundation (the research reported on in Chapter 17 was supported in part by Grant SBR-9410612). Special thanks are due to Antony Unwin for providing Figure 1 on Manet, and to Noel Cressie and Jürgen Symanzik for providing Figures 4 and 5 on ArcView-XGobi.

Richard J Aspinall would like to thank Peter Aspinall, Simon Aspinall, David Balharry, Dick Birnie, Marianne Broadgate, Marsailidh Chisholm, Roy Haines-Young, Matt Hare, Rachel Harvey, Ann Humble, Brian Lees, Kim Lowell, David Maguire, Jeff Maxwell, Elaine McAlister, David Miller, Julia Miller, Diane Pearson, Jonathan Raper, Allan Sibbald, Neil Veitch, Paul Walker, and Joanna Wherrett for their many and varied contributions to his thinking on the topics discussed in Chapter 69.

Mike Barnsley wishes to acknowledge the Natural Environment Research Council for data used to construct Plates 21 and 22.

Barry Boots would like to thank Michael Tiefelsdorf who provided the results in Table 1 and who drew Figures 11, 12, and 13 in Chapter 36; and Atsuyuki Okabe and Narushige Shiode who drew Figure 16 using software package PLVOR created by Toshiyuki Imai of the University of Tokyo.

Thomas Cova would like to thank Michael Goodchild for the invitation to contribute to this book and David Maguire for helpful comments on earlier drafts.

Susan Elshaw Thrall and **Grant Ian Thrall** would like to thank Mr Mark McLean of the Department of Geography at University of Florida for his comments on the layout of Table 1 in Chapter 23, and for his comments on the section on 'ready-made maps'.

Manfred Fischer gratefully acknowledges a research grant provided by the Austrian Ministry for Science, Research and Art (EZ 308.937/2 – W/3/95).

Peter Fisher's chapter was completed when the author was Visiting Fellow at the Department of Geomatics, University of Melbourne. The use of facilities, and the kind invitation to visit are both gratefully acknowledged. Figure 3 in Chapter 13 is reproduced with the kind permission of Taylor and Francis. The assistance of Alan Strachan and Paul Longley is also gratefully acknowledged.

Anthony Gatrell and **Martyn Senior** are grateful to the following for providing, or allowing them to modify, illustrations: Dr Anders Schaerstrom, Dr Danny Dorling, Professor Gerry Rushton, Professor Graham Moon, and Dr Andy Jones.

Art Getis would like to thank Judy Getis, Stuart Phinn, and Serge Rey for reviewing his chapter.

Michael Goodchild acknowledges the support of the National Science Foundation for the National Center for Geographic Information and Analysis (SBR 88–10917 and SBR 96–00465) and the Alexandria Digital Library (IRI 94–11330).

Gerard Heuvelink would like to thank Dr J Bouma and Dr A Stein (Agricultural University Wageningen) for permission to use the Allier dataset.

Mark Horn: acknowledgement is due to Eamonn Clifford and Christine Hansford at the Office of the Surveyor-General of New South Wales, Australia, who produced the illustrations for Chapter 67.

Michael Hutchinson and **John Gallant** gratefully acknowledge the assistance of Tingbao Xu and Janet Stein in the production of the figures in Chapter 9.

Dave MacDevette, Richard Fincham, and **Greg Forsyth** would like to thank Adele Wildsehut of the Centre for Rural and Legal Studies, Stellenbosch for permission to reproduce Figure 3 in Chapter 65 from Larry Zietsman's original.

David Mark's paper is a result of research at the US National Center for Geographic Information and Analysis, supported by a grant from the National Science Foundation (SBR-88-10917); support by NSF is gratefully acknowledged.

Robin McLaren wishes to thank the Ministry of Agriculture in Hungary for granting permission to use the cadastral map of Budapest (Figure 5 in Chapter 61).

Jeffery R Meyers wishes to express his gratitude for the invaluable research, editorial, and narrative review assistance provided by Christine M Condit in the preparation of his chapter. Without Ms Condit's efforts, the work would have suffered, and quite possibly not have been completed at all.

Lubos Mitas and Helena Mitasova wish to acknowledge that data for Plates 26 and 27 were supplied by K Auerswald of the Technische Universität München and S Warren of the US Army Construction Engineering Research Laboratories. Data for Plate 28 were supplied by L A K Mertes, Department of Geography, University of California Santa Barbara; data for plate 29 by L Iverson, USDA Forest Service, Delaware, Ohio; and data for Plates 30 and 31 are from US EPA Chesapeake Bay Program Office. The research in GIS applications of spline interpolation methods was supported in part by Strategic Environmental Research and Development Program (SERDP).

Harlan Onsrud's chapter is based upon work partially supported by the National Center for Geographic Information and Analysis (NCGIA) under National Science Foundation grant No. SBR 88-10917. Any opinions, findings, and conclusions are those of the author and do not necessarily reflect the views of the National Science Foundation.

Stan Openshaw and Seraphim Alvanides wish to acknowledge that Cray T3D time was provided by EPSRC under Grant GR/K43933. The 1991 Census data and boundary files are provided by ESRC and JISC and the resulting maps are all Crown Copyright.

Donna Peuquet's work was supported by National Science Foundation grant no. FAW 90-27. Portions of this work were previously published in Donna Peuquet 1988 'Representations of geographic space: toward a conceptual synthesis' in *Annals of the Association of American Geographers* 78: pages 375–94.

John Pickles draws on Chapter 1 of *Ground Truth*, the founding proposal and progress reports of Initiative 19 (I-19) of the National Center for Geographic Information Analysis (NCGIA), as well as his article 'Tool or science? GIS, techno-science, and the theoretical turn' in *Annals of the Association of American Geographers*. In particular, the chapter owes a great deal to the writings of, and discussion with, a small group of colleagues working in the liminal (and at times uncomfortable) spaces between GIS and social theory: Nick Chrisman, Michael Curry, Jon Goss, Carol Hall, Trevor Harris, Ken Hillis, Bob McMaster, David Mark, Patrick McHaffie, Roger Miller, Harlan Onsrud, Eric Sheppard, Paul Schroeder, Dalia Varanka, Dan Weiner; and at the University of Kentucky, Oliver Froehling, Eugene McCann, and Steve Hanna. The chapter draws heavily on the work of this group in I-19, especially the discussions and presentations of the planning group and participants at the Friday Harbor workshop on Geographic Information and Social Theory (1993), the Koinonia Workshop on the Representation of Space, People, and Nature in GIS (1996), the planning group of I-19 (Helen Couclelis, Michael Curry, Trevor Harris, Bob McMaster, David Mark, Eric Sheppard, and Dan Weiner), and the participants in the Critical Social History of GIS Workshop in Santa Barbara (1996) (Michael Curry, Jon Goss, David Mark, Patrick McHaffie, Roger Miller, and Dalia Varanka). Parts of section 4 lean heavily on the founding proposal for I-19 written by the author, Michael Curry, Trevor Harris, Bob McMaster, David Mark, Roger Miller, Eric Sheppard, and Dan Weiner. The summary of GIS-2 was adapted from I-19 discussions presented by Paul Schroeder and Harlan Onsrud. The salient points governing the new systems for a GIS-2 have been abstracted from the results of I-19 and the Public Participation Project, and these can be found at *http://ncgia.maine.edu/ pgis/ppgishom.html*. A discussion list for this issue has been set up at *http://ncgia.spatial.maine.edu/ webforum.html*. None of the above are responsible for any egregious errors, misinterpretations, or outrageous claims.

Writing the chapter was aided immensely by the opportunity to present these ideas to the national postgraduate programme in geography at the University of Turku in Finland. For this opportunity the author is indebted to Harri Anderson of the Department of Geography at Turku and students in the course.

Jonathan Raper would like to acknowledge that Figure 1 in Chapter 5 was developed by John Walker (*http://www.fourmilab.ch*) – the image is based on the Global Topographic Map from the Marine Geology and Geophysics Division of the National Geophysical Data Center, Boulder, Colorado, USA.

David Rhind wishes to thank Ray Harris and Ian Masser for sight of pre-publication versions of their important books cited in Chapter 56. He also gratefully acknowledges his debt to Nancy Tosta's published work on the US National Spatial Data Infrastructure.

Nigel Waters would like to acknowledge comments, suggestions, ideas and references from Shelley Alexander, Chad Anderson, Robert Arthur, Stefania Bertazzon, Murray Rice, Terry Woods, and Clarence Woudsma (all of the Department of Geography at the University of Calgary); to Tim Nyerges for supplying copies of his seminal papers; and to Harvey Miller for copies of his most recent papers. Finally, he would like to thank Howard Slavin, President, Caliper Corporation, for providing technical documentation on the TransCAD package and for other support.

Robert Weibel and Geoffrey Dutton wish to thank Frank Brazile for helping with the preparation of illustrations. A number of people have generously provided illustrations or helped with the compilation of figures, including Dietmar Grünreich and Brigitte Husen of the University of Hanover, Corinne Plazanet and Anne Ruas of IGN France, and Chris Jones of the University of Glamorgan. Partial support from the Swiss NSF through project 2100-043502.95/1 is gratefully acknowledged.

John Wilson acknowledges the following permissions for reproduction: Plates 60–62 are reprinted with permission from Hutchinson, Nix, McMahon, and Ord *Africa: a topographic and climatic database (version 1)* © 1995 by Australian National University, Canberra, Australia; Plates 63 and 64 are reprinted with permission from Corbett and Carter 'Using GIS to enhance agricultural planning: the example of inter-seasonal rainfall variability in Zimbabwe' *Transactions in GIS 1*: 207–18 © 1997 by GeoInformation International, Cambridge, UK; Figures 1 and 2 in Chapter 70 are reprinted with permission from Bell, Cunningham, and Havens 'Soil drainage class probability using a soil landscape model' *Soil Science Society of America Journal* 58: 464–70 © 1997 by Soil Science Society of America, Madison, Wisconsin; Figure 3 is reprinted with permission from Usery, Pocknee, and Boydell 'Precision farming data management using geographic information systems' *Photogrammetric Engineering and Remote Sensing* 61: 1383–91 © 1995 by American Society for Photogrammetry and Remote Sensing, Falls Church, Virginia.

The editors and contributors are grateful to the following for permission to reproduce copyright figures and tables:

Atsuyaki Okabe and Narushige Shiode for Figure 16 in Chapter 36; *Computing and Statistics* magazine for Figures 4 and 5 in Chapter 17; Garmin (Europe) Ltd for Figures 1 and 2 in Chapter 33; Georgia Tech Virtual GIS project for Figure 2 in Chapter 39; IGN France for Figures 13 and 14 in Chapter 10, courtesy of C Plazenet; Institute of Geography, University of Hanover for Figure 15 in Chapter 10; John Wiley & Sons Inc. for permission to reproduce Figures 1 and 2 in Chapter 15; MEGRIN for Table 2 in Chapter 47; Michael Tiefelsdorf for Figures 11, 12, and 13 in Chapter 36; NASA for permission to reproduce Table 2 in Chapter 45 and Table 2 in Chapter 48; New University of Lisbon for Figure 5 in Chapter 39; Oracle Corporation 1996 for Figure 2 in Chapter 29; Swiss Federal Office of Topography, DHM25©1997, 1263a for Figure 16 in Chapter 10; Tables 1 and 2 in Chapter 43 Courtesy of the State of Washington, Department of Natural Resources; Taylor and Francis, London for Figure 2 in Chapter 8 which appeared in *Time in GIS* by Gail Langran (1992) and for Table 1 in Chapter 44; Trimble Navigation Ltd for Figures 3 and 4 in Chapter 33.

We are grateful to the following for permission to reproduce copyright photographs:

A P Jones for Plate 56; American Society for Photogrammetry and Remote Sensing for permission to reproduce Plate 9; Combined Universities Collection of Air Photographs for Plate 19; Figure 2 in Chapter 29 © Caliper Corporation 1996; Figure 5 in Chapter 61 © Department of Lands and Mapping, Ministry of Agriculture, Hungary; ESRI Inc. for Plate 49; Georgia Tech Virtual GIS project for Plate 37; John Wiley & Sons, Chichester, for Plate 56; Kendall Publishing Co. for Figure 2 in Chapter 54; taken from Morgan J M et al (1996) *Directory of Academic GIS Education*; Longman for Plate 7 which appeared in Kraak and Ormeling *Cartography, visualisation of spatial data*, 1996; Microsoft *Encarta World Atlas* for Plate 8; New University of Lisbon for Plate 40; Office of the Surveyor-General of NSW for Figure 1 in Chapter 67; Plate 57 © NSW Department of Land and Water Conservation 1997; Space Imaging for permission to use Plate 18; Swiss Federal Office of Topography, DHM25©1997, 1263a for Figure 16 in Chapter 10; Taylor and Francis, London, for Figure 4 in Chapter 47; The Caliper Corporation for permission to reproduce Plate 50; Trimble Navigation Ltd for Plates 23 and 24; UCL 3D Image Maker Plate 20.

While every effort has been made to trace the owners of copyright material, in a few cases this has proved impossible and we take this opportunity to offer our apologies to any copyright holders whose rights we may have unwittingly infringed.

1

Introduction

P A LONGLEY, M F GOODCHILD, D J MAGUIRE, AND D W RHIND

Every day in different parts of the world people pose questions just like these:

Politician: 'What is the population of the Sedgefield parliamentary constituency?'

Farmer: 'What are the characteristics of the soils in the Lobley Plantation?'

Retailer: 'Where should I locate my next clothing outlet store?'

Gas engineer: 'Where should I dig up the road to gain access to the gas main?'

Health practitioner: 'How can my authority best respond to the needs of those single parent families with low income and poor housing?'

Climatologist: 'How has the hole in the ozone layer changed in the past 10 years?'

Geologist: 'Are there any trends in the pattern of earthquakes in Italy which could help predict future quakes?'

Planner: 'How has the distribution of urban and rural population changed between the past two censuses?'

Military commander: 'If I deploy my equipment and personnel here who will be able to see me and shoot at me?'

Home delivery service manager: 'What is the shortest route I can use to deliver all these refrigerators to the homes of new customers?'

City accountant: 'What is the total value of the land and property assets which the city has sold in the last 12 months?'

Forester: 'If a fire were to start here on a breezy day, in which direction would it spread and how much timber would be lost?'

Hydrologist: 'A large quantity of a pollutant has been introduced into this well: where will it spread and which customers will be affected?'

All of these questions and many more like them are concerned with geographical patterns and processes on the surface of the Earth. As practitioners of these fields know only too well, answering such questions requires access to geographical information which is characterised by its multidimensional nature (x,y,z coordinates and time), its large volume and high processing cost. To answer apparently simple geographical questions requires that data from several sources be integrated into a consistent form. The art, science, engineering, and technology associated with answering geographical questions is called Geographical Information Systems (GIS). GIS is a generic term denoting the use of computers to create and depict digital representations of the Earth's surface.

From humble beginnings in the 1960s, GIS has developed very rapidly into a major area of application and research, and into an important global business. In 1997 GIS was being taught in over 1500 universities and over 1000 schools, it had over 500 000 regular users (plus innumerable casual map users), and was a global business worth over US $12 billion. It has moved from being an esoteric academic field to being recognised as part of the information technology (IT) mainstream. Today GIS is a vibrant, active and rapidly expanding field which generates considerable public and private interest, debate, and speculation.

1 A BRIEF HISTORY OF GIS

The phenomenon – no other word seems quite as appropriate – now known as 'GIS' has many roots,

1

Table 1 Major GIS textbooks. Note only core text books are included here.

Antenucci J, Brown K, Croswell, Kevany M 1991 *Geographic information systems: a guide to the technology*. New York, Van Nostrand Reinhold

Aronoff S 1989 *Geographic information systems: a management perspective*. Ottawa, WDL Publications

Bernhardsen T 1992 *Geographic information systems*. Arendal, Norway, Viak IT and Norwegian Mapping Authority Cambridge (UK), GeoInformation International

Bonham-Carter G F 1994 *Geographic information systems for geoscientists: modeling with GIS*. New York, Pergamon Press

Burrough P A, McDonnell R A 1997 *Principles of geographical information systems*, 2nd edition. Oxford, Oxford University Press

Cassettari S 1993 *Introduction to integrated geo-information management*. London, Chapman and Hall

Chrisman N R 1997 *Exploring geographic information systems*. New York, John Wiley & Sons Inc.

Clarke K C 1997 *Getting started with geographic information systems* Englewood Cliffs, Prentice-Hall

Dale P F, McLaughlin J D 1989 *Land information management: an introduction*. Oxford, Oxford University Press

Davis B E 1996 *GIS: a visual approach*. Santa Fe, Onword Press

DeMers M N 1996 *Fundamentals of geographic information systems*. New York, John Wiley & Sons Inc.

Huxhold W E 1991 *An introduction to urban geographic information systems*. New York, Oxford University Press

Huxhold W E, Levinsohn A G 1995 *Managing geographic information system projects*. New York, Oxford University Press

Jones C 1997 *Geographical information systems and computer cartography*. Harlow, Longman

Laurini R, Thompson D 1992 *Fundamentals of spatial information systems*. London, Academic Press

Maguire D J, Goodchild M F, Rhind D W 1991 *Geographical information systems: principles and applications*. Harlow, Longman/New York, John Wiley & Sons Inc.

Martin D S 1996 *Geographic information systems: socioeconomic applications*, 2nd edition. London, Routledge

Peuquet D J, Marble D F 1990 *Introductory readings in geographic information systems*. London, Taylor and Francis

Star J L, Estes J E 1990 *Geographic information systems: an introduction*. Englewood Cliffs, Prentice-Hall

Worboys M F 1995 *GIS: a computing perspective*. London, Taylor and Francis

and it is impossible to do justice to all of them in a brief history. The first edition of this 'Big Book of GIS' (Maguire et al 1991) included a full chapter on GIS history; a book on the history of GIS edited by Foresman appeared early in 1998 (Foresman 1998) and many introductory texts include short histories (see Table 1). Rather than attempt to summarise, the emphasis here is on the diversity of GIS's roots, and on updating the story with a brief account of major events and trends since 1991 (when the first edition of this book appeared).

1.1 GIS as data analysis and display tools

The history of GIS is in many (but not all) ways the history of using digital computers to handle and analyse mapped data. Early computers were literally 'number crunchers', not handlers of the complex forms of information found on maps, and were designed to perform a task – the manipulation of numbers – that had no obvious applications in the world of map production and use. Thus it was many years after the development and deployment of the first electronic computers that uses for the new technology for handling maps began to emerge. It is now generally accepted that the British Colossus computer of the early 1940s, used to break the German Enigma codes, was probably the first electronic computer, although an electro-mechanical

one had operated in Harvard a few years earlier. By the 1950s (Rhind 1998), Swedish meteorologists were producing weather maps with the aid of computers. Shortly afterwards, Terry Coppock was geographically analysing agricultural data by computer. At the end of the 1950s, he analysed about half a million records from the Agricultural Census using an early computer in London University. The programmes summarised the data records and classified them ready for mapping by hand. Though the potential value of computer mapping was clearly appreciated at the time, the limitations of machine performance and output devices rendered such automation impossible (Coppock 1962). His work may be the earliest substantive 'GIS-based research'. Working in Canada, Roger Tomlinson (see also section 7 below) is rightly credited with seeing the need for computers to perform certain simple but enormously labour-intensive tasks associated with the Canada Land Inventory in the mid 1960s, and with being the father of the Canada Geographic Information System (CGIS), itself widely acknowledged to be the first real GIS. Tomlinson saw that if a map could be represented in digital form, then it would be easy to make measurements of its basic elements, specifically the areas assigned to various classes of land use. At that time, normal practice involved laborious and tedious hand-measurement of area by counting dots on transparent overlays of known dot density. Tomlinson's cost–benefit analysis showed that computerisation would be cost effective, despite the enormous costs and primitive nature of the computers of the time.

It is, however, important to note that many other pioneers, often working alone, also played a very significant role: for instance, many of the same technical tools were also devised in Australia, while at Northwestern University in the USA, Duane Marble and colleagues became interested in using geographical information technologies to solve transportation and other urban problems.

1.2 GIS as map-making tools

A second and quite distinct history of GIS stems from the benefits of automating the map production process. Once information of any kind is in digital form, it is much easier to manipulate, copy, edit, and transmit. The primary GIS innovator in this context was David Bickmore: at his urging, Ray Boyle invented the 'free pencil' digitiser and, by 1964, Bickmore and Boyle had set up the Oxford system for high quality digital cartography (Rhind 1988). At that time, major mapping agencies – including the US and other military bodies – began the lengthy and often rocky process of automation. The complexity of the issues involved in doing this are confirmed by the fact that even today major map-producing agencies employ a sometimes awkward mix of manual and automated techniques (for a sense of some of the reasons behind this continuing difficulty, see Weibel and Dutton, Chapter 10). Widespread achievement of the benefits of automated cartography had to await the development of suitable mechanisms for input, display, and output of map data, but the necessary devices – map digitiser, interactive graphics display device and plotter, respectively – had become available at reasonable cost by the early to mid 1970s and from then onwards an increasing number of organisations set out to convert all their maps into computerised form.

1.3 Other roots of GIS

A third root of GIS lies in landscape architecture and environmentally sensitive planning. In the 1960s, a view of planning emerged that saw the world as composed of a set of largely independent layers, each representing some component of the environment, and thus some set of environmental concerns. These layers might include groundwater, natural vegetation, or soil. McHarg (1969; 1996) was the foremost proponent of this view, and his group at the University of Pennsylvania applied it in a long series of exemplary studies. Although the initial idea was strictly manual, the computerisation of these ideas in a layer-based raster GIS was a simple step, and many systems owe their origins to McHarg's simple model (e.g. Tomlin 1990).

GIS also has urban and demographic roots. Efforts to automate national population censuses go back to Hollerith and the very early days of office automation, and the mechanical card sorters that predate digital computing. A census is inherently geographical, requiring the tabulation and publication of statistics for a range of geographical units, with complex hierarchical relationships in space (see Martin, Chapter 6). The cost of these aggregations, and the notion that they could be performed automatically from a single representation at the most detailed level, had by the late 1960s driven the US Bureau of the Census to introduce the dual independent map encoding (DIME) system – a primitive GIS representation of the urban street network with simple topology. Interestingly, part of the rationale for the use of this approach to encoding – which initially contained no coordinates – was to permit automated checking of data consistency because the data collection process was spread over many offices. Many of these ideas were reapplied at even more detailed scales in cities in support of such urban functions as infrastructure maintenance, and the Urban and Regional Information Systems Association (URISA) was founded at about this time to foster further development.

Finally, GIS has roots in the stimulus provided by the development of remote sensing, again in the late 1960s and early 1970s, as a potentially cheap and ubiquitous source of Earth observations. While many of the techniques for processing images are highly specialised, more general GIS techniques become important in order to combine information from remote sensing with other information (Star et al 1997). Today, many GIS include extensive functionality for image processing, and all types of remote sensing are increasingly the data source of choice, particularly for detection of landscape change (see Barnsley, Chapter 32; Estes and Loveland, Chapter 48).

1.4 GIS as a coherent, multi-purpose 'thing'

If GIS has so many apparently independent roots, what brought them together, and why has the umbrella term 'GIS' become so widely accepted? First, there are obvious commonalties. For example, the representation of topology invented for the DIME system at the US Bureau of the Census is almost identical to that incorporated in CGIS and in Australian work; the methods of raster processing and storage used in remote sensing systems are almost identical conceptually to those used by systems that have implemented McHarg's multi-layer view of the world. Second, it was easy from the viewpoint of the software engineering paradigms of the 1970s and 1980s to integrate functions around common representations. Once a raster or vector data model had been established, functions that process that data model in different ways were easy to add – thus it was possible, for example, to build large-scale integrations of image processing functions around a common raster representation. By the end of the 1970s, the term 'GIS' had emerged in recognition both of common technical requirements and of the opportunity to build systems that could potentially satisfy all of these applications. It took rather longer for the 'raster GIS' of the McHarg and remote sensing roots to merge with the 'vector GIS' of the CGIS, mapping, urban, and census roots. Debates on whether one or the other was 'better' were commonplace in the 1970s and 1980s, with hybrids like the 'vaster' structure emerging. To some extent this remains a cleavage in GIS to this day, exacerbated by the many variants on the basic raster and vector options (see the various contributions on representational issues in the 'Space and time in GIS' Section of the Principles Part of this volume).

When the first edition of this book was assembled, between 1989 when the project started and 1991 when the book finally appeared, the prevailing view of GIS was this notion of large-scale software integration around a common data model. Since GIS made it possible to store many coverages, software development was seen as providing a large number of functions to operate on those layers, as well as basic housekeeping functions for input, storage, and output. Extending the data model, for example by adding an option to order layers as a temporal sequence, would allow even more functions to be added. Progress in GIS was for a time measured by such additions to the richness of its data models, and associated additions of functionality – all within a monolithic and often proprietary software environment.

This view began to crumble in the early 1990s. First, the demarcation that it implied between geographical and other types of data became less valid. It became possible, for example, to handle an image within a relational database environment or a statistical package; or to make a map from a simple spreadsheet. Second, while such monolithic and expensive packages optimised the overall use of available computer power, this did not necessarily mean that individual GIS operations were performed in the most efficient manner. Third, there was growing resistance in the marketplace to solutions that required all customers to acquire all functions, regardless of need. Finally, customers became increasingly frustrated with the direct and indirect costs of monolithic proprietary solutions.

As we discuss below, today's GIS is in the process of being reinvented. There is much less emphasis on 'system', with all that is implied in that term – a clearly demarcated, monolithic, probably proprietary solution. The 'open GIS' movement, most clearly seen in the Open GIS Consortium (but by no means restricted to it), is driven by a vision of GIS as a collection of interoperable modules, under common standards (Sondheim et al, Chapter 24). The growth of electronic communications networks and associated applications means that it is no longer necessary for the data, the software, and the user to be in the same place at the same time – in the late 1990s vision the activities associated with the term 'GIS' are increasingly distributed (Coleman, Chapter 22). In time these technical innovations are likely to be reflected in institutional changes, as the field moves further from its societal roots. The advent of powerful PCs has provided substantial GIS functionality, shrink-wrapped and relatively stable and easy to use, on the individual desktop. Perhaps most important of all, the advent of the World Wide Web (WWW) has facilitated the routinisation of database linkage (Pleuwe 1997). Since GIS software systems built by many different vendors and running on different hardware in different countries can now be linked routinely together and the data used in combination, the old concepts of GIS are totally dead. This is explored in much greater detail later in this chapter.

In 1980 the GIS collective was dominated by the disciplines that gave it its impetus – landscape architecture, urban and regional planning, geography, cartography, and remote sensing, among others. With the rapid growth of GIS in the 1980s came new alliances, notably with computer science and many of its sub-fields – computer graphics, computational geometry, and database theory. Interest in making GIS easier to use led to alliances with cognitive science and environmental psychology (see Mark, Chapter 7). Increasingly, GIS is seen as a specialised sub-field of information technology and information science, and there are links of growing importance with the library science community (see Adler and Larsgaard, Chapter 64). Perhaps as a result of all this, the large, national and general-purpose GIS conferences popular in the 1980s have begun to lose attendance. They are being replaced in popularity by regional and local general-purpose conferences and by vertical market ones (e.g. GIS appears in utility company conferences).

It is difficult to identify specific individual events in the past seven or so years that have been particularly significant in redirecting GIS. The founding of the Open GIS Consortium may be one, along with the events and trends in the wider information technology arena of 'open systems' that preceded it. Certain moves by GIS vendors – new products, changes of direction, adoption of standards – have also had trend-setting significance, as have various failures, demises, and terminations in the industry. The 1990s marked the final victory of commercial off-the-shelf (COTS) software over the public-sector software development efforts that had characterised earlier decades, and had persisted well into the 1990s in the case of GRASS. It marked very significant moves by major software vendors – Microsoft, Oracle, and Autodesk among them – to establish positions in the geographical information marketplace. It also saw moves by GIS vendors into the consumer software market – an alliance between Intergraph and Egghead, for example, and new consumer GIS products from ESRI (for more on consumer GIS, see Elshaw Thrall and Thrall, Chapter 23). Arguably, however, it is the advent of the WWW that has been the single most important development affecting GIS in the last 20 years.

2 DEFINITION AND CLASSIFICATION OF GIS

Geographical information is information about geography, that is, information tied to some specific set of locations on the Earth's surface (including the zones immediately adjacent to the surface, and thus the sub-surface, oceans, and atmosphere). 'Spatial' is often used synonymously with, or even in preference to, 'geographical' in this context, although in principle it might be taken to include information that is tied to frames other than the Earth's surface, such as the human body (as in medical imaging) or a building (as in architectural drawings). Because of this difficulty, the term 'geospatial' has become popular recently, notably in the context of the US National Spatial Data Infrastructure, the Canadian National Geospatial Infrastructure, and the UK National Geospatial Data Framework. In this book, the terms 'geographical' and 'geospatial' are used interchangeably.

2.1 GIS, GI, and maps

Goodchild (1992a; see also Peuquet, Chapter 8; Gatrell and Senior, Chapter 66) identifies two distinct primitive types of geographical information: field information, in which geography is conceived as a set of spatially continuous functions, each having a unique value everywhere in space; and information about discrete entities, where the world is conceived as populated by geometric objects that litter an otherwise empty space and are characterised by attributes, such that any point in space may lie in any number of discrete entities. The field/object dichotomy underlies many areas of GIS, including its data models, data quality, analysis, and modelling (e.g. Burrough and Frank 1996; see also Raper, Chapter 5; Martin, Chapter 6).

Over the years the vision of a GIS has shifted significantly, but has always included the notion of processing geographical information within an integrated environment. It has been argued that the environment need not be digital, and that the principles of GIS can certainly be taught outside the digital environment, but today's world is increasingly digital and GIS is now almost always associated with digital computing in one form or another. It has also been argued (e.g. Maguire 1991) that the definition of GIS should include much more than the digital environment – in this conception the people who interact with it are also part of the system. Finally, GIS has been defined by its objectives, as in Cowen's definition of a GIS as a spatial decision-support system (Cowen 1988).

Today, the term GIS tends to be applied whenever geographical information in digital form is manipulated, whatever the purpose of that manipulation. Thus using a computer to make a map is as likely to be described as 'GIS' as is using the same computer to analyse geographical information and to make future forecasts using complex models of geographical processes. At the same time, there are significant exceptions. The Earth images collected by remote-sensing satellites are geographical data, but the systems that process them are not likely to be called GIS as long as they remain specialised to this particular form of data – in such cases, 'GIS' tends to be reserved for systems that integrate remotely-sensed data with other types, or process data that have already been cleaned and transformed. Similarly, an atmospheric scientist or oceanographer will tend to associate 'GIS' with systems used more for multidisciplinary work and policy studies, and will use other software environments for modelling and analysis within the confines of his or her own discipline. In short, because GIS implies a generalised software environment that is exclusive to geographical information there is a tendency for it to be most strongly associated with multidisciplinary, integrative work and applications; in more narrowly-defined environments less general solutions may be adequate.

Moreover, there is a persistent – albeit unfortunate and misleading – tendency for 'GIS' to be associated with the digital representation of the kind of geographical information that has traditionally been shown on paper maps, rather than geographical information conceived more generally. While maps may appear to place few restrictions on their compilers and users, in reality they can be highly constrained in the ways they represent the Earth's surface. Traditionally (although with notable and celebrated exceptions) paper map information has typically been:

- static, favouring the representation of fixed aspects of the Earth's surface, because once made, a paper map cannot be changed;
- 2-dimensional, and unable to show many diverse attributes of 3-dimensional socioeconomic systems such as cities, or physical environments such as the subsurface, oceans, or atmosphere;
- flat, because the curved surface of the Earth must be projected in order for it to be shown on a sheet of paper – or a regular solid like a globe;
- apparently exact, because there have been few applications of cartographic techniques for showing uncertainty in mapped information;
- unconnected to other information that may be available about the same set of places, but cannot be shown on the same map (and possibly cannot even be physically stored in the same place).

Because of its roots in mapping in general, and traditional cartographic practice in particular, much of GIS practice and application has remained similarly shackled to these limitations, unable to move beyond the metaphor of the traditional paper map (but see the Epilogue for a prospective view).

Wright et al (1997) define several different interpretations of what it means, in today's parlance, to be 'doing GIS'. One interpretation might simply be the **application** of a particular class of software, having chosen it from among the classes available today by considering various pros and cons, in order to gain insight, learn more about the world, support some kind of **management** decision-making, etc. In a more general sense, 'doing GIS' might involve applying the **principles** of GIS, including its particular ways of representing the world, and thus operating within a 'GIS paradigm'. Or it might involve furthering GIS **technology** by developing new capabilities. Finally, GIS might provide the medium for studying one or more of the fundamental issues that arise in using digital information technology to examine the surface of the Earth. Wright et al argue that only in the last instance is one necessarily 'doing science' when 'doing GIS'.

This argument, and others related to it, has led to a search for new terms that encompass activities that are less dependent on the particular nature of today's software offerings. Goodchild (1992b) has argued that this can be done by decoding the familiar acronym as geographical information science (GISc), and this idea is reflected in the recent establishment in the USA of the University Consortium for Geographic Information Science (UCGIS), an organisation of the principal GIS research institutions (see *http://www.ucgis.org*). The term geomatics has also gained some popularity, particularly in Europe and Canada and in the surveying engineering and geodetic science communities (see for instance, *http://www.geocan. nrcan.gc.ca*). Geocomputation also has similar connotations, although here the modelling of process may be more important than the modelling of information per se. Forer and Unwin (Chapter 54) have suggested no fewer than three decodings of GIS: GISy for the systems, GISc for the science, and GISt when the focus is on studies of GIS, particularly in the context of society and its institutions.

2.2 Is spatial 'special'?

Ultimately, the continued existence of GIS relies on the belief that there is some value in dealing with geographical information as a special case – that there is 'something special about spatial' (unfortunately there seems to be no available English term to complete the more appropriate 'something . . . about geographical' – 'magical', 'fanatical' don't quite serve the purpose). In the past, the case was argued on several grounds, including:

- the nature of geographical queries, potentially combining topological, geometric, and attribute elements, all with some fuzziness embedded;
- the special data structures, indexing systems, and algorithms needed for efficient processing of geographical information;
- the multi-dimensional nature of geographical information (x,y,z,n. . .);
- the voluminous nature of much geographical information;
- the fundamental inability to create a perfect representation of the Earth's surface, forcing users of GIS to deal with problems of data quality, accuracy, and uncertainty;
- the isolated nature of traditional production arrangements for geographical data, including the existence of public sector mapping agencies in most countries;
- the need for special standards for geographical information;
- the combination of distinct legal and economic contexts of geographical information, including copyright laws, liability, privacy protection, freedom of information laws, and costs of acquisition, that vary markedly from one country to another.

Recently, however, much of this basis for demarcation has diminished, if not disappeared altogether. In today's software environments, the special structures needed for handling geographical data are largely invisible to the user. The size of a single remotely-sensed image from a sensor like Landsat no longer seems formidable when personal computers often include gigabytes of storage. And debates about the legal and economic contexts of GIS are increasingly embedded within much broader debates about information policy and practice in general. Moreover, several recent technical developments have reduced the need to maintain distinctions within today's computing environments. Open standards like Microsoft's Object Linking and Embedding/Component Object Model (OLE/COM) and Object Management Group's Common Object Request Broker Architecture (CORBA) allow information of different types to be passed between environments, suitably enclosed in 'wrappers' (interfaces) that describe the type to the host. Thus it is increasingly possible to hold geographical information within an environment designed for processing text – that is, a familiar word processor. In effect, these technologies decouple the handling of a container of information from the nature of its contents, treating all information as 'bags of bits'. Structured Query Language (SQL) and other query languages have been extended recently to handle the special cases of geographical information and geographical queries, and extensions like Oracle's SDO increasingly allow geographical information to be handled within the frameworks of mainstream database management systems.

2.3 Geographical Information is special

Unlike GIS software, geographical information *is* special in many ways, but some of the more fundamental of these have little to do with its manipulation in digital systems. Anselin (1989) has argued that 'spatial is special' in two crucial respects. The first is expressed in Tobler's famous 'First Law of Geography' (Tobler 1970): 'all things are related but nearby things are more related than distant things'. This property of spatial dependence, or at least autocorrelation, is endemic to geographical data, violates the principle of independence that underlies much of classical statistics, and is the basis on which any representation of the infinite complexity of the Earth's surface is even approximately possible.

Anselin's second special characteristic is spatial heterogeneity, the propensity of geographical data to 'drift' such that conditions at one place are not the same as conditions elsewhere. Statistically, this concept corresponds to non-stationarity, and is well-known in geostatistics (e.g. Isaaks and Srivastava 1989). Practically, it means that the results of any analysis are always dependent on how the boundaries of the study are drawn – whereas it is often (erroneously) assumed that a geographical study area is analogous to a sample in statistics, drawn from the set of all possible study areas by some random process, and thus that the choice of study area has minimal effect on the results. Many of the arguments that emerge from this point can be found in the fractal literature (e.g. Mandelbrot 1982). More recently, Fotheringham (1997), Getis and Ord (1992), and others have argued for a new approach to geographical analysis based on the need to determine the local characteristics of places, rather than universal generalities (see also Getis, Chapter 16).

To these two might be added a third, which is particularly apposite in the context of GIS. The idea of expressing geography as a series of layers suggests that each layer captures something unique to it; statistically, that each layer makes an independent contribution to the total picture of geographical variability. In practice, however, geographical layers are almost always highly (if variably) correlated. It is very difficult to imagine that two layers representing different aspects of the same geographical area would not somehow reveal that fact through similar patterns. For example, a map of rainfall and a map of population density would often clearly have *some* similarities: population could be dependent on agricultural production and thus rainfall (or irrigation!), or might tend to avoid steep slopes and high elevations where rainfall was also highest. Of course, these correlations are often indirect, with other controlling variables and cultural features and inertia playing important roles.

These special characteristics of geographical data are undoubtedly important, but often not unique. Dependence is also endemic in time series; non-stationarity occurs in many contexts. While there is every reason for users of GIS to be aware of the ecological fallacy (Robinson 1950) and the Modifiable Areal Unit Problem (Openshaw 1984) – and these themes are explored at greater length in the chapters on spatial analysis later in this volume

(e.g. Openshaw and Alvanides, Chapter 18) – it is difficult to argue that they justify the demarcation of GIS from other types of software.

One final characteristic is worth discussion, because it appears to be of increasing significance as the information society moves to reliance on a world of distributed computing. Society's arrangements for production, storage, and use of information depend critically on how interest in that information is determined. In the case of detailed geographical information, interest tends to be highly localised – interest in a street map of Manchester is clearly of greater importance to users located in Manchester than it is to users in Paris. Traditionally, this has been reflected in the pattern of availability of that information in libraries, bookshops, etc. In a world in which information is distributed over a myriad of servers accessible through tools such as the Web it is of critical importance to know where a particular set of information can be found. That issue is resolved in the case of textual information through the existence of search engines, which use Web crawlers to find and catalogue text by key word. But no comparable mechanism yet exists for geographical information though embryonic Web-based geographical services already exist. In developing new geographical data search engines, the new world of distributed computing is likely to find new ways in which 'spatial is special'.

3 CURRENT TRENDS IN GIS

3.1 The evolving GIS environment

GIS is a young area of technological innovation and application. It is also a very rapidly changing one. Without doubt, developments in computer technology have been a major contributor to the rapid advances of GIS. Thus in exploring the world of GIS it is appropriate to begin by charting the main relevant technological advances of recent years and seeking to gauge their impact on GIS.

Perhaps the root cause of all technological advances, as far as GIS is concerned, is improvement in computer hardware. Twenty years ago Gordon Moore, co-founder of the microprocessor company Intel, suggested that computer hardware performance would double and price would halve every 18 months. In the intervening years this prediction, subsequently dubbed 'Moore's Law', has held true and it appears that for the foreseeable future hardware will continue to improve at this rate. In mid 1997, however, after many years of close adherence to Moore's Law, announcements by IBM and Intel predicted that the rate of growth of processor speed would be even faster in the next few years. IBM announced a technique to replace aluminium connections on microprocessors with copper (which has greater conductivity), and Intel announced 'flash' technology, which allows two or even more bits to be processed by each processor element instead of one.

As a result of these developments, not only have hardware systems become faster and cheaper, but their physical size has also decreased. Notebook and field portable computers, for example, are now very commonly used in GIS applications. Yet the full implications of improvements in computer processor speed have yet to be fully recognised in GIS applications. Perhaps inevitably, hardware bottlenecks do remain in today's computers, notably with respect to the internal communication bus and the speed of disk access. Some of the hardware performance increases have been soaked up by the development of ever more sophisticated graphical user interfaces (GUIs), while the emphasis in spatial analysis has been to use enhanced hardware performance to support visualisation and data exploration rather than data modelling as more traditionally conceived.

Only a few years ago, the engineering workstation with its UNIX operating system was the dominant platform for delivering GIS. Since then, there has been the shift towards the personal computer, the innovation of desktop computing, and the gradual domination of Microsoft (the Windows operating system) and Intel's microprocessors (the 'Wintel' combination). By 1997 the Wintel combination had become the system of choice for GIS applications on the desktop. For server machines and specialist applications, UNIX remains a credible and important alternative. But Windows has become so widely adopted in GIS applications because of its widespread use in general applications, its (comparative) ease of use, its ability to run both GIS and non-GIS applications, and its low cost. As a consequence, the major GIS software systems have a remarkably similar 'look and feel'.

As we saw in the opening paragraphs of this introduction, one of the fundamental characteristics of GIS applications has been their use of large and very large quantities of multi-dimensional data (i.e. x,y,z coordinates) and the need for multi-user access to spatially continuous databases. The early GIS software systems used binary flat files to store data and specialist data management routines for data organisation and access. Fairly quickly, with the rapid growth of relational database management system (RDBMS) technology, many software developers began to manage non-geometric data using RDBMS. Today, the issues of performance, multi-user access, and data compression have largely been resolved and it is the norm for GIS software systems to store both geometric and non-geometric data in an RDBMS. With the development of Object-Relational DBMS and their capability for extension so that they can manage complex data types, like spatial, these are expected quickly to become the standard.

Most early GIS were individual isolated islands of technology. Since then, the rise in importance of network technology has had a profound impact on GIS. The words of Scott McNealy, President of Sun Microsystems, 'the computer is the network, the network is the computer', clearly state the importance of networks. In the late 1980s there was a move to connect machines together using local area network technology. More recently, wide area network (WAN) technology has been of interest to users. None of these can really compare, however, to the growth in interest and rapid uptake of the Internet as network-based technology.

The Internet is the world's largest public network. It is a multi-faceted mosaic of computer servers supplying information upon request to multiple clients. The Internet is unified by common use of the Internet Protocol (IP). This communication standard allows heterogeneous hardware to communicate in a simple, but effective, fashion. The WWW is a popular application which operates over the Internet. The Web is a distributed collection of sites (servers) composed of multimedia documents. These are linked together using the hypertext transmission protocol (http) and are spatially referenced using a uniform resource locator (URL). Web use has increased at a truly incredible rate in recent years, establishing new standards for many types of GIS application. Those focusing on data publishing, simple display, and query have been most successfully implemented.

While the Internet is almost certainly the technological innovation that is exerting the greatest external influence upon GIS at the present time, its impacts are all the more far-reaching because of contemporaneous developments within GIS. Central to these developments has been the establishment of the Open GIS Consortium (OGC) in August 1994. This is an international consortium of more than 100 corporations, government agencies, and universities. The OGC has put considerable effort into the development of 'interoperable' software using OpenGIS (Open Geodata Interoperability Specification) to build links between different proprietary systems (Sondheim et al, Chapter 24). Allied with the development of the Internet, open object standards and object brokers have been used to support distributed computing. The CORBA and OLE/COM standards allow 'objects', or packages of digital information, to be passed freely between different software environments, and make the contents of objects understandable to systems. More recently, the Java language has provided a means for sending program modules over the Internet as well as data, allowing one system to send a process for another system to execute. Other fragments of programs known as 'applets', 'plug-ins' and 'add-ons' are now routinely distributed from one system to another. Each of these developments is contributing to a new Internet-based computing environment in which it is as common to distribute the ability to process as it is to distribute the subject of processing – that is, the data. This increasing fragmentation of programs is extending the GIS environment ever further beyond its self-contained, monolithic roots.

The combined effect of the application of these technologies is that GIS software is breaking up into reusable 'plug-and-play' modules, which can be assembled and used through the Internet. It is also leading to the development of packages of software modules and data for use as so-called 'desktop GIS' (Elshaw Thrall and Thrall, Chapter 23): some observers view this as a transitory phase on the way towards use of the Internet as the principal platform for GIS.

Each of these advances in technology has, of course, been designed to improve the ability to store, manage, manipulate, display, and query geographical data. Together they have also profoundly changed the way that computing is carried out, as the practice of a user interacting with a file server becomes supplemented by 'peer-to-peer' computing in which every user is potentially both a client and a server – both a source and a destination for computation.

3.2 Our digital world

There have also been a number of significant changes in the way data are used and disseminated which have additionally influenced GIS applications. Spatial referencing is by definition essential to any GIS application, yet application-specific thematic layers alone rarely create a readily-recognisable view of the world – as anyone who has been presented with a choropleth map of an unfamiliar area will testify. Important developments are taking place in the provision of digital 'framework data' for GIS (Rhind 1997b). Framework data provide information pertaining to the location of topographic and other key features in the natural, built, or cultural landscape, which may be used as a backcloth to application-specific thematic data. Since the first edition of this book, such data have been created by a number of national mapping, cadastral, and census agencies and these present officially sanctioned views of the surface of the Earth, to a range of emergent data standards (Salgé, Chapter 50). 'Unofficial' sources of framework data also exist in the form of classified high-resolution satellite images, obtained from the new generation of high-resolution remote sensing satellites or from the new radar sources (which are less limited by cloud).

Each of these sources of framework data has become increasingly commercialised during the 1990s – on the one hand, national mapping and census agencies in many parts of the world are developing commercial datasets in order to meet their cost recovery targets; while, on the other, the break up of the former Soviet Union and the launch of new commercial satellites has done much to multiply the number of sources of remote sensing imagery. The latter commercial developments have become of wider import to GIS given recent technical developments in softcopy photogrammetry and pattern recognition. These are leading to the widespread creation of new products such as digital orthophoto maps and elevation models (DEMs) at much lower cost than has previously been the case.

With the general proliferation of digital datasets it has become increasingly difficult for the GIS user to know what datasets exist, what quality they are, and how they might be obtained. Allied to the development of the Internet, an important current development is the creation of on-line metadata – data about data – services, a number of which are designed for use with geographical location as a primary search criterion. An interesting development in 1997 was the creation of comparatively low cost intelligent data products containing functionality and metadata which allow fast direct access by GIS software packages. More generally, the development of whole digital libraries of geographical information is becoming feasible, and there is growing interest in using the metaphor of libraries to support geographical information management and data sharing (Adler and Larsgaard, Chapter 64).

Just as it is becoming easier for GIS users to find out exactly which digital data exist, so it is also becoming easier for them to collect their own digital data. Although many of the bottlenecks of digitising data from old hardcopy sources remain, much new data are now collected using the global positioning system (GPS) technology that has developed rapidly during the 1990s (Lange and Gilbert, Chapter 33). Low cost hand-held or mounted GPS receivers are suitable for many (but by no means all) field data collection purposes, and record geographical location routinely to quite high levels of precision (40–100 metres for civilian 'selective availability' applications and 10–32 metres for military applications) by reference to the US NAVigation Satellite Timing And Ranging Global Positioning System (NAVSTAR GPS) or its Russian equivalent (GLONASS). Much higher resolutions are obtainable using differential GPS and post-processing. This technology has revolutionised data collection for a wide swath of applications, particularly as receivers have been developed which also permit input of aspatial attribute data during the data collection phase.

Even in 1991 it was clear that information in general and geographical information in particular were becoming both a tradable commodity and a strategic resource. Nowhere in GIS has this continuing trend become more apparent than in business applications of GIS, where a huge value added reseller (VAR) and consultancy industry has developed to service business client needs. The data for most business applications have hitherto largely been obtained by combining census variables into composite 'geodemographic' indicators, which experience has shown bear an identifiable correspondence with observed consumer behaviour. More recently, the proliferation of digital customer records, allied to the collection of data from new customer loyalty programmes, is leading to the creation of more and more 'lifestyles' databases.

These are not as geographically comprehensive as conventional geodemographics, but are much more frequently updateable and contain data which might be judged more pertinent to prediction of customer behaviour than those from conventional censuses.

3.3 Scientific trends and research directions

Elsewhere in this book we will explore the broader scientific trends in GIS: the current emphasis on the big questions of geographic information science (GISc) over the small technical questions; the growth of interest in human cognition that should make GIS easier to use (Mark, Chapter 7); the shift in emphasis towards data modelling and ontological issues (Raper, Chapter 5; Martin, Chapter 6); and the development of new strong links to mainstream computer science (e.g. Worboys, Chapter 26; Oosterom, Chapter 27). These and many other interesting developments and research directions are discussed at length throughout the book, and particularly in the first two sections.

4 WHAT WAS WRONG LAST TIME

The message of all of this is that GIS continues to be a vibrant and fast-changing area of business, application development, and research. From its origins in the 1970s, through its rapid growth phase in the 1980s, GIS has rapidly expanded and matured into a general-purpose information technology that is capable of solving the widest range of problems in a geographical context. Although its disciplinary heart lies in academic geography (Coucelis, Chapter 2; Johnston, Chapter 3), its continued growth and vitality is much more broadly-based than this – GIS is at least as much grounded in people's enduring fascination with maps, and the ease of spatial expression and reasoning that maps allow, as in any particular disciplinary matrix.

The first edition of this book (Maguire et al 1991) attempted not just to set out the whole panoply of GIS circa 1991, but also to anticipate the directions in which its inherent dynamism would move it. If book sales and patterns of academic citations are anything to go by, the first edition certainly provided an accessible and comprehensive snapshot of the state of GIS at the time of its publication, but it is only now with the benefit of hindsight that we can identify the respects in which it failed to anticipate the direction and strength of change.

Perhaps the most glaring omission is the complete failure of the book to anticipate the growth of the Internet and the World Wide Web into a massive global computer. It follows that there was far too little discussion of the technologies required to support distributed databases, distributed processing, and above all distributed users, together with the emergent role of the Internet in supporting vast numbers of servers and clients.

Second, in retrospect, there is the sense throughout the book that the most important technical problems had all been solved and that the big remaining ones concerned GIS management and institutional usage. While there is undoubtedly truth in the latter, it is clear in hindsight that very big technical issues still remain, whilst in the related area of methodology the emergence of GISc and geocomputation suggests that spatial analytical elements may not have been afforded sufficient prominence last time.

Third, there was a sense in the first edition of a quest for the Holy Grail of an 'all-singing, not all-dancing' GIS which would permit the fullest range of analytical operations to be performed. Even from the brief discussion of current trends contained in the previous section, it should be clear that a strong counter-trend has been the break-up of GIS software into packaged components, and that data components are often of similar importance to analytical functions in such systems. The Internet has had the opposite effect in allowing software to converge across different domains, and as a result users have been able to assemble task-oriented systems at will and as needs dictate – particularly given that the drive towards interoperability has meant that component software modules need not all originate from a single source. Neither trend has fostered the development of a single integrated GIS software system. Indeed the emphasis upon the development of analytical functions proved to be a distraction from the under-played information management functions of GIS, development of which has subsequently been key to the wider dissemination and adoption of GIS.

Fourth, passages of the first edition are redolent of a rather more technocentric view of the world – a sentiment which also characterises most of the first generation of GIS textbooks. This sense of mechanistic manipulation has subsequently dissipated somewhat, with the advent of social critiques of GIS and the wider realisation that GIS can be as much an empowering technology as it is a technology of control. The reasons for this emphasis in the first edition probably lie in the then prohibitively high cost of GIS software systems (at a time prior to licensing deals for higher education and government usage, for example) and a fascination with the implications of plummeting costs of computation for analytical functionality rather than the far wider distribution of PC and networked computer technology. The technocentric view is epitomised by the amount of space devoted to the promise of artificial intelligence – a theme which requires surrender of power to the machine rather than encouraging user empowerment, and which subsequent experience suggests cannot deliver much of its early promise.

Finally, there is a recurring sense throughout the first edition that because 'spatial is special' the GIS industry would continue to comprise a set of isolated, proprietary, specialised vendors. Most of those have subsequently disappeared, although two of the early market leaders (ESRI and Intergraph) retain large market shares. The new entrants to the industry are the IT heavyweights Microsoft, Autodesk, and Oracle – as we will discuss further in the next section.

5 THE WORLD OF GIS

There are several encouraging signs that in recent years GIS has reached new levels of popularity, respectability and maturity, and here we will provide something of the flavour of the state of GIS in the late 1990s. It is impossible to be comprehensive in summarising the state of GIS. Quite apart from anything else, space – even in a book at large as this – does not permit it. Rather the approach we will take is to review some of the major strands of development and current interest.

A key sign of the maturity of any discipline or business area is the development of coordinating bodies and academic and professional societies. GIS now has these in abundance. In the USA, the best known include: ACSM (American Congress on Surveying and Mapping), the GIS speciality group of the AAG (Association of American Geographers), AM/FM (Automated Mapping and Facilities Management: also in Europe), ASPRS (American Society of Photogrammetry and Remote Sensing), UCGIS (University Consortium for Geographic

Information Science), and URISA (Urban and Regional Information Systems Association). In other parts of the world comparable organisations include: AGI (the UK Association for Geographic Information), EUROGI (the European GI organisation), AGILE (Association of Geographic Information Laboratories in Europe), CPGIS (Chinese Professionals in GIS), GISRUK (GIS Research – UK) and UDMS (the Urban Data Management Society in Europe). These and many other bodies regularly organise society meetings featuring conferences and exhibitions. Together with a parallel set of meetings organised by private companies and public agencies (notably under the auspices of the OGC, discussed in section 3.1 above), GIS events often feature several thousand participants and provide close interaction between vendors, users, consultants, and researchers.

OGC, through OpenGIS, has brought forward standards for the interoperability of GIS software. The initial standard is based on the straightforward exchange of simple features (points, lines, and polygons) between commercial systems. Comparable international standards bodies that are focusing effort on developing *de jure* standards for GIS include ISO (the International Standards Organisation) and CEN (Comité Européen de Normalisation: Salgé, Chapter 50). ISO is an international body with representatives in many countries and CEN is a European umbrella organisation. These and other organisations are seeking to standardise almost all aspects of GIS, from metadata to database interfaces. If these standards are complementary and are widely adopted then they should further stimulate the growth of GIS.

One of the interesting aspects of GIS is the close involvement of software vendors in the continued evolution. Two of the earliest and most successful vendors – Environmental Systems Research Institute Inc. (ESRI) and Intergraph Corporation – remain the GIS market leaders. However, the increasing use of GIS on the desktop has led to new market entrants such as Mapinfo Corporation, while the movement of GIS to the Web and the ever closer relationships between computer-aided design (CAD) and GIS software has brought firms like Autodesk and Bentley into the GIS market. At the same time, IBM Corporation, Informix Corporation and Oracle Corporation have extended their respective

DBMS to incorporate spatial data. In late 1997 the value of the global software market was estimated to be worth between US$627 and $904 million, depending upon whether a narrow or broad definition of GIS was used, with ESRI and Intergraph having market shares of about 33 per cent each (using the narrow definition) or 20 per cent each (using the broad definition) (Crockett 1997). Each of the market leaders is diversifying into emergent market niches and data-related products. Smallworld Systems maintains a strong position in utilities. After a period of rationalisation (because of takeovers and bankruptcies) GIS has become dominated by just a handful of vendors. By 1997, the GIS software market was probably worth about $1 billion worldwide.

Overall, expenditures on GIS are much higher than simply those on software. The US Office of Management and Budget (OMB) found in 1993 that total expenditures on digital geographical information in Federal agencies amounted to over US$4 billion. Adding the effects of activities at the state and local levels, and the activities of the private sector and non-governmental organisations leads to estimates of between $10 billion and $14 billion for the total value of the digital geographical information industry in the USA, although this is almost certainly an underestimate. Precise estimates of the total number of GIS users are similarly difficult to ascertain. A conservative estimate is that there are about 100 000 highly technical or professional GIS users in the world. When the 500 000 desktop users and one million casual viewers are added, the total becomes about 1.6 million. This is well in excess of the 250 000 or so predicted by the editors of the first edition of this book (Maguire et al 1991). At the current rate of expansion there could be eight million GIS users worldwide by the year 2000.

Just as the number of users has grown, so has the interest and involvement of academics. Education in GIS began in the universities, but has spread over the years to include significant efforts in training colleges and vocational programs, secondary schools, and even elementary schools. These are largely complementary to the training programs offered by major GIS vendors. Recently there has been much interest in distance learning, to address what is perceived to be a lack of educational opportunities for professionals in mid-career, and the UNIGIS consortium now offers distance

13

learning through a network of institutions in several countries. University-based research has been stimulated in many countries by major funding for centres. In the USA the National Center for Geographic Information and Analysis (NCGIA) was established in 1988, with funding from the National Science Foundation, as a consortium of three institutions. In the UK, the Regional Research Laboratories stimulated the development of a network of universities committed to GIS-based research, funded by the Economic and Social Research Council between 1987 and 1991. Similar national research programmes exist in Korea, the Netherlands, France, Japan, and many other countries. The University Consortium for Geographic Information Science (UCGIS) was established in the USA in 1995 as a network of major research universities, and now has nearly 50 members. The European Science Foundation's GISDATA program coordinated and stimulated GIS research in a network of European countries between 1993 and 1997.

6 GIS: PRINCIPLES, TECHNIQUES, MANAGEMENT, AND APPLICATIONS

Just about the only thing that has not changed about GIS during the 1990s is its inherent dynamism. It is seven years since the first edition of this 'Big Book of GIS' appeared, and the editors of this second edition find themselves dealing with a subject which has developed and expanded enormously – not least in the range of geographical realities that GIS used to represent and the wider range of media through which digital representations of that reality may be constructed. Since the first edition was published the scale and pace of human interactions with computers has accelerated, and the provision and use of digital geographical information has provided one means of navigating through a geographical reality that we understand to be ever more detailed and complex. What, in the face of these remarkable upheavals, are the prospects for recreating a GIS reference work that is as relevant in terms of content and coverage as its forebear?

It is perhaps best to begin with a view of what this book is not. First, in these two volumes we have not sought to revisit all of the principles expounded in the first edition, since much of this material has

completed the transition from application-led research and practice to standard textbook material. Table 1 on page 2 lists some of the general GIS textbooks that are available. Even in a work of this length, it is impossible to cover everything in GIS from first principles, given the vast expansion of the field since the first edition. Second, neither is it possible to cover the entire range of GIS applications, and our aim here has been to review those applications from operational and strategic GIS practice which we judge to be of key importance in understanding the breadth of the field. Applications of GIS are truly legion and the detail of practice is as fast-changing as the field of GIS itself. For this reason, readers with particular application interests should instead consult any of the range of GIS journals and professional magazines, listed in Table 2, which contain periodic reports of the experience of a wide range of GIS applications – many of these are targeted at national or supranational markets, which adds further specificity to the experience that is reported. Third, it is not just an extended guide to the latest research in GIS by academics – various monographs (notably the GISDATA and Innovations in GIS series, and the books arising out of the NCGIA initiatives) exist to document these rapid developments and changes.

Table 2 Major GIS journals and magazines

(a) Journals

Cartography and Geographic Information Systems
Computers and Geosciences
Computers, Environment, and Urban Systems
Earth Observation Science
Geographical Analysis
Geoinformatica
International Journal of Geographical Information Science
Journal of the Urban and Regional Information Systems
    Association
Photogrammetric Engineering and Remote Sensing
Transactions in GIS

(b) Magazines

Geo Info Systems
GIM International: Geomatics Info Magazine
GIS Africa
GIS Asia Pacific
GIS Europe
GIS World
Mapping Awareness

Instead we have attempted to produce a work which is focused towards 'frontiers in GIS' and which discusses and explains the issues and practices important to everybody who comes into contact with GIS. Thus we have tried to summarise existing state-of-the-art knowledge and best practice, to explain recent developments, and to anticipate possible future ones. We have sought to cross-reference related themes and to provide pointers to other textbooks, research papers, and consultancy reports wherever appropriate. We hope that readers will find this new edition at least as comprehensive, readable and well-illustrated, and as thoroughly up-to-date as the first edition. In short, we have attempted to create a hybrid of relevant pedagogy and research and development, produced by the leading writers in the GIS field. The result looks very different to the first edition, but this is only fitting given the transformation of GIS itself over the last seven years.

In producing a second edition of what we hope will remain the definitive GIS reference book ('Big Book Two') we began essentially from scratch. At an early stage in our deliberations we recognised that we should separate our discussion of *technical issues* from underlying *principles* in order to reflect different interests among our readership. Due recognition of the wider *management* functions that GIS now has would require that a separate section be devoted to such issues. Finally, a new range of *applications* would be used in order to illustrate the ways in which theory, technique, and management map into a representative range of operational and strategic situations in practice. **Principles** and **Technical Issues** are discussed in the first volume of this set, and **Management Issues** and **Applications** in the second.

Of course it is not just the world of GIS that has changed so profoundly during the 1990s, but also those many aspects of the real world that GIS seeks to abstract and to model. At its simplest, if we recognise that the world is not the same as it was, then we should not be surprised if the ways in which we order it are not the same either. Science is also changing, as many of the old certainties are breaking down in response to the challenges of relativism. We thus begin the wholly rewritten **Principles** Part of this book with a review and reappraisal of the central role of GIS in structuring our geographical understanding of the world, including the arguments, debates, and dialogues that have developed since the first edition was published. New chapters also chart developments in the representation and visualisation of spatial phenomena. Data quality, error, and uncertainty are also given new and extended treatments, and an expanded group of contributions on spatial analysis present a contemporary view of the usefulness of GIS in analysing spatial distributions.

As we have seen, the technological setting to GIS has been transformed since the publication of the first edition – so our new **Technical Issues** Part traces the emergence of new technologies such as the development of networked and 'open' GIS and the introduction of GIS for the desktop. New techniques of spatial database management receive extensive attention, as does data capture through the latest remote sensing and GPS technologies. Finally in this section, a range of techniques for transforming and linking geographical data are discussed, notably in the context of terrain modelling, hydrographical analysis, and the creation of virtual GIS environments.

As GIS comes to play an important role in an ever-wider range of organisations, so management issues such as the choice between different commercial GIS, data availability and operational management become of importance to increasing numbers of people. These issues are addressed in the all-new **Management Issues** Part of the book. Information managers also need to be aware of legal liability issues in the provision and use of GIS, as well as data pricing and availability, and issues of privacy and confidentiality. This Part provides comprehensive introductions to these important emergent topics in GIS usage.

In many respects applications are the most important aspect of GIS since the only real point of working with GIS is to solve substantive real-world problems. Diverse though the range of GIS applications is, many nevertheless share common themes. In the **Applications** Part of this book we have selected a range of operational ('nitty gritty') and more strategic social and environmental applications. The former generally focus on practical issues such as cost effectiveness, service provision, system performance, competitive advantage, and database creation/access/use; while the latter are often more concerned with model sophistication, the social and environmental consequences of results, and the precision and accuracy of the findings.

In the Epilogue the editors draw some conclusions and indulge in some speculation as to what the future holds for GIS. We hope that readers will judge the end result to be an authoritative, comprehensive, and up-to-date statement of all that is relevant and interesting about GIS.

7 SOME INDEPENDENT VIEWS ON THE STATE, RELEVANCE, VALUE, OR FUTURE OF GIS

The act of producing a book, even one as large and diverse as this, is liable to force some degree of homogeneity on the contributions. Each author is honour-bound to report the latest trends or research findings in his or her field and assess these in a rational way; the editors need to ensure balance and provide cross-links between chapters. We considered this and agreed that a small number of iconoclastic, individual and personal views could add materially to the book. This would be especially true if they were written by individuals known to be incapable of being seduced by editorial or other blandishments

and who had worked in the furnace at the centre of some major GIS developments.

As a consequence, we invited five contributions from well-known figures, with use of the first person to emphasise this personal viewpoint. Their brief was to write about the state, relevance, value, or future of GIS. We suggested that they might use 'major historical events', 'GIS in a societal context', 'future trends', 'how has GIS changed the way we live today?', 'a personal story about becoming involved in GIS' or 'what are the remaining challenges to GIS?' as the basis for their contributions, but no restrictions were placed on comments.

What follows represents some of the wider strands of thinking about GIS worldwide.

GIS as the national Majlis
by Sheik Ahmed Bin Hamad Al-Thani
Centre for Geographic Information Systems, Doha, Qatar

The Majlis, an informal village meeting to discuss community issues and resolve differences, is an ancient tradition known throughout the Middle East. Even as a child, I wondered at the ease with which this simple, open forum prompted inquiry, discussion, analysis, and resolution.

As a member of the Qatari government I faced, with others, the challenge of establishing methods of master planning and the redevelopment of our cities in a systematic way that would rectify the make-or-break construction projects of the past and provide a definitive guide for future development.

In the late 1980s I saw, by chance, my first demonstration of GIS. It was as if a beacon, or guiding light, was suddenly sighted and I realised that this technology was the key that would provide the framework for developing an information infrastructure for our entire country.

As with all computer-based technologies, compatibility was the central issue. If we were to implement a successful national GIS, standardisation would be critical. With the authority of the senior

members of our government, I was able to establish a National GIS Steering Committee responsible for developing and maintaining national standards and the Centre for GIS which was tasked with implementing these standards. Today Qatar enjoys a unique, nationwide GIS in which all participating government agencies are connected by a high-speed optic fibre network. Each agency can access the data of all others but the responsibility for maintaining the data rests with the individual data custodians, the different agencies. As a result of all this, Qatar now has a GIS that will facilitate intragovernmental cooperation and coordination for many generations to come.

It is clear to me that, for successful implementation of a national GIS, those in the highest levels of government must understand the benefits of the technology and must actively support its implementation. GIS provides an easy method of standardising and sharing a wide variety of information amongst all levels of government. Like the Majlis, it fosters cooperation, interaction, analysis, and well-considered decisions, solving real problems in real time – from which a society can only benefit.

Technology changes everything

by John O'Callaghan
Cooperative Research Centre for Advanced Computational Systems,
The Australian National University, Canberra, Australia

I think the opportunities for GIS in the current age of 'convergence' are really exciting. We have now entered the age where the integration of computing, communications, and content is providing an information infrastructure which is fuelling the widespread use of GIS by government, industry, and the community.

GIS have built on the rapid advances in information technology and, since the 1960s, have exhibited typical stages of growth towards maturity: the experimentation with GIS technologies, the demonstration of GIS on practical applications, the consolidation of the geographical data infrastructure, and the realisation of benefits from operational GIS.

My own country – Australia – has been an early adopter of information technology and this, coupled with our coordinated approach to land ownership, our large geographical size and our dependence on natural resources, has resulted in Australia playing a leading role in the development and application of GIS.

Today, the most obvious demonstration of 'convergence' is the Internet, which is revolutionising the way we access data, interact with systems, and communicate with people. For GIS, the Internet is enabling the rapid deployment and widespread dissemination of geographical information services.

My group's research is now focused on enriching the user interfaces to these kinds of services: on-line navigation and analysis of large and distributed geographical databases; 3-dimensional modelling and visualisation of geographical data using 'immersive' display and haptic devices; and cooperative working on geographically-based simulations at several locations. We expect the results of this research to be adopted rapidly through the information infrastructure of the Internet and to contribute to the huge opportunities for GIS in this age of convergence.

How it all began and the importance of bright people

by Roger F Tomlinson
Tomlinson Associates, Ottawa, Ontario, Canada

The Canadian contribution to the development of GIS centres around the idea of using *computers to ask questions of maps*. This idea stemmed from the need for multiple map overlay and analysis facing Spartan Air Services, an Ottawa company working in Kenya in 1960. Later, in 1962, the approach was proposed by Spartan Air Services to the federal government of Canada, who adopted it for the Canada Land Inventory then planning to generate thousands of new maps to describe current and potential land use in Canada. This very successful federal-provincial programme funded the development of

GIS in Canada for the next decade. From the basic idea came the concept that many maps in digital form could be linked across Canada to form a continent-wide map database to be permanently available for analysis, and further, that these digital maps could be linked intelligently to digital databases of statistics (particularly the Census of Canada) so that a wide range of spatial questions could be answered.

I directed the development of the Canada Geographic Information System from its conception until 1969. During that time over 40 people were involved in the work and there are many who deserve great credit. Lee Pratt was the young head of the Canada Land Inventory who, as a civil servant, took the entire risk of funding the new ideas. D R Thompson of IBM designed and built the first 48 x 48 cartographic scanner for primary map input. A R Boyle, then working for Dobbie McInnes (Electronics) Ltd in Scotland, designed and built the first 48 x 48 high precision free cursor digitising tables used to input point data. Guy Morton designed the continent-wide data structure incorporating a brilliant tessellation schema (the Morton Matrix) that allowed many maps to be handled by the tiny (in terms of speed and capacity) computers of the time. Don Lever was central to most of the logic of converting scanner data to topologically coded map format. It was the first use of the arc-node concept of line encoding incorporated in a GIS. Bruce Sparks and Peter Bédard made major contributions to the automatic map sheet edge match capability, which topologically matched polygons and contents seamlessly over a continent. Art Benjamin played a major part in designing the automatic topological map error recognition capability and in designing the links between map data and statistical data. Bob Kerneny developed the essential map data compaction methods using eight-directional codes originated by Galton and later called Freeman codes. Frank Jankaluk devised the reference coordinate system and made the calculations of error in calculation algorithms. Bob Whittaker designed the system for error correction and updating. Also incorporated in the system were map projection change, rubber sheet stretch, scale change, line smoothing and generalisation, automatic gap closing, area measurement, dissolve and merge, circle generation and new polygon generation, all operating in the topological domain.

The computer command language that recognised geographical analysis terms used to pose spatial questions, and that could be understood by a wide range of potential users, was a very important part of the system. Peter Kingston was responsible for the overall design of this data retrieval system and particularly for the efficient polygon-on-polygon overlay process. He also designed the command language, together with Ken Ward, Bruce Ferrier, Mike Doyle, John Sacker, Frank Jankaluk, Harry Knight, and Peter Hatfield.

Our most useful links to the academic world were through Waldo Tobler and Duane Marble in the USA, and Terry Coppock in the United Kingdom. In Canada the principal initiatives came from within private industry and government rather than academia. The links to work in the UK were through David Bickmore of the Oxford Cartographic System who, in the early 1960s, was responsible for many of the ideas for using computers to make maps. We disagreed on almost everything in the early days, but eventually our paths converged and we became firm friends.

The 1960s in Canada were exciting years, and I am happy to have been part of that excitement. While we all worked extremely hard, there was a spirit of adventure and the feeling that if you could imagine it you could make it. In those days, a few key individuals – many of them mentioned above – really counted. In the process I described, the first GIS was born and the field was named. We still call them the Champagne years.

GIS, politics, and technology

by Nancy Tosta
Director of Forecasting and Growth Strategy, Puget Sound Regional Council, Seattle, Washington, USA

In 1978, I tried to convince the Director of the California Department of Forestry that pixels were good for him and his agency. In those days, appointed and elected officials were highly suspicious of any form of geospatial technology. Their fears were justified. The price tags were huge and no one had proved that spending all those dollars to digitise data would pay off. I remember him asking why there were all those little squares on the map/image. Why didn't it look like the maps he usually used? How could the data be used? Now, writing in the early months of 1997, I would be hard pressed to find an elected official who does not know the meaning of GIS and who does not have a story to tell about how GIS was used to clarify or solve a problem. I knew that we had crossed a watershed in political acceptance of the technology in 1994 when President Clinton signed Executive Order 12906: 'Coordinating Geographic Data Acquisition and Access: The National Spatial Data Infrastructure'. While labouring in the preparation of that order, I was astounded at the lack of questions from the White House and others about the technology. The assumption was that GIS was valuable and that data should be coordinated and shared to use the technology more effectively. Other nations have used Clinton's Order to generate political support for their GIS data efforts. The local elected officials I interact with today may not know about Federal Executive Orders, or exactly how much has been expended to develop their GIS, or what the software does, but they accept that the technology works. What more do we need to make a difference?

It's all about money, stupid!

by Joe Lobley

Lobley Associates, Santatol, Southlands, USA

Much rubbish has been talked about the special value of GIS. Even more rubbish has been heard about the essential contributions of academic research and the role of government in GIS. These two groups have made almost no contribution to the evolution of GIS to date nor will they greatly influence its future. Government talks a lot, produces lots of paper, and consumes our taxes. Other than spasmodic politically correct initiatives to 'modernise' itself, government is as moribund as ever it was (and will be). Academics are supposed to exist to question what is taken for granted but when did we ever see anything really critical or new come out of the geographers at least? Technically, it was probably in the mid 1960s. Since then we have spent loads of money on fancy research centres to little effect except airline revenues. Maybe some social geographers have hit something interesting in this ethics business but their posturing and soul-bearing seems a mite contrived to me (and has no real effect other than to cause more trees to be felled for their precious publications, read only by themselves).

No, the mainspring of everything important that has happened in GIS is business and the profit motive.

Nothing of any significance started until the first commercial GIS became available. The growth in use of GIS has been fuelled by the decrease in cost of technology, driven in turn by commercial competition and salesmanship. Unlike most academics, some government data producers have a potentially important role simply because they hold valuable data assets. It's just a pity that they are typically complacent and act on geological timescales; the only way to jolt them out of all this is to contract out many of their activities. So far as access to software, hardware, and data are concerned – if people won't pay for software, data, and services, they don't really need them. If we pay for software and hardware from the commercial sector, why should we not pay for data from it – and why should government be involved at all?

The moral is obvious. Official history is created by those with the luxury of time to write and claim the credit. But the real achievers are those who have put their money on the line and built a business worldwide. I don't expect this situation to change much in future and I don't really care. But don't forget who really makes GIS happen!

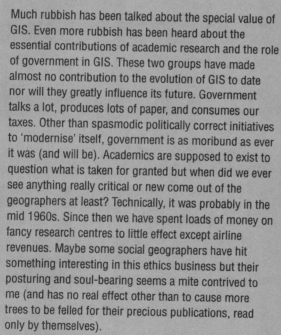

References

Anselin L 1989 *What is special about spatial data? Alternative perspectives on spacial data analysis.* Technical paper 89-4. Santa Barbara, NCGIA

Burrough P A, Frank A U (eds) 1996 *Geographic objects with indeterminate boundaries.* London, Taylor and Francis

Collins M, Rhind J 1997 Developing global environmental databases: lessons learned about framework information. In Rhind D W (ed.) *Framework for the world.* Cambridge (UK), GeoInformation International: 120–9

Coppock J T 1962 Electronic data processing in geographical research. *Professional Geographer* 14: 1–4

Cowen D J 1988 GIS versus CAD versus DBMS: what are the differences? *Photogrammetric Engineering and Remote Sensing* 54: 1551–4

Crockett M 1997 GIS companies race for market share. *GIS World* 10 (4): 54–7

Foresman T W (ed.) 1998 *The history of geographic information systems: perspectives from the pioneers.* Upper Saddle River, Prentice-Hall

Fotheringham A S 1997 Trends in quantitative methods 1: stressing the local. *Progress in Human Geography* 21: 88–96

Getis A, Ord J K 1992 The analysis of spatial association by use of distance statistics. *Geographical Analysis* 24: 189–206

Goodchild M F 1992a Geographic data modeling. *Computers and Geosciences* 18: 401–8

Goodchild M F 1992b Geographical information science. *International Journal of Geographical Information Systems* 6: 31–46

Isaaks E H, Srivastava R M 1989 *Applied geostatistics.* New York, Oxford University Press

Maguire D J 1991 An overview and definition of GIS. In Maguire D J, Goodchild M F, Rhind D W (eds) *Geographical information systems: principles and applications.* Harlow, Longman/New York, John Wiley & Sons Inc. Vol. 1: 9–20

Maguire D J, Goodchild M F, Rhind D W 1991 *Geographical information systems: principles and applications.* Harlow, Longman/New York, John Wiley & Sons Inc.

Mandelbrot B B 1982 *The fractal geometry of nature.* San Francisco, Freeman

McHarg I L 1969 *Design with nature.* New York, Natural History Press

McHarg I L 1996 *A quest for life.* New York, John Wiley & Sons Inc.

Openshaw S 1984 *The modifiable areal unit problem.* Concepts and Techniques in Modern Geography Vol. 38. Norwich, GeoBooks

Pleuwe B 1997 *GIS online: information retrieval, mapping and the Internet.* Santa Fe, Onword Press

Rhind D W 1988 Personality as a factor in the development of a discipline. *American Cartographer* 15: 3277–89

Rhind D W 1997b *Framework for the world.* Cambridge (UK), GeoInformation International

Rhind D J 1998 The incubation of GIS in Europe. In Foresman T W (ed.) *The history of geographic information systems.* Upper Saddle River, Prentice-Hall: 293–306

Robinson G K 1950 Ecological correlation and the behavior of individuals. *American Sociological Review* 15: 351–7

Star J L, Estes J E, McGwire K C 1997 *Integration of geographic information systems and remote sensing.* New York, Cambridge University Press

Tobler W R 1970 A computer movie: simulation of population change in the Detroit Region. *Economic Geography* 46: 234–40

Tomlin C D 1990 *Geographic information systems and cartographic modeling.* Englewood Cliffs, Prentice-Hall

Wright D J, Goodchild M F, Proctor J D 1997 Demystifying the persistent ambiguity of GIS as 'tool' versus 'science'. *Annals of the Association of American Geographers* 87: 34–62

SECTION 1(a): SPACE AND TIME IN GIS

Introduction

THE EDITORS

The term GIS is fundamentally about the use of digital data to represent space and time, and few of the readers of this book will be unfamiliar with the standard sequence of operations that GIS invoke to create such representations – data input, storage, manipulation, and output. For many users of GIS, this simple chronology of operations has provided an adequate framework for understanding what GIS is about. Yet reality is infinitely complex in its totality, and our digital representations are inevitably simplifications or 'models' of it. With experience, and perhaps the demands of wider domain and strategic applications, many GIS users will begin to get a feel (from the 'bottom up') for the nature of the simplifying assumptions, or 'transformations' (Martin 1996) which are inherent in reducing the myriad complexities of geographical reality to digital computer records. From a quite different perspective, the fundamental ('top down') views of social science and science held by some academics bring into question the very validity of GIS-based representations of the real world. The opening five chapters of this book seek to set out the context to GIS, as a contribution towards reconciling philosophy and science with practice, concepts with application, analytical capability with social context. As such, and although avowedly academic in emphasis, they contain material of relevance to everyone who has considered using GIS to formulate and analyse problems in the real world. Successive chapters begin to translate these abstract notions and ideas into firmer guiding principles of GIS, in order that principles in turn might coalesce into operational guidelines for implementation.

In the opening chapter to this Section, Helen Couclelis traces the disciplinary origins of interest in representing space and time to the disciplines of geography, mathematics, philosophy, and physics. The traditional paper map subsequently emerged as the dominant paradigm of spatial representation, with its goal of depicting spatial phenomena using established and recognisable schemes of representation. The more recent innovation of GIS has sought to develop and enhance such analogue models of the world using computer hardware, software, and digital data. More detailed and sophisticated than paper maps they may be, but most GIS-based maps remain similarly constrained – they must present a world that has been projected onto a flat plane; they must be static and 2-dimensional; they depict the world as if it were known perfectly, or at least as accurately as the scale of the map allows; and they must present the world at a uniform scale or level of geographical detail. These are all examples of simplifications of reality, yet the GIS medium is fundamentally more capable of relaxing these assumptions, constraints, and conventions than paper mapping. Thus it is with some confidence that Couclelis sees GIS as rising to the challenge of achieving 'the seamless integration of space and time, the representation of relative and non-metric spaces, the representation of inexact geographical entities and phenomena, and the accommodation of multiple spatio-temporal perspectives to meet a variety of user purposes and needs'.

Of the different disciplines that have sought to represent space and time, it has been geography that has identified itself most closely with the innovation of GIS – although (as a number of the contributors to this section note) geography has not been central to its technological development. Ron Johnston (Chapter 3) uses the debates that have developed within geography to explore the implications of the fundamentally empiricist view of the world that GIS provides – that is, a view founded upon the philosophical belief that there is a separate objective world that is outside and independent of any individual observer. Empirical scientific approaches have come to be viewed with disdain by some academics working in human geography, yet they remain the predominant *modus operandi* in physical geography. Johnston concedes that a pragmatic application-led science, couched in the world of appearances, is not universally attractive to all geographers (although it has undoubtedly enhanced the status of their discipline), yet his closing remarks suggest that the very richness of digital media no longer need constrain GIS in this way. If, as Couclelis asserts, GIS has already come a long way since the era of early computer cartography, then the rapidity and pace of current developments should in turn now begin to suggest ways in which GIS might inform other, non-empiricist, approaches to social science.

In Chapter 4, John Pickles develops the critique of GIS from a more functionalist perspective: that is, how its approach to science has impacted upon its technological capacities and social uses. The early 1990s critique of GIS within academic geography was fundamentally one of empiricism – that is, the approach to science in which, in Johnston's words, 'facts speak for themselves'. As such, it was to some extent a re-run of the critiques of quantitative geography that had developed in the 1970s and 1980s. This time, however, the detailed critique developed on two quite different fronts. First, what was different for some this time was the power of the technology, and the drive towards data-rich depictions of geographical reality capable of eroding privacy and increasing (social, political, military . . .) control. Second, empiricist approaches (in contrast to other social science approaches such as social and critical theory) were deemed most unlikely to shed light on questions of valid and intrinsic academic interest, and thus infusion of GIS into the discipline of geography would never create more than a diversion and irrelevant distraction. Although perhaps contradictory (if a technology truly is capable of eroding privacy, then it surely is a worthy focus of academic concern), these two perspectives shared the common sentiment that GIS has introduced a technological distraction to legitimate academic discourse and thus has reinvigorated an approach to social science in geography which by the 1990s many had thought discredited. Some of the later chapters in the Management Issues and Applications sections air these issues in much greater detail. Pickles' important contribution, here as well as elsewhere (Pickles 1995), has been to open up these issues to constructive dialogue between the 'top down' views of the best practice of science and the 'bottom up' empirical experience of GIS users. Such dialogue is likely to lead to GIS applications breaking free from notions of 'objective' reality and

(echoing the views of Johnston) may also lead to supplementation of quantifiable attributes and characteristics of geographical reality with measures of local knowledge, place-based information and other qualitative considerations.

There are lessons here for even the most unequivocal advocate of GIS – namely the need for cognisance of the philosophical background and context to analysis, and the inherent subjectivity of even the most apparently 'objective' models of reality that are abstracted within GIS. In Jonathan Raper's view this relativist honing of GIS to context is not restricted to social science applications that embrace human agency, as suggested in his review of scientific representation (Chapter 5). Physical science lies much more uncontroversially in the empiricist domain than social science, and hence it might be taken as axiomatic that there is a strong correspondence between increased richness of digital information and the accurate and orderly depiction of real-world systems. Raper formalises the empiricist conception of GIS in natural scientific applications as representing a 'bridge' between scientific theory and the real world. Yet even within natural science the way in which this structure is fashioned is profoundly influenced by our information sources and scientific conventions. 'Scientific conventions' are the ways in which we define and give significance to geographical phenomena, and the ways that we identify phenomena in space and time within GIS (when is a sand dune not a sand dune? what are its boundaries? what sort of time increments should be used to represent and model its dynamics?).

Raper's view is that the spatial and temporal context to natural science representations in GIS should be specified inductively rather than deductively, in a context-sensitive manner and in a spirit of humility rather than conviction – a view developed later in this section with respect to digital terrain models by Hutchinson and Gallant. As such, empirical science should refocus more of its efforts on the ways in which simplified representations of an infinitely complex reality are developed within GIS, and greater detail and volume of information is seen as but one ingredient of an enhanced approach to model-building. Spatial representation is thus an intrinsic component of scientific method, and must be related to theory about the way the world works and how geographical reality is structured. This mapping of reality into model is seen as being accomplished through geographic information science (Goodchild 1992) – that is, the development of formal conventions and rules for the appropriate representation of phenomena within conventional and unconventional (3-dimensional models, video and multimedia) GIS representations.

The issues and problems that Raper identifies are at least as problematic as many in the socioeconomic realm, yet here there are additional problems arising out of the strictures of confidentiality (and consequent areal aggregation of data), the ways in which boundaries are imposed around continuous spatial features, and the ways in which time is discretised during data collection. These and other problems create some differences in the definition and handling of geographical objects of study between natural and social science applications. David Martin reviews these differences here (Chapter 6). Spatial boundaries and temporal intervals pose more than an analytical inconvenience, since they lie outside of the control of the GIS analyst and cannot be changed. Moreover, there are conceptual difficulties in assigning precise geographical coordinates to human individuals and describing their activity patterns. Martin describes how these provide additional challenges to effective representation, analysis, and display within socioeconomic GIS, and discusses how representational strategies may be used to contain ecological fallacy and modifiable areal unit effects.

Taken together, the contribution of these chapters is to steer GIS users towards a more relativist conception of reality and its representation within GIS, and thence to identify how GIS data structures and architectures can be developed to accommodate the widest range of information sources. This provides a general framework for enhancing GIS-based representations of reality which are tailored to the perceptions and needs of the many. In short, it presents a broad canvas to GIS applications: the remaining chapters in this section investigate a range of topics that can further improve representation within GIS.

One socially significant facet of this critique is that if GIS has been rendered accessible only to the scientific community, then successful users have been placed in the role of experts. As a consequence, GIS is intolerant of diversity of viewpoint. Cognitive interest in GIS stems from a desire to make it easier to use, by making its user interfaces and

representations more compatible with the ways people naturally think and reason about the world around them. David Mark's chapter (7) reviews the current state of cognitive research in GIS, and discusses some of the issues that are raised by this line of reasoning. He begins by appraising the correspondence between the ways in which humans perceive real-world phenomena and the ways they are represented within GIS. His approach is avowedly empiricist in approach, and sets out to examine the possible mismatches between objective measurement and cognitive models of reality. He shares some of Johnston's optimism that GIS may provide a suitable medium through which realist models of geographical reality might be built. The detail of his empirical analysis of cognitive categories substantiates the views of Raper, Martin, Pickles, and Couclelis that boundaries are *de facto* often indistinct, fuzzy, and graduated. He also describes how distance, direction, reference frames, and topology are all subjectively manipulated in common parlance, and how variations may be compounded by natural language differences (when is a lake not a lake but a pond? why are 'lodge' water bodies apparently confined to northern England?).

As Raper has already intimated, our inability to deal with time, and thus to represent the dynamic elements of the geographical world, is perhaps the most compelling of the inadequacies of maps and traditional GIS. A technology that is forced to represent the world as static inevitably favours the static aspects of the world. Thus (as Martin describes) our maps show the locations of buildings and roads rather than people, and the same biases have been inherited by GIS. The chapter by Donna Peuquet (8) reviews the state of the art in the representation of time in GIS, and specifically the problems that arise out of the inadequate definition and representation of events and timeframes within GIS. She assesses the merits of different data structures for representing time in GIS – for example, as raster coverage-based snapshots, variable length (raster) pixels, entity-based (vector) representations and geographical objects. Peuquet then reviews modes of exploring and visualising space-time interactions using GIS. It is clear from this that query languages for identifying temporal change are much better developed for aspatial, rather than spatial, database management systems (DBMS), because of the reduced dimensionality of the queries. Nevertheless, in an upbeat conclusion, she anticipates considerable improvements in the representational power and analytical capabilities of GIS in this regard.

The remaining chapters share a pragmatic emphasis on the ways in which data models may be used to fulfil a variety of end-uses. Much of the conceptual debate surrounding GIS has arisen simply because the user of today's GIS is faced with many more options in representation. Choices must be made between different scales, or levels of geographical detail; between raster and vector options; and between various approaches to representing change. Nowhere are these choices more apparent than in the representation of topography, or the form of the Earth's surface. Data are available at various scales, and in three major representational schemes: the meshes of triangles known as triangulated irregular networks (TINs); grids of regularly spaced sample elevations (digital elevation models or DEMs); or digitised contour lines. But many more complex and subtle issues exist in finding accurate and useful representations.

Michael Hutchinson and John Gallant's main concern (Chapter 9) is with identification of the guiding principles for generating digital terrain models, and an area in which (firmly in the empiricist tradition) spatial analysis of form is frequently used to draw inference about environmental process. Digital terrain models are also used in the conceptualisation (cf. Raper) and display. Accuracy and extent of spatial coverage are of importance here, of course, but there is also a sense of the recursive relationship between the way that relevant phenomena are identified and defined (ontology) and the ways in which they are subsequently analysed. Additionally, the representation of 3-dimensional structures creates a potentially vast increase in the amount of data that might be stored within GIS, many of which are likely to be redundant: clear thinking, coupled with appropriate choice of analytical technique are thus required in order to create realistic yet manipulable models of real-world 3-dimensional structures. This raises a wide range of considerations in making the choice of data model: how to anticipate/manage errors associated with GPS data capture (Lange and Gilbert, Chapter 33), whether priorities favour capturing surface variability through use of variable point densities across a surface or by representing local properties of curvature, whether and how grids may be adapted to local terrain structure, the range

of scales relevant to the end-user, and the ways in which features are defined and parameterised. As the previous chapters in this section imply, the model-building process does not then terminate with a single pass through the data, and Hutchinson and Gallant describe further recursive stages of data quality assessment and model interpretation.

Robert Weibel and Geoffrey Dutton (Chapter 10) broaden this analysis, looking at the generalisation of geographical objects and cartographic features. They develop a typology of motivations for generalisation within GIS, ranging from data storage, through improved data robustness to optimising visual communication. All of the preceding contributions have emphasised that good environmental and social science data models are sensitive to context in what, through abstraction, they retain and discard. Yet such reflection is clearly not practicable where a multitude of routine decisions must be made, or where the outcome of data modelling is a cartographic product for visual display. In such circumstances, sensitivity to context may nevertheless be achieved using a range of automated and semi-automated knowledge-based methods, such as generalisation algorithms and methods for structure/shape recognition, and further methods for evaluating the 'quality' of generalisations. The principles underpinning such methods are seen as an automated development of traditional map-making conventions, in which the cartographer was always to some extent the arbiter, even architect, of cartographic form. That said, progress towards automated generalisation of digital maps has apparently been rather slower than was anticipated in the first edition of this book: however, Weibel and Dutton provide evidence of encouraging prospects in this regard, suggesting that automated generalisation may not be a 'holy grail'.

The final chapter in this section (11), by Menno-Jan Kraak, extends the discussion of the theme of scientific visualisation – that is, the presentation, analysis, and exploration of geographical phenomena. Although we naturally tend to think of visual displays of GIS databases as the digital equivalent of making paper maps, there are significant differences and opportunities. The design of a paper map is permanent, but visual displays can be manipulated and transformed freely. Scale, for example, takes on a different and more interesting meaning in a world of zoomable displays.

GIS displays can be animated, raising a host of new issues for the user's ability to perceive and understand geography through visual display. Although computer display screens are approximately flat, their use allows us to re-examine the significance of map projections, and to ask whether they are actually necessary in a digital geographical world, since there are no flat surfaces in a digital computer that are as constraining as the inevitable flatness of paper. There are also echoes here of many of the previous contributions in the description of the 'overlay model' as a simplified, error-prone depiction of reality, and a review of the effective use of symbolisation and other cartographic conventions to present spatial distributions – as well as to interpret data reliability and quality. As such, visualisation is considered an important adjunct to explanation, which helps through query, re-expression, multiple views, dynamics, animation and changes in dimensionality. Kraak's view (cf. Raper) is that GIS provides a bridge between the map and the database 'text', and he anticipates some of the ways in which video and other multimedia are set to develop and enhance the links between digital models of reality and their visual front ends.

Other issues of representation are addressed elsewhere in this book. The representation of uncertainty, another missing element in traditional maps, is taken up in the next section; and object-oriented issues in representation are discussed later as Technical Issues. But research on data modelling is proceeding at such a pace as to make it impossible to achieve a complete coverage in the space available here. This is why we have presented an extended overview of guiding principles rather than fast-changing practices. The interested reader is referred to the references in each chapter, to recent collections (for example, Molenaar and Hoop 1994), and to the continual stream of new research papers appearing in the journals of the field.

References

Goodchild M F 1992b Geographical information science. *International Journal of Geographical Information Systems* 6: 31–45

Martin D J 1996 *Geographic information systems: socioeconomic applications,* 2nd edition. London, Routledge

Molenaar M, Hoop S de 1994 *Advanced geographic data modelling.* Publications on Geodesy, New Series No. 40. Delft, Netherlands Geodetic Commission

Pickles J (ed.) 1995a *Ground truth: the social implications of geographic information systems.* New York, Guilford Press

3

Geography and GIS

R J JOHNSTON

Geographers study three major concepts – environment, space, and place – but their approaches to these have varied considerably over recent decades, incorporating three very different conceptions of science and views of the world: all of these can be encapsulated into 'geography as spatial science' and 'geography as social theory'. As reviewed here, most applications of GIS to date have fallen within the first of these categories: their utility in the latter is also discussed.

1 INTRODUCTION

There are very strong links between GIS and the academic discipline of geography, which extend well beyond the commonality of titles. The academic discipline has been the 'home' for much of the research, development, and training for GIS practitioners, and some see GIS as a major element in the discipline's 'survival package' for the foreseeable future, in a period of considerable pressures on public sector funding for academic work (see Gober et al 1995, on employment prospects for geographers in the USA). Nevertheless, the discipline embraces a great deal more than GIS and, as Openshaw (WWW 1996) has argued, much of what is done with GIS has very little to do with that discipline as generally practised: over 90 per cent of all applications, he claims, are 'of no significant consequence to people and society. They ... are concerned with the management of the physical infrastructure ... [and] involve little more than a digital replacement for various large-scale paper map-making, recording and handling industries'. This chapter looks at the context within which much GIS work has been, and continues to be, nurtured, while recognising that the 'child' has grown immensely in stature and influence beyond the 'academic nest' in the last two decades.

Significantly, different views on the practice of geography have been advanced on several occasions over the last half-century, but core beliefs regarding the discipline's *raison d'être* have remained constant. There is little difference, for example, between Hartshorne's (1939) classic definitional statements:

'... geography is a science that interprets the realities of areal differentiation of the world as they are found, not only in terms of the differences in certain things from place to place, but also in terms of the total combination of phenomena in each place, different from those in every other place'

so that '... geography is concerned to provide accurate, orderly and rational description and interpretation of the variable character of the Earth's surface' which

'... seeks to acquire a complete knowledge of the areal differentiation of the world, and therefore discriminates among the phenomena that vary in different parts of the world only in terms of their geographic significance – i.e. their relation to the total differentiation of areas. Phenomena significant to areal differentiation have areal expression – not necessarily in terms of physical extent over the ground, but as a characteristic of an area of more or less definite extent'.

through its updating (Hartshorne 1959):

'... geography is that discipline that seeks to describe and interpret the variable character from place to place of the earth as the world of man' and '... geography is primarily concerned to describe... the variable character of areas as formed by existing features in interrelationships'.

and McDowell's (1995) modern rendering, that geography is a discipline '... whose *raison d'être* is the explanation of difference and diversity' (1995: 280) with geographers needing '...theoretical perspectives that not only permit the elucidation of the main outlines of difference and diversity, of the contradictory patterns of spatial differentiation in an increasingly complex world, where ever-tighter global interconnections coexist with extreme differences between localities; but perspectives that also allow us to say something about the significance of these differences.' Geography remains the study of differences across the Earth's surface.

Throughout this period, geographical research and writing have focused on three main concepts:

1 *Environment* – or nature
2 *Space*
3 *Place*

although the relative stress placed on each, and on their interactions, has varied somewhat. Geographers are concerned with the where, how, and why of the physical environment on which material life is based, the spatial organisational structures erected and operated by human societies to sustain and promote their material well-being, and the nature of the places which they have created within those structures. In pursuing those interests, they have engaged with a variety of approaches to science, and applied their findings in a number of separate ways.

2 APPROACHES TO SCIENCE

Discussions of the nature of science reflect not only the changing worlds in which scientists live and work but also their own varying conceptions of what comprises knowledge, how it can be obtained, and to what purposes it should be used. This literature is simplified into a typology containing just three categories, each of which has been used and argued about within geography in recent decades.

The first type (that most associated with the term 'science') is the *empirical*, which contains several sub-categories. Knowledge is acquired through direct experience, especially visual: its production involves accurate observation, and its propagation calls for unambiguous reportage. Within empirical science, therefore, facts speak for themselves. The scientist, having decided what to study, is a neutral observer-reporter, presenting material using accepted languages and categories (and, if necessary, proposing extensions to them). The acid test of the validity of a piece of science is that when replicated it produces the same outcome.

Many empirical scientists seek not only to describe accurately but also to explain: geographers aim both to show what is where and to indicate why. Positivist empirical science explains through generalisation: an event is accounted for as an occurrence of the operation of one or more general laws, whose identification is the scientists' goal. (Logical positivist science claims that there is no other route to knowledge.) Scientific laws can be used to predict, and that power is associated with the power to change, applying scientific findings to achieve certain ends.

The empirical sciences are founded on a belief in a separate world outside and independent of any individual observer, and some claim that its methods are equally valid in the social as in the natural sciences – in human as well as physical geography. The *hermeneutic* sciences challenge that: in them, nothing exists outside the observer, because any perception of an empirical event involves interpretation, using human constructs (especially language) to give meaning to what is being recorded. Individuals ascribe meanings to the worlds they live in, and act according to those meanings, so understanding an action involves appreciation of the meanings on which it was based. Natural scientists operate in this way as well as social scientists, but whereas the latter study the creation and transmission of meanings by human actors the natural scientist's subject matter comprises phenomena lacking the distinguishing features of 'humanness' – memory and reason.

The hermeneutic social sciences study people continuously interacting with their natural and human environments – individuals learning and using their powers of reasoning to interpret their observations, in often-changing situations. Explanation through generalisation is not possible, therefore, since humans cannot be equated with machines which always respond in the same way to given stimuli: they may react differently because of their learning processes (which may involve reinterpretation of the stimulus itself), or because the stimuli and the contexts in which they are encountered are changed. Instead of explanation, therefore, hermeneutic science advances understanding, appreciation of why people acted in particular ways: this promotes awareness of the past and the present, and provides a guide to the future, but no more.

The final type of science – the *critical* – promotes explanation, but not necessarily through generalisation. Underpinning these sciences is a belief that observation is insufficient to appreciate the world, so that theories are needed which can account for the hidden structures involved in creating what is recorded by the senses – the theories must be consistent with the outcomes of those hidden processes. The 'law of gravity' exemplifies this. Gravity cannot be observed: all that can be recorded is behaviour which is consistent with its assumed operation. The law is accepted as valid because of its successful predictive power in repeated experiments under controlled conditions; if it should fail, however, then its validity in similar circumstances must be reconsidered.

Application of the rules of the hidden structures in the subject matter studied by social scientists involves human interpretation and action, as argued in the hermeneutic sciences. Biological drives demand food and liquid sustenance in humans, for example, but what foods and liquids is not determined: people individually and collectively decide how to meet those needs, not by making *ad hoc* decisions every time that they are thirsty and hungry but by developing strategies to meet their requirements on a regular basis. Those strategies vary over time and space, reflecting the conditions within which they are developed (such as the environmental context), the processes of learning, and the lessons which are passed on (as part of a society's cultural inheritance). Their outcomes may be inscribed in the landscape – as with land-use patterns which reflect food-raising strategies. The strategies may occasionally be reconsidered – either as a matter of urgency (because of a crop failure following an environmental disaster, for example) or more slowly (as population pressure on environmental resources builds).

The critical sciences identify three 'domains of interest':

1 The *real*, comprising the underlying, unobservable, mechanisms;
2 The *actual*, involving the operation of those mechanisms;
3 The *empirical*, which is the outcome of those operations.

In the natural sciences, the law of gravity occupies the first level: it is put into operation at the second level when rain falls, for example; and the empirical outcome is the rain's impact on the landscape. In the social sciences, both the biological drives necessary to human survival and the mechanisms put in place to sustain them socially occupy the domain of the *real* – as with the capitalist and socialist modes of production which have evolved to ensure individual and collective well-being, and improving material standards. The domain of the *actual* involves implementing those mechanisms, the myriad decisions made by 'knowing individuals' applying the fruits of their own learning and acquired meanings within the 'rules'; and the domain of the *empirical* contains the outcomes – which become part of the context within which future decisions on operating the mechanisms are made.

In both natural and social science it is possible to observe the outcomes and the procedures by which they are generated, but not the underlying mechanisms – one can no more 'see' capitalism than one can 'see' the law of gravity, so that the nature of each has to be inferred (or theorised) and the validity of that thought process evaluated against reality. Beyond this similarity there is a major difference between the two, however. The law of gravity is assumed to be invariant, and nothing that humans can do will remove it: it is an enduring fact. The mechanisms studied in the social sciences are not invariant: not only are they human creations which can be destroyed by human action but they are also continually being changed, as societies respond to situations in the light of previous decisions and their consequences. Hence, whereas a critical realist approach to the natural sciences underpins the search for explanation via generalisation (since the mechanisms are invariant and all that can alter is how they are put into operation) such a strategy is not viable in the social sciences, since the same conditions may never be repeated: events in the natural sciences may be unique; in the social sciences they may be singular (Johnston 1989).

The three types of science – empirical, hermeneutic, and critical – represent different world-views, different conceptions of how the world can be understood. They are incommensurate and cannot be evaluated for their validity against some external criterion: choosing a world-view involves determining one's conception of what is involved in achieving understanding. Each world-view contains a number of separate approaches (sometimes termed disciplinary matrices), different strategies accepted by scientists for achieving their desired ends. In turn, each of those strategies has its own detailed exemplars, paradigm instances of how the science should be done. When becoming a scientist, therefore, an individual adopts a world-view, a general orientation for her or his work. The next decision is which community to join, whose strategy to adopt. Finally, which detailed exemplar(s) to follow has to be determined. (These three decisions and scales are readily equated with Kuhn's paradigm model of scientific practice: Kuhn 1962 – and for its application to geography: Johnston 1997.) These decisions are not taken in isolation, of course: they are made during the process of becoming a scientist, of being socialised into its culture and accepting its norms, the would-be researcher already having decided his or her scientific subject-matter – what discipline to join. And they can be revisited, with different exemplars being adopted as the models to follow, different strategies being entertained, and even different world-views canvassed.

Those decisions are not irrevocable. Change is most likely at the level of the exemplar, or *modus operandi*: as science progresses so new ways of achieving its goals are identified and adopted by practitioners. It may occasionally occur at the intermediate level, when a scientist decides that one community's strategy is better than another's. Change is rare at the highest level, however, since this involves altering one's world-view which cannot be done 'objectively' against predetermined criteria: it involves deciding that one's approach to science to date has been entirely wrong, and that an alternative offers a 'better' path to understanding – and that decision is at root a subjective one, a 'leap of faith' from one cosmology to another. A number of geographers have made such leaps in recent decades, and many more have switched their preferred disciplinary matrix.

3 GEOGRAPHY AND THE SCIENCES

Over recent decades geographers have encountered, experimented with, and debated the relative merits of a range of approaches to science, at each scale. At the largest, whereas physical geographers have largely remained within the empirical world-view (Gregory 1985; see also Aspinall, Chapter 69), human geographers have embraced all three – with some individuals shifting conceptions (see Sheppard 1995) and many more being socialised into the one proving particularly attractive to their generation. Within each world-view, different strategies have attracted varying support – as exemplified among physical geographers by debates over the relative merits of logical positivism and Popperian critical rationalism (Haines-Young and Petch 1985) and among human geographers by the debates on the relevance of the 'economic man' model (see Barnes 1996; Johnston 1997). Within each strategy, the importance of different exemplars has waxed and waned: technical developments in the analysis of spatial data generated a major shift in methodological orientation between the two editions of Haggett's *Locational analysis in human geography*, for example (Haggett 1965; Haggett et al 1977). A third edition, prepared in the wake of developments in GIS since, would undoubtedly be a very different book yet again.

The history of geography in recent decades has been charted in a number of volumes and will not be repeated here (Gregory 1985; Johnston 1997). Instead, focus is on broad categorisations of how geographers approach their discipline – and why. Some have simply divided geographers into two camps, with separate world-views – the spatial analysts and the social theorists (Sheppard 1995): all physical geographers fall into the former, but human geographers are divided between the two (see Bennett 1989). In slightly more detail, Buttimer (1993) has identified four 'root metaphors' at the heart of different approaches to the discipline and four separate rationales for its study.

3.1 The root metaphors

Metaphor is basic to many approaches to science – it provides a way of moving from the known to the unknown, via analogy (Barnes and Duncan 1992; Barnes 1996). Root metaphors provide analogies for use in a wide range of circumstances.

Buttimer's (1993) first metaphor – *the world as a mosaic of forms* – involves the 'recognition of forms and patterns, similarities and differences, among people, places and events'. These are described – verbally or through other representational forms, such as maps in geography – using 'language' which stresses their similarity (as with the centre/periphery model applied in a variety of contexts within human geography and the analogies between river systems and settlement patterns identified by Woldenberg and Berry 1967; see also Haggett and Chorley 1969, on networks). The approach provides a way of summarising the vast volumes of geoinformation now available: its main concern is description, saying what is where rather than why it is.

The second metaphor – *the world as a mechanical system* – emphasises the mechanisms that provide an explanatory base for the study of forms, as in developments in physical geography from the late 1960s when a 'systems approach' offered an integration of studies of processes and forms (Chorley and Kennedy 1971; and Bennett and Chorley 1978 promoted its application in human as well as physical geography).

Buttimer's third metaphor – *the world as organic whole* – focuses on the interrelationships among society and nature, using organic analogies to argue that the whole is greater than the sum of its parts. Whereas the mechanistic analogy promotes the study of order, rationality, and certainty, the organic concentrates on human individuality, identity, and ingenuity. The organic metaphor was initially applied to the identification of regions, territorial units in which society and milieu interacted to form identifiable niches in the spatial organisation of urban as well as rural worlds (on which see Gregory 1994; and Livingstone 1992).

Finally, *the world as arena of events* focuses on 'Events, in all their complexity, possible uniqueness, and contingency...; one looks for holistic understanding of particular events rather than ways of fitting them into some *a priori* schema of form, process, or organic whole.' (Buttimer 1993: 187). In its search for explanation this has similarities with the mechanistic vision (rather than the hermeneutic goal of the organic), but it rejects the search for universal truths and order underpinning the Enlightenment project that informs the mechanistic approach (Barnes 1996). It stresses pragmatic decision-making in the critical science mould, drawing conclusions within an embracing context of interpretations, as in the Marxist and realist approaches promoted by Harvey (1982) and Sayer (1992) respectively. This metaphor embraces a wide range of disciplinary matrices, including those such as postmodernism which promotes a relativistic

form of knowledge and denies the existence of universal truths; all accounts are incommensurate (having been produced within different contexts and from different positions) and thus are of equal value.

3.2 The metaphors in use

These four metaphors have varied in their application within the practice of geography. In physical geography, for example, the promotion of studying form rather than process led to its relative demise within the discipline in the United States for several decades (see Marcus 1979) – a situation not repeated in the UK (Johnston and Gregory 1994); the mechanical system metaphor now dominates. Human geography, on the other hand, has experienced considerable, continuing debate over their validity and significant variations in their relative importance (as chronicled in Johnston 1997). These variations have reflected changing emphases on the role of science and social science within society (on which see Taylor 1985). Buttimer (1993) identified four roles, or vocational meanings:

1 *Poesis* encourages curiosity about the relationships among individuals, society and nature by evoking geographical awareness;
2 *Paideia* promotes self-education through reflective practice regarding life and landscape;
3 *Logos* emphasises the search for order and generalisation via analytical rigour;
4 *Ergon* involves a focus on appropriate behaviour (applied geography) in the face of contemporary societal and environmental problems.

As a broad generalisation, the first two are associated with the organic and arena metaphors whereas the others are linked with form and mechanism.

Human geographers have long debated the validity of these metaphors and roles, characterised by Sheppard (1995) as a contest between spatial science and social theory. While that has continued, discussions have taken place within each camp regarding which disciplinary matrix to adopt and which exemplars to follow. Progress within each matrix has led to the replacement of some exemplars by others. With regard to GIS, this has occurred almost entirely within the spatial science world-view, involving the mechanism and form metaphors (and the virtual rejection of organism and arena) and the *logos* and *ergon* roles.

4 GIS AND GEOGRAPHY AS SPATIAL SCIENCE

Progress within geography as spatial science has been substantially linked to technical advances in the collection, collation, display, and analysis of data; computers with greater size, speed, and power have enabled the expansion of existing approaches and the introduction of new.

Several topics illustrate how GIS has enabled analyses which were previously technically extremely time-consuming, if not impossible.

- In the analysis of spatial form, maps of different phenomena are overlaid to identify correlations. This may involve the straightforward linking of point and area patterns, as in Ravenhill's (1955) investigations of the pattern of Celtic settlements in Cornwall. It may, however, involve attempts to link data from incommensurate sources, such as the classic paper by Robinson et al (1961) on areal correlations which related point (rainfall amounts) to area (population density) data through interpolation processes involving manually-created hexagonal tesselations.[1]
- In analyses of spatial diffusion using simulation techniques, time-consuming methods were employed to generate changing patterns over time using mean information fields (as in Hägerstrand's 1968 classic and works by his followers, such as Morrill 1965a, 1965b) – so much so that some simulations were based on a single iteration only (Robinson 1981).
- Although the ultimate focus of much research was on individual decision-making and behaviour, a great deal of the available data related to spatially-defined aggregates. Their analysis involved not only the ecological fallacy of inferring from a population (usually defined as those living in a particular area) to its individual members but also the aggregation and scale problems which exacerbated the difficulties of drawing valid conclusions. The simulation exercises involving what became known as the modifiable areal unit problem (MAUP: Openshaw 1977; Openshaw and Alvanides, Chapter 18) were further examples of time-consuming procedures which significantly constrained evaluation not only of the technical issues but also of the underlying substantive concerns (Wrigley 1995).[2]

In each of these, GIS technology has increased the efficiency of research endeavours many-fold,

illustrated for the first topic by Openshaw's proto-GIS work on the location of nuclear facilities (Openshaw 1986) and his 'geographical analysis machine' for the evaluation of significant clusters of disease outbreaks (Openshaw et al 1987). Their development does not eliminate decision-making regarding technical issues, as illustrated by the concerns of GIS researchers regarding how to integrate datasets compiled on incommensurate spatial frameworks (indeed in some ways they may extend them), but they make experiments much more easy to conduct.

In addition to improving the efficiency and effectiveness of existing research approaches, technological advances open opportunities for work in previously-closed areas. This is exemplified by 3-dimensional modelling of landscapes (Hutchinson and Gallant, Chapter 9) – for which the proto-technology (the Harvard mapping package) was extremely limited and precluded interactive experiments involving, for example, altering perspective and changing various components of a map format.

The introduction of GIS to geographical research thus provided new exemplars for the conduct of original investigations and assisted progress in spatial science approaches to both human and physical geography: use of the form and machine metaphors was significantly enhanced. These changes occurred during a period of very considerable debate over the use to which science should be put and the priorities over investment in its enhancement. In the early–mid 1980s, for example, the funding of new staff positions in British Universities under a competitive 'new blood' scheme focused substantially on investments in 'enabling technologies': of the 11.5 posts allocated to Departments of Geography, five were for work on remote sensing and digital mapping and three for mathematical modelling (see Smith 1985). More recently, investigations of the nature of posts vacated and filled in US university geography departments indicate a net growth in GIS specialists (Gober et al 1995; Miyares and McGlade 1994).

Of the various roles for geography-as-spatial-science, *ergon* has been promoted over *logos*, increasingly so in an economic environment which stresses the material gains to be derived from educational investment (Johnston 1995), on the grounds that successful technical applications are more likely to bring political (and thus financial) recognition for the discipline than are improvements in explanatory power. Thus established location–allocation models are used in 'spatial decision-making systems' applied to questions regarding the optimal location of service facilities (Birkin et al, Chapter 51; Clarke and Clarke 1995). Some argue that the vast amounts of data available to geographers should be dredged in an empiricist fashion rather than addressed in a theoretically-informed context (Openshaw 1989: 73, refers to 'data-driven computer modelling in an information economy'; and see Openshaw and Alvanides, Chapter 18): their expertise with spatially-referenced data allows them to 'add value' to them, thereby making that expertise more saleable (see also Rhind 1989), as exemplified by the growing field of geodemographics which uses computer-generated classifications of ecological data to identify target markets for 'niche products' (Batey and Brown 1995; Birkin 1995). Openshaw (1994: 202) has argued that a combination of three recent developments – in large-scale parallel computing, the use of artificial intelligence, and what he terms the 'GIS revolution' – is stimulating the emergence of an 'IT State' characterised by:

'large scale, benign, universal data capture covering most aspects of modern life, the computerisation of virtually all of the management and control systems on which societies and economies depend, the linkage of separate computer systems, the dissolution of technical obstacles to systems integration by the emphasis on open systems, and the increased reliance on computers at all levels and all scales for the continuation of life on Earth'.

This offers the potential of a new mode of 'computational geography ... sufficiently broadly defined, generic in its technology, and flexible in philosophical outlook so as to encompass most, if not all, areas of human and physical geography' (208–9; see also the discussions in Openshaw and Alvanides, Chapter 18 and Longley et al, Chapter 72). Despite that claim, however, Openshaw has yet to convince many (mainly human) geographers committed to organism and arena, poesis, and paideia that GIS has much to offer them and their research agenda.

5 GIS AND GEOGRAPHY AS SOCIAL THEORY

The division within geography between spatial scientists and social theorists, the association of GIS with the former, and the intensity of some of the debate between the two camps (including over GIS: see Johnston 1997) have largely blocked substantial exploration of the potential for GIS (and the 'IT State' more generally) within the latter camp. Gilbert (1995) argued that:

'The use of computers in geography has been stunted by identification with ... GIS, by both its proponents and critics. By and large geographers have come to regard computers as analytical machines, and have ignored their growing potential as a distinctively new mode of expression.'

Computers are much more than ever-faster and more-powerful calculating machines, but within geography the continuity of personnel from spatial science into GIS has created a barrier to this realisation among social theorists, despite the increasingly widespread use of multi- and hyper-media within the humanities.

This has come about despite the early use of a proto-GIS in one of the pioneer hypermedia educational tools – the BBC Domesday project of 1986 (Openshaw et al 1986). This integrated textual and visual material in novel ways, since very significantly enhanced. Spatial scientists increasingly employ hypermedia in decision-support systems, whereby a range of digitally-coded spatial data (remotely-sensed images, maps, other images, video etc.) is combined with textual and other material to aid a range of projects (Shiffer, Chapter 52; Raper and Livingstone 1995). Such usages have been subject to substantial critiques (as in several of the contributions to Pickles 1995) and the role of GIS in creating new images of the world is increasingly appreciated (Roberts and Schein 1995),[3] but the technology's positive potential has been submerged under the weight of this (usually valid) assessment of likely negative impacts.

A central argument of geographers attracted to the postmodernist disciplinary matrix is its emphasis on variety in both space and time that cannot be readily captured by conventional literary forms and styles – contemporaneity is extremely difficult to express (see the attempt by Pile and Rose 1992). Hypermedia make this much more possible, allowing what Gregory (1989) refers to as alternative textual strategies 'which do not attempt to reduce differences and fragmentations to a single overarching account, particularly where there is a concern to convey the spatial simultaneity of different experiences' (Gilbert 1995: 8) which allow 'changes in perspective, jump-cuts and cross-cuts between scenes, dislocations of chronology and composition, commentaries on the construction of the text by author and reader, and so on'. (See also Crang 1992, on polyphony.) The ability to integrate textual and other materials (including moving images with sound) would enable exploration of many of the issues raised by adherents to the postmodernity approach (as well as, more prosaically, allowing readers to refer to illustrative material alongside the relevant text, and to access cited material without relying on the – potentially biased – selection made by another author; they need not depend on the world as represented to them by someone else, but can access the world directly themselves).

6 CONCLUSION

There have been many changes in the practice of geography over recent decades, but the discipline has sustained its core concern with spatial variations in the nature of and the interactions among environment, space, and place. Four separate world-views, each with a number of associated disciplinary matrices and their paradigm exemplars, have been called on to address these concerns – with four separate 'applied goals'. These have been reduced in this brief discussion to just two major world-views – spatial science and social theory – with their associated applications.

Over the last decade, GIS has been associated with the former world-view – largely because of its perceived strong links with the technical apparatus and empiricist/positivist aims of spatial science. It has both increased the efficiency and efficacy of established research directions therein and opened up other possibilities. By contrast, applications within geography-as-social-theory have been few, and its potential within hypermedia and other applications largely ignored; the challenges thus remain massive and should be extremely fruitful.

Endnotes

1 The hexagons were not regular, because of an error by one of the researchers (personnel information).

2 As in several other areas, Stan Openshaw's creative abilities as computer programmer partially overcame these (Openshaw and Taylor 1979).

3 As exemplified by the NCGIA's Initiative 19 – 'GIS and society: the social implications of how people, space and environment and represented in GIS'. This addressing the

forms of representation adopted and rejected in GIS applications, the level of empowerment which user groups obtain, the ethical and regulatory issues raised by GIS applications, and the potential for using GIS in democratic resolutions of social and environmental conflicts.

References

Barnes T J 1996 *Logics of dislocation: models, metaphors, and meanings of economic space*. New York, Guilford Press

Barnes T J, Duncan J S 1992 Introduction: writing worlds. In Barnes T J, Duncan J S (eds) *Writing worlds: discourse, text & metaphor in the representation of landscape*. London, Routledge: 1–17

Batey P, Brown P 1995. In Longley P A, Clarke G (eds) *GIS for business and service planning*. Cambridge (UK), GeoInformation International

Bennett R J 1989 Whither models and geography in a post-welfarist world. In MacMillan W (ed.) *Remodelling geography*. Oxford, Basil Blackwell: 373–90

Bennett R J, Chorley R J 1978 *Environmental systems: philosophy, analysis and control*. London, Methuen

Birkin M 1995 Customer targeting, geodemographics, and life-style approaches. In Longley P A, Clarke G (eds) *GIS for business and service planning*. Cambridge (UK), GeoInformation International

Buttimer A 1993 *Geography and the human spirit*. Baltimore, Johns Hopkins University Press

Chorley R J, Kennedy B A 1971 *Physical geography: a systems approach*. Englewood Cliffs, Prentice-Hall

Clarke M, Clarke G 1995 The development and benefits of customised spatial decision support systems. In Longley P A, Clarke G (eds) *GIS for business and service planning*. Cambridge (UK), GeoInformation International: 227–46

Crang P 1992 The politics of polyphony: reconfigurations in geographical authority. *Environment and Planning D: Society and Space* 10: 527–49

Gilbert D 1995 Between two cultures: geography, computing, and the humanities. *Ecumene* 2: 1–13

Gober P A, Glasmeier A K, Goodman J M, Plane D A, Stafford H A, Wood J S 1995 Employment trends in geography. *The Professional Geographer* 47: 317–46

Gregory D 1989 Areal differentiation and postmodern human geography. In Gregory D, Walford R (eds) *Horizons in human geography*. London, Macmillan: 67–96

Gregory D 1994 *Geographical imaginations*. Oxford, Blackwell

Gregory K J 1985 *The nature of physical geography*. London, Edward Arnold

Hägerstrand T 1968 *Innovation diffusion as a spatial process*. Chicago, University of Chicago Press

Haggett P, Cliff A D, Frey A E 1977 *Locational analysis in human geography,* 2nd edition. London, Edward Arnold

Haggett P 1965 *Locational analysis in human geography*. London, Edward Arnold

Haggett P, Chorley R J 1969 *Network models in geography.* London, Edward Arnold

Haines-Young R, Petch J R 1985 *Physical geography: its nature and methods.* London, Harper and Row

Hartshorne R 1939 *The nature of geography: a critical survey of current thought in the light of the past.* Association of American Geographers: 21, 462–3

Hartshorne R 1959 *Perspective on the nature of geography.* Chicago, Rand McNally/London, John Murray

Harvey D 1982 *The limits to capital.* Oxford, Basil Blackwell

Johnston R J 1989 Philosophy, ideology, and geography. In Gregory D, Walford R (eds) *Horizons in human geography.* London, Macmillan: 48–66

Johnston R J 1995 The business of British geography. In Cliff A D, Gould P R, Hoare A G, Thrift N J (eds) *Diffusing geography: essays for Peter Haggett.* Oxford, Blackwell: 317–41

Johnston R J 1997 *Geography and geographers: Anglo-American human geography since 1945,* 5th edition. London, Edward Arnold

Johnston R J, Gregory S 1994 The United Kingdom. In Johnston R J, Claval P (eds) *Geography since the Second World War, an international survey.* London, Croom Helm: 107–31

Kuhn T S 1962 *The structure of scientific revolutions.* Chicago, University of Chicago Press.

Livingstone D N 1992 *The geographical tradition: episodes in the history of a contested enterprise.* Oxford, Blackwell

Marcus M G 1979 Coming full circle: physical geography in the twentieth century. *Annals of the Association of American Geographers* 69: 521–32

McDowell L 1995 Understanding diversity: the problem of/for theory. In Johnston R J, Taylor P J, Watts M J (eds) *Geographies of global change: remapping the world in the late twentieth century.* Oxford, Blackwell: 280–94

Miyares I M, McGlade M S 1994 Specialisation in 'Jobs in Geography' 1980–1993. *The Professional Geographer* 46: 170–7

Morrill R L 1965a *Migration and the growth of urban settlement.* Lund, C W K Gleerup

Morrill R L 1965b The negro ghetto. *The Geographical Review* 55: 339–61

Openshaw S 1977 A geographical study of scale and aggregation problems in region-building, partitioning, and spatial modelling. *Transactions of the Institute of British Geographers* NS2: 459–72

Openshaw S 1986 *Nuclear power: siting and safety.* London, Routledge

Openshaw S 1989 Computer modelling in human geography. In MacMillan W (ed.) *Remodelling geography.* Oxford, Blackwell: 70–88

Openshaw S 1994 Computational human geography: exploring the geocyberspace. *The Leeds Review* 37: 201–20

Openshaw S 1996b GIS and society: a lot of fuss about very little that matters and not enough about that which does! In Harris T, Weiner D (eds) *GIS and society: the social*

implications of how people, space and environment are represented in GIS. Scientific report for the Initiative–19 Specialist Meeting, 2–5 March 1996, Koinia Research Center, South Haven, NCGIA Technical Report 96–7: D54–D58. Santa Barbara, NCGIA

Openshaw S, Taylor P J 1979 A million or so correlation coefficients: three experiments on the modifiable areal unit problem. In Bennett R J, Thrift N J, Wrigley N (eds) *Statistical applications in the spatial sciences.* London, Pion

Openshaw S, Wymer C, Charlton M 1986 A geographical information and mapping system for the BBC Domesday optical discs. *Transaction of the Institute of British Geographers* 11: 296–304

Openshaw S, Charlton M, Wymer C, Craft A W 1987 A Mark I geographical analysis machine for the automated analysis of point datasets. *International Journal of Geographical Information Systems* 1: 335–58

Pickles J (ed) 1995 *Ground truth: the social implications of geographic information systems.* New York, Guilford Press

Pile S, Rose G 1992 All or nothing? Politics and critique in the modernism–postmodernism debate. *Environment and Planning D: Society and Space* 10: 123–36

Raper J, Livingstone D 1995 The development of a spatial data explorer for an environmental hyperdocument. *Environment and Planning B: Planning and Design* 22: 679–87

Ravenhill W L D 1955 The settlement of Cornwall during the Celtic period. *Geography* 40: 237–48

Rhind D W 1989 Computing, academic geography, and the world outside. In MacMillan W (ed.) *Remodelling geography.* Oxford, Blackwell: 177–90

Roberts S M, Schein R H 1995 Earth shattering: global imagery and GIS. In Pickles J (ed.) *Ground truth: the social implications of geographic information systems.* New York, Guilford Press: 171–95

Robinson A H, Lindberg J B, Brinkmann L W 1961 A correlation and regression analysis applied to rural farm densities in the Great Plains. *Annals of the Association of American Geographers* 51: 211–21

Robinson V B 1981. In Jackson P, Smith S J (eds) *Social interaction and ethnic segregation.* London, Academic Press

Sayer A 1992 *Method in social science: a realist approach,* 2nd edition. London, Routledge

Sheppard E S 1995 Dissenting from spatial analysis. *Urban Geography* 16: 283–303

Smith D M 1985 The 'new blood' scheme and its application to geography. *Area* 17: 237–43

Taylor P J 1985 The value of a geographical perspective. In Johnston R J (ed.) *The future of geography.* London, Methuen: 92–110

Woldenberg M J, Berry B J L 1967 Rivers and central places: analogous systems? *Journal of Regional Science* 7: 129–40

Wrigley N 1995 Revisiting the modifiable areal unit problem and the ecological fallacy. In Cliff A D, Gould P R, Hoare A G, Thrift N J (eds) *Diffusing geography: essays for Peter Haggett.* Oxford, Blackwell: 49–71

4

Arguments, debates, and dialogues: the GIS–social theory debate and the concern for alternatives

J PICKLES

The chapter provides an historical overview of the emergence of the debate in geography about the disciplinary and social implications of GIS: its theory of science, technological capacities, and social uses. It describes the epistemological, methodological and at times political positions that have guided both practitioners and theorists of GIS on the one hand, and those who have expressed concern about these positions and the social implications on the other. The chapter charts these debates from opposition arguments through lively debates to recent critical engagements and joint projects between scholars of GIS and social theory. The chapter ends with discussion of Research Initiative 19 of the US National Center for Geographic Information and Analysis (NCGIA) on 'GIS and Society' and some of its contributions.

1 GIS AND SOCIETY

In recent years GIS practitioners have begun to argue for the importance of building a more flexible, open, and theoretical science of geographic information systems and geographic information – a geographic information science (Goodchild 1992, 1993, 1995; Openshaw 1991, 1992, 1996; Wright et al 1997). This theoretical turn has emerged as GIS itself has changed from an enterprise involving the development and testing of software and hardware, to the application of GIS and the study of data structures and visualisation techniques, to a field that has become so generalised in everyday life and in academic research that the specific role of any single discipline – especially one with a special relationship to GIS (geography) has to be rethought (Pickles 1997; Wright et al 1997). This chapter maps out the parallel evolution of responses to these phases of GIS development in geography, and geographers' attempts to come to grips with the changing possibilities and problems that GIS has brought to the discipline and the wider society (see also Forer and Unwin, Chapter 54; Martin, Chapter 6).

Specifically, the chapter seeks to locate the GIS social theory debate in geography (and their respective claims to method, science, and

knowledge) in terms of a decade of changing technological and institutional ensembles, discourses, and practices which have brought about different responses and forms of engagement. We seek to capture something of the dynamism in the debate that occurred in the transition from the mid 1980s to the mid 1990s. This debate ranged from GIS as a research tool and scholarly practice (and the epistemological grounds on which these battles were fought), to debate about its fundamental assumptions and transformative capacities, to dialogue about alternative pathways for a technology that is increasingly realising both its utopian and dystopian possibilities.

The chapter outlines briefly the nature of the opposition arguments that emerged as a result of the disciplinary impacts wrought by GIS in the 1980s. It then shows how these opposition arguments – while they still continued in some quarters – gradually began to take the form of a constructive debate about the real material and intellectual effects of GIS. We go on to show how this debate is currently leading to experiments in dialogue among individuals and groups with quite distinct goals and perhaps different conceptions of GIS as technology, practice, and body of ideas. A different understanding of the possibilities and constraints of GIS as a tool, and of the study of GIS as a social practice, emerges from these engagements (Gilbert 1995).

The primary goals of this kind of work should be spelled out briefly, given the suspicion about the critique that has emerged within the field:

- to contribute to a theory of GIS which is neither technical nor instrumental, but locates GIS as an object, set of institutions, discourses, and practices that have disciplinary and societal effects;
- to show how these disciplinary and societal effects operate;
- to push against the limits of GIS and its unacknowledged conditions and unintended consequences of development and practice (e.g. corporate influence, epistemological assumptions, and understanding of appropriate applications);
- to ask whether GIS could have been different, or in what ways it may be made different in the future.

2 GIS AND GEOGRAPHY: NEW SCIENCE OR OLD WINE?

GIS is not only big business, it is becoming bigger and bigger business with every passing year. In the 1980s and 1990s GIS and related spatial data handling and imaging systems became central elements in the restructuring of economic activity, the modernisation of the state, and the administration of social life by public and private organisations (Cowen 1995). In the 1960s most geographers would probably have welcomed such changes and lauded as progressive the rationalisation of planning. In the 1990s these matters have given rise to deep divisions within the discipline about the role and function of social engineering and the information revolution that makes it possible in new forms. Although a substantial part of the discipline cannot understand why the geographic profession displays such distrust of the developments in GIS and why it remains sceptical about motives, potential value, and political consequences of adoption, another part of the discipline cannot understand why these questions have not yet been asked within the GIS community, how practitioners cannot see the problems raised by corporate control, proprietary systems, limitations on available data, and the uses to which GIS has been put in recent years. For some the revolution in spatial data processing and digital

imaging systems offers new opportunities for constructing 'informed' societies and pursuing rational and efficient social planning; for others the new systems of knowledge engineering and social engineering raise serious questions about freedom, civil society, and democratic practice (for further discussion see Curry 1994, 1995, 1996, 1997; Goss 1995a, 1995b; Harris et al 1995; Lake 1993; Miller 1995; Pickles 1991, 1992, 1993, 1995, 1997; Sheppard 1995; Sui 1994). Thus, as GIS has become a more significant element in restructuring public and private life, it becomes crucial to ask what impacts these technologies and applications have on the ways in which people interact with one another.

Until recently, discussions of the social impacts of GIS have been limited mainly to an internal analysis of technique and methodology: improving accuracy, extending capabilities, and widening the scope of applications that are possible. Little attention has been given to the broader discussions in geography about the interests that influence scientific research, the socially constituted nature of objects, categories, and concepts, the gendering of science, or the differing commitments of empiricist, hermeneutic, and critical epistemologies (Johnston, Chapter 2; Gregory 1978, 1994). Instead, much of the discussion has taken the form of a theoretical advocacy and an almost evangelical celebration of the possibilities offered by GIS to save geography – from its marginal economic position in universities, from its weak professional status in areas of public policy, from its underdeveloped technical capacities in applied fields, and from its humpty-dumpty like fragmentation in the discipline (Abler 1993; Openshaw 1991, 1992). In each of these domains GIS, it is claimed, offers rigorous science, useful technique, and universal possibilities for application. An objectivist epistemology and a pragmatic politics combine to reject any broader theorising of the consequences of this form of knowledge production and management.

Other geographers sometimes disagreed. Jordan's 1988 Presidential column in the newsletter of the Association of American Geographers (Jordan 1988) signalled the first reaction on the part of the old guard in the discipline to what was perceived as the pretension of GIS and its claim to intellectual standing: GIS was, in his view, merely a technical field without intellectual vigour or promise. Moreover, the need for large investments in capital equipment, personnel, training, and recurring costs for maintenance and upgrade was not matched initially by quality output and clear results. Indeed, the points of contact with GIS for most geographers in the 1980s were requests for budget and faculty lines on the one hand, and faint, Cubist- and Futurist-like map images on the other hand. As a result, many Realists in the discipline greeted the emergence of GIS with quiet resistance and the knowing scepticism of the bourgeois critic, comfortable in the assurance that the fad would pass.

One unfortunate and unnecessary side-effect of such opposition positions has been the tendency of one side or the other to reject as 'unreasonable' the arguments of the other. The result has been a closing down of constructive and open debate on both sides, and the emergence of 'cultures of indifference' on both sides. Since the personal, institutional, and social stakes are high this is not unexpected, but what was lost in this opposition was any serious debate with some important issues on each side. Where a fuller engagement with the ideas and claims of each has occurred, the result has been an 'energising' of both communities and an opening of new avenues of research.

3 EPISTEMOLOGICAL CRITIQUE: DEBATING THE ASSUMPTIONS

The first serious engagements between GIS and social theory occurred over issues related to the politics of knowledge and the social impacts of use (Lake 1993; Miller 1995; Pickles 1991, 1995; Sheppard 1995; Sui 1994). In his trenchant critique of GIS as the new imperialist geography, Taylor (1990) suggested that GIS emerged as a two-part strategy on the part of unreconstructed 'quantifiers' who had 'bypassed' the critiques levied against the empiricism of spatial analysis, and at the same time captured the rhetorical ground of a progressive modernism by readily accepting the switch from knowledge to information:

'Knowledge is about ideas, about putting ideas together into integrated systems of thought we call disciplines. Information is about facts, about separating out a particular feature of a situation and recording it as an autonomous observation . . . The positivist's revenge has been to retreat to information and leave their knowledge problems – and their opponents – stranded on a foreign shore. But the result has been a return of the very worst sort of positivism, a most naive empiricism.' (Taylor 1990: 211–212)

In this (re)turn the geographical is defined as the study of anything that is spatial:

'GIS is a technological package that can treat any systematic collection of facts that are individually identified spatially. These facts may be medical statistics, remote-sensing images, crime files, land-use data, population registers or whatever. In terms of the package, spatial patterns can be produced irrespective of what the information is about . . . Such quantifiers can produce a maverick geography dealing with crime one week, bronchitis the next, and so on.' (Taylor 1990: 212)

The colonising aspirations of such claims are, for Taylor, transparent. But many practitioners of GIS saw these claims as exaggerated at best and false at worst, or, as Openshaw (1991) argued they represent reductionist assertions and derogatory and confrontational language; 'knockabout stuff' that emerges from a reactionary desire to protect a particular system of order and power. Thus, for Openshaw the crisis to which Taylor points is redefined as 'contrived' and should be replaced by a notion of 'creative tensions' between at times complementary, at times competing, but equally productive intellectual projects. In place of any narrow delimitation of the possibilities of GIS, Openshaw (1991) offered an expansive vision of emerging GIS practice:

'A geographer of the impending new order may well be able to analyse river networks on Mars on Monday, study cancer in Bristol on Tuesday, map the underclass of London on Wednesday, analyse groundwater flow in the Amazon basin on Friday. What of it? Indeed, this is only the beginning.' (Openshaw 1991: 624)

This new order geography needs GIS in order to 'put the pieces of geography back together again to form a coherent scientific discipline':

'It would appear then that GIS can provide an information system domain within which virtually all of geography can be performed. GIS would emphasise an holistic view of geography that is broad enough to encompass nearly all geographers and all of geography. At the same time it would offer a means of creating a new scientific look to geography, and confer upon the subject a degree of currency and relevancy that has, arguably, been missing.' (Openshaw 1991: 626)

In this view, GIS has an overreaching technology and approach broad enough to allow any geographer to pursue his or her research questions: GIS offers the epistemological and methodological flexibility to the creative researcher to be adapted to any practical circumstance.

The divide is not, in this sense, between GIS and social theory, but between a social theory and notion of science rooted in empiricism (in which theory is that which accounts for the outcome of model testing) and social theory in which theory is the precondition for any understanding and analysis in the first place.

For these reasons, several commentators have argued strongly against the particular view of the discipline, of science, and of research practice and application that ties the development of GIS to the 'resurrection' of a rational model of planning and a positivist epistemology (Lake 1993; Sui 1994):

'. . . the unrelenting embrace of the rational model by planning and applied geography is not adequately described merely in terms of the tenacity and inertia of convenient and familiar practices. The rational model has been actively resurrected and rehabilitated by the ascendance of GIS to a position near to or at the core of both planning and geography.' (Lake 1993: 404)

In the 1980s, human geography developed strong critiques of the reductionist ontology of spatialism and turned to questions of contextual knowledge, contingency and necessity, society, space, and Nature, the (social/political/gendered) construction of space, and the production of scale, each of which in various ways problematised aspects of Cartesian science and the ontology of spatial analysis. These approaches questioned the overemphasis on pattern, challenged geographers to rethink the meaning of space, problematised the dominance of natural science method in the study of social phenomena, and raised questions about the underlying ontology of objects, location, and application on which spatial analysis was predicated. Yet, in his 1993 review of the field, Lake found few publications on the part of GIS proponents which consider these epistemological, political and ethical critiques of positivism, or any serious engagement with what he terms the 'fundamental disjuncture growing at the core of the disciplines'.

By the decade of the 1990s, social theorists within the discipline began to take aim at what they saw as the transformative capacities of GIS, both in disciplinary and broader social terms. The author's own 1991 essay on 'The Surveillant Society', Gregory's (1994) claim that GIS positivists represented the 'new Victorians', and Smith's (1992) charge that the war against Iraq – the Gulf War – represented the first GIS war, incensed many practitioners and theorists of GIS. How could these neophytes and outsiders levy such charges, particularly against the only part of the discipline that really exercised rigour in its work and power in regard to other disciplines and funding agencies?

Such concern turned to outrage as more GIS practitioners interpreted claims about GIS and its origins in surveillance and battlefield logistical needs in the military (Pickles 1991) as a direct attack on their own credibility and commitments, and 'GIS über Alles' (the title of the first section in *Real wars, theory wars*: Smith 1992) and the purposeful ambiguity in the first two sentences: 'The war against Iraq in 1990–91 was the first full-scale GIS war. It put geography on the public agenda in a palpable if unpalatable way as it claimed an estimated 200 000 Iraqi lives' as suggesting that GIS, fascism, and imperialistic warmongering were somehow synonymous.

There was a double irony here. First, in the *declarations of war* against social theory and the expressions of the need to *mobilise in defence* of GIS against this onslaught that ensued in various gatherings of geographers. Second, that these responses occurred at the very time when, for example, Dangermond was bringing Ralph Nader to speak to the ESRI Users' Conference to argue for a 'vigorous GIS' – that is, one that recognised its current embeddedness within the institutions of government, military, and corporate interests, and instead sought to foster democratic access and public participation. This at a time when Openshaw (1991, 1992) was calling for a more open, *flexible* GIS, and when Goodchild (1992) was arguing for the need for a geographical information science that would address the impacts, as well as the possibilities, of the use of GIS. In one sense, social theorists and theorists of GIS had reached similar conclusions but by different paths.

In essence, the speed at which the technology was changing, the breadth of adoption and use, and the depth of the impacts of contemporary GIS had changed the terrain on which the discussion would occur. The opposition logics of the 1980s were no longer practically helpful. User-friendly software had increased the number of GIS users. GIS had grown institutionally strong and – with its own conferences, journals, and funding sources – no longer took the arguments of disciplinary theorists seriously. At the same time, few critics followed sufficiently closely the emerging capacities and the new applications to understand the changes they wrought. In particular, few understood that while the instrumental logics and positivist justifications they attacked were being ever more deeply ingrained, they were also being fundamentally challenged by new practices and notions of space, object, and science that did not fit within such positivist frameworks: GIS itself was beginning to experience contradictions in its own claims and practices.

Despite last ditch efforts on the part of the traditionalists (Jordan 1988), GIS could not be wished away, nor could the hard resource decisions be avoided by departments and individuals in their research and teaching. When the Chancellor of the University and Manager of the Office of Facilities Planning both pull up GIS for the day-to-day management of their campus, when city planners are digitising every street in the city, when city engineers are GPSing every waterline and powerline they manage, when new forms of red-lining using GIS maps have become second nature to insurance companies, and when the US Department of Defense solves complex peace negotiations over delimiting territorial borders in Bosnia with digital terrain models and repeated flyovers for negotiators, there can be no question that geographers must take GIS seriously as a set of tools, institutions, ideas, and practices that are shaping our lives and landscapes, and that are transforming the possibilities for certain types of research in the discipline. How to ask these questions was the crucial issue.

4 GIS AND SOCIETY: DIALOGUE AND ENGAGEMENT

In response to the sterile binaries of uncritical support and outright denial, Brian Harley and the author decided (following two sessions of 'GIS and Society' held at the Annual Conference of the Association of American Geographers in 1991) to edit a book of essays that attempted to theorise the impacts of GIS in the discipline and in the wider society as a means of stimulating students to begin to think about alternatives to the rather sterile pro and con positions that dominated discussions at the time. To our

surprise, the 'idea of the book' *Ground truth* (Pickles 1995) achieved some of these goals prior to publication. This 'idea of the book' began to circulate on list servers like GIS-L, and concerns were expressed that such a book could undermine the growing position of GIS in the field: the book was somehow to be thought of as a dangerous attack on GIS.

One outgrowth of these discussions was an NCGIA-sponsored workshop 'GIS and Society'. The workshop addressed the kinds of questions that needed to be asked to understand the growing influence and social implications of GIS development and use, to consider how and in what ways such questioning might be sustained, and to investigate the possibilities for future critical engagements among GIS and social theory (Poiker 1993).

Perhaps the single issue that causes confusion in geography over the possibilities and limits of GIS use is what generally is represented as the debate about positivism, a term that has served geographers as a recurring moment for mobilisation or vilification. The concept itself stands as a signifier for something broader and it is here that the problem needs to be located. The apparent incommensurability between GIS and social theory critiques has its origin, perhaps, in how one understands the appropriate scope for inquiry (see also Johnston, Chapter 2). Most discussion of GIS operates within a very circumscribed understanding of the appropriate domain of inquiry, and this bounding of the field has variously been criticised as technicist, instrumentalist, and positivist. Social theorists have gradually broadened their own understanding of the appropriate scope within which inquiry must be situated, and currently any single social theoretic critique might operate at any one scale ranging from theories of geography, and science, to theories of society and technology (including the role of commerce, planning, and strategic thinking), to theories of modernity (including political theories of liberalism and critiques of masculinism, imperialism, and observer epistemologies) to Enlightenment thought itself. For each of these domains distinct literatures and languages have been carefully developed to enable critical thought. *Ground truth* aimed to locate discussions of GIS in a variety of these possible interpretative frameworks, and thereby to provide illustrations that might lead others to deepen the analysis of the intellectual and practical commitments and impacts of GIS.

This was also the goal for the 1995 special issue of *Cartography and Geographic Information Systems* – 'GIS and Society' – edited by Sheppard (1995). In his introduction Sheppard argued that the opposition nature of the debates occasioned by the emergence of GIS was full of heroic images and cruel caricatures, and that supporters and critics of GIS could learn from each other. Sheppard demonstrated how the origins of GIS affect the ways of thinking that can be employed. First, the dependence of GIS on digital computing (as opposed to analog computers, for example) constrains GIS by the structure and logic of the Turing machine, which employs deductive, Aristotelian logic. Second, the link between GIS and computers means that GIS is embedded in a broader set of social relations within which the computer is deployed:

'A major theme of the post-war era, in both the first and second worlds, has been extending the ability of both public and private institutions to control and organise the production and delivery of goods and services effectively. The principles of operations research as a methodology for optimally achieving well-defined goals, so effectively demonstrated in the armed services during the second world war, have been promoted as facilitating the rationality of both private enterprise operating in a free market and of public planning in a welfare or socialist state.' (Sheppard 1995: 8)

The result is that 'large institutional actors favour, and finance, those developments meeting their needs' (Sheppard 1995), and thus influence the development of computing and the directions taken by applications such as GIS. Since these large institutional actors have primarily been corporate, military, or public administration institutions, it should be no surprise that applications that favour surveillance, private sector interests, and control functions have been more common than those favouring public participation, data access, and community-defined goals. Such biases may be unproblematic for some, but for others they present a serious challenge to the possibility of a critical and rigorous science. Either way, GIS is a product of such technological and social constraints and its capacities have been influenced and delimited by these constraints.

5 GIS AND SOCIETY – NCGIA INITIATIVE 19 (I-19)

Following the Friday Harbor workshop, and in part stimulated by it, a group of interested individuals proposed that the US National Center for Geographic Information and Analysis dedicate one of its research initiatives to the issues now known as 'GIS and society'. A proposal was submitted to the NCGIA Board of Directors and approved, and a meeting of specialists convened in early 1996 (Harris and Weiner 1996). This section examines the continuing work of the initiative in some detail.

What marks the Friday Harbour and I-19 workshops as unique and important in the emerging theory of GIS is that Friday Harbour ended with, and I-19 began with, a set of assumptions that have been absent from debates about GIS until recently. Questions of origins, epistemology, data selection and data access, forms of representation, and the politics and ethics of information have generally been seen as marginal to the more technical questions of systems development and application (Martin, Chapter 6; see also Raper, Chapter 5). At these meetings they were seen as essential for any discussion of GIS and society. GIS is thus seen as a set of institutionalised systems of data handling and imaging technologies and practices situated within particular economic, political, cultural, and legal structures. They can thus be thought of as spatial data institutions (Curry 1995) and sociotechnological ensembles (Latour 1993). Understanding GIS as both a set of social practices and institutions embedded in a particular discourse is, perhaps, unique in the history of the engagement between GIS and social theory. Certainly, such social constructionist, genealogical or post-positivist theoretical frameworks have been virtually absent until recently in the debates over GIS.

Deploying such frameworks has been an important part of an emerging theory of GIS and society in which description (of the development of particular logics, systems, and uses of GIS), analysis (of the limits of access, range of diffusion, and effects of use) and critique (focused on the epistemological assumptions embedded in systems and use, conceptions of language in use, and logics and representations) are all present.

5.1 Critical social history of GIS

The written history of GIS is quite limited and few detailed case studies have appeared in print (Coppock and Rhind 1991; Goodchild 1988; Petchenik 1988). But it is vital to any critical field of inquiry that its practitioners know about the origins of the choices made and those rejected in defining and delimiting the field. In particular, it is vital that the technical, logical, and epistemological constraints on what GIS does, and the ways in which particular logics and visualisation techniques, values systems, forms of reasoning, and ways of understanding the world have been incorporated into existing GIS techniques are understood. It is equally important that practitioners and theorists understand the ways in which alternative forms of representation have been filtered out.

In the first instance, this has to do with the development paths taken within GIS and the possible alternatives that were not chosen but were available.

- Accepting that scientific knowledge is socially produced and rejecting any linear path of technical development, what were the debates and decisions leading to certain system choices and foundational logics rather than others within GIS over the past 30 years?
- Second, if alternatives were not pursued or accepted at the time, what were these and what were the conditions under which they were rejected or not pursued?
- Third, if there are always choices being made in the design and implementation of any technology and research tool, can alternative cultural and social conceptions of objects (property, land, resource relational values, historical meaning) be incorporated within GIS, and what are the actual possibilities for extending GIS to incorporate new ways of understanding the world?
- Fourth, since system and procedural choices have already been made and are now rooted in place through technical, financial, and practical inertia, what are the limits on what present-day GIS can do and what any reformed GIS might achieve?

There is a broader context that is also relevant here. This has to do with the issue of historical antecedents. GIS does not spring full blown or completely new into our world (Coppock and Rhind 1991; Goodchild 1988; Petchenik 1988). It emerges out of systems of land surveying, mapping, and data collection each with long heritages, and each having been centrally placed in the systematising and formalising of social life under capitalism. It is a constant surprise to social theorists in geography that the published histories of GIS tend to be what Livingstone (1992) referred to as 'internalist' and 'hagiographic', and do not deal with these historical antecedents, the ways in which GIS developed and diffused (who funded development, what options were considered and rejected, what institutional and intellectual linkages were forged in the development of GIS, etc.) and the patterns of production, marketing, and use that emerge in different cultures and settings. This would seem to be vitally important for any area of science in assessing the effectiveness, value, and limitation of its own technical and theoretical practices. Moreover, such questions locate the study of GIS at the heart of contemporary geographical issues (Wright et al 1997).

Recognising that GI comprises a series of institutions, discourses, and practices (as well as a set of tools) means that any theory of GIS must account for its origins and effects. In other words, GIS as a socially embedded and historically produced set of practices must account for its own history. It is to this question that the Critical History of GIS (CHGIS) Group, an activity initiated under I-19, has recently turned its attention.

Attempting to write a history of GIS that is not internalist or hagiographic, the CHGIS Project aims to bring a variety of theoretical perspectives from contemporary social theory to bear on the question of GIS as social practice. It also attempts to contextualise GIS in its social, political, and economic context, to locate GIS in terms of a broader history of science and technology than heretofore – and specifically to do so through an engagement with the systems and logics that were developed, the paths that were not taken, and the institutional linkages that provided the context for that which emerged.

5.2 Marginalised groups and the politics of access, exclusion, and control

In recent years, new technological capacities and an expansion of the scope of their application in many areas of social life have made it increasingly important to think about the ways in which the logics, systems, and representations deployed by contemporary GIS support particular types of social practice and inhibit others. What effects are GIS having? If GIS has been influenced by the demands of their developers and funders, many of them tied to large institutional and corporate interests and high-cost applications, what forms of access to information do these systems promote and deny? Specifically, how has the proliferation and dissemination of databases associated with GIS, as well as differential access to these databases, influenced the ability of different social groups to gain access to and use this information for their own purposes (see Rhind, Chapter 56)? Second, what types of knowledge and forms of reasoning are not well represented within GIS and what are the consequences of their exclusion (Onsrud 1992a, 1992b)?

A theory of GIS and society must address the impacts of these limits and impediments on groups or individuals where unequal access to software, hardware, and technical skills present real barriers to use, and seriously affect the types of outcome that result from the use of GIS in making decisions.

Differential access to databases is, clearly, becoming one of the central issues facing scholars and users of GIS and all forms of electronic data. As spatial data handling capabilities increase in power, the social impacts become more important. Geodemographic spatial data handling, for example, is already raising serious questions about privacy and access to databases (Curry, Chapter 55; 1997; Goss 1995a, 1995b). Until very recently, the primary sites at which GIS have been developed have been at national and local (in the USA, the state) level. In Britain, GIS has been used for land-use applications related to zoning, long-term planning, and the like. But the increasing availability and ease of use of GIS, accelerated by the development and deployment of global positioning systems and remote sensing systems, now constitute a powerful means of systematically tracking a wide range of natural and social phenomena, and in particular of developing monitoring systems for tracking populations (Graham 1997; Pickles 1991). The development of

these systems raises a wide range of questions about the types of assumption, data, and representation that are incorporated in any GIS. Who decides which data are to be collected? Who decides how those data are collected, which categories (of race, gender, species and so on) are to be used? How will the accuracy and validity of those data be measured and guaranteed, not in the technical sense of data error, but in a political sense of data appropriateness? Finally, because state agencies are both users and regulators of software, hardware, and data, questions arise concerning the ways in which these agencies adjudicate their sometimes competing responsibilities of protecting citizens and promoting use. (See Goodchild and Longley, Chapter 40, for a discussion of the technical implications of some of these issues.) In summary, how is the balance between rights to access and rights to privacy currently being struck (Curry 1995)?

The emergence of geodemographic information systems (GDIS) as targeted marketing strategies has already pointed to the emergent dangers of the use of GIS to further the commodification of everyday life (Curry 1997; Goss 1995a, 1995b). In the case of GDIS the issues go beyond the increasing efficiency of marketing agencies to target consumers with particular tastes and purchasing habits. They involve questions about the constitution of identity. GDIS consumer profiles, are aggregate profiles based on neighbourhood level data from which individual profiles are constructed. The targeting of commercial, political, and public service materials to individuals based on neighbourhood-derived profiles in turn 'produces' new identities (in that it channels and restricts the information individuals in that neighbourhood receive). Thus, even beyond questions of access and privacy, GDIS raises fundamental questions about the ethics of using information systems in ways that presuppose (and in turn contribute to the development of) socially homogeneous neighbourhoods.

There is a basic paradox in using GIS to address issues of land-use planning of any sort. On the one hand, conflicts over the use of space typically involve competing sets of values, assumptions, and interests. Not unexpectedly, the representations incorporated in GIS models of landuse conflicts tend to reflect the views, values and interests of dominant sectors of society. Ethnic, racial, and sexual minorities whose values and interests differ from those of culturally or economically dominant groups may be doubly disadvantaged when attempts to resolve conflict involve a significant GIS component (see Fisher, Chapter 13). Not only are their interests not intrinsic to the models on which technical solutions to complex problems are based, but they may lack access to the tools used by planners and politicians in making their decisions (Aitken and Michel 1995; Lake 1993; Miller 1992; Yapa 1991).

5.3 Ways of knowing

Beyond questions of access and exclusion is a related set of issues having to do with the ways in which knowledge and information are represented. An interesting change in the thinking of geographers seems to have occurred as GIS has been applied to more and more questions of this sort. Geographical information is increasingly assumed to refer to that which is captured or could be captured by GIS. Since GIS typically assume a universal set of objectifiable and 'self-evident' components of the processes they model (Sheppard 1995), GIS representations are often based on the assumption that there is a single version of reality to be modelled, and that land-use planning and conflict resolution principally involve the discovery of the most efficient solution to this objectifiable location problem. The use of GIS in locational conflict resolution has, in one important sense, poorly served the interests of those whose viewpoints and values differ from those incorporated in GIS models. Other forms of geographical information: place-based information, local knowledge, historical memory of land-use struggles, past events etc., are being marginalised as subjective information, *doxa*, or opinion (Curry 1996, 1997; but see Fisher, Chapter 13, and Veregin, Chapter 12, for discussions of uncertainty and data quality, respectively).

One example already addressed in literature is the case of the use of GIS to revisit claims of North Americans whose lands were ceded to the government in the nineteenth century, and whose abrogated treaty rights are now a basis for re-evaluation of that land alienation process. A basic problem emerges in the fact that GIS is far better at incorporating certain types of variable than others (Fischer, Chapter 13; Poiker 1993). Clearly, the variables incorporated in GIS representations are

not always tangible: for instance, both physical forest resources and conceptual property boundaries are included in GIS databases used in adjudicating land disputes. However, intangible factors related to competing value systems are not usually present in such analyses. How factors such as emotional attachments and the sacredness of place, the role of place in creating and maintaining community, use rights versus property ownership rights, and alternative views of nature are incorporated adequately into the GIS analysis of such conflicts has a huge impact on the types of claims and decisions that can be made (Rundstrom 1991). Rundstrom (1995) has even gone so far as to ask whether decisions should be based on GIS analysis at all in cases where such calculi are not amenable to incorporation into GIS models.

It is not yet clear how any technical systems can deal with alternative knowledge systems in cross-cultural settings. Some ways of knowing are privileged in existing GIS approaches, but it is not clear how different types of knowledge and information can be included. Nor is it clear whether the apparent technical and epistemological limitations of present systems could incorporate different ways of knowing without reducing one to the other, or whether new, different system logics, configuration, and practices need to be developed. The possibilities and the difficulties involved in these efforts are well documented by Harris et al (1995) and Weiner et al (1995).

With the inclusion of locationally fuzzy knowledge many issues arise as to how the multi-objective goals, based on multiple criteria, and using spatially imprecise and possibly conflicting data might actually achieve what is assumed to be consensus decision-making. Perhaps one reason why GIS has achieved such astounding 'success' to date in decision-making support roles is that it is based on only one seemingly non-contradictory perception of reality. Collaborative spatial decision-making is a complex issue even among participants with similar world views and knowledge. In the absence of this commonality the difficulties are qualitatively greater. But these difficulties are also opportunities; they arise as such partly because of new technical capacities for handling large datasets and displaying and disseminating spatial images. What a 'pluralistic GIS' (one containing multiple views of resource value, potentially fuzzy, and conflicting information) would look like and what it would imply for the ways in which GIS can be used in collaborative decision-making remain open questions.

5.4 Public participation and GIS-2

If it is the case that the systems and logics that underpin much GIS emerged in response to the requirements and influence of large institutional supporters (be they public or private), then issues of surveillance, ownership, and control raise questions about the possibility of access, participation, and community-based involvement in GIS. This is even more pressing if one is not willing to reduce such issues of access and participation to the logics already present in existing systems. If GIS has emerged in its present form as a result of influences from a variety of financial and institutional interests, and if it does operate (through its technical demands, cost structure, types of data, and differential access) as a top-down technology and practice, can it be democratised? The democratisation of GIS means that the emerging possibilities of the technology must be considered. What must also be considered is how the types of systems and logics emerged within contemporary GIS and whether they can be changed.

If these forms of embeddedness do function as real constraints on public participation, can alternative social relations, ways of knowing, and marginalised groups be represented or given access in ways that do not reduce their own positions and logics to those of current GIS practice? How can the knowledge, needs, desires, and hopes of marginalised social groups be represented adequately as input to a decision-making process, and what are the possibilities and limitations of GIS as a way of encoding and using such representations?

If contemporary GIS can be thought of as predicated on the computerisation of the cartographic industry (GIS-1), can alternatives (GIS-2) be thought of which might range from 'knowledge creation environments' (Goodchild 1995) to public access centres and which address these issues? Also, how should people, space, and nature be represented? Who should have the right to speak on the nature of the representations that are created (Latour 1993)? What criteria might govern the emergence of such a GIS-2?

This question was raised and discussed at Friday Harbor and has become a central focus of I-19 research. The issue of system design is being addressed in the public participation project at the University of Maine, headed by Schroeder and Onsrud. Questions of legal and ethical conditions that enable and prevent intrusion are being addressed in a joint project between UCLA and the University of Minnesota – specifically by Curry, Sheppard, and Miller. The nature of geographical information in situations involving social conflict, and its relationship to the present capacities of GIS, is being addressed variously in projects in Minnesota and UCLA, and at the University of Kentucky and West Virginia University.

These efforts are aimed at asking what GIS-2 might look like. It would certainly have to be cheaper, more accessible, and sufficiently flexible to be of use to a wider range of users. But it would also have to address public concerns about privacy and access to information. Such a public GIS would have to guard against the reduction of multiple ways of knowing to a single logic and the premature resolution of differences. Instead, it will have to develop ways to represent different conceptions of space or Nature, and preserve contradiction, inconsistency, and disputes. Finally, a more flexible and accessible GIS-2 needs to be capable of integrating all data components, such as WWW, data archives, parallel and counter texts in diverse media, standard maps and datasets, and sketch map and field notes, all from one interface (Harris and Weiner 1996; *http://ncgia.spatial.maine.edu/ppgis/ppgishom.html*).

6 CONCLUSIONS

What are the results of the engagements described above? In the first place, these are early days in each of these projects and concrete research results are limited. Several conceptual advances have, however, been made.

- The relationship between the speed of developments and depth of the impacts of GIS technology, theory, and practice can now be seen in the context of a field that has been reticent to acknowledge the conditions of its own production, that has been lax in building its own archive, and that has by and large failed to develop sustained and detailed critical reflection upon its own practices. The discussion around GIS until the late 1980s remained focused largely on technical issues, unreflective in nature, and theoretical only insofar as theory referred to either empirical findings or internal technical concerns (but see, for example, Chrisman 1987a, 1991a; Coppock and Rhind 1991; and Goodchild 1995).

- The debate thus far has broadened discussion of GIS practice so that it now encompasses the social impacts of GIS. This is particularly important as new cyberspaces emerge and new forms of geographical information are finding a home through which important reconfigurations of material life are being affected.
- The need to think of GIS as a social object with its own institutional contexts, discourses, and practices has been demonstrated. This is not, however, an argument for a form of exceptionalism or professionalising of GIS as a discipline. Instead it calls for the necessity of locating those institutions, discourses, and practices in terms of broader debates in social theory about science/technology/society, theories of science, and the political economy of informatics on the one hand, and the recontextualising of GIS practice within the broader debates about geography on the other hand.
- The engagement has rectified one important absence within GIS communities (the legitimacy of 'GIS and society' questions and the availability of sites and groups among whom such discussion can continue).

The emergence of critical dialogue between GIS and social theory offers great promise for the emergence of a critical GIS aware of its own effects and striving to open its capacities to the needs, questions, and ways of knowing of broader and different 'publics'.

References

Abler R F 1993 Everything in its place: GPS, GIS, and geography in the 1990s. *The Professional Geographer* 45: 131–9

Aitken S C, Michel S M 1995 Who contrives the 'real' in GIS? Geographic information, planning, and critical theory. *Cartography and Geographic Information Systems* 22: 17–29

Chrisman N R 1987a Directions for research in GIS. *Proceedings, IGIS Symposium* 1: 101–12

Chrisman N R 1991a A geography of geographic information: placing GIS in cultural and historical context. *Mimeo*

Coppock J T, Rhind D W 1991 The history of GIS. In Maguire D J, Goodchild M F, Rhind D W (eds) *Geographical information systems: principles and applications*. Harlow, Longman/New York, John Wiley & Sons Inc. Vol 1: 21–43

Cowen D J 1995 The importance of GIS for the average person. *GIS in government: the federal perspective 1994*. Proceedings First Federal Geographic Technology Conference, 26–28 September 1994. Washington DC, GIS World Inc: 7–11

Curry M R 1994 Images, practice, and the hidden impacts of GIS. *Progress in Human Geography* 18: 441–59

Curry M R 1995 Rethinking rights and responsibilities in GIS: beyond the power of imagery. *Cartography and Geographic Information Systems* 22: 58–69

Curry M R 1996a Digital people, digital places: rethinking privacy in a world of geographic information. Paper presented at the conference on Technological Assaults on Privacy, Rochester Institute of Technology, Rochester (USA), 18–19 April

Curry M R 1997a Geodemographics and the end of the private realm. *Annals of the Association of American Geographers* 87: 681–99

Gilbert D 1995 Between two cultures: geography, computing, and the humanities. *Ecumene* 2: 1–13

Goodchild M F 1988 Stepping over the line: technological constraints and the new cartography. *The American Cartographer* (special issue on the history of GIS) 15: 311–22

Goodchild M F 1992 Geographical information science. *International Journal of Geographical Information Systems* 6: 31–45

Goodchild M F 1993 Ten years ahead: Dobson's automated geography in 1993. *The Professional Geographer* 45: 444–5

Goodchild M F 1995a GIS and geographical research. In Pickles J (ed.) *Ground truth: the social implications of GIS*. New York, Guilford Press: 31–50

Goss J 1995a Marketing the new marketing: the strategic discourse of geodemographic information systems. In Pickles J (ed.) *Ground truth: the social implications of GIS*. New York, Guilford Press: 130–70

Goss J 1995b 'We know who you are and we know where you live': the instrumental rationality of geodemographic information systems. *Economic Geography* 71: 171–98

Graham S 1997 Surveillant simulation and the city: telematics and the new urban control revolution. *Environment and Planning D: Society and Space*

Gregory D 1978 *Ideology, science, and human geography*. London, Hutchinson

Gregory D 1994 *Geographical imaginations*. Cambridge (USA), Blackwell

Harris T M, Weiner D 1996 *GIS and society: the social implications of how people, space, and environment are represented in GIS*. Scientific report for the Initiative-19 Specialist Meeting, 2–5 March, Koininia Retreat Center, South Haven. NCGIA Technical Report 96–7. Santa Barbara, NCGIA

Harris T M, Weiner D, Warner T A, Levin R 1995 Pursuing social goals through participatory GIS. In Pickles J (ed.) *Ground truth: the social implications of GIS*. New York, Guilford Press: 196–222

Jordan T 1988 The intellectual core: President's column. *AAG Newsletter* 23: 1

Lake R W 1993 Planning and applied geography: positivism, ethics, and GIS. *Progress in Human Geography* 17: 404–13

Latour B 1993 *We have never been modern*. Cambridge (USA), Harvard University Press

Livingstone D N 1992a *The geographical imagination.* Cambridge (USA), Blackwell

Miller B 1992 Collective action and rational choice: place, community, and the limits of individual self-interest. *Economic Geography* 68: 22–42

Miller B 1995 Beyond method, beyond ethics: integrating social theory into GIS and GIS into social theory. *Cartography and Geographic Information Systems* 22: 98–103

Onsrud H J 1992a In support of open access for publicly held geographic information. *GIS Law* 1: 3–6

Onsrud H J 1992b In support of cost recovery for publicly held geographic information. *GIS Law* 1: 1–7

Openshaw S 1991a A view on the GIS crisis in geography, or using GIS to put Humpty Dumpty back together again. *Environment and Planning A* 23: 621–8

Openshaw S 1992 Further thoughts on geography and GIS: a reply. *Environment and Planning A* 24: 463–6

Openshaw S 1996b GIS and society: a lot of fuss about very little that matters and not enough about that which does! In Harris T, Weiner D (eds) *GIS and society: the social implications of how people, space, and environment are represented in GIS*. Scientific report for the Initiative-19 Specialist Meeting, 2–5 March 1996, Koininia Retreat Center, South Haven. NCGIA Technical Report 96–7: D.54–D.58. Santa Barbara, NCGIA

Petchenik B B (ed.) 1988 Special issue on the history of GIS. *The American Cartographer* 15: 249–322

Pickles J 1991 Geography, GIS, and the surveillant society. *Papers and Proceedings of Applied Geography Conferences* 14: 80–91

Pickles J 1992 Review of D Martin 'GIS and their socio-economic applications'. *Environment and Planning D: Society and Space* 10: 597–606

Pickles J 1993 Discourse on method and the history of discipline: reflections on Jerome Dobson's 1983 'Automated geography'. *The Professional Geographer* 45: 451–5

Pickles J (ed.) 1995 *Ground truth: the social implications of geographic information systems*. New York, Guilford Press

Pickles J 1997 Tool or science? GIS, technoscience, and the theoretical turn. *Annals of the Association of American Cartographers* 87: 363–72

Poiker T (ed.) 1993 *Proceedings 'Geographic Information and Society' workshop, Friday Harbor, 11–14 November*. Santa Barbara, NCGIA

Rundstrom R 1991 Mapping, postmodernism, indigenous people, and the changing direction of North American cartography. *Cartographica* 28: 1–12

Rundstrom R 1995 GIS, indigenous peoples, and epistemological diversity. *Cartography and Geographic Information Systems* 22: 45–57

Sheppard E (ed.) 1995 Special issue: GIS and society. *Cartography and GIS* 22 (1)

Smith N 1992 Real wars, theory wars. *Progress in Human Geography* 16: 257–71

Sui D Z 1994 GIS and urban studies: positivism, post-positivism, and beyond. *Urban Geography* 15: 258–78

Taylor P 1990 Editorial comment: GKS. *Political Geography Quarterly* 9: 211–12

Weiner D, Warner T A, Harris T M, Levin R M 1995 Apartheid representations in a digital landscape: GIS, remote sensing and local knowledge in Kierpersol, South Africa. *Cartography and Geographic Information Systems* 22: 30–44

Wright D J, Goodchild M F, Proctor J D 1977 Demystifying the persistent ambiguity of GIS as 'tool' versus 'science'. *Annals of the Association of American Geographers* 87: 34–62

Yapa L 1991 Is GIS appropriate technology? *International Journal of Geographic Information Systems* 5: 41–58

11

Visualising spatial distributions

M-J KRAAK

Maps are an integral part of the process of spatial data handling. They are used to visualise spatial data, to reveal and understand spatial distributions and relations. Recent developments such as scientific visualisation and exploratory data analysis have had a great impact. In contemporary cartography three roles for visualisation can be recognised. First, visualisation may be used to present spatial information where one needs function to create well-designed maps. Second, visualisation may be used to analyse. Here functions are required to access individual map components to extract information, and functions to process, manipulate, or summarise that information. Third, visualisation may be used to explore. Functions are required to allow the user to explore the spatial data visually, for instance by animation or linked views.

1 INTRODUCTION

Developments in spatial data handling have been considerable during the past few decades and it seems likely that this will continue. GIS have introduced the integration of spatial data from different kinds of sources, such as remote sensing, statistical databases, and recycled paper maps. Their functionality offers the ability to manipulate, analyse, and visualise the combined data. Their users can link application-based models to them and try to find answers to (spatial) questions. The purpose of most GIS is to function as decision support systems in the specific environment of an organisation.

Maps are important tools in this process. They are used to visualise spatial data, to reveal and understand spatial distributions and relations. However, maps are no longer the final products they used to be. Maps are now an integral part of the process of spatial data handling. The growth of GIS has changed their use, and as such has changed the world of those involved in cartography and those working with spatial data in general. This is caused by many factors which can be grouped in three main categories. First, technological developments in fields such as databases, computer graphics,

multimedia, and virtual reality have boosted interest in graphics and stimulated sophisticated (spatial) data presentation. From this perspective it appears that there are almost no barriers left. Second, user-oriented developments, often as an explicit reaction to technological developments, have stimulated scientific visualisation and exploratory data analysis (Anselin, Chapter 17). Also, the cartographic discipline has reacted to these changes. New concepts such as dynamic variables, digital landscape models, and digital cartographic models have been introduced. Map-based multimedia and cartographic animation, as well as the visualisation of quality aspects of spatial data, are core topics in contemporary cartographic research.

Tomorrow's users of GIS will require a direct and interactive interface to the geographical and other (multimedia) data. This will allow them to search spatial patterns, steered by the knowledge of the phenomena and processes being represented by the interface. One of the reasons for this is the switch from a data-poor to a data-rich environment, but it is also because of the intensified link between GIS and application-based models. As a result, an increase in the demand for more advanced and sophisticated visualisation techniques can be seen.

The developments described here have led to cartographers redefining the word 'visualisation' (Taylor 1994; Wood 1994). In cartography, 'to visualise' used to mean just 'to make visible', and as such incorporated all cartographic products. According to the newly established Commission on Visualisation of the International Cartographic Association it reflects '. . . modern technology that offers the opportunity for real-time interactive visualisation'. The key concepts here are interaction and dynamics.

The main drive behind these changes has been the development in science and engineering of the field known as 'visualisation in scientific computing' (ViSC), also known as scientific visualisation. During the last decade this was stimulated by the availability of advanced hardware and software. In their prominent report McCormick et al (1987) describe it as the study of 'those mechanisms in humans and computers which allow them in concert to perceive, use and communicate visual information'. In GIS, especially when exploring data, users can work with the highly interactive tools and techniques from scientific visualisation. DiBiase (1990) was among the first to realise this. He introduced a model with two components: 'private visual thinking' and 'public visual communication'. Private visual thinking refers to situations where Earth scientists explore their own data, for example. Cartographers and their well-designed maps provide an example of public visual communication. The first can be described as geographical or map-based scientific visualisation (Fisher et al 1993; MacEachren and Monmonier 1992). In this interactive 'brainstorming' environment the raw data can be georeferenced resulting in maps and diagrams, while other data can result in images and text. By the publication of two books, *Visualization in geographical information systems* (Hearnshaw and Unwin 1994) and *Visualization in modern cartography* (MacEachren and Taylor 1994) the spatial data handling community clearly demonstrated their understanding of the impact and importance of scientific visualisation on their discipline. Both publications address many aspects of the relationships between the fields of cartography and GIS on the one hand, and scientific visualisation on the other. According to Taylor (1994) this trend of visualisation should be seen as an independent development that will have a major influence on cartography. In his view the basic

aspects of cognition (analysis and applications), communication (new display techniques), and formalism (new computer technologies) are linked by interactive visualisation (Figure 1).

Three roles for visualisation may be recognised:

- First, visualisation may be used to present spatial information. The results of spatial analysis operations can be displayed in well-designed maps easily understood by a wide audience. Questions such as 'what is?', or 'where is?', and 'what belongs together?' can be answered. The cartographic discipline offers design rules to help answer such questions through functions which create proper well-designed maps (Kraak and Ormeling 1996; MacEachren 1994a; Robinson et al 1994).

- Second, visualisation may be used to analyse, for instance in order to manipulate known data. In a planning environment the nature of two separate datasets can be fully understood, but not their relationship. A spatial analysis operation, such as (visual) overlay, combines both datasets to determine their possible spatial relationship. Questions like 'what is the best site?' or 'what is the shortest route?' can be answered. What is required are functions to access individual map components to extract information and functions to process, manipulate, or summarise that information (Bonham-Carter 1994).

- Third, visualisation may be used to explore, for instance in order to play with unknown and often

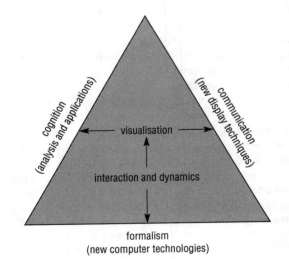

Fig 1. **Cartographic visualisation (Taylor 1994).**

raw data. In several applications, such as those dealing with remote sensing data, there are abundant (temporal) data available. Questions like 'what is the nature of the dataset?', or 'which of those datasets reveal patterns related to the current problem studied?', and 'what if . . .?' have to be answered before the data can actually be used in a spatial analysis operation. Functions are required which allow the user to explore the spatial data visually (for instance by animation or by linked views – MacEachren 1995; Peterson 1995).

These three strategies can be positioned in the map use cube defined by MacEachren (1994b). As shown in Figure 2, the axes of the cube represent the nature of the data (from known to unknown), the audience (from a wide audience to a private person) and the interactivity (from low to high). The spheres representing the visualisation strategies can be positioned along the diagonal from the lower left front corner (present: low interactivity, known data, and wide audience) to the upper right back corner (explore: high interactivity, unknown data, private person). Locating cartographic publications within the cube would reveal a concentration in the lower left front corner. However, colouring the dots to differentiate the publications according to their age would show many recent publications outside this corner and along the diagonal.

The functionality needed for these three strategies will shape this chapter. Each of them requires its own visualisation approach, described in turn in the

following three sections. The first section provides some map basics. It will briefly explain cartographic grammar, its rules and conventions. Depending on the nature of a spatial distribution, it will suggest particular mapping solutions. This strategy has the most developed tools available to create effective maps to communicate the characteristics of spatial distributions. When discussing the second strategy, visualisations to support analysis, it will be demonstrated how the map can work in this environment, and how information critical for decision-making can be visualised. In a data exploration environment, the third strategy, it is likely that the user is unfamiliar with the exact nature of the data. It is obvious that, compared to both other strategies, more appropriate visualisation methods will have to be applied. Specific visual exploration tools in close relation to 'new' mapping methods such as animation and hypermaps (multimedia) will be discussed. It is this strategy that will benefit most from developments in scientific visualisation.

2 PRESENTING SPATIAL DISTRIBUTIONS

Maps are uniquely powerful tools for the transfer of spatial information. Using a map one can locate geographical objects, while the shapes and colours of its signs and symbols inform us about the characteristics of the objects represented. Maps reveal spatial relations and patterns, and offer the user insight into the distribution of particular phenomena. Board (1993) defines the map as 'a representation or abstraction of geographical reality' and 'a tool for presenting geographical information in a way that is visual, digital or tactile'. Traditionally cartographers have concentrated most of their research efforts on enhancing the transfer of spatial data. This knowledge is very valuable, although some additional new concepts need to be introduced as illustrated in Figure 3. The traditional paper map functioned not only as an analogue database but also as an information transfer medium. Today a clear distinction is made between the database and presentation functions of the map, known respectively as the Digital Landscape Model (DLM) and Digital Cartographic Model (DCM). A DLM can be considered as a model of reality, based on a selection process. Depending on the purpose of the database, particular geographical objects have been selected from reality, and are represented in the

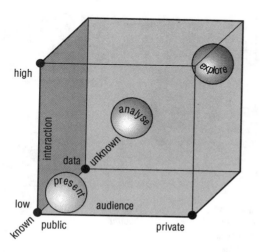

Fig 2. The three visualisation strategies plotted in MacEachren's (1994) map use cube.

database by a data structure (see Dowman, Chapter 31; Martin, Chapter 6). Multiple DCMs can be generated from the same landscape model, depending on the output medium or map design. To visualise data in the form of a paper map requires a different approach to an onscreen visualisation, and a road map for a vehicle navigation system will look different from a map designed for a casual tourist. Both, however, can be derived from the same DLM.

Next in importance to its contents, the usefulness of a map depends on its scale. For certain GIS applications one needs very detailed large-scale maps, while others require small-scale maps. Figure 4 shows a small-scale map (on the left) and a large-scale map. Traditionally maps have been divided into topographic and thematic types. Topographic maps portray the Earth's surface as accurately as possible subject to the limitations of the map scale. Topographic maps may include such features as houses, roads, vegetation, relief, geographical names, and a reference grid. Thematic maps represent the distribution of a particular phenomenon. In Plate 6 the upper map shows the topography of the peak of Mount Kilimanjaro in Africa. The lower thematic map shows the geology of the same area. As can be noted, the thematic map contains information also found in the topographic

map, since to be able to understand the theme represented one needs to be able to locate it as well. The amount of topographic information required depends on the map theme. A geological map will need more topographic data than a population density map, which normally only needs administrative boundaries. The digital environment has diminished the distinction between the two map types. Often both the topographic and the thematic maps are stored in layers, and the user is able to switch layers on or off at will.

The design of topographic maps is mostly based on conventions, of which some date back to the nineteenth century. Examples are representing water in blue (see MacDevette et al, Chapter 65), forests in green, major roads in red, urban areas in black, etc. The design of thematic maps, however, is based on a set of cartographic rules, also called cartographic grammar. The application of the rules can be translated in the question 'how do I say what to whom?'. 'What' refers to spatial data and its characteristics – for instance whether they are of a qualitative or quantitative nature. 'Whom' refers to the map audience and the purpose of the map – a map for scientists requires a different approach to a map on the same topic aimed at children. 'How' refers to the design rules themselves.

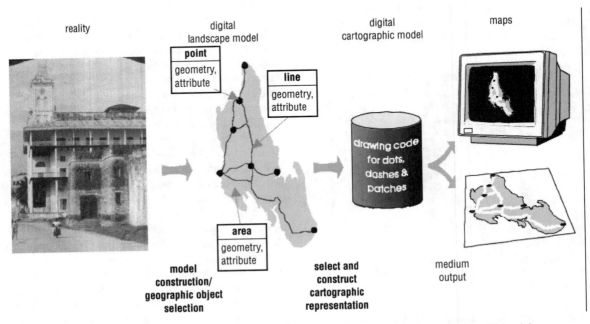

Fig 3. Spatial data characteristics: from reality to the map via a digital landscape model and a digital cartographic model.

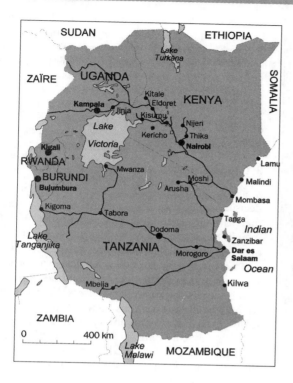 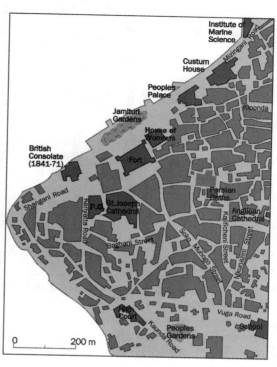

Fig 4. A small-scale map of East Africa, and a large-scale map of Stone Town (Zanzibar).

To identify the proper symbology for a map one has to conduct cartographic data analysis. The objective of such analysis is to access the characteristics of the data components in order to find out how they can be visualised. The first step in the analysis process is to find a common denominator for all of the data. This common denominator will then be used as the title of the map. Next the individual component(s) should be accessed and their nature described. This can be done by determining the measurement scale, which can be nominal, ordinal, interval, or ratio (see Martin 1996 for a discussion of geographical counterparts to these). Qualitative data such as land-use categories are measured on a nominal scale, while quantitative data are measured on the remaining scales. Qualitative data are classified according to disciplinary convention, such as a soil classification system, while quantitative data are grouped together by mathematical method.

When all the information is available the data components should be linked with the graphic sign system. Bertin (1983) created the base of this system. He distinguished six graphical variables: size, value, texture (grain), colour, orientation, and shape (Plate 7). Together with the location of the symbols in use these are known as visual variables. Graphical variables stimulate a certain perceptual behaviour with the map user. Shape, orientation, and colour allow differentiation between qualitative data values. Size is a good variable to use when the purpose of the map is to show the distribution amounts, while value functions well in mapping data measured on an interval scale. The design process results in thematic maps that are instantly understandable (for example newspaper maps and simple maps such as the one in Figure 5), and maps which may take some time to study (for example road maps or topographic maps – Plate 6(a)). A final category includes maps which require additional interpretative skills on the part of the user (for example geological or soil maps – Plate 6(b)).

Figure 6 presents an overview of some possible thematic maps. They represent different mapping

53

Fig 5. Zanzibar Town: appealing map design by visual hierarchy and the use of fonts.

font used for the wording of 'Zanzibar Town' has been chosen to express its oriental Arabic atmosphere. However, to be effective the text must be placed in an appropriate position with respect to the relevant symbols.

3 VISUAL ANALYSIS OF SPATIAL DISTRIBUTIONS

3.1 Introduction

Since one of a GIS's major functions is to act as a decision support system, it seems logical that the map as such should play a prominent role. With the map in this role one can even speak of visual decision support. The maps provide a direct and interactive interface to GIS data. They can be used as visual indices to the individual objects represented in the map. Based on the map, users will get answers to more complex questions such as 'what relationship exists?' This ability to work with maps and to analyse and interpret them correctly is one very important aspect of GIS use. However, to get the right answers the user should adhere to proper map use strategies.

Figure 7 demonstrates that this is not easy at all. The map displayed shows the northern part of the Netherlands. It is a result of a GIS analysis executed by an insurance company which wanted to know if it would make sense to initiate a regional operation. A first look at the map, which shows the number of traffic accidents for each municipality, would indeed suggest so. The eastern region seems to have worse drivers than the western region. However, a closer look at the map should make one less sure. First, the geographical units in the western part are much larger than the average units in the east; because each unit has a symbol the map looks much denser in the east. Second, when looking at the legend it can be seen that the small squares can represent from 1 to 99 accidents; the map shows some small squares representing only one accident, while others represent over 92. The west could therefore still have far more accidents then the east. The example illustrates not only that care is required when interpreting maps, but also that access to the map's single objects and the database behind the objects is a necessity. Additional relevant information such as the number of cars and the length of road should be available as well.

methods, many of which are found in the cartographic component of GIS software. In addition to the measurement scale, it is also important to take into account the distribution of the phenomenon, whether continuous or discontinuous, whether boundaries are smooth or not, and whether the data refer to point, line, area, or volume objects. The maps in Figure 6 are ordered in a matrix with the (dis)continuous nature along one side and qualitative/quantitative nature along the other side. From the above it will be clear that each spatial distribution requires a unique mapping solution depending on its character (see also Elshaw Thrall and Thrall, Chapter 23).

However, if all rules are applied mechanically the result can still be quite sterile and uninteresting. There is an additional need for a design that is appealing as well. Figure 5 provides an example of good design. Here information is ordered according to importance and is translated into a visual hierarchy. The urban area of Zanzibar Town is the first item on the map that will catch the eye of the map user. The map also shows some other important ingredients needed, such as an indication of the map scale and its orientation. Placement and style of text can be seen to play a prominent role too. Text can be used to convey information additional to that represented by the symbols alone, and the graphical

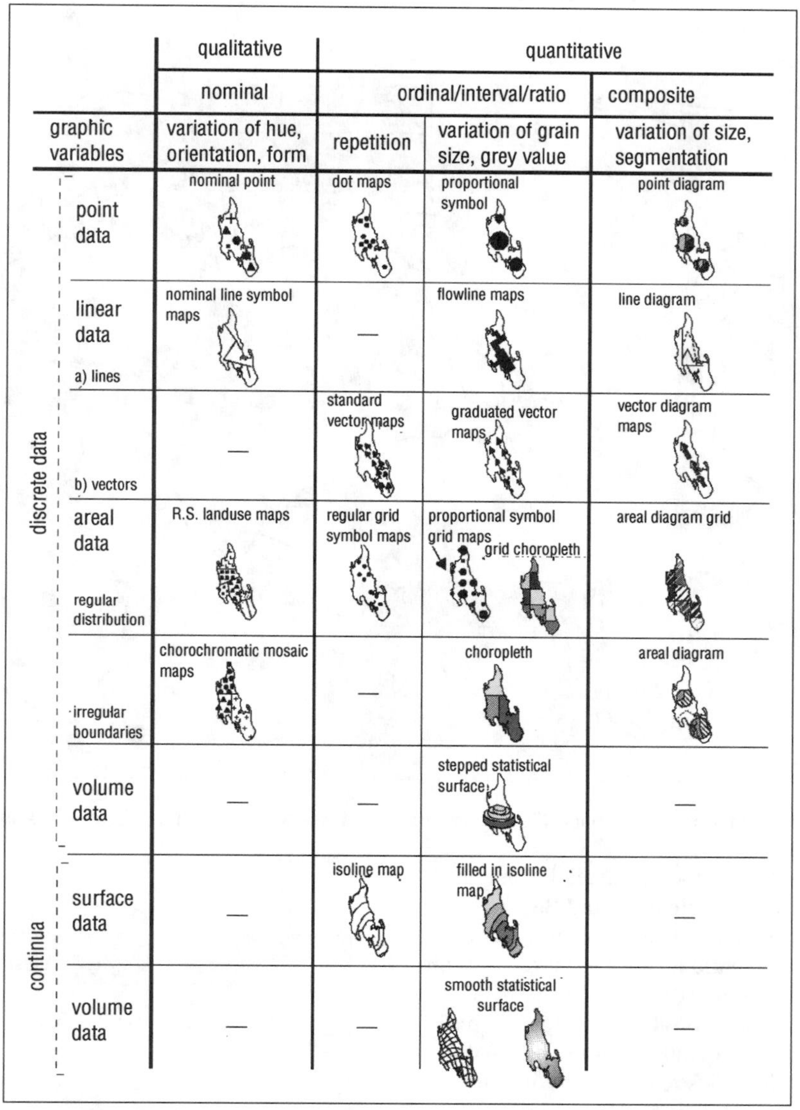

	qualitative	quantitative		
	nominal	ordinal/interval/ratio		composite
graphic variables	variation of hue, orientation, form	repetition	variation of grain size, grey value	variation of size, segmentation
point data	nominal point	dot maps	proportional symbol	point diagram
linear data a) lines	nominal line symbol maps	—	flowline maps	line diagram
b) vectors	—	standard vector maps	graduated vector maps	vector diagram maps
areal data regular distribution	R.S. landuse maps	regular grid symbol maps	proportional symbol grid maps / grid choropleth	areal diagram grid
irregular boundaries	chorochromatic mosaic maps	—	choropleth	areal diagram
volume data	—	—	stepped statistical surface	—
surface data	—	isoline map	filled in isoline map	—
volume data	—	—	smooth statistical surface	—

discrete data (spanning point data through irregular boundaries)

continua (spanning volume data, surface data, volume data)

Fig 6. A subdivision of thematic map types, based on the nature of the data (after Kraak and Ormeling 1996).

How can map tools help with the visual analysis of spatial distributions (Armstrong et al 1992)? Little is known about how people make decisions on the basis of map study and analysis. Giffin (1983) found that the strategies followed by individuals vary widely in relation to map type and complexity as well as according to individual characteristics. From the example above, it becomes clear that the user needs to have access to the appropriate spatial data in order to solve spatial tasks. Compared to the mapping activities in the previous section, the link between map and database (DCM and DLM) as well as access to the tools to describe and manipulate the data are of major importance. A key word here is *interaction*.

In order to make justifiable decisions based on spatial information, its nature and its quality (or reliability) must be known (Beard and Buttenfield,

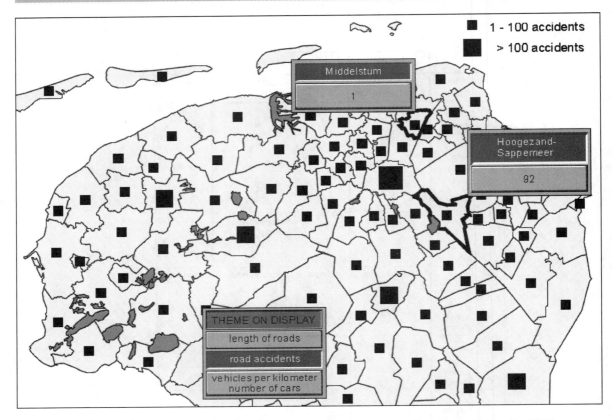

Fig 7. Maps and decision-making: traffic accident in the Netherlands and insurance policy (from Kraak and Ormeling 1996).

Chapter 15; Fisher, Chapter 13; Heuvelink, Chapter 14; Veregin, Chapter 12; Buttenfield and Beard 1994). Whether the data are fit for use is a complex matter, especially where combinations of different datasets are used. Visual decision support tools can help the user to make sensible spatial decisions based on maps. This is the most efficient way of communicating information about spatial reliability. This requires formalisation which can be done by providing functionality for data integration, standardisation (e.g. exchange formats), documentation (e.g. metadata), and modelling (e.g. generalisation and classification). This will lead to insights into the quality of the data on which the user will base spatial decisions. This is necessary because GIS is very good at combining datasets; notwithstanding the fact that these datasets might refer to different survey dates, different degrees of spatial resolution, or might even be conceptually unfit for combination: the software will not mind, but instead will happily combine them and present the results. The schema shown in Figure 8, and described extensively by Kraak et al (1995), summarises this approach.

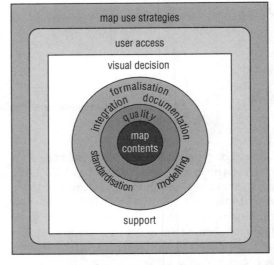

Fig 8. Visual decision support for spatio-temporal data handling (from Kraak et al 1995).

While working with spatial data in a GIS environment one commonly has to deal with 'where?', 'what?', and 'when?' queries. In a spatial analysis operation the queries will result in the manipulation of geometric, attribute, or temporal data components, separately or in combination. However, just looking at a map that displays the data already allows an evaluation of how certain phenomena vary in quantity or quality over the mapped area. Often one is not just interested in a single phenomenon but in multiple phenomena. For some aspects analytical operations are required, but sometimes a visual comparison will reveal interesting patterns for further study. Spatial, thematic, and temporal comparisons can be distinguished (Kraak and Ormeling 1996).

3.2 Comparing spatial data's geometric component

Comparing two areas seems to be relatively easy while focusing on a single theme – for example, hydrology, relief, settlements, or road networks. However, to make a sensible comparison the maps under study should have been compiled according to the same methods. They should have the same scale and the same level of generalisation or adhere to the same classification methods. For instance, if one is comparing the hydrological patterns in two river basins the individual rivers should be represented at the same level of detail in respect to generalisation and order of branches.

Figure 9 shows a comparison of the islands of Zanzibar and Pemba. They have been isolated from their original location and positioned next to each other. The coastline, reefs, road network, and villages are displayed, all derived from the Digital Chart of the World. It can be seen that Pemba, the island on the right, has a typical north–south settlement pattern, while Zanzibar, slightly larger, has a more evenly spread settlement with a larger urban area on the west coast (see Openshaw and Alvanides, Chapter 18, for a discussion of the analysis of geographically averaged data).

3.3 Comparing the attribute components of spatial data

If two or more themes related to a particular area are mapped according to the same method, it is possible to compare the maps and judge similarities or differences. However, not all mapping methods

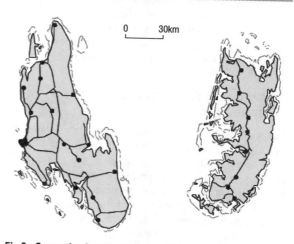

Fig 9. Comparing location: Zanzibar (Unguja) and Pemba.

are easy to compare. Choropleth maps are the simplest to compare, at least as long as the administrative units are the same in both maps. Isoline maps can be compared by measuring values in each map at the same locations.

Figure 10 compares a chorochromatic map (a soil map, right) with an isoline map (precipitation, left). At first sight it appears that low precipitation corresponds with a soil type that dominates the eastern part of the island, and that high precipitation results in a wider diversity of soils. Those familiar with Earth science in general will know that there is no necessary link between the two topics, but the above visual map analysis could be

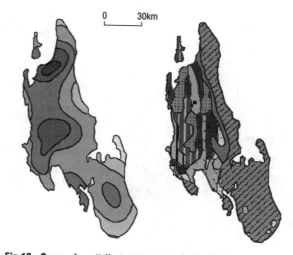

Fig 10. Comparing attributes: precipitation and soils.

57

true. It shows that only the expert can do the real analytical work, but comparing or overlaying two datasets can be done by anybody – but whether the operation makes sense remains unanswered.

3.4 Comparing the temporal components of spatial data

Users of GIS are no longer satisfied with analysis of snapshot data but would like to understand and analyse whole processes. A common goal of this type of analysis is to identify typical patterns in space-time. Change can be visually represented in a single map. Understanding the temporal phenomena from a single map will depend on the cartographic skills of both the map maker and map user, since these maps tend to be relatively complex. An alternative is the use of a series of single maps each representing a moment in time. Comparing these maps will give the user an idea of change. The number of maps is limited since it is difficult to follow long series of images. Another, relatively new alternative is the use of *dynamic* displays or animation (Kraak and MacEachren 1994). Change in the display over time provides a more direct impression of change in the phenomenon represented.

Figure 11 visualises the growth of the population of Zanzibar. From the maps it becomes clear that there is growth, and that growth in the urban district is faster than in the other parts of the island.

4 VISUAL EXPLORATION OF SPATIAL DISTRIBUTIONS

4.1 Introduction

Keller and Keller (1992) identify three steps in the visualisation process: first, to identify the visualisation goal; second, to remove mental roadblocks; and third, to design the display in detail. In cartography the first step is summarised by the phrase 'how do I say what to whom?', which was addressed in section 2. In the second step, the authors suggest removing oneself some distance from the discipline in order to reduce the effects of traditional constraints and conventional wisdom. Why not choose an alternative mapping method? For instance, one might use an animation instead of a set of single maps to display change over time; show a video of the landscape next to a topographic map; or change the dimension of the map from 2 dimensions to 3 dimensions. New, fresh, creative graphics could be the result, would probably have a greater and longer lasting impact than traditional mapping methods, and might also offer different insight. During the third step, which is particularly applicable in an exploratory environment, one has to decide between mapping data or visualising phenomena. An example of the mapping of the amount of rainfall may be used to clarify this distinction (Figure 12). Experts exploring rainfall patterns would like to distinguish between different

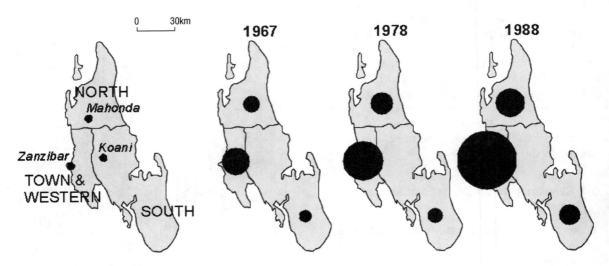

Fig 11. Comparing time: population growth.

precipitation classes, by using different colours for each class, such as blue, red, yellow, and green. A wider television audience might prefer a map showing areas with high and low precipitation. This can be realised using one colour, for instance blue, in different tints for all classes. Making dark tints correspond with high rainfall and light tints with low rainfall would result in an instantly understandable map. When exploring, data visualisation might be favoured; while presenting, phenomena visualisation may be preferred.

This approach to visualisation requires that a flexible and extensive functionality be available. The keywords 'interaction' and 'dynamics' were mentioned before. Compared with the presentation and analytical visualisation strategies these are clearly the extras. However, options to visualise the third dimension as well as temporal datasets should also be available. When exploring their data, users can work with the highly interactive tools and techniques from scientific visualisation. How are those tools implemented in a geographical exploratory visualisation environment?

Work is currently underway to develop tools for this exploratory environment (DiBiase et al 1992; Fisher et al 1993; Kraak 1994; Monmonier 1992; Slocum et al 1994). In 1990 Monmonier introduced the term 'brushing', as illustrated in Figure 13. It is

about the direct relationship between the map and other graphics related to the mapped phenomenon, like diagrams and scatter plots. The selection of an object in the map will automatically highlight the corresponding elements in the other graphics. Depending on the view in which the object is selected, the options are with geographical brushing (clicking in the map), attribute brushing (clicking in the diagram), and temporal brushing (clicking on the time line). Similar experiments on classification of choropleth maps have been made by Egbert and Slocum (1992). MacDougall (1992) followed a similar approach, while Haslett et al (1990) developed the Regard package as an interactive graphic approach to visualising statistical data. Other applications are discussed by DiBiase et al (1992), and Anselin (Chapter 17). Dykes (1995) has built a prototype of what he calls a cartographic data visualiser (CDV) which has much exploratory functionality. The system consists of a set of linked widgets, such as slide bars, buttons, and labels.

The illustrations in Figure 14 show some of the important functions that should be available to execute an exploratory visualisation strategy. The following functions are discussed in the works referred to above:

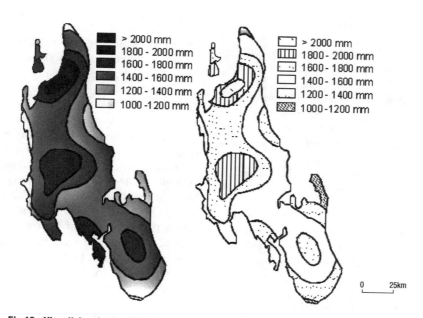

Fig 12. Visualising the classification: phenomena (left) or data (right).

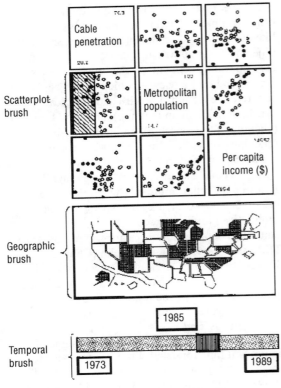

Fig 13. Geographic, attribute, and temporal brushing (Monmonier 1990).

- *Query*: an elementary function, that should always be available whatever the strategy. The user can query the map by clicking a symbol, which will activate the database. Electronic atlases incorporate this functionality (Figure 14(a): see also Elshaw Thrall and Thrall, Chapter 23).
- *Re-expression*: this function allows the same data, or part of the data, to be visualised in different ways. A time series of earthquakes could be reordered by the Richter scale instead, which could reveal interesting spatial patterns; or the classification method followed could be changed and the grey tints inverted as well – as can be seen in Figure 14(b).
- *Multiple views*: this approach could be described as interactive cartography. The same data could be displayed according to different mapping methods. Population statistics could be visualised as a dot map, a proportional circle map or a diagram map as shown in Figure 14(c).
- *Linked views*: this option is related to Monmonier's brushing principle. Selecting a

geographical object in one map will automatically highlight the same object in other views. For instance clicking a geographical unit in a cartogram would change the colour of the same unit in a geometrically correct map. In Figure 14(d), clicking the diagram showing clove production reveals a photograph of a clove plant and a map with the distribution of clove plantations in that particular year. This type of functionality allows one to introduce the multimedia components which will be discussed later in this section.

- *Animation*: the dynamic display of (temporal) processes is best done by animation. As will be explained in the next section interaction is a necessary add-on to animation (Figure 14(e)).
- *Dimensionality*: to view 3-dimensional spatial data one should be able to position the map in 3-dimensional space with respect to the map's purpose and the phenomena mapped (see Hutchinson and Gallant, Chapter 19). This means that all kinds of interactive geometric transformation functions to scale, translate, rotate, and zoom should be available, because it may be that the features of interest are located behind other features in the image (Figure 15).

4.2 Animation

Maps often represent complex processes which can be explained expressively by animation. To present the structure of a city, for example, animations can be used to show subsequent map layers which explain the logic of this structure (first relief, followed by hydrography, infrastructure, and land-use, etc.). Animation is also an excellent way to introduce the temporal component of spatial data, as in the evolution of a river delta, the history of the Dutch coastline, or the weather conditions of last week. An interesting example is ClockWork's Centennia (previously Millennium: ClockWork 1995; *http://www.clockwk.com*), a historical electronic atlas which presents an interactive animation of Europe's boundary changes between the years 1000 and 1995. This type of product can be used to explore or analyse the history of Europe.

The need in the GIS environment to deal with processes as a whole, and no longer with single time-slices, also influences visualisation. It is no longer

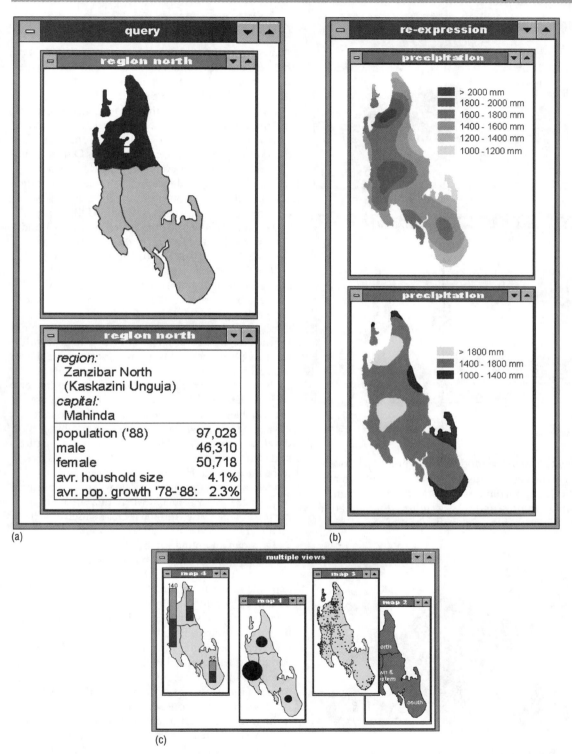

Fig 14. Data exploration: (a) query; (b) re-expression; (c) multiple views; (d) linked views; (e) animation (overleaf).

61

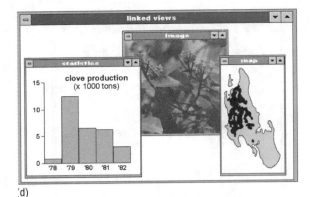

(d)

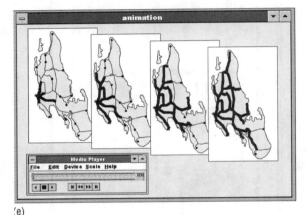

(e)

Fig 14. (cont.)

efficient to visualise models or planning operations using static paper maps. However, the onscreen map does offer opportunities to work with moving and blinking symbols, and is very suitable for animation. Such maps provide a strong method of visual

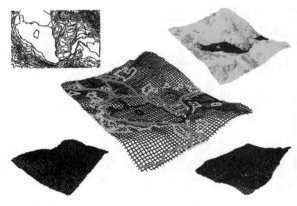

Fig 15. Working with the third dimension (from Kraak and Ormeling 1996).

communication, especially as they can incorporate real data, as well as abstract and conceptual data. Animations not only tell a story or explain a process, but also have the capability to reveal patterns or relationships which would not be evident if one looked at individual maps.

Attempts to apply animation to visualise spatial distributions date from the 1960s (see, for example, Thrower 1961; Tobler 1970) although only non-digital cartoons were possible initially. During the 1980s technological developments gave a second impulse to cartographic animation (see Moellering 1980). A third wave of interest in animation has developed, driven by interest in GIS (DiBiase et al 1992; Langran 1992; Monmonier 1990; Weber and Buttenfield 1993). Historic overviews are given by Campbell and Egbert (1990) and Peterson (1995).

The field of (cartographic) animation is about to change. Peterson (1995) expresses this as 'what happens between each frame is more important then what exists on each frame'. This should worry cartographers since their tools were developed mainly for the design of static maps. How can we deal with this new phenomenon? Is it possible to provide the producers of cartographic animation with sets of tools and rules to create 'good' animation, in the form 'If your data are . . ., and your aim is . . ., then use the variables . . .'? To be able to do so, and to take advantage of knowledge of computer graphics developments and the 'Hollywood' scene, the nature and characteristics of cartographic animations have to be understood. However, the problem is that 'understanding' animations alone will not be of much help, since the environment where they are used, the purpose of their use, and the users themselves will greatly influence 'performance'.

How can an animation be designed to make sure the viewer indeed understands the trend or phenomenon? The traditional graphic variables, as explained earlier, are used to represent the spatial data in each individual frame. Bertin, the first to write on graphic variables, had a negative approach to dynamic maps. He stated in his work (1967): '. . . however, movement only introduces one additional variable, it will be dominant, it will distract all attention from the other (graphic) variables'. Recent research, however, has demonstrated that this is not the case. Here we should remember that technological opportunities offered at the end of the 1960s were limited compared with those of today. Koussoulakou and Kraak (1992) found that the viewer of an animation would not necessarily get a better or worse understanding of the contents of the animation when compared with static maps. DiBiase et al (1992) found that movement would give the traditional variable new energy.

In this context DiBiase introduced three so-called dynamic variables: duration, order, and rate of change. MacEachren (1994b) added frequency, display time and synchronisation to the list:

- Display time – the time at which some display change is initiated.
- Duration – the length of time during which nothing in the display changes.
- Frequency – the same as duration: either can be defined in terms of the other.
- Order – the sequence of frames or scenes.
- Rate of change – the difference in magnitude of change per unit time for each of a sequence of frames or scenes.
- Synchronisation – (phase correspondence) refers to the temporal correspondence of two or more time series.

In the animation literature the so-called animation variables have surfaced (Hayward, 1984). They include size, position, orientation, speed of scene, colour, texture, perspective (viewpoint), shot (distance), and sound. The last of these is not considered here but can have an important impact (see Krygier 1994). These variables are shown in Figure 16 in relation to the graphic and dynamic variables. From this figure it can be seen that Bertin's graphic variables each have a match with one of the animation variables. From the dynamic variables only order and duration have a match, but they are the strongest in telling a story. Research is currently under way to validate and elaborate the new dynamic variables.

4.3 Maps and multimedia components

This section presents a cartographic perspective on multimedia. The relationship between the map and the individual multimedia components in relation to visual exploration, analysis, and presentation will be discussed (see Plate 8).

Maps supported with sound to present spatial information are often less interactive than those created to analyse or explore. In some electronic atlases pointing to a country on a world map plays the national anthem of the country (Electronic World Atlas; Electromap 1994). In this category one can also find the application of sound as background music to enhance a

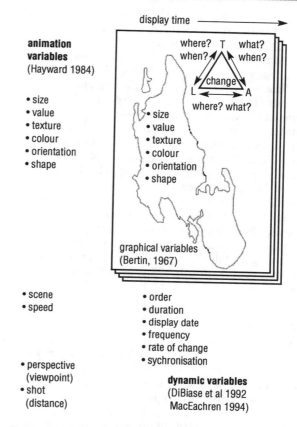

animation variables
(Hayward 1984)

- size
- value
- texture
- colour
- orientation
- shape

- scene
- speed

- perspective (viewpoint)
- shot (distance)

- order
- duration
- display date
- frequency
- rate of change
- sychronisation

dynamic variables
(DiBiase et al 1992
MacEachren 1994)

Fig 16. Cartographic animation and variable types.

mapped phenomenon, such as industry, infrastructure, or history. Experiments with maps in relation to sound are known on topics such as noise nuisance and map accuracy (Fisher 1994; Krygier 1994). In both cases the location of a pointing device in the map defines the volume of the noise. Moving the pointer to a less accurate region increases the noise level. The same approach could be used to explore a country's language – moving the mouse would start a short sentence in a region's dialect.

GIS is probably the best representation of the link between a map and text (the GIS database). As shown in Plate 8 the user can point to a geographical unit to reveal the data behind the map. Electronic atlases often have various kinds of encyclopaedic information linked to the map as a whole or to individual map elements. It is possible to analyse or explore this information. Country statistics can be compared. However, multimedia has more to offer. Now scanned text documents, such as those that describe the ownership of parcels, can be included. Text in the format of hypertext can be used as a lead to other textual information or other multimedia components.

Maps are models of reality. Linking video or photographs to the map will offer the user a different perspective on reality. Topographic maps present the landscape, but it is also possible to present, next to this map, a non-interpreted satellite image or aerial photograph to help the user in his or her understanding of the landscape. The analysis of a geological map can be enhanced by showing landscape views (video or photographs) from characteristic spots in the area. A real estate agent could use the map as an index to explore all houses for sale on company file. Pointing at a specific house would show a photograph of the house, the construction drawings, and a video would start showing the house's interior. New opportunities in the framework are offered by the application of virtual reality in GIS.

References

Armstrong M P, Densham P J, Lolonis P, Rushton G 1992 Cartographic displays to support locational decision making. *Cartography and Geographic Information Systems* 19: 154–64

Bertin J 1983 *Semiology of graphics*. Madison, University of Wisconsin Press (original in French, 1967)

Board C 1993 Spatial processes. In Kanakubo T (ed.) *The selected main theoretical issues facing cartography: report of the ICA Working Group to Define the Main Theoretical Issues in Cartography*. Cologne, International Cartographic Association: 21–4

Bonham-Carter G F 1994 *Geographical information systems for geo-scientists: modelling with GIS*. New York, Pergamon Press

Buttenfield B P, Beard M K 1994 Graphical and geographical components of data quality. In Hearnshaw H, Unwin D J (eds) *Visualisation in geographic information systems*. Chichester, John Wiley & Sons: 150–7

Campbell C S, Egbert S L 1990 Animated cartography: thirty years of scratching the surface. *Cartographica* 27: 24–46

ClockWork Software 1995 *Centennia*. PO Box 148036, Chicago 60614, USA

DiBiase D 1990 Visualization in earth sciences. *Earth and Mineral Sciences*, Bulletin of the College of Earth and Mineral Sciences, Pennsylvania State University 59: 13–18

DiBiase D, MacEachren A M, Krygier J B, Reeves C 1992 Animation and the role of map design in scientific visualisation. *Cartography and Geographic Information Systems* 19: 201–14

Dykes J 1995 Cartographic visualisation for spatial analysis. *Proceedings, Seventeenth International Cartographic Conference, Barcelona*: 1365–70

Egbert S L, Slocum T A 1992 EXPLOREMAP: an exploration system for choropleth maps. *Annals of the Association of American Geographers* 82: 275–88

Fisher P F 1994a Randomization and sound for the visualization of uncertain spatial information. In Hearnshaw H, Unwin D J (eds) *Visualization in geographic information systems*. Chichester, John Wiley & Sons: 181–5

Fisher P, Dykes J, Wood J 1993 Map design and visualisation. *The Cartographic Journal* 30: 136–42

Giffin T L C 1983 Problem-solving on maps – the importance of user strategies. *The Cartographic Journal* 20: 101–109

Haslett J, Willis G, Unwin A 1990 SPIDER: an interactive statistical tool for the analysis of spatially distributed data. *International Journal of Geographical Information Systems* 4: 285–96

Hayward S 1984 *Computers for animation*. Norwich, Page Bros

Hearnshaw H M, Unwin D J (eds) 1994 *Visualization in geographical information systems*. Chichester, John Wiley & Sons

Keller P R, Keller M M 1992 *Visual cues, practical data visualization*. Piscataway, IEEE Press

Koussoulakou A, Kraak M J 1992 Spatio-temporal maps and cartographic communication. *The Cartographic Journal* 29: 101–8

Kraak M J 1994 Interactive modelling environment for 3-D maps, functionality and interface issues. In MacEachren A M, Taylor D R F (eds) *Visualization in modern cartography*. Oxford, Pergamon: 269–86

Kraak M J, MacEachren A M 1994b Visualization of spatial data's temporal component. In Waugh T C, Healey R G (eds) *Advances in GIS research – Proceedings Fifth Spatial Data Handling Conference*. London, Taylor and Francis: 391–409

Kraak M-J, Ormeling F J 1996 *Cartography, visualization of spatial data*. Harrow, Longman

Kraak M-J, Ormeling F J, Müller J-C 1995 GIS-cartography: visual decision support for spatio-temporal data handling. *International Journal of Geographical Information Systems* 9: 637–45

Krygier J 1994 Sound and cartographic visualization. In MacEachren A M, Taylor D R F (eds) *Visualization in modern cartography*. Oxford, Pergamon: 149–66

Langran G 1992 *Time in geographical information systems*. London, Taylor and Francis

MacDougall E B 1992 Exploratory analysis, dynamic statistical visualization, and geographic information systems. *Cartography and Geographic Information Systems* 19: 237–46

MacEachren A M 1994a *Some truth with maps: a primer on design and symbolization*. Washington DC, Association of American Geographers

MacEachren A M 1994b Visualization in modern cartography: setting the agenda. In MacEachren A M, Taylor D R F (eds) *Visualization in modern cartography*. Oxford, Pergamon: 1–12

MacEachren A M 1995 *How maps work*. New York, Guilford Press

MacEachren A M, Monmonier M 1992 Geographic visualization: introduction. *Cartography and Geographic Information Systems* 19: 197–200

MacEachren A M, Taylor D R F (eds) 1994 *Visualization in modern cartography*. Oxford, Pergamon

Martin D J 1996 *Geographic information systems: socioeconomic applications*. London, Routledge

McCormick B, DeFanti T A, Brown M D 1987 Visualization in scientific computing. *ACM SIGGRAPH Computer Graphics* 21 special issue.

Moellering H 1980 The real-time animation of 3-dimensional maps. *The American Cartographer* 7: 67–75

Monmonier M 1990 Strategies for the visualization of geographic time-series data. *Cartographica* 27: 30–45

Monmonier M 1992 Authoring graphic scripts: experiences and principles. *Cartography and Geographic Information Systems* 19: 247–60

Peterson M P 1995 *Interactive and animated cartography*. Englewood Cliffs, Prentice-Hall

Robinson A H, Morrison J L, Muehrcke P C, Kimerling A J, Guptill S C 1994 *Elements of cartography*, 6th edition. New York, John Wiley & Sons Inc.

Slocum T A , Egbert S, Weber C, Bishop I, Dungan J, Armstrong M, Ruggles A, Demetrius-Kleanthis D, Rhyne T, Knapp L, Carron J, Okazaki D 1994 Visualization software tools. In MacEachren A M, Taylor D R F (eds) *Visualization in modern cartography*. Oxford, Pergamon: 91–122

Taylor D R F 1994 Perspectives on visualization and modern cartography. In MacEachren A M, Taylor D R F (eds) *Visualization in modern cartography*. Oxford, Pergamon: 333–42

Thrower N 1961 Animated cartography in the United States. *International Yearbook of Cartography*: 20–8

Tobler W R 1970 A computer movie: simulation of population change in the Detroit region. *Economic Geography* 46: 234–40

Weber R, Buttenfield B P 1993 A cartographic animation of average yearly surface temperatures for the 48 contiguous United States: 1897–1986. *Cartography and Geographic Information Systems* 20: 141–50

Wood M 1994 Visualization in a historical context. In MacEachren A M, Taylor D R F (eds) *Visualization in modern cartography*. Oxford, Pergamon: 13–26

Introduction

THE EDITORS

A recurrent theme throughout the first part of this book is that the inherent complexity of the geographical world makes it virtually impossible for any digital representation to be complete, however limited its scope. Although some exceptions exist (we can, for example, create a perfect digital representation of the latitude of the Equator, or a line on the Earth's surface that is by definition straight), there will otherwise be differences between the database contents and the phenomena they represent. Various terms are used to describe these differences, depending on the context. Differences can exist because of errors of measurement, while the term 'uncertainty' seems more appropriate if the digital representation is simply incomplete. More generally, one might simply refer to the 'quality' of the representation.

If data quality is an important property of almost all geographical data, then it must affect the decisions made with those data. In general, the poorer the quality of the data, the poorer the decision. Bad decisions can have severe consequences, as when an ambulance is sent to the wrong location, or a school is inadvertently built over an abandoned storage facility for hazardous waste. Geographical data are often used for regulatory purposes, or to resolve disputes: the custodians of such data are clearly exposed to potential liability if the data are shown to be in error.

Despite what appear to be obvious arguments in favour of explicit treatment of data quality in GIS, and despite substantial research into appropriate methods, much GIS practice continues to proceed as if data were perfect. Results of GIS analysis – whether in the form of tables, maps, or displays – rarely show estimates of confidence, or other indicators of the effects of data quality. In part, such attitudes have been inherited from cartographic practice, since it is often difficult to determine the quality of mapped information. In part, they may also reflect a general tendency to give computers more credit than they deserve – to believe that because numbers or maps have emerged as if by magic from digital black boxes, they must necessarily be reliable.

This section contains four chapters that together represent the state of the geographical data quality art, or, more accurately, science. Howard Veregin presents in Chapter 12 an overview of the components of data quality; their interactions and dependencies; and the efforts that have been made in recent years to embed them in standards. From the perspective of the data producer, quality refers to the difference between the actual characteristics of the product and the relevant specifications that define it, or the claims made about it. Information on quality is immensely useful in managing the production process, particularly if the results of quality analysis point back to suspect sources. On the other hand, details of the production process may be of only marginal interest to a potential user of the data, who is concerned solely with whether the data meet particular requirements. Data quality can thus vary from user to user, depending on respective needs; and the effective measurement and documentation of data quality against needs that are often poorly defined can be an immensely complex and frustrating process.

The problems of determining data quality have been further complicated in recent years by the growth of new communication technologies. These have made it far easier for data to be found, accessed, and shared. The user of a geographical dataset may now be many steps removed from the producer. User and producer may be from entirely different backgrounds, with very little in the way of shared terminology or culture. Even if the data are well documented, the lack of effective systems for documentation, in the form of metadata, may leave the user with an incomplete or incorrect understanding of the meaning of the data. For example, if the units of measurement of a variable are not documented, or if the documentation is not transferred to the user, then from the perspective of the user the data are now subject to a further source of inaccuracy. To the user at the end of a long chain of communication, data quality is most appropriately defined as a measure of the difference between the database's contents and the user's understanding of their true values. The same collection of bits can have different levels of quality, both increasing and decreasing, as it passes from one custodian to another.

In Chapter 13, Peter Fisher discusses alternative models of uncertainty. The traditional scientific concept of measurement error, which accounts for differences between observers or measuring instruments, turns out to be far too simple as a framework for understanding quality in geographical data. Many geographical concepts are incompletely specified – as for example when population density at a point is reported, without specifying the area over which the density was measured – and such incompleteness of specification is an appropriate component of data quality. Many concepts are poorly defined, leading to understandable disagreement between observers. In this context it is useful to distinguish, as Fisher does, between such terms as 'vague', 'fuzzy', and 'probable'. Both fuzzy set theory and probability theory have been found to be useful in modelling uncertainty in geographical data, although their axioms differ in several key respects.

If agreement can be reached on how to measure and express data quality, then such information should be made available to users, preferably by storing it as part of, or in conjunction with, the database. Quality measures that are true of the entire contents are conveniently stored as part of the database metadata, the digital equivalent of documentation. But other quality measures may be true only of parts of the database, such as classes of objects, or individual objects, or even parts of objects, or regions of the study area. In such cases, it is necessary to have 'slots' in the database available to store data quality information in appropriate, meaningful ways. Such slots might take the form of additional data quality attributes of objects, or components of an object class's description; or it might even be necessary to create a complete map of data quality, showing how quality varies across the study area. Thus data quality becomes a significant part of the representation itself.

With adequate information available on data quality, it is possible to determine its effects on the results of GIS analysis, and for decisions made with GIS to reflect the uncertainties present in the base data. This topic of error propagation is the subject of Chapter 14, by Gerard Heuvelink. Several general strategies for error propagation are proposed, at least one of which will be valid in any context. Typically the error propagation is hidden from the user, who sees only a standard GIS function, such as 'compute slope', but is presented with results that include both the requested estimates of slope, and measures of confidence or uncertainty in the results. Software for error propagation is increasingly available in the GIS world, often in the form of specialised 'add-ons' written in a GIS's macro or scripting language.

In the final chapter in this section, Kate Beard and Barbara Buttenfield review techniques that have been developed for visualising uncertainty, issues raised by their use, and problems requiring additional research. Traditional cartographic practice includes remarkably few methods for visualising uncertainty; whether this is because it is difficult to do so within the constraints of map-making, or whether it reflects a human desire to see the world as simpler than it really is, remains a subject of debate. What is beyond doubt, however, is that the continuation of such practices in the world of GIS is both technically and ethically indefensible. The digital world is far more flexible, and Beard and Buttenfield illustrate many of the methods that have been proposed and implemented by the research community. There have been experiments with sound, animation, and use of the third dimension, each with attendant advantages and disadvantages.

Despite such progress on the research front, the issues of dealing with uncertainty remain. GIS has been adopted by individuals and agencies who see its benefits in terms that often include increased accuracy compared to previous methods; yet the data stored in GIS are in most cases no more accurate. Suppose, for example, that the research community were to suggest, on theoretically defensible grounds, that the only effective method for visualising uncertainty would be to present the user with several equally likely versions of how the world might actually look. Uncertainty in soil mapping could be presented by showing ten alternative, equally likely maps of the same area. While this might make perfect sense from the perspective of error theory, it would be almost completely alien to a culture raised on single, apparently exact maps. The problems of coping with uncertainty, and of introducing its effective treatment from a managerial perspective, are the subject of Chapter 45 by Gary Hunter.

13

Models of uncertainty in spatial data

P F FISHER

Spatial information is rife with uncertainty for a number of reasons. The correct conceptualisation of that uncertainty is fundamental to the correct use of the information. This chapter attempts to document different types of uncertainty – specifically error, vagueness, and ambiguity. Examples of these three types are used to illustrate the classes of problems which arise, and to identify appropriate strategies for coping with them. The first two categories are well documented and researched within the GIS field, and are now recognised in many varied contexts. The third has not been so widely researched. Cases are also identified where uncertainty is deliberately introduced into geographical information in order to anonymise individuals. Examples are given where both error and vagueness can be applied to the same phenomenon with different understandings and different results. Methods to address the problems are identified and are explored at length.

1 INTRODUCTION

'The universe, they said, depended for its operation on the balance of four forces which they identified as charm, persuasion, uncertainty and bloody-mindedness.'

Terry Pratchett (1986)

acuracy *n.* An absence of erors. 'The computer offers both speed and acuracy, but the greatest of these is acuracy' (*sic*)

Kelly-Bootle (1995)

The handling of large amounts of information about the natural and built environments, as is necessary in any GIS, is prone to uncertainty in a number of forms. Ignoring that uncertainty can, at best, lead to slightly incorrect predictions or advice and at worst can be completely fatal to the use of the GIS and undermine any trust which might have been put in the work of the system or operator. It is therefore of crucial importance to all users of GIS that awareness of uncertainty and error should be as widespread as possible. Fundamental to such understanding is the nature of the uncertainty, in its different guises. This is the subject of this chapter. A minimal response should be that users of the GIS be aware of the possible complications to their analysis caused by uncertainty, and at best present the user of the analysis with a report of the uncertainty in the final results together with a variety of plausible outcomes. A complete response to uncertainty is to present the results of a full modelling exercise which takes into account all types of uncertainty in the different data themes used in the analysis. It seems that neither response is widespread at present, and in any case the tools for doing the latter are currently the preserve only of researchers.

This chapter explores the developing area of the conceptual understanding (modelling) of different types of uncertainty within spatial information. These are illustrated in Figure 1. At the heart of the issue of uncertainty is the problem of defining both the class of object to be examined (e.g. soils) and the individual object (e.g. soil map unit) – the so-called problem of definition (Taylor 1982). Once the conceptual modelling identifies whether the class of objects to be described is well or poorly defined the nature of the uncertainty as follows:

1 If both the class of object and the individual are well defined then the uncertainty is caused by errors and is probabilistic in nature;

2 If the class of object or the individual is poorly
    defined then additional types of uncertainty may
    be recognised. Some have been explored by GIS
    researchers and others have not:
    a If the uncertainty is attributable to poor
       definition of class of object or individual object,
       then definition of a class or set within the
       universe is a matter of *vagueness*, and this can
       conveniently be treated with fuzzy set theory.
    b Uncertainty may also arise owing to ambiguity
       (the confusion over the definition of sets
       within the universe) owing, typically, to
       differing classification systems. This also takes
       two forms (Klir and Yuan 1995), namely:
       i Where one object or individual is clearly
          defined but is shown to be a member of two
          or more different classes under differing
          schemes or interpretations of the evidence,
          then *discord* arises;
       ii Where the process of assigning an object to
          a class at all is open to interpretation, then
          the problem is *non-specificity*.

In the context of spatial databases, only vagueness
as expressed by fuzzy set theory and error as
represented by probability theory have been
researched, and these are the primary focus of the
discussion below. The list is necessarily not
exhaustive: however, the volume of research and the
amount of interest in this area continues to increase.

If a chapter had been written in this form for the
first edition of this book, it would have focused on
only one variety of uncertainty, namely error
(Chrisman 1991). A few years later there are two
equally important strands to be discussed. Although
the strands discussed here seem to explain the
majority of the long-recognised causes of
uncertainty in spatial information, it is already
possible to identify other types of uncertainty that
should be addressed in future research.

2 THE PROBLEM OF DEFINITION

The principal issue of geographical uncertainty is
the understanding of the collector and user of the
data as to the nature of that uncertainty. There are
three facets to this, namely uncertainty in
measurement of attributes, of space, and of time. In
order to define the nature of the uncertainty of an
object within the dimensions of space and time, a
decision must be made as to whether or not it is

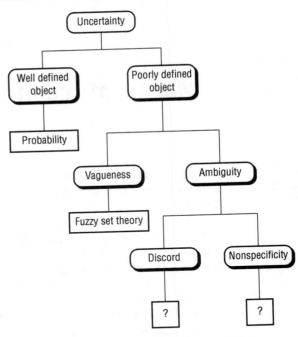

Fig 1. A conceptual model of uncertainty in spatial data
(adapted from Klir and Yuan 1995: 268).

clearly and meaningfully separable from other
objects in whichever dimension is of interest –
ideally it will be separable in both. This is a complex
intellectual process, one which draws on the history
and the critical appraisal of subject-specific
scientists. This conceptual model has been
complicated and muddied by conventions which
influence the perception of geographical
information. Foremost among these is the historical
necessity of simplification of information for map
production; what Fisher (1996) denotes the
paradigm of 'production cartography'. Equally
important are the concepts of classification,
commonly based on hierarchies, in which objects
must fall into one class or another, and of computer
database models in which objects are treated as
unique individuals and form the basis to analysis.

If a spatial database is to be used, or to be created
from scratch, then investigators or users have to ask
themselves two apparently simple questions:

1 Is the class of objects to be mapped (e.g. soils,
   rocks, ownership, etc.) clearly separable from
   other possible classes?
2 Are the geographical individuals within the class of
   objects clearly and conceptually separable from
   other geographical individuals within the same class?

If it is possible to separate unequivocally the phenomenon to be mapped into mappable and spatially distinct objects using the spatial distribution of some individual attribute or collection of attributes, at a given time, then there is no problem of definition. A phenomenon which is well defined should have diagnostic properties for separating individuals into classes based on attributes and into spatially contiguous and homogenous areas.

If it is not possible to define the spatial extent of an object to be mapped or analysed, there is a problem of definition, and it can be said to be 'vague' (Williamson 1994). In this circumstance, while specific properties may be measured and these measurements may be precise, no combination of properties allows the unequivocal allocation of individual objects to a class, or even the definition of the precise spatial extent of the objects. Most spatial phenomena in the natural environment share this problem of definition to some extent. Error analysis on its own does not help with the description of these classes, although any properties which are measured may be subject to errors just as they are in other cases.

2.1 Examples of well-defined geographical objects

In developed countries *census geographies* tend to be well defined; even in less developed countries the geographical concepts are generally well defined, if less clearly implemented. They usually consist of a set of regions each with precise boundaries within which specific attributes are enumerated (Openshaw 1995). The areas at the lowest level of enumeration (city blocks, enumeration districts, etc.) are grouped with specific instances of other areas at the same level to make up higher level areas, which in turn are grouped with other specific areas to form a complete and rigid hierarchy (e.g. see Martin, Chapter 6). The attributes to be counted within the areas are typically based on property units, individuals, and households: although the definitions of 'household' may differ between different surveys (Office of National Statistics 1997) and there is rarely any perfect correspondence between households and property units (e.g. houses in multiple occupation), each definition is nevertheless quite transparent and unambiguous. The data collection process in the western world relies on a certain level of cooperation and literacy amongst those being counted, and while there are frequently legal sanctions for non-cooperation these cannot easily be enforced if people are reluctant to

cooperate. The primary errors associated with the US Census of Population arise out of underenumeration of groups such as illegal immigrants and the homeless (Bureau of the Census 1982).

A second example of a well-defined geographical phenomenon in western societies is *land ownership*. The concept of private ownership of land is fundamental to these societies; therefore the spatial and attribute interpretation of that concept is normally quite straightforward in its spatial expression. The boundary between land parcels is commonly marked on the ground, and marks an abrupt and total change in ownership. In point of fact, at least in the UK, the surveyed boundary is only deemed indicative of the actual position of the boundary, and so any property boundary has a defined uncertainty in position, otherwise it would require resurveying every time the boundary marker is rebuilt (Dale and McLaughlin 1988). Even in instances of collective ownership in which two groups may own two adjoining parcels and one person may belong to both groups, the question of ownership and responsibility remains clear in law.

Well-defined geographical objects are essentially created by human beings to order the world they occupy. They exist in well-organised and established political and legal realms. Some other objects in our built and natural environments may seem to be well defined, but they tend to be based on a single measurement, and close examination frequently shows the definition to be obscure. For example, the land surface seems well defined, and it should be possible to determine its height above sea level rigorously and to specified precision. But even the position of the ground under our feet is being brought into question. This is caused by the increasing availability of elevation models derived from photogrammetry to sub-centimetre precision, when the actual definition of the land surface being mapped must come into question, and whether the field was ploughed or the grass was cut, become serious issues in defining the so-called land surface. Most, if not all, other geographical phenomena are similarly poorly defined to some extent.

2.2 Examples of poorly-defined geographical objects

In aboriginal societies the concept of ownership is much less clear than in western society. There are many different native cultures, but many have a conception of the land owning the people, and responsibility for nurturing the land is a matter of common trust within a group (Native North Americans and Australians, for example: Young 1992). Areas of responsibility are less well defined, with certain core areas for which a group or an individual may be responsible (e.g. the sacred sites of the Australian Aborigines: Davis and Prescott 1992), and other regions for which no one is actually responsible but many groups may use (so-called 'frontier zones'). Among both North American and Australian native groups, the spatial extents of these core and peripheral areas have been shown to be well known to the groups concerned, although they may not be marked, precisely located, or fixed over time (Brody 1981; Davis and Prescott 1992). There are therefore acknowledged divisions of space, but the spatial location of the divider may be uncertain. The extent of the zones of uncertainty can be resource dependent, so that when resources are plentiful there may be relatively precise boundaries, and when scarce there may be very diffuse frontiers (Davis and Prescott 1992; Young 1992). Alternatively, ties of kinship between groups may create less specific frontiers, and lack of kinship hard boundaries (Brody 1981). These aboriginal territories have much in common with the documented 'behavioural neighbourhoods' of western individuals. Such neighbourhoods are also poorly defined both spatially and temporally: they may be discontinuous and will inevitably overlap with others, and while possibly unique to an individual or family, may nonetheless make up part of a geographical region that is occupied by a group.

Complexity is also inherent in the mapping of vegetation (Foody 1992). The allocation of a patch of woodland to the class of oak woodland, for example – as opposed to any other candidate woodland type – is not necessarily easy. It may be that in that region a threshold percentage of trees need to be oak for the woodland to be considered 'oak', but what happens if there is one per cent less than that threshold? Does it really mean anything to say that the woodland needs to be classed to a different category? Indeed, the higher level classification to woodland at all has the same problems. Mapping the vegetation is also problematic since in areas of natural vegetation there are rarely sharp transitions from one vegetation type to another, rather an intergrade zone or ecotone occurs where the dominant vegetation type is in transition (Moraczewski 1993). The ecotone may occupy large tracts of ground. The attribute and spatial assignments may follow rules, and may use indicator species to assist decisions, but strict deterministic rules may trivialise the classification process without generating any deeper meaning.

In discussion of most natural resource information we typically talk about central concepts and transitions or intergrades. Figure 2 shows a scatter plot of some remotely-sensed (LANDSAT) data from Band 3 and Band 5 (which record the amounts of reflected electromagnetic radiation in the wavelength ranges 0.63–0.69 and 1.55–1.75 μm, respectively). This is part of the information used in the assignment of pixels in an image to land covers. The conceptualisation of the land covers is as Boolean objects (discussed below), and yet it is clear from Figure 2 that there are no natural breaks in the distribution of points in the 2-dimensional space shown. This is typical of satellite imagery. Although LANDSAT actually records information in seven spectral bands which can give identification to some natural groups of pixels, the number of identifiable groups very rarely corresponds with the number of land cover types being mapped (Campbell 1987). The classification process involves the identification of prototypical values for land cover types, and the extension of that mapping from the attribute dimensions shown to the spatial context. Conceptually, the same basic process is executed in almost all traditional mapping operations, and the problem of the identification of objects is fundamental. It is apparent from Figure 2 that the intergrades (all possible locations in attribute space which are between the prototype or central concepts) are more commonly and continuously occupied than the prototypical classes.

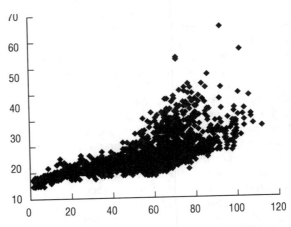

Fig 2. A scatterplot of Bands 3 and 5 of a LANDSAT TM image.

The problem of identification may be extended into locations. Figure 3 shows a soil map of part of the Roujan catchment in France with numbers indicating soil map units (soil types) and the shading indicating the extent of boundary intergrades between types. The width of intergrades is based on the knowledge of soil surveyors who prepared the map (Lagacherie et al 1996).

Within natural resource disciplines the conceptualisation of mappable phenomena and the spaces they occupy is rarely clear cut, and is still more rarely achieved without invoking simplifying assumptions (see also Veregin, Chapter 12). In forestry, for example, tree stands are defined as being clearly separable and mappable; yet trees vary within stands by species density, height, etc., and often the spatial boundary between stands is not well defined (Edwards 1994). Although theorists may recognise the existence of intergrades, the conceptual model of mapping used in this and other natural resource disciplines accepts the simplification and places little

importance on them, although the significance has not been assessed. In other areas, such as soil science and vegetation mapping, some of the most interesting areas are at the intergrade, and these are rightly a focus of study in their own right (Burrough 1989; Burrough et al 1992; Lagacherie et al 1996). The interest in intergrades as boundaries is not a preserve of natural resouce scientists, however, and in discussion of urban and political geography considerable attention is paid to these concepts (Prescott 1987; Batty and Longley 1994).

3 ERROR

If an object is conceptualised as being definable in both attribute and spatial dimensions, then it has a *Boolean* occurrence; any location is either part of the object, or it is not. Yet within GIS, for a number of reasons, the assignment of an object or location to the class may be expressed as a probability. There are

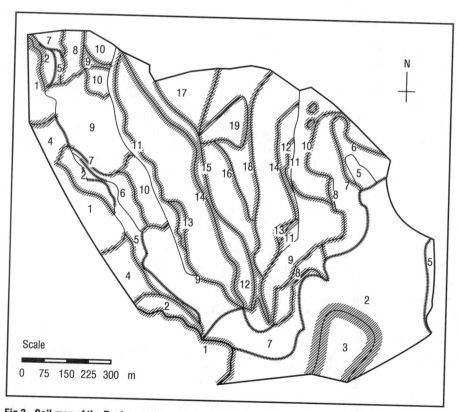

Fig 3. Soil map of the Roujan catchment in France showing the extent of soil intergrades (after Lagacherie et al 1996: 281).

any number of reasons why this might be the case. Three are briefly discussed here:

1 probability owing to error in the measurement;
2 probability because of the frequency of occurrence;
3 probability based on expert opinion.

Errors occur within any database, and for any number of reasons; some reasons are given in Table 1. They are given more complete treatment by Fisher (1991b) and Veregin (Chapter 12). The simplest to handle are those associated with measurement, because well-advanced error analysis procedures have been developed (Heuvelink, Chapter 14; Heuvelink et al 1989; Heuvelink and Burrough 1993; Taylor 1982). If the true value of a property of an object were precisely known, then it would be possible to identify the distribution of 'real world' measurement error by making repeated measurements of the property (which would each differ from the true value by a variable measurement error). It would then be possible to estimate the distribution of the error in its measurement, and thus to develop a full error model of the measurement error. This is, in fact, the basis of the 'root mean square' reporting of error in digital elevation models (see also Beard and Buttenfeld, Chapter 15). Yet there are many instances in which such reductionist measures of error are over-simplistic and aspatial, failing to identify the spatial distribution of the error in GIS-based modelling (Monckton 1994; Walsby 1995).

A further means of describing aspatial error is to create a confusion matrix which shows the cover-type actually present at a location crosstabulated against the cover-type identified in the image classification process. Typically the matrix is generated for a complete image. It reports errors in the allocation of pixels to cover types (Campbell 1987; Congalton and Mead 1983). However, the confusion matrix is of limited use if the precise interpretation of either the classification process or the ground information is not clear cut.

A different view of probability is based on the frequency of the occurrence of a phenomenon. The classic applications of probability in this area include weather and flood forecasting. Floods of a particular height are identified as having a particular return period which translates as a particular probability of a flood of that level occurring.

A third view of probability is as a manifestation of subjective opinion, where an expert states a 'gut feeling' of the likelihood of an event occurring. Much

Table 1 Common reasons for a database being in error.

Type of error	Cause of error
Measurement	Measurement of a property is erroneous
Assignment	The object is assigned to the wrong class because of measurement error by field, or laboratory scientist, or by surveyor
Class generalisation	Following observation in the field and for reasons of simplicity, the object is grouped with objects possessing somewhat dissimilar properties
Spatial generalisation	Generalisation of the cartographic representation of the object before digitising, including displacement, simplification, etc. (see Weibel and Dutton, Chapter 10)
Entry	Data are miscoded during (electronic or manual) entry to a GIS
Temporal	The object changes character between the time of data collection and of database usage
Processing	In the course of data transformations an error arises because of rounding or algorithm error

geological and soil mapping is actually the result of Boolean classification of subjective probability, since it is impracticable to observe directly either of these phenomena across the entire countryside: rather inference is made using sampled points such as outcrops and auger borings. Between those locations it is expert opinion as to what is there; so long as a Boolean model of soil and rock occurrence is applied, the map is implicitly a matter of the expert's maximum probability (Clarke and Beckett 1971).

Probability has been studied in mathematics and statistics for hundreds of years. It is well understood, and the essential methods are well documented. There are many more approaches to probability than the three described here. Probability is a subject that is on the syllabus of almost every scientist qualified at degree level, and so it pervades the understanding of uncertainty through many disciplines. It is not, however, the only way to treat uncertainty.

4 VAGUENESS

In contrast with error and probability which are steeped in the mathematical and statistical literature, vagueness is the realm of philosophy and logic and has been described as one of the fundamental challenges to those disciplines (Williamson 1994; Sainsbury 1995). It is relatively easy to show that a concept is 'vague', and the classic pedagogic exposition uses the case of the 'bald' man. If a person with no hair at all is considered bald, then is a person with one hair bald? Usually, in any working definition of 'bald', the answer to this would be 'yes'. If a person with one hair is bald, then is a person with two hairs bald: again, 'yes'. If you continue the argument, one hair at a time, then the addition of a single hair never turns a bald man into a man with a full head of hair. On the other hand, you would be very uncomfortable admitting that someone with plenty of hair was bald, since this is illogical (Burrough 1992; Burrough 1996; Zadeh 1965). This is known as the *Sorites Paradox* which, little by little, presents the logical argument that someone with plenty of hair is bald! A number of resolutions to the paradox have been suggested, but the most widely accepted is that the logic employed permits only a Boolean response ('yes' or 'no') to the question. A graded response is not acceptable. And yet there is a degree to which a person can be bald. It is also possible that the initial question is false, because 'bald' would normally be qualified if we were examining it in detail, so we might ask whether someone was 'completely bald', and we might define that as someone with no hair at all. Can we ever be certain that individuals have absolutely no hair on their heads? Furthermore, where on their neck and face is the limit of the head such that we can judge whether there is any hair on it? You are eventually forced to admit that by incremental logical argument, it is impossible to specify whether someone is 'completely', 'absolutely', 'partially', or 'not at all' bald, given a count of hairs on their head, even if the count is absolutely correct. So no matter the precision of the measurement, the allocation to the set of people is inherently vague.

The Sorites Paradox is one way which is commonly used to define vague concepts. If a concept is 'Sorites susceptible', it is vague. Many geographical phenomena are 'Sorites susceptible', including concepts and objects from the natural and built environments (e.g. see Band, Chapter 37). When, exactly, is a house a house; a settlement, a settlement; a city, a city; a podsol, a podsol; an oak woodland, an oak woodland? The questions always revolve around the threshold value of some measurable parameter or the opinion of some individual, expert or otherwise.

Fuzzy set theory was introduced by Zadeh (1965) as an alternative to *Cantor (Boolean) sets*, and built on the earlier work of Kaplan and Schott (1951). Membership of an object to a Cantor set is absolute, that is it either belongs or it does not, and

membership is defined by integer values in the range {0,1}. By contrast, membership of a fuzzy set is defined by a real number in the range [0,1] (the change in type of brackets indicates the real and integer nature of the number range). Definite membership or non-membership of the set is identified by the terminal values, while all intervening values define an intermediate degree of belonging to the set, so that, for example, a membership of 0.25 reflects a smaller degree of belonging to the set than a membership of 0.5. The object described is less like the central concept of the set.

Fuzzy memberships are commonly identified by one of two methods (Robinson 1988):

1 the *Similarity Relation Model* is data driven and involves searching for patterns within a dataset similarly to traditional clustering and classification methods, the most widespread method being the Fuzzy c Means algorithm (Bezdek 1981). More recently, fuzzy neural networks have been employed (Foody 1996);
2 the *Semantic Import Model*, in contrast, is derived from a formula or formulae specified by the user or another expert (Altman 1994; Burrough 1989; Wang et al 1990).

Many studies have applied fuzzy set theory to geographical information processing. There are several good introductions to the application of fuzzy sets in geographical data processing, including books by Leung (1988) and Burrough and Frank (1996) – see also Eastman (Chapter 35).

Fuzzy set theory is now only one of an increasing number of soft set theories (Pawlak 1982), in contrast to hard, Cantor sets. However, a number of authorities consider that fuzzy set theory is mistakenly used for problems which more correctly fall within the realm of subjective probability (Laviolette and Seaman 1994). They have, however, primarily addressed fuzzy logic rather than fuzzy sets, and illustrated their arguments with Boolean conditions and decisions. As such, they have failed to address the nature of the underlying set and any inherent vagueness which may be present, as Zadeh (1980) has shown. Moreover, Kosko (1990) has argued that fuzzy sets are a superset of probability.

5 AMBIGUITY

The concepts and consequences of ambiguity (Figure 1) in geographical information are not well researched. Ambiguity occurs when there is doubt as to how a phenomenon should be classified because of differing perceptions of it. Two types of ambiguity have been recognised, namely *discord* and *non-specificity*. In other areas of study some partial solutions have been suggested, but they are not reviewed here because of the lack of specific research with geographical information.

Within geography the most obvious form of discord through ambiguity is in the conflicting territorial claims of nation states over particular pieces of land. History is filled with this type of ambiguity, and the discord which results. Examples in the modern world include intermittent and ongoing border conflicts and disagreements in Kashmir (between India and China) and the neighbouring Himalayan mountains (between China and India). Similarly, the existence or non-existence of a nation of Kurds is another source of discord. All represent mismatches between the political geography of the nation states and the aspirations of people (Horn, Chapter 67; Prescott 1987; Rumley and Minghi 1991).

As has already been noted, many if not most phenomena in the natural environment are also ill-defined. The inherent complexity in defining soil, for example, is revealed by the fact that many countries have slightly different definitions of what a soil actually constitutes (cf. Avery 1980; Soil Survey Staff 1975), and by the complexity and the volume of literature on attempting to define the spatial and attribute boundaries between soil types (Webster and Oliver 1990; Lagacherie et al 1996). Furthermore, no two national classification schemes have either the same names for soils or the same definitions if they happen to share names. This causes many soil profiles to be assigned to different classes in different schemes, as shown in Table 2 (see also Isbell 1996; Soil Classification Working Group 1991; Soil Survey Staff 1975). Within a single country this is not a problem, yet ambiguity arises in the international efforts to produce supra-national or global soil maps. The individual national classifications cause considerable confusion in the process and the classification scheme becomes part of the national identity within the context. There is also rarely a one-to-one correspondence between classification systems (soil type x in this classification corresponds to soil type a in that), but rather a many-to-many classification (soil types a and b correspond broadly to soil type x, but some profiles of soil type a are also soil types y and z). This leads to different placement of soil boundaries in both attribute and spatial dimensions, and generates considerable problems in mapping soils across international and interstate boundaries (FAO/UNESCO 1990; Campbell et al 1989), as has been exemplified in the creation of the Soil Map of the European Communities (Tavernier and Louis 1984).

Several measures of social deprivation have been suggested which are based upon information from the UK Census of Population (Table 3). Enumeration areas are assigned to one class or another, and the classes have been used in the allocation of resources for a range of social and economic programmes. The fact that there are different bases to the measurement of deprivation means that enumeration areas may be afforded special policy status using one indicator, but not using another, and this is a source of potential discord.

Ambiguity through non-specificity can be illustrated from geographical relations. The relation 'A is north of B' is itself non-specific, because the concept 'north of' can have at least three specific meanings: that A lies on exactly the same line of longitude and towards the north pole from B; that A lies somewhere to the north of a line running east to west through B; or, in common use, that A lies in the sector between perhaps north-east and north-west, but is most likely to lie between north-north-east and north-north-west of B. The first two definitions are precise and specific, but equally valid. The third is the natural language concept which is itself vague. Any lack of definition as to which should be used means that uncertainty arises in interpreting 'north of'.

Arguably, soil classification is a process whereby modern schemes have removed the problem of non-specificity which was inherent in earlier schemes and replaced it by supposedly objective, globally applicable diagnostic criteria. The remaining problems arise out of creating Boolean boundaries in a vague classification environment and the problem of discord.

None of this should be taken to imply that ambiguity is inappropriate or intrinsically 'wrong'. The England and Wales soil classification scheme at the scale of England and Wales is possibly the most relevant classification scheme for the soils in that country. Similarly, the United States Department of Agriculture scheme (Soil Taxonomy) was the best scheme for the US when it was finalised in 1975 (although it does claim a global application). The problem of ambiguity arises when we move to a higher level, and data from the British Soil Survey

Table 2 Alternative soil classification schemes for global and national use.

US Classification (Soil Survey Staff 1975)	Australian Classification (Isbell 1996)	Soil Map of the World (FAO/UNESCO 1990)		British Soil Classification (Avery 1980)
Entisol	Anthroposol	Fluvisol	Kastanozem	Terrestrial raw soil
Inceptisol	Organosol	Gleysol	Chernozem	Hydric raw soil
Spodosol	Podsol	Regosol	Phaeozem	Lithomorphic soil
Mollisol	Hydrosol	Lithosol	Greyzem	Pelosol
Oxisol	Kurosol	Arenosol	Cambisol	Brown soil
Ultisol	Sodosol	Rendzina	Luvisol	Podzolic soil
Alfisol	Chromosol	Ranker	Podzoluvisol	Ground-water gley soil
Aridisol	Calcarosol	Andosol	Podzol	Surface-water gley soil
Histosol	Ferrosol	Vertisol	Planosol	Man-made soil
Vertisol	Dermosol	Solonchak	Acrisol	Peat soil
	Kandosol	Solonetz	Nitosol	
	Rudosol	Yermosol	Ferrasol	
	Tenosol	Xerosol	Histosol	

Table 3 Measures of social deprivation used in the UK, with the associated census variables used in their calculation (Openshaw 1995).

Variable	Jarman	Townsend	Department of the Environment
Unemployment	X	X	X
No car		X	
Unskilled	X		
Overcrowding (more than 1 person per room)	X	X	X
Lacking amenities			X
Not owner occupied		X	
Single-parent household	X		X
Children under 5 years old	X		
Lone pensioners	X		
Ethnic minorities	X		

within the UK, housing indicators of deprivation replace ethnic indicators in Wales.) Ambiguity nevertheless does come into play in the allocation of social and economic programme resources, and can lead to contention between local, national, and (in the case of EU programmes) supra-national, politicians over the issue of the basis to financial support.

6 CONTROLLED UNCERTAINTY

Many agencies distribute and allow access to spatial information which is degraded deliberately through creating uncertainty. Two examples of this are discussed (see also Heuvelink, Chapter 14, and Hunter, Chapter 45, for a discussion of the management of uncertainty).

If the exact locations of rare or precious objects such as nesting sites of endangered birds or archaeological sites are recorded in a dataset, any more widely distributed versions may introduce a systematic or random error introduced into the locational component. This may be done by only reporting information for large areal aggregations (e.g. 4 km^2 in the county flora of Leicestershire, England: Primavesi and Evans 1988; and 100 km^2 grid in the state flora of Victoria, Australia, distributed on CD-ROM: Viridians 1996). In some cases both systematic and random elements are introduced in order to protect the phenomenon reported, and although the error may be inconvenient, the consequences of not introducing it may be worse.

have to be fused with data from neighbouring countries or countries further afield. In preparing the Soil Map of the European Community, for example, the FAO/UNESCO classification was employed with some amendments.

In a like manner, there is nothing wrong with there being three different methods of defining deprived regions in Britain. Deprivation is a social construct and any quantitative index can only be an approximation which is deemed relevant and acceptable within its own terms of reference. If the constituent attributes of a particular index happen not to be measured in another country, that index simply ceases to have international application. (In fact, with regard to the use of the Jarman Index

Uncertainty is also deliberately introduced into census data in order to preserve confidentiality. If only a few people living within any one enumeration area have a particular characteristic – for example high income – and incomes are reported, it may be very easy to identify exactly which person that is. This is not socially acceptable, and so most census organisations withhold or falsify small counts. For example, in the USA, data for areas with small counts are withheld (Bureau of the Census 1982), whereas in the UK small counts have had a random value between +1 and -1 added (Dewdney 1983).

7 DISTINGUISHING BETWEEN VAGUENESS AND ERROR

Appropriate conceptualisation of uncertainty is a prerequisite to its modelling within GIS. In this section two areas of previous study are examined, and the reasons for the use of either vague or error models of uncertainty are discussed.

7.1 Viewshed

The viewshed is a simple operation within many current GIS, which, in its usual implementation, reports those areas in a landscape which are in view and those which are not (coded 1 and 0 respectively), whether in a triangulated grid or dataset (De Floriani and Magillo, Chapter 38; De Floriani et al 1986; Fisher 1993). Fisher (1991a) has shown how, for a variety of reasons, the visible area is very susceptible to error in the measurement of elevations in the Digital Elevation Model (DEM). (While Fisher used a rectangular grid in his 1991 study, the same would be true for a triangulated model.) The database error is propagated into the binary viewshed because of error in the elevation database (Fisher 1991a) and uncertainty in determination of visibility because of variation between different algorithms (Fisher 1993). Fisher (1992, 1993, 1994) has proposed that it is possible to define the error term from the Root Mean Squared Error (RMSE) for the DEM such that the error has a zero mean and standard deviation equal to the RMSE. This is not in fact true and provides insufficient description of the error for a fully justifiable error model since the mean error may be biased (non-zero) and must have spatial structure. Spatial structure of the error may be identified

through spatial autocorrelation measures or full specification of the variogram of the error field (Journel 1996). If the error field is generated using this method then it can be added to the DEM, yielding a revised DEM which includes the known error. If the viewshed is determined over that DEM with error, then a version of the Boolean viewshed is generated. If the process is repeated, then a second version of the Boolean viewshed, a third, a fourth, and so on are generated. If each Boolean viewshed image is coded as 0 and 1 indicating areas which are out-of-sight and in-sight, then using map algebra to find the sum of Boolean viewsheds, a value between 0 and the number of realisations will be found for all locations depending on the number of realisations in which that location is visible. Dividing by the number of realisations will then give an estimate of the probability of that location actually being within the viewshed. The probability of any pixel being visible from the viewing point, or the probability of the land rising above the line of sight somewhere between the viewer and the viewed is given by:

$$p(x_{ij}) = \frac{\sum_{k=1}^{n} x_{ijk}}{n} \tag{1}$$

where

$p(x_{ij})$ is the probability of a cell at row i and column j in the raster image being visible; and x_{ijk} is the value at the cell of the binary-coded viewshed in realisation k such that k takes values 1 to n.

This is illustrated in Plate 9.

In contrast, using a Semantic Import Model it is possible to define a number of different fuzzy viewsheds (Fisher 1994; note that the term is used incorrectly by Fisher 1992) from a family of equations relating the distance from the viewer to the viewed to the fuzzy membership function (Plate 10). Any number of different circumstances can be described, and two are included here: Equation 2 represents normal atmospheric conditions, and Equation 3 describes the visibility through fog.

$$\mu(x_{ij}) = \begin{cases} 1 & \text{for } d_{vp} \to ij \leq b_1 \\ \dfrac{1}{\left(1 + \left(\dfrac{d_{vp} \to ij - b_1}{b_2}\right)^2\right)} & \text{for } d_{vp} \to ij > b_1 \end{cases} \tag{2}$$

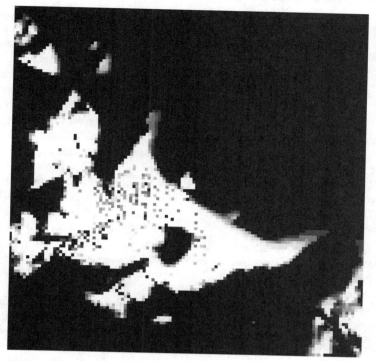

Fig 4. Probable viewshed based on Equation 1.

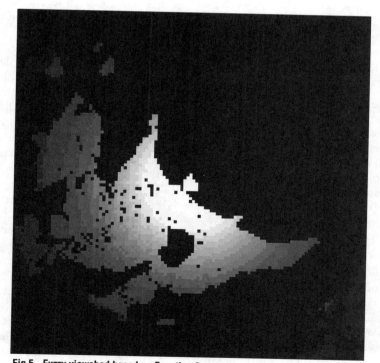

Fig 5. Fuzzy viewshed based on Equation 2.

$$\mu(x_{ij}) = \begin{cases} 1 & \text{for } d_{vp} \to ij \le b_1 \\ 0 & \text{for } d_{vp} \to ij > b_1 + 2 \bullet b_2 \\ \sin\left(\left(\dfrac{d_{vp} \to ij - b_1}{2 \bullet b_2}\right)\dfrac{\pi}{2}\right) & \text{for } d_{vp} \to ij > b_1 \end{cases} \qquad (3)$$

where

$\mu(x_{ij})$ is the fuzzy membership at the cell at row i, column j;

$d_{rp} \to_{ij}$ is the distance from the viewpoint to row i, column j;

b_1 is the radius of the zone around the viewpoint where the clarity is perfect, and the target object can be seen at the defined level of detail;

b_2 is the distance from b_1 to fuzzy membership of 0.5, sometimes called the cross-over point.

The distinction between fuzzy and probable viewsheds is that the first describes the probability of a location being visible, while the second portrays the degree to which objects can be distinguished. Thus there is an objective definition of the first, and only subjective versions of the second which may describe group or even personal circumstances.

7.2 Remote sensing

Classification of remotely-sensed data has been a major source of land cover and land-use information for GIS. The basic methods, based on a number of discriminant functions from numerical taxonomy, are well known and widely documented (Campbell 1987). The assumptions implicit in this approach are threefold:

1 the cover type itself is a well-defined phenomenon with clear breaks reflected by there being more similarity within cover types than between them;
2 the digital numbers recorded in the original satellite image allow the discrimination of land cover/use types, mapping on a one-to-one basis between reflectance and cover type;
3 the area of the pixel on the ground can be identified as having a single cover type (that area can be assigned to one and only one land-cover or land-use).

From these assumptions it is possible to allow the conceptualisation of both the spatial extent of the pixel and the land cover attributes to be determined as Boolean concepts. Therefore uncertainties can be described by probability, and functional methods such as the maximum likelihood classifier are applicable. Unfortunately, all the assumptions are made for the convenience of the operator, and none matches the actual situation either pragmatically or theoretically.

It is a fact of life that the spatial extent of geographical objects is not coincident with the image pixel, hence the class types on the ground are often hard to define precisely (many are Sorites susceptible), and the digital numbers do not show greater similarity within cover type than between. Therefore, arguably, fuzzy set theory (as an expression of concepts of vagueness) is a more appropriate model for working with satellite imagery and has been the subject of a number of explorations (Foody 1992, 1996; Fisher and Pathirana 1991; Goodchild et al 1994). Both Foody (1992, 1996) and Fisher and Pathirana (1991) have shown that the fuzzy memberships extracted from digital images can be related to the proportion of the cover types within pixels. This can be seen as a step towards a full interpretation of the fuzzy memberships derived from the imagery, since in the work reported the land covers analysed are still well-defined Boolean concepts; the vagueness is introduced by the sensor characteristics (Fisher and Pathirana 1991; Foody 1996). On the other hand, Foody (1992) uses the fuzzy sets to examine a zone of intergrade between vegetation communities, where both the communities and the intergrade are vague concepts.

The confusion between land cover and land use is also problematic (see also Barnsley, Chapter 32). Land use has a socioeconomic dimension to it, which cannot be sensed from satellites. Land cover, on the other hand, pertains to directly observable physical properties of the Earth's surface, and so can be classified directly. Indeed, one reason for the poor results of classification accuracy is the confusion in the conceptualisation of this transformation, and the opacity of the relationship between the surface reflectance of land covers and land use. The most successful attempts at land-use mapping from satellite imagery have adopted rule-based (Wang et al 1991) or graph theoretic approaches to the problem (Barnsley, Chapter 32; Barr and Barnsley 1995), and a combination of fuzzy set theory with these other methods may well further improve the results.

Within remote sensing, it can therefore be seen that the conceptualisation of the problem is the controlling influence. If the assumptions as to the spatial and attribute discrimination of land cover within a pixel noted above are accepted, then there is a Boolean mapping between land cover and digital number which can be extracted by classification, and uncertainty can be expressed probabilistically. If they cannot be accepted, then the Sorites susceptibility of the subjects of mapping indicates their vagueness, and so fuzzy set theory is a more appropriate approach to analysis. A clear conceptualisation of the nature of the phenomenon to be mapped and the approach to be taken is essential to the successful analysis of satellite imagery.

8 CONCLUSION: UNCERTAINTY IN PRACTICE

Through citing a number of different examples, this chapter has argued that within geographical information there are a number of different causes of and approaches to uncertainty. Anyone using uncertain information (i.e. the overwhelming majority of GIS users) needs to think carefully about the possible sources of uncertainty, and how they may be addressed. Uncertainty is a recurrent theme throughout many of the chapters of this book (e.g. Hunter, Chapter 45; Martin, Chapter 6; Raper, Chapter 5); the particular contribution of this chapter is to relate our conceptualisation of the nature of uncertainty to GIS-based data models. Analysis without accommodating data uncertainty (both error and vagueness) can quite severely limit its usefulness. Yet an appropriate conceptualisation of uncertainty and the application of related analytical methods creates a rich analytical environment where decision making based on spatial information is facilitated not only by objective orderings of alternatives but also by giving confidence in those alternatives. New analytical products are beginning to appear as a result of processing, and not ignoring, uncertainty (Burrough 1989; Burrough et al 1992; Davidson et al 1994; Wang et al 1990).

It is crucial to the correct use of geographical information systems that all aspects of uncertainty should be accommodated. This can only be achieved through awareness of the issues and a thorough and correct conceptualisation of uncertainty. The subject of uncertainty in spatial information has developed rapidly, and is still changing, particularly with the increasing use and exploration of alternative, soft set theories (Pawlak 1982).

References

Altman D 1994 Fuzzy set theoretic approaches for handling imprecision in spatial analysis. *International Journal of Geographical Information Systems* 8: 271–89

Avery B W 1980 *Soil classification for England and Wales (higher categories)*. Harpenden, Soil Survey Technical Monograph 14

Barr S, Barnsley M 1995 A spatial modelling system to process, analyse, and interpret multi-class thematic maps derived from satellite sensor images. In Fisher P F (ed) *Innovations in GIS 2*. London: Taylor and Francis: 53–65

Batty M, Longley P 1994 *Fractal cities: a geometry of form and function*. London/San Diego, Academic Press

Bezdek J C 1981 *Pattern recognition with fuzzy objective function algorithms*. New York, Plenum Press

Brody H 1981 *Maps and dreams; Indians and the British Columbia frontier*. Harmondsworth, Penguin

Bureau of the Census 1982 *Census of Population and Housing*. Washington DC, US Department of Commerce

Burrough P A 1989 Fuzzy mathematical methods for soil survey and land evaluation. *Journal of Soil Science* 40: 477–92

Burrough P A 1992a Are GIS data structures too simple minded? *Computers & Geosciences* 18: 395–400

Burrough P A 1996 Natural objects with indeterminate boundaries. In Burrough P A, Frank A U (eds) *Geographic objects with indeterminate boundaries*. London, Taylor and Francis: 3–28

Burrough P A, Frank A U (eds) 1996 *Geographic objects with indeterminate boundaries*. London, Taylor and Francis

Burrough P A, MacMillan R A, Deursen W van 1992 Fuzzy classification methods for determining land suitability from soil profile observations and topography. *Journal of Soil Science* 43: 193–210

Campbell J B 1987 *Introduction to remote sensing*. New York, Guilford Press

Campbell W G, Church M R, Bishop G D, Mortenson D C, Pierson S M 1989 The role for a geographical information system in a large environmental project. *International Journal of Geographical Information Systems* 3: 349–62

Chrisman N R 1991b The error component in spatial data. In Maguire D J, Goodchild M F, Rhind D W (eds) *Geographical information systems: principles and applications*. Harlow, Longman/New York, John Wiley & Sons Inc. Vol. 1: 165–74

Clarke G P, Beckett P 1971 *The study of soils in the field*, 5th edition. Oxford, Clarendon Press

Congalton R G, Mead R A 1983 A quantitative method to test for consistency and correctness in photointerpretation. *Photogrammetric Engineering and Remote Sensing* 49: 69–74

Dale P F, McLaughlin J D 1988 *Land information management*. Oxford, Oxford University Press

Davidson D A, Theocharopoulos S P, Bloksma R J 1994 A land evaluation project in Greece using GIS and based on Boolean fuzzy set methodologies. *International Journal of Geographical Information Systems* 8: 369–84

Davis S L, Prescott J R V 1992 *Aboriginal frontiers and boundaries in Australia*. Melbourne, Melbourne University Press

De Floriani L, Falcidieno B, Pienovi C, Allen D, Nagy G 1986 A visibility-based model for terrain features. *Proceedings, Second International Symposium on Spatial Data Handling*. Columbus, International Geographical Union: 235–50

Dewdney J G 1983 Census past and present. In Rhind D W (ed.) *A census user's handbook*. London, Methuen: 1–15

Edwards G 1994 Characteristics and maintaining polygons with fuzzy boundaries in geographic information systems. In Waugh T C, Healey R G (eds) *Advances in GIS research: Proceedings Sixth International Symposium on Spatial Data Handling*. London, Taylor and Francis: 223–39

FAO/UNESCO 1990 *Soil map of the world: revised legend*. FAO, Rome, World Soil Resources Report 60

Fisher P F 1991a First experiments in viewshed uncertainty: the accuracy of the viewable area. *Photogrammetric Engineering and Remote Sensing* 57: 1321–7

Fisher P F 1991b Data sources and data problems. In Maguire D J, Goodchild M F, Rhind D W (eds) *Geographical information systems: principles and applications*. Harlow, Longman/New York, John Wiley & Sons Inc. Vol. 1: 175–89

Fisher P F 1992 First experiments in viewshed uncertainty: simulating the fuzzy viewshed. *Photogrammetric Engineering and Remote Sensing* 58: 345–52

Fisher P F 1993 Algorithm and implementation uncertainty in the viewshed function. *International Journal of Geographical Information Systems* 7: 331–47

Fisher P F 1994a Probable and fuzzy models of the viewshed operation. In Worboys M (ed.) *Innovations in GIS 1*. London, Taylor and Francis: 161–75

Fisher P F 1996a Concepts and paradigms of spatial data. In Craglia M, Couclelis H (eds) *Geographic information research: bridging the Atlantic*. London, Taylor and Francis: 297–307

Fisher P F, Pathirana S 1991 The evaluation of fuzzy membership of land cover classes in the suburban zone. *Remote Sensing of Environment* 34: 121–32

Foody G M 1992 A fuzzy sets approach to the representation of vegetation continua from remotely-sensed data: an example from lowland heath. *Photogrammetric Engineering and Remote Sensing* 58: 221–5

Foody G M 1996 Approaches to the production and evaluation of fuzzy land cover classification from remotely-sensed data. *International Journal of Remote Sensing* 17: 1317–40

Goodchild M F, Chi-Chang L, Leung Y 1994 Visualising fuzzy maps. In Hearnshaw H M, Unwin D J (eds) *Visualisation in geographical information systems*. Chichester, John Wiley & Sons: 158–67

Heuvelink G B M, Burrough P A 1993 Error propagation in cartographic modelling using Boolean logic and continuous classification. *International Journal of Geographical Information Systems* 7: 231–46

Heuvelink G B M, Burrough P A, Stein A 1989 Propagation of errors in spatial modelling with GIS. *International Journal of Geographical Information Systems* 3: 303–22

Isbell R F 1996 *The Australian soil classification*. Australian Soil and Land Survey Handbook 4, CSIRO, Collingwood

Journel A 1996 Modelling uncertainty and spatial dependence: stochastic imaging. *International Journal of Geographical Information Systems* 10: 517–22

Kaplan A, Schott H F 1951 A calculus for empirical classes. *Methodos* 3: 165–88

Kelly-Bootle S 1995 *The computer contradictionary*, 2nd edition. Cambridge (USA), MIT Press

Klir G J, Yuan B 1995 *Fuzzy sets and fuzzy logic: theory and applications*. Englewood Cliffs, Prentice-Hall

Kosko B 1990 Fuzziness vs probability. *International Journal of General Systems* 17: 211–40

Lagacherie P, Andrieux P, Bouzigues R 1996 The soil boundaries: from reality to coding in GIS. In Burrough P A, Frank A U (eds) *Geographic objects with indeterminate boundaries*. London, Taylor and Francis: 275–86

Laviolette M, Seaman J W 1994 The efficacy of fuzzy representations of uncertainty. *IEEE Transactions on Fuzzy Systems* 2: 4–15

Leung Y C 1988 *Spatial analysis and planning under imprecision*. New York, Elsevier Science

Monckton C G 1994 An investigation into the spatial structure of error in digital elevation data. In Worboys M (ed.) *Innovations in GIS 1*. London, Taylor and Francis: 201–11

Moraczewski I R 1993 Fuzzy logic for phytosociology 1: syntaxa as vague concepts. *Vegetatio* 106: 1–11

Office of National Statistics 1997 *Harmonised concepts and questions for government social surveys*. London, Her Majesty's Stationery Office

Openshaw S (ed.) 1995a *Census users' handbook*. Cambridge (UK), GeoInformation International

Pawlak Z 1982 Rough sets. *International Journal of Computer and Information Sciences* 11: 341–56

Pratchett T 1986 *The light fantastic*. Gerrards Cross, Colin Smythe

Prescott J R V 1987 *Political frontiers and boundaries*. London, Allen and Unwin

Primavesi A L, Evans P A 1988 *Flora of Leicestershire*. Leicester, Leicestershire County Museum Service

Robinson V B 1988 Some implications of fuzzy set theory applied to geographic databases. *Computers, Environment, and Urban Systems* 12: 89–98

Rumley D, Minghi J V (eds) 1991 *The geography of border landscapes*. London, Routledge

Sainsbury R M 1995 *Paradoxes*, 2nd edition. Cambridge (UK), Cambridge University Press

Soil Classification Working Group 1991 *Soil classification, a taxonomic system for South Africa*. Memoirs on Agricultural Natural Resources of South Africa 15, Pretoria

Soil Survey Staff 1975 *Soil taxonomy: a basic system of soil classification for making and interpreting soil surveys*. USDA Agricultural Handbook 436. Washington DC, Government Printing Office

Tavernier R, Louis A 1984 *Soil map of the European Communities*. Luxembourg, Office of Offical Publications of the European Communities

Taylor J R 1982 *An introduction to error analysis: the study of uncertainties in physical measurements*. Oxford, Oxford University Press/Mill Valley, University Science Books

Viridians 1996 *Victorian flora database CD-ROM*. Brighton East, Victoria, Viridians Biological Databases

Walsby J C 1995 The causes and effects of manual digitising on error creation in data input to GIS. In Fisher P F (ed.) *Innovations in GIS 2*. London, Taylor and Francis: 113–22

Wang F, Hall G B, Subaryono 1990 Fuzzy information representation and processing in conventional GIS software: database design and application. *International Journal of Geographical Information Systems* 4: 261–83

Wang M, Gong P, Howarth P J 1991 Thematic mapping from imagery: an aspect of automated map generalisation. *Proceedings of AutoCarto 10*. Bethesda, American Congress on Surveying and Mapping: 123–32

Webster R, Oliver M A 1990 *Statistical methods in soil and land resource survey*. Oxford, Oxford University Press

Williamson T 1994 *Vagueness*. London, Routledge

Young E 1992 Hunter-gatherer concepts of land and its ownership in remote Australia and North America. In Anderson K, Gale F (eds) *Inventing places; studies in cultural geography*. Melbourne, Longman: 255–72

Zadeh L A 1965 Fuzzy sets. *Information and Control* 8: 338–53

Zadeh L A 1980 Fuzzy sets versus probability. *Proceedings of the IEEE* 68: 421

83

14

Propagation of error in spatial modelling with GIS

G B M HEUVELINK

Most GIS users are now well aware that the accuracy of GIS results cannot naively be based on the quality of the graphical output alone. The data stored in a GIS have been collected in the field, have been classified, generalised, interpreted or estimated intuitively, and in all these cases errors are introduced. Errors also derive from measurement errors, from spatial and temporal variation, and from mistakes in data entry. Consequently, errors are propagated or even amplified by GIS operations. But exactly how large are the errors in the results of a spatial modelling operation, given the errors in the input to the operation? This chapter describes the development, application, and implementation of error propagation techniques for quantitative spatial data. Techniques considered are Taylor series approximation and Monte Carlo simulation. The theory is illustrated using a case study.

1 INTRODUCTION

One of the most powerful capabilities of GIS, particularly for the earth and environmental sciences, is that it permits the derivation of new attributes from attributes already held in the GIS database. For example, elevation data in the form of a digital elevation model (DEM) can be used to derive maps of gradient and aspect (Hutchinson and Gallant, Chapter 9); or digital maps of soil type and gradient can be combined with information about soil fertility and moisture supply to yield maps of suitability for growing maize (Burrough 1986). The many basic types of function used for derivations of this kind are often provided as standard functions or *operations* in many GIS, under the name of 'map algebra' (Burrough 1986; Tomlin 1990).

In practice, many GIS operations are used in sequence in order to compute an attribute that is the result of a (computational) model. For instance, the channel flow at the outlet of a watershed can be computed after the relevant hydrological processes have been translated into mathematical equations, thus after reality has been approximated by a suitable computational model. Using GIS for the evaluation of computational models is identified here by the term *spatial modelling* within GIS.

To date, most work on spatial modelling with GIS has been concentrated on the business of deriving computational models that operate on spatial data, on the building of large spatial databases, and on linking computational models with the GIS. However, there is an important additional aspect that has long received too little attention. This concerns the issue of data quality and how errors in spatial attributes propagate through GIS operations.

1.1 The propagation of errors through GIS operations

It can safely be said that no map stored in a GIS is truly error-free. Note that the word 'error' is used here in its widest sense to include not only 'mistakes' or 'blunders', but also to include the statistical concept of error meaning 'variation' (Burrough 1986). An extensive account of important error sources in GIS has been given in a previous chapter (Veregin, Chapter 12).

When maps that are stored in a GIS database are used as input to a GIS operation, then the errors in the input will *propagate* to the output of the operation. Therefore the output may not be sufficiently reliable for correct conclusions to be drawn from it. Moreover, the error propagation continues when the output from one operation is used as input to an ensuing operation. Consequently, when no record is kept of the accuracy of intermediate results, it becomes extremely difficult to evaluate the accuracy of the final result.

Although users may be aware that errors propagate through their analyses, in practice they rarely pay attention to this problem. Perhaps experienced users know that the quality of their data is not reflected by the quality of the graphical output of the GIS, but they cannot truly benefit from this knowledge because the uncertainty of their data still remains unknown. No professional GIS currently in use can present the user with information about the confidence limits that should be associated with the results of an analysis (Burrough 1992; Forier and Canters 1996; Lanter and Veregin 1992).

The purpose of this chapter is to present a methodology for handling error and error propagation in (quantitative) spatial modelling with GIS. Note that this chapter mainly deals with the propagation of *quantitative attribute* errors in GIS, where in addition it is assumed that spatially referenced data are represented as fields, not as objects (Goodchild 1992). However, many of the results presented in this chapter can be generalised and are thus valuable for the general problem of error propagation in GIS. For instance, Wesseling and Heuvelink (1993) have applied the same methodology to spatial objects and the propagation of positional errors can also be studied using a similar approach (Griffith 1989; Keefer et al 1991; Stanislawski et al 1996). The propagation of categorical errors is more difficult because in such circumstances error probability distributions cannot easily be reduced to a few parameters. Some recent work in this area is given by Forier and Canters (1996), Goodchild et al (1992), Lanter and Veregin (1992), and Veregin (1994, 1996). Recent applications of error propagation in spatial modelling are described by Finke et al (1996), Haining and Arbia (1993), Heuvelink and Burrough (1993), Leenhardt (1995), Mowrer (1994), and Woldt et al (1996).

2 DEFINITION AND IDENTIFICATION OF A STOCHASTIC ERROR MODEL FOR QUANTITATIVE SPATIAL ATTRIBUTES

Before considering the propagation of error one must first give a suitable definition of error. An 'error' in a quantitative attribute can be conveniently

defined as the difference between reality and our representation of reality (i.e. the map). For instance, if the nitrate concentration of the shallow groundwater at some location equals 68.6 g/m^3, while according to the map it is 62.9 g/m^3, then there will be no disagreement that in this case the error is $68.6 - 62.9 = 5.7$ g/m^3. Generalising this example, let the true value of a spatial attribute at some location x be $a(x)$, and let the representation of it be $b(x)$. Then, according to the definition, the error $v(x)$ at x is simply the arithmetical difference $v(x) = a(x) - b(x)$.

We consider the situation in which the true value $a(x)$ is unknown, because if it were known, then error could simply be eliminated by assigning $a(x)$ to $b(x)$. What is known exactly is the representation $b(x)$, because this is the estimate for $a(x)$ that is available from the map. The error $v(x)$ is also not known exactly, but we should have some idea about the range or distribution of values that it is likely to take. For instance, we may know that the chances are equal that $v(x)$ is positive or negative, or we may be 95 per cent confident that $v(x)$ lies within a given interval.

Knowledge about the error $v(x)$ is thus limited to specifying a range or distribution of possible values. This type of information can best be conveyed by representing the error as a *random variable $V(x)$*. Note that notation using capitals is introduced here, in order to distinguish random variables from deterministic variables. Typically, a random variable is associated with the outcome of a probabilistic experiment, such as the throw of a die or the number drawn in a lottery. But a random variable is equally suited to model the concept of *uncertainty* (Fisher, Chapter 13). For instance, since we do not know the true nitrate concentration of the shallow groundwater, we may think that it is a value drawn from a large set of values that surround the estimated value of 62.9 g/m^3. Although we are aware that the attribute has only one fixed, deterministic value $a(x)$, our uncertainty about $a(x)$ allows us to treat it as the outcome of some random mechanism $A(x)$. We must then proceed by specifying the rules of this random mechanism, by saying how likely each possible outcome is. This will be done more formally in the next section.

2.1 Definition of the stochastic error model

Consider a quantitative spatial attribute

$$A(.) = \{A(x) \mid x \in D\}$$

that is defined on the spatial domain of interest D. Refer to the value of $A(.)$ at some location $x \in D$ as $A(x)$. The error model introduced in the previous section thus becomes:

$$A(x) = b(x) + V(x) \quad \text{for all } x \in D \qquad (1)$$

where $A(x)$ and $V(x)$ are random variables and $b(x)$ is a deterministic variable. Note that $A(.)$ and $V(.)$ are not random variables but random fields, in the geostatistical literature also termed random functions (Cressie 1991; Journel and Huijbregts 1978).

Let us first consider the error at location x only. Denote the mean and variance of $V(x)$ by $E[V(x)] = \mu(x)$ and $Var(V(x)) = \sigma^2(x)$. The mean $\mu(x)$ is often referred to as the systematic error or bias, because it says how much $b(x)$ systematically differs from $A(x)$. The standard deviation $\sigma(x)$ of $V(x)$ characterises the non-systematic, random component of the error $V(x)$. In standard error analysis, it is often assumed that errors follow the normal (Gaussian) distribution (Taylor 1982), but this is not always sensible. For instance, in geology, hydrology, and soil science, many attributes are skewed and the errors associated with them may be described more adequately using a lognormal distribution.

Next consider the spatial and multivariate extension of the error model. Although a complete characterisation of the error random field $V(.)$ would require its entire finite-dimensional distribution (Cressie 1991: 52), here we only define its first and second moments, which are assumed to exist. Let x and x' be elements of D. The (spatial auto-) correlation $\rho(x,x')$ of $V(x)$ and $V(x')$ is defined as:

$$\rho(x, x') = \frac{R(x, x')}{\sigma(x)\,\sigma(x')} \qquad (2)$$

where $R(x,x')$ is the covariance of $V(x)$ and $V(x')$. Clearly, when $x = x'$ then covariance equals variance, so $R(x,x) = \sigma^2(x)$ and $\rho(x,x) = 1$ for all $x \in D$.

When there are multiple attributes $A_i(x)$ and errors $V_i(x)$, $i = 1, \ldots, m$, then for each of the attributes an error model $A_i(x) = b_i(x) + V_i(x)$ is defined, where the error $V_i(x)$ follows some distribution with mean $\mu_i(x)$ and variance $\sigma_i^2(x)$. Let $\rho_{ij}(x,x')$ be the (spatial cross-) correlation of $V_i(x)$ and $V_j(x')$, defined as:

$$\rho_{ij}(x,x') = \frac{R_{ij}(x, x')}{\sigma_i(x)\,\sigma_j(x')} \qquad (3)$$

where $R_{ij}(x,x')$ is the covariance of $V_i(x)$ and $V_j(x')$. The cross-covariance function $R_{ij}(.\,,.)$ thus defines the covariance of different attribute errors, possibly at different locations.

To illustrate that errors in spatial attributes are often correlated, consider the example of soil pollution by heavy metals, such as is the case in the river Geul valley, in the south of the Netherlands (Leenaers 1991). Consider the concentration of lead and cadmium in the soil, maps of which are obtained from interpolating point observations. In this case the interpolation errors $V_{lead}(x)$ and $V_{cadmium}(x)$ are likely to be positively correlated, because *unexpectedly* high lead concentrations will often be accompanied by *unexpectedly* high cadmium concentrations. Unforeseen low concentrations will also often occur simultaneously. Heuvelink (1993) derives these error correlations mathematically for geostatistical interpolation.

The observation that errors in spatial attributes are often correlated is important because in what follows we will see that presence of non-zero correlation can have a marked influence on the outcome of an error propagation analysis.

2.2 Identification of the error model

To estimate the parameters of the error random field $V(.)$ in practice, certain stationarity assumptions have to be made (Cressie 1991: 53). This can be done in various ways. The most obvious way is to impose the assumptions directly on $V(.)$. This is acceptable when inference on $V(.)$ is based solely on observed errors at test points. For instance, to assess the error standard deviation of an existing DEM it may be sensible to assume that $\sigma(.)$ is spatially invariant, so that it can be estimated by the root mean squared error (RMSE), computed from the differences between the DEM and the true elevation at the test points (Fisher 1992). In addition, it may be sensible to assume that the spatial autocorrelation $\rho(x,x')$ is a (decreasing) function of only the distance $|x-x'|$. If sufficient test points are available (say 60 or more), then $\rho(.)$ can be estimated using geostatistical tools (Cressie 1991; Pannatier 1996).

However, in many situations it is not very sensible to impose the stationarity assumptions directly on the error map $V(.)$. In many situations $V(.)$ is the residual from mapping an attribute from point observations, and where the spatial variability of the attribute has been identified prior to, and has been incorporated in, the mapping. In order to avoid inconsistencies, the error model parameters should then be derived from the spatial variability of the attribute and the mapping procedure used. The spatial variability of the attribute may be characterised using a discrete, continuous, or mixed model of spatial variation, but in all three cases the mapping and error identification will involve some form of Kriging (Heuvelink 1996). The advantage of Kriging is that it not only yields interpolated values but that it also quantifies the interpolation error. For a discussion of Kriging, see Cressie (1991) or Mitas and Mitasova (Chapter 34).

3 THE THEORY OF ERROR PROPAGATION

The error propagation problem can now be formulated mathematically as follows. Let $U(.)$ be the output of a GIS operation $g(.)$ on the m input attributes $A_i(.)$:

$$U(.) = g(A_1(.), ..., A_m(.)) \qquad (4)$$

The operation $g(.)$ may be one of various types, such as a standard filter operation to compute gradient and aspect from a gridded DEM (Carter 1992), a pedotransfer function to predict soil hydraulic properties from basic soil properties (Finke et al 1996), or a complex distributed runoff and soil erosion model (De Roo et al 1992). The objective of the error propagation analysis is to determine the error in the output $U(.)$, given the operation $g(.)$ and the errors in the input attributes $A_i(.)$. The output map $U(.)$ also is a random field, with mean $\xi(.)$ and variance $\tau^2(.)$. From an error propagation perspective, the main interest is in the uncertainty of $U(x)$, as contained in its variance $\tau^2(.)$.

It must first be observed that the error propagation problem is relatively easy when $g(.)$ is a linear function. In that case the mean and variance of $U(.)$ can be directly and analytically derived. The theory on functions of random variables also provides several analytical approaches to the problem for non-linear $g(.)$, but few of these can be resolved by simple calculations (Helstrom 1991). In practice, these analytically-driven methods nearly always rely on numerical methods for a complete evaluation. Thus for the general situation analytical methods are not very suitable. In this context, two alternative methods will now be discussed.

For practical purposes the discussion hereafter will be confined to point operations, i.e. GIS operations that operate on each spatial location x separately. This is no real restriction because non-point operations can be handled by minor modification (Heuvelink 1993). For notational convenience, the spatial index x will be dropped. It will also be assumed that the errors V_i have zero mean. This is because unbiasedness conditions are usually included in the mapping of the A_i.

3.1 Taylor series method

The idea of the Taylor series method is to approximate $g(.)$ by a truncated Taylor series centred at $\bar{b}=(b_1,...,b_m)$. In case of the first order Taylor method, $g(.)$ is linearised by taking the tangent of $g(.)$ in $\bar{b}$. The linearisation greatly simplifies the error analysis, but only at the expense of introducing an approximation error.

The first order Taylor series of $g(.)$ around $\bar{b}$ is given by:

$$U = g(\bar{b}) + \sum_{i=1}^{m}(A_i - b_i)g_i'(\bar{b}) + \text{remainder} \qquad (5)$$

where $g_i'(.)$ is the first derivative of $g(.)$ with respect to its i-th argument. By neglecting the remainder in Equation 5 the mean and variance of U are given as (Heuvelink et al 1989):

$$\xi \approx g(\bar{b}) \qquad (6)$$

$$\tau^2 \approx \sum_{i=1}^{m}\sum_{j=1}^{m}\rho_{ij}\sigma_i\sigma_j g_i'(\bar{b})g_j'(\bar{b}) \qquad (7)$$

Thus the variance of U is the sum of various terms, which contain the correlations and standard deviations of the A_i and the first derivatives of $g(.)$ at $\bar{b}$. These derivatives reflect the sensitivity of U for changes in each of the A_i. From Equation 7 it also appears that the correlations of the input errors can have a marked effect on the variance of U. Note also that Equation 7 constitutes a well known result from standard error analysis theory (Burrough 1986: 128–31; Taylor 1982).

To decrease the approximation error invoked by the first order Taylor method, one option is to extend the Taylor series of $g(.)$ to include a second order term as well (Heuvelink et al 1989). This is particularly useful when $g(.)$ is a quadratic function, in which case the second order method is free of approximations and the first order method is not. The case study provides an example.

Another method comparable to the first order Taylor method has been proposed by Rosenblueth (1975). This method estimates ξ and τ^2 from 2^m function values of $g(.)$, evaluated at all 2^m corners of a hyperquadrant in m-dimensional space. Unlike the Taylor method, this method does not require that $g(.)$ is continuously differentiable.

3.2 Monte Carlo method

The Monte Carlo method (Hammersley and Handscomb 1979; Lewis and Orav 1989) uses an entirely different approach to analyse the propagation of error through the GIS operation (Equation 4). The idea of the method is to compute the result of $g(a_1,...,a_m)$ repeatedly, with input values a_i that are randomly sampled from their joint distribution. The model results form a random sample from the distribution of U, so that parameters of the distribution, such as the mean ξ and the variance τ^2, can be estimated from the sample.

The method thus consists of the following steps:

1 repeat N times:
   a generate a set of realisations a_i, $i=1,...,m$
   b for this set of realisations a_i, compute and store the output $u=g(a_1,...,a_m)$
2 compute and store sample statistics from the N outputs u.

A random sample from the m inputs A_i can be obtained using an appropriate random number generator (Lewis and Orav 1989; Ross 1990). Note that a conditioning step will have to be included when the A_i are correlated. One attractive method for generating realisations from a multivariate Gaussian distribution uses the Cholesky decomposition of the covariance matrix (Johnson 1987).

Application of the Monte Carlo method to error propagation with non-point operations requires the simultaneous generation of realisations from the random fields $A_i(.)$. This implies that spatial correlation will have to be accounted for. Various techniques can be used for stochastic spatial simulation, an attractive one being the sequential Gaussian simulation algorithm (Deutsch and Journel 1992).

The accuracy of the Monte Carlo method is inversely related to the square root of the number of runs N. This means that to double the accuracy, four times as many runs are needed. The accuracy thus slowly improves as N increases.

3.3 Evaluation and comparison of error propagation techniques

The main problem with the Taylor method is that the results are only approximate. It will not always be easy to determine whether the approximations involved using this method are acceptable. The Monte Carlo method does not suffer from this problem, because it can reach an arbitrary level of accuracy.

The Monte Carlo method brings with it other problems, however. High accuracies are reached only when the number of runs is sufficiently large, which may cause the method to become extremely time consuming. This will remain a problem even when variance reduction techniques such as Latin hypercube sampling are employed. Another disadvantage of the Monte Carlo method is that the results do not come in an analytical form.

As a general rule it seems that the Taylor method may be used to obtain crude preliminary answers. These should provide sufficient detail to be able to obtain an indication of the quality of the output of the GIS operation. When exact values or quantiles and/or percentiles are needed, the Monte Carlo method may be used. The Monte Carlo method will probably also be preferred when error propagation with complex operations is studied, because the method is easily implemented and generally applicable.

3.4 Sources of error contributions: the balance of errors

When the error analysis reveals that the output of $g(.)$ contains too large an error then measures will have to be taken to improve accuracy. When there is a single input to $g(.)$ then there is no doubt where the improvement must be sought, but what if there are multiple inputs to the operation? Also, how much should the error of a particular input be reduced in order to reduce the output error by a given factor? These are important questions that will now be considered.

To obtain answers to the questions above, consider Equation 7 again, which gives the variance of the output U using the first order Taylor method. When the inputs are uncorrelated, this reduces to:

$$\tau^2 \approx \sum_{i=1}^{m} \sigma_i^2 \, (g_i{}'(\bar{b}))^2 \tag{8}$$

Equation 8 shows that the variance of U is a sum of parts, each to be attributed to one of the inputs A_i. This *partitioning property* allows one to analyse how much each input contributes to the output variance. Thus from Equation 8 it can directly be seen how much τ^2 will reduce from a reduction of σ_i^2. Clearly the output will mainly improve from a reduction in the variance of the input that has the largest contribution to τ^2. Note that this need not necessarily be the input with the largest error variance, because the sensitivity of the operation $g(.)$ for the input is also important. Note also that Equation 8 is derived under rather strong assumptions. When these assumptions are too unrealistic it may be advisable to derive the error source contributions using a modified Monte Carlo approach (Jansen et al 1994).

In the introduction, it was noted that a GIS operation is often in effect a computational model. Consequently, not only will *input error* propagate to the output of a GIS operation, but *model error* will as well. In practice, model error will often be a major source of error and should therefore be included in the error analysis. Ignoring it would severely underestimate the true uncertainty in the model output. Model error can be included by assigning errors to model coefficients or by adding a residual error term to the model equations.

If a reduction of output error is required, it will not necessarily be sensible to improve the input with the highest error contribution. This is because the cost of reducing input error may vary from attribute to attribute. However, in many cases it will be most rewarding to strive for a *balance of errors*. When the error in an attribute has a marginal effect on the output, then there is little to be gained from mapping it more accurately. In that case, extra sampling efforts can much better be directed to an input attribute that has a larger contribution to the output error. For instance, if a pesticide leaching model is sensitive to soil organic carbon and less so to soil bulk density, then it is more important to map the former more accurately (Loague et al 1989).

The example of the pesticide leaching model draws attention to the fact that a balance of errors must also include model error. It is clearly unwise to spend much effort on collecting data if what is gained is immediately thrown away by using a poor model. On the other hand, a simple model may be as good as a complex model if the latter needs lots of data that cannot be accurately obtained. This is why researchers in catchment hydrology have raised the question of whether there is much benefit to be gained from developing ever more complex models when the necessary inputs cannot be evaluated in the required spatial and temporal resolution (Beven 1989; Grayson et al 1992).

4 APPLICATION TO MAPPING SOIL MOISTURE CONTENT WITH LINEAR REGRESSION FOR THE ALLIER FLOODPLAIN SOILS

As part of a research study in quantitative land evaluation, the World Food Studies (WOFOST) crop simulation model (Diepen et al 1989) was used to calculate potential crop yields for floodplain soils of the Allier river in the Limagne rift valley, central France. The moisture content at wilting point (Θ_{wp}) is an important input attribute for the WOFOST model. Because Θ_{wp} varies considerably over the area in a way that is not linked directly with soil type, it was necessary to map its variation separately to see how moisture limitations affect the calculated crop yield.

Unfortunately, because Θ_{wp} must be measured on samples in the laboratory, it is expensive and time-consuming to determine it for a sufficiently large number of data points for Kriging. An alternative and cheaper strategy is to calculate Θ_{wp} from other attributes which are cheaper to measure. Because the moisture content at wilting point is often strongly correlated with the moisture content at field capacity (Θ_{fc}) and the soil porosity (Φ), both of which can be measured more easily, it was decided to investigate how errors in measuring and mapping these would work through to a map of calculated Θ_{wp}. The following procedure was used to obtain a map of the mean and standard deviation of Θ_{wp}.

The properties Θ_{wp}, Θ_{fc} and Φ were determined in the laboratory for 100 cc cylindrical samples taken from the topsoil (0–20 cm) at 12 selected sites shown as the circled points in Figure 1.

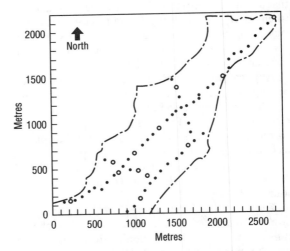

Fig 1. The Allier study area showing sampling points. Circled sites are those used to estimate the regression model.

These results were used to set up a pedotransfer function, relating Θ_{wp} to Θ_{fc} and Φ, which took the form of a multiple linear regression:

$$\Theta_{wp} = \beta_0 + \beta_1 \Theta_{fc} + \beta_2 \Phi + \varepsilon \qquad (9)$$

The coefficients β_0, β_1, and β_2 were estimated using standard ordinary least squares regression. The estimated values for the regression coefficients and their respective standard deviations were $\hat{\beta}_0 = -0.263 \pm 0.031$, $\hat{\beta}_1 = 0.408 \pm 0.096$, $\hat{\beta}_2 = 0.491 \pm 0.078$. The standard deviation of the residual ε was estimated as 0.0114. The correlation coefficients of the regression coefficients were $\rho_{01} = -0.221$, $\rho_{02} = -0.587$, $\rho_{12} = -0.655$. The regression model explains 94.8 per cent of the variance in the observed Θ_{wp}, indicating that the model is satisfactory. Note that presence of spatial correlation between the observations at the 12 locations was ignored in the regression analysis.

Sixty-two measurements of Θ_{fc} and Φ were made in the field at the sites indicated in Figure 1. From these data experimental variograms were computed. These were then fitted using the linear model of coregionalisation (Journel and Huijbregts 1978). For the purposes of this study the input data for the regression were mapped to a regular 50×50 m grid using block co-Kriging with a block size of 50×50 m. The block co-Kriging yielded raster maps of means and standard deviations for both Θ_{fc} and Φ, as well as a map of the correlation of the block co-Kriging prediction errors. Figure 2 displays these maps. Note that there are clear spatial variations in the correlation between the block co-Kriging errors.

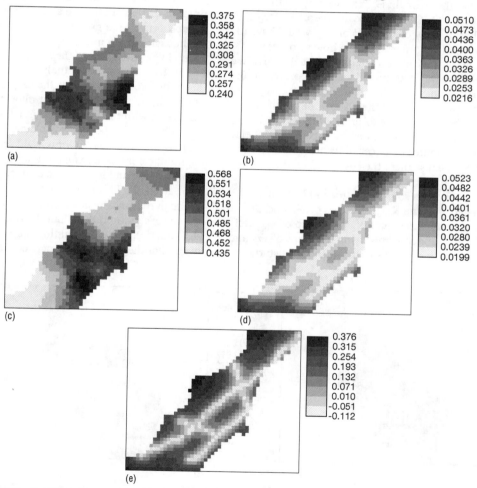

Fig 2. Kriging results for the Allier study area (50 x 50 m grid): (a) block mean and (b) standard deviation of Θ_{fc} (cm³/cm³), (c) block mean and (d) standard deviation of Φ (cm³/cm³), (e) correlation of co-Kriging prediction errors of Θ_{fc} and Φ.

The maps of Θ_{fc} and Φ were substituted in the regression Equation 9 yielding maps of the attribute Θ_{wp} and the associated error. The operation is a quadratic function and therefore the second order Taylor was considered the most appropriate error propagation technique. Because the model coefficients and the field measurements were determined independently, the correlation between the $\hat{\beta}_i$ and the co-Kriging prediction errors was taken to be zero. The results are given in Figure 3. The accuracy of the map of Θ_{wp} is reasonable: the standard deviation in Θ_{wp} rarely exceeds 25 per cent of the mean. These maps could be used as the basis of a subsequent error propagation analysis in the WOFOST crop yield model.

If an uncertainty analysis with WOFOST would show that the errors in Θ_{wp} cause errors in the output of WOFOST that are unacceptably large, then the accuracy of the map of Θ_{wp} would have to be improved. In order to decide how to proceed in such a situation, the contribution of each individual error source was determined using the partitioning property discussed in the previous section. Figure 4 presents the results and these show that both Θ_{fc} and Φ form the main source of error. Only in the immediate vicinity of the data points is the model a meaningful source of uncertainty, as would be expected because there the co-Kriging variances of Θ_{fc} and Φ are the smallest.

Thus the main source of error in Θ_{wp} is that associated with the Kriging errors of Θ_{fc} and Φ. Improvement of the quality of the map of Θ_{wp} can thus best be done by improving the maps of Θ_{fc} and Φ, by taking more measurements over the study area. The variograms of Θ_{fc} and Φ could be used to assist in optimising sampling (McBratney et al 1981). This technique would allow one to judge *in advance* how much improvement is to be expected from the extra sampling effort.

5 DISCUSSION AND CONCLUSIONS

Error propagation in spatial modelling with GIS is a relevant research topic because rarely if ever are the data stored in a GIS completely error-free. In this chapter several methods were described for analysing the propagation of errors. None of these methods is perfect: some do not apply to all types of operations, others are extremely time consuming or involve large approximation errors. However, in practice there will often be at least one method that is appropriate for a given situation. Thus the methods are in a sense complementary, and as a group in almost all cases they enable one to carry out an error propagation analysis successfully.

Unfortunately, at present the majority of GIS users still has no clear information about the errors associated with the attributes that are stored in the GIS. This is an important problem because an error propagation analysis can only yield sensible results if the input errors have realistic values. Often there will only be crude and incomplete estimates of input error available. This lack of information is perhaps the main reason why error propagation analyses are still the exception rather than the rule in everyday GIS practice. It is essential that map makers become aware that they should routinely convey the accuracy

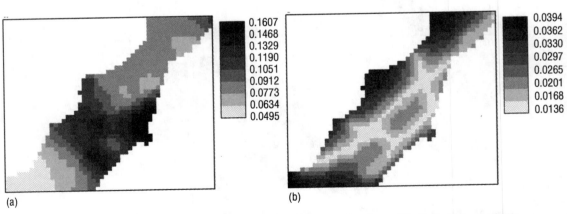

(a)

(b)

Fig 3. Results of the error propagation: (a) block mean and (b) standard deviation of Θ_{wp} (cm³/cm³) as obtained with the regression model.

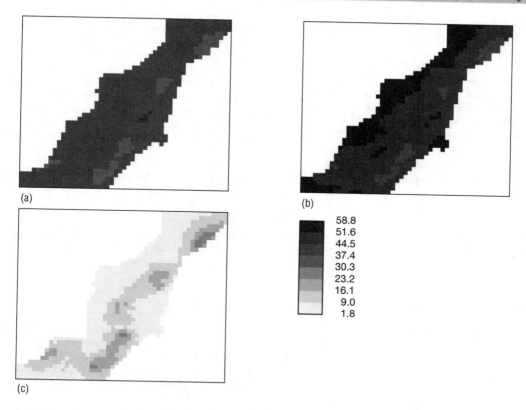

Fig 4. Maps showing the relative contributions (per cent) of the different input errors to the variance of Θ_{wp}: (a) due to Θ_{fc}, (b) due to Φ, and (c) due to the regression model.

of the maps they produce, even when accuracy is less than expected. At the same time, it is important that GIS manufacturers increase their efforts to add error propagation functionality to their products.

It is important to note that an error analysis offers much more than the computation of output error. The partitioning property of an error analysis allows one to determine how much each individual input contributes to the output error. Information of this sort may be extremely useful, because it allows users to explore how much the quality of the output improves, given a reduction of error in a particular input. Thus the improvement foreseen due to intensified sampling can be weighed against the extra sampling costs.

The partitioning property can also be used to compare the contributions of input and model error. With the advent of GIS, and the many computational models that often come freely with it, there is an increased risk of disturbing the balance between input and model error. When there is no protection against improper use then ignorant users will be tempted to apply models to improper scales, use them for purposes for which they were not developed, or combine them with data that are too uncertain (Heuvelink 1998). These problems can only be tackled when users become more aware of the issue of spatial data quality and when error propagation analysis becomes a routine instrument available to the GIS community.

References

Beven K 1989 Changing ideas in hydrology – the case of physically-based models. *Journal of Hydrology* 105: 157–72

Burrough P A 1986 *Principles of geographical information systems for land resources assessment.* Oxford, Clarendon Press

Burrough P A 1992b Development of intelligent geographical information systems. *International Journal of Geographical Information Systems* 6: 1–11

Carter J R 1992 The effect of data precision on the calculation of slope and aspect using gridded DEMs. *Cartographica* 29: 22–34

Cressie N A C 1991 *Statistics for spatial data*. New York, John Wiley & Sons Inc.

De Roo A P J, Hazelhoff L, Heuvelink G B M 1992 Estimating the effects of spatial variability of infiltration on the output of a distributed runoff and soil erosion model using Monte Carlo methods. *Hydrological Processes* 6: 127–43

Deutsch C V, Journel A G 1992 *GSLIB: geostatistical software library and user's guide*. New York, Oxford University Press

Diepen C A van, Wolf J, Keulen H van, Rappoldt C 1989 WOFOST: a simulation model of crop production. *Soil Use and Management* 5: 16–24

Finke P A, Wösten J H M, Jansen M J W 1996 Effects of uncertainty in major input variables on simulated functional soil behaviour. *Hydrological Processes* 10: 661–9

Fisher P F 1992 First experiments in viewshed uncertainty: simulating fuzzy viewsheds. *Photogrammetric Engineering and Remote Sensing* 58: 345–52

Forier F, Canters F 1996 A user-friendly tool for error modelling and error propagation in a GIS environment. In Mowrer H T, Czaplewski R L, Hamre R H (eds) *Spatial accuracy assessment in natural resources and environmental sciences*. Fort Collins, USDA Forest Service General Technical Report RM-GTR-277: 225–34

Goodchild M F 1992a Geographical data modeling. *Computers and Geosciences* 18: 401–8

Goodchild M F, Sun G, Yang S 1992 Development and test of an error model for categorical data. *International Journal of Geographical Information Systems* 6: 87–104

Grayson R B, Moore I D, McMahon T A 1992 Physically based hydrologic modelling: 2. Is the concept realistic? *Water Resources Research* 28: 2659–66

Griffith D A 1989 Distance calculations and errors in geographic databases. In Goodchild M F, Gopal S (eds) *Accuracy of spatial databases*. London, Taylor and Francis: 81–90

Haining R P, Arbia G 1993 Error propagation through map operations. *Technometrics* 35: 293–305

Hammersley J M, Handscomb D C 1979 *Monte Carlo methods*. London, Chapman and Hall

Helstrom C W 1991 *Probability and stochastic processes for engineers*. New York, Macmillan

Heuvelink G B M 1993 'Error propagation in quantitative spatial modelling: applications in geographical information systems'. PhD thesis. Utrecht, Netherlands Geographical Studies 163

Heuvelink G B M 1996 Identification of field attribute error under different models of spatial variation. *International Journal of Geographical Information Systems* 10: 921–36

Heuvelink G B M 1998 Uncertainty analysis in environmental modelling under a change of spatial scale. *Nutrient Cycling in Agro-Ecosystems* 50: 257–66

Heuvelink G B M, Burrough P A 1993 Error propagation in cartographic modelling using Boolean methods and continuous classification. *International Journal of Geographical Information Systems* 7: 231–46

Heuvelink G B M, Burrough P A, Stein A 1989 Propagation of errors in spatial modelling with GIS. *International Journal of Geographical Information Systems* 3: 303–22

Jansen M J W, Rossing W A H, Daamen R A 1994 Monte Carlo estimation of uncertainty contributions from several independent multivariate sources. In Grasman J, Straten G van (eds) *Predictability and non-linear modelling in natural sciences and economics*. Dordrecht, Kluwer: 334–43

Johnson M E 1987 *Multivariate statistical simulation*. New York, John Wiley & Sons Inc.

Journel A G, Huijbregts C J 1978 *Mining geostatistics*. London, Academic Press

Keefer B J, Smith J L, Gregoire T G 1991 Modeling and evaluating the effects of stream mode digitizing errors on map variables. *Photogrammetric Engineering and Remote Sensing* 57: 957–63

Lanter D P, Veregin H 1992 A research paradigm for propagating error in layer-based GIS. *Photogrammetric Engineering and Remote Sensing* 58: 825–33

Leenaers H 1991 Deposition and storage of solid-bound heavy metals in the floodplains of the river Geul (The Netherlands). *Environmental Monitoring and Assessment* 18: 79–103

Leenhardt D 1995 Errors in the estimation of soil water properties and their propagation through a hydrological model. *Soil Use and Management* 11: 15–21

Lewis P A W, Orav E J 1989 *Simulation methodology for statisticians, operations analysts, and engineers* Vol. 1. Pacific Grove, Wadsworth & Brooks/Cole

Loague K, Yost R S, Green R E, Liang T C 1989 Uncertainty in pesticide leaching assessment in Hawaii. *Journal of Contaminant Hydrology* 4: 139–61

McBratney A B, Webster R, Burgess T M 1981 The design of optimal sampling schemes for local estimation and mapping of regionalized variables: 1. Theory and method. *Computers and Geosciences* 7: 331–4

Mowrer H T 1994 Monte Carlo techniques for propagating uncertainty through simulation models and raster-based GIS. In Congalton R G (ed) *Proceedings of the international symposium on spatial accuracy of natural resource databases*. Washington, American Society for Photogrammetry and Remote Sensing: 179–88

Pannatier Y 1996 *VARIOWIN: software for spatial data analysis in 2D*. New York, Springer

Rosenblueth E 1975 Point estimates for probability moments. *Proceedings of the National Academy of Sciences of the United States of America* 72: 3812–14

Ross S M 1990 *A course in simulation*. New York, MacMillan

Stanislawski L V, Dewitt B A, Shrestha R L 1996 Estimating positional accuracy of data layers within a GIS through error propagation. *Photogrammetric Engineering and Remote Sensing* 62: 429–33

Taylor J R 1982 *An introduction to error analysis: the study of uncertainties in physical measurement*. Oxford, Oxford University Press/Mill Valley, University Science Books

Tomlin C D 1990 *Geographic information systems and cartographic modeling*. Englewood Cliffs, Prentice-Hall

Veregin H 1994 Integration of simulation modelling and error propagation for the buffer operation in GIS. *Photogrammetric Engineering and Remote Sensing* 60: 427–35

Veregin H 1996 Error propagation through the buffer operation for probability surfaces. *Photogrammetric Engineering and Remote Sensing* 62: 419–28

Wesseling C G, Heuvelink G B M 1993 Manipulating quantitative attribute accuracy in vector GIS. In Harts J, Ottens H F L, Scholten H J (eds) *Proceedings EGIS 93*. Utrecht, EGIS Foundation: 675–84

Woldt W, Goderya F, Dahab M, Bogardi I 1996 Consideration of spatial variability in the management of non-point source pollution to groundwater. In Mowrer H T, Czaplewski R L, Hamre R H (eds) *Spatial accuracy assessment in natural resources and environmental sciences*. Fort Collins, USDA Forest Service General Technical Report RM-GTR-277: 49–56

Introduction

THE EDITORS

It is clear from the discussion of GIS principles thus far that we now live in a data-rich world in which a vast and increasing array of geographical phenomena are represented in digital form. GIS-based data models are by definition selective abstractions and the data used to build them are error prone, yet they can lay the foundation for legitimate context-sensitive inputs to generalisable analysis and forecasting. The contributions to this section set out to identify how GIS allows us to summarise the properties of spatial distributions, inductively solve spatial problems, and contribute towards spatial decision-making.

The established paradigm for quantitative description and generalisation about geographical phenomena has been to use spatial statistics. What makes spatial statistics distinct from its parent discipline is its concern with observations which are located near to one another in space and which, as a consequence, tend to share similar attribute values – in Anselin's words (Chapter 17): 'the phenomenon where locational similarity (observations in spatial proximity) is matched by value similarity (correlation)'. Spatial and geostatisticians have developed a range of specialised methods and techniques for dealing with such cases. The emergence and chronological development of spatial statistics in the pre-GIS era is the first theme considered in the contribution by Art Getis (Chapter 16).

An important emergent debate within GIS has been the continuing relevance of spatial statistical approaches. Briefly, the key arguments may be summarised as: first, spatial statistics developed in what all of the contributors to this section would recognise as a 'data-poor' era, in which statistics were based upon few (by present-day standards) observations; second, this paucity of data made computation a fairly straightforward procedure; and, third, the geography of areal units was fixed (and usually coarse), and not itself subjectable to the

range of transformations and sensitivity analyses that have been outlined in previous contributions to this Principles part of the book. There is evident consensus among the contributors to this section that spatial analysis has received far too little attention in the development of GIS, yet they have some different views as to whether and how spatial statisticians can continue to contribute practical spatial analysis skills to GIS. On the one hand, Getis (Chapter 16) and Luc Anselin (Chapter 17) set out some of the enduring contributions that spatial statistics is making to GIS, especially in the area of exploratory spatial data analysis (ESDA). On the other hand, the views of Stan Openshaw and Seraphim Alvanides (Chapter 18) and Manfred Fischer (Chapter 19) lean towards the view that the changes associated with the development of GIS require a more root-and-branch reappraisal of the practice of spatial analysis in GIS. Richard Church (Chapter 20) presents a review of the ways in which GIS is being applied to locational analysis problems in GIS.

The review by Getis (Chapter 16) charts the considerable progress that has been made in developing ESDA spatial statistics through the media of GIS. However, he laments the relative lack of progress in developing spatial hypothesis testing within GIS. By implication, he seems to sound a warning that the media of GIS are in danger of overwhelming the message of spatial statistical analysis as conventionally understood. Indeed, developments in scientific visualisation and ESDA appear to have contributed little to our incomplete theoretical and statistical understandings of the ways that observations should be differentially weighted across space (i.e. the effects of distance) and the effects of boundaries and edges on the results of spatial statistical analysis.

To dwell upon such (possibly unresolvable) statistical issues might be seen as admonishing failure, when the media of GIS have been demonstrably effective in exploring locational scenarios and visualising spatial outcomes. Data exploration within GIS is the domain of ESDA techniques – defined by Anselin (1994) as being used 'to describe and visualise spatial distributions, identify atypical locations (spatial outliers), discover patterns of spatial association (spatial clusters) and suggest different spatial regimes and other forms of spatial instability or spatial non-stationarity'. Anselin's contribution to this volume develops the views that GIS has become data rich but theory poor, and that ESDA statistics can be used to structure, visualise, and explain a wide array of geographical data. His comprehensive review builds upon Getis' conceptions of spatial autocorrelation (i.e. geostatistical and spatial weights formulations), and in this context goes on to identify important domains of ESDA as pertaining to identification of local patterns of spatial association within global patterns.

For Openshaw and Alvanides, by contrast, the dominant impression is that we are seemingly unable to structure and analyse the vast quantities of spatial data that are now available, and that this failure reflects our continued adherence to the 'pre-GIS' spatial statistical analysis paradigm. Given that the development of GIS and of modern databases was substantially technology-led, it is at least intuitively plausible that analysis might be developed through the same guiding force. Thus for Openshaw and Alvanides (Chapter 18) the way forward lies through broad-based techniques of 'geocomputation' – that is, 'the adoption of a large-scale computationally-intensive approach to the problems of physical and human geography in particular, and the geosciences in general'. This kind of approach has undoubtedly had a profound impact through demonstration and exploration of modifiable areal unit effects, and the geocomputational paradigm has clear application in allowing spatial analysts to create zone designs that satisfy particular constellations of constraints. Building upon this, Openshaw and Alvanides see generic solutions emerging from the use of 'intelligent' pattern-seeking techniques which might become integral to GIS. The implication is that new computational techniques may be used to search for new theories and to generate new knowledge using applied, problem-solving approaches.

Fischer (Chapter 19) takes a wider perspective on the emergence of what he terms 'computational intelligence' (CI) technologies in relation to classical spatial statistics. He shares many of the expressed doubts of Openshaw and Alvanides that conventional spatial statistics can be adapted to accommodate the richness of the GIS environment, and instead advocates a CI paradigm (involving artificial life, evolutionary computation, and neural networks) based essentially upon geocomputation. He develops an extended case study involving the use of a neural net model for satellite image classification, and shows how spatial analysis proceeds through model specification, estimation, and testing phases. His exposition is lucid and non-technical, and he is at pains to dispel suspicion about the 'mystique and metaphorical jargon' hitherto associated with CI techniques. Whether the widest GIS audience will share his confidence that use of the 'universal language of mathematics' alone will dispel such scepticism is a moot point: indeed, this raises important issues about the gulf between 'machine-intelligent' spatial analysis of digital abstractions and scientific theory and reasoning as conventionally understood.

We have seen in the first section to this part of the book how ontology ('meta-theory') prescribes particular detailed approaches to analysis, and that no stage in scientific reasoning can be considered in isolation. The emphasis in much of the second part was on demonstrating how digital data provide only imperfect, incomplete, and error-prone representations of reality. Together this would seem to require that choice of spatial analytical method is rational, informed, and sensitive to context, rather than being data led in a naive empiricist way. Just as Goodchild and Longley (Chapter 40) argue later in the Technical issues section, that volume of data is not a substitute for scientific rigour, so the paradigm of geocomputation needs to demonstrate why and in what circumstances spatial analysts should have 'confidence' (both broadly and narrowly defined) in its substantive findings. Openshaw and Alvanides cite a number of high-profile and celebrated case studies which have adopted what has come to be described as a geocomputational approach (notably in the identification of clusters of diseases), yet none appears wholly to have withstood scientific scrutiny: as such, the jury must still be out regarding the advisability of wholesale reliance upon geocomputational approaches. It is beyond doubt that the world has never been as data rich but, in the realm of spatial analysis, there have been a number of false dawns before – as our comments in the Introduction on the coverage of artificial intelligence in the first edition of this book testifies. Thus while Openshaw and Alvanides conjecture that 'it may be possible to compute our way out of the data swamp', it remains a moot point as to how and why we may have become lured into it in the first place. It remains to be seen whether and how far the protagonists of computational approaches are able to rebuff concerns that they are indulging in uninformed pattern-seeking empiricism in the absence of clear theoretical guidance as conventionally understood. Elsewhere, Openshaw and Openshaw (1997) have begun to demystify new ways of viewing our digital world although, at the other end of the spectrum, Curry (1995: 82) has concluded that 'to develop an understanding of the data adequate to a resolution of the problems which arise in the production of a GIS would very likely render those systems irrelevant'. There is still some way to go, both with regard to demystifying technique and to understanding data.

An oft-rehearsed but nevertheless resonant theme running through all of these contributions concerns the integration of spatial analytic functionality into proprietary GIS. Neither spatial statistical models, nor geocomputational methods, nor refined ESDA techniques form part of proprietary GIS. This is in part because of user ignorance about the range of simplifying assumptions that routine usage brings, and in part because (as a consequence) vendors are unlikely to prioritise functionality for which there are no strong user demands. Given the very small likelihood of fully integrated spatial analytical GIS in the foreseeable future, Getis, Openshaw and Alvanides, and Anselin each explore a number of options for the close and loose coupling of GIS to specialist spatial analysis packages, as well as the potential role of the Internet as a platform for integrating GIS and spatial analysis.

Some of the earliest applications of spatial analysis involved the calculation of statistical moments and distributions for the classic locational models of geography, models which were used in the pre-GIS era to identify the best location for industrial and service facilities. The enduring relevance of locational modelling to GIS is the focus of the contribution by Church (Chapter 20). A conventional facet to this problem has been the use of GIS to locate single facilities with respect to spatial patterns of demand and, through use of the overlay model, to identify corridors linking the different sites involved in activities in the most cost-efficient way. In recent years, progress has been made towards solving multiple-site location problems, most notably instances in which sites may need to be relocated in response to very short-term changes in demand (as in the relocation of 'on call' ambulances to cover for vehicles which are already attending emergencies). This echoes Getis' sentiment of continuity of approach in spatial analysis, but what has changed here with the innovation of GIS is the richness of the data which can be brought to bear on site location problems, the computational support for new and complex location–allocation algorithms, and the visual quality of the computer environment for data exploration, investigation of scenarios and decision support. Location modelling also provides a good exemplar of the wider problems that remain on the spatial analysis agenda, namely: the compatibility of data structures/data quality issues; the representation of spatial patterns of demand, and the screening process used to identify sites; the size and scale of the elemental units used to specify location–allocation problems; and the ways in which errors are created and propagated in formulation of problems. GIS is clearly having far-reaching impacts upon the specification, estimation, and testing of spatial relationships, and the Applications part of this book (in Volume 2) provides evidence of the practical relevance of these techniques (e.g. Cova, Chapter 60; Gatrell and Senior, Chapter 66).

References

Anselin L 1994a Exploratory spatial data analysis and geographic information systems. In Painho M (ed.) *New tools for spatial analysis*. Luxembourg, Eurostat: 45–54

Curry M R 1995a GIS and the inevitability of ethical inconsistency. In Pickles J (ed.) *Ground truth: the social implications of geographic information systems*. New York, Guilford Press

Openshaw S, Openshaw C A 1997 *Artificial intelligence in geography*. Chichester, John Wiley & Sons

18

Applying geocomputation to the analysis of spatial distributions

S OPENSHAW AND S ALVANIDES

Developments in IT and GIS have combined to create an enormously data-rich world. The need now is to develop GIS-relevant spatial analysis tools that will assist endusers in making good use of their spatial information. The problems are: (a) the absence of many appropriate tools; and (b) the new types of analysis required as a consequence of the innovation of GIS. The hope is that a combination of GIS databases, high performance computers, artificial intelligence and a geocomputation paradigm will together provide a generic workable solutions strategy.

1 INTRODUCTION

The GIS revolution has created an immense wealth of spatial information in a large number of different application areas. The emphasis in GIS upon database creation and systems building will soon have to be replaced by a new concern for applications using spatial analysis and modelling. The development of GIS can be largely regarded as the computerisation of pre-existing manual procedures and established technologies that were already fairly mature in research terms. Thus no great innovation was required in order to underpin the GIS revolution, although doubtless much did occur as the software systems developed. However, when the focus of attention switches to spatial analysis and modelling then it is a very different story. Frequently there have been no existing useful manual procedures to computerise, and over the last 20 years there has been very little relevant new research focused specifically on the special needs of GIS. Moreover, an increasing number of the emerging analysis tasks are novel and have not previously attracted much or any attention – for example, the exploration of very large spatial datasets for completely unknown patterns and relationships, or the real-time analysis of live spatial databases for emerging patterns, 'hot spots' (see Getis, Chapter 16), and anomalies of interest.

The quest for improved analysis of spatial distributions is predicated upon three considerations:

1 a general desire to make use of the information resources created by GIS;
2 attempts to gain competitive advantages or other benefits from investments in information technology (of which GIS is a component);
3 hardware developments that are trivialising the costs of computation and are hence creating new ways of analysing spatial distributions.

One might argue that it is scandalous that so many key databases are not currently being properly analysed – be they pertaining to morbidity, mortality, cancer, or crime incidence – while commercial concerns and government agencies probably waste many millions (if not billions) of dollars or ECUs by poor spatial data management and inefficient locational decisions (Openshaw 1994d). Yet perhaps the users cannot be blamed for not using tools that are unavailable! The problem is essentially a longstanding failure to evolve distinctly geographical-data-appropriate tools and styles of analysis and modelling – albeit with a small number of exceptions. The legacy of statistical methods (Getis, Chapter 16) may not be helpful, at least partly because its inherent limitations need to be properly understood. In essence, no amount of apparent statistical sophistication should be allowed to hide the fact that much of spatial statistics is very limited in what it can do, and is even more limiting in its view of spatial information and the handling of their special properties. It also needs to be appreciated that the 'post GIS-revolution' world of the late 1990s is quite different from the primeval (by contemporary standards) computing and data-poor environments in which many of the existing spatial analysis and modelling technologies were developed. Are these old legacy technologies and their latter-day offspring still appropriate, or is a new period of basic research and development needed to create the spatial analysis tools likely to be required in the late 1990s and beyond? There is no denial that advances have been made in improving some spatial statistical methods (Getis, Chapter 16), but these advances – such as developing local versions of global statistics – are of limited usefulness in practice.

This chapter attempts to address some of these concerns by focusing on the changes in the computational environment that has occurred during the 1990s. It argues that large-scale computation can now be used as a paradigm for solving some of the major spatial analysis problems that are relevant to GIS. However, if computational power is to be useful, then there also has to be a clear understanding of what the requirements and the user needs are. This leads us to a brief typology of alternative approaches and a brief illustrative case study based on one method in which high performance computing is combined with GIS data and artificial intelligence (AI) tools to develop better ways of engineering zoning systems as a decision support, analysis, modelling, and data management tool.

2 ADVANCES IN 'HIGH PERFORMANCE COMPUTING'

There is now considerable excitement in many traditional sciences about developments in supercomputing or high performance computing (HPC). Computation is now regarded as a scientific tool of equal importance to theory and experimentation, since fast computers have stimulated new ways of doing science via large-scale computer-based experimentation, simulation, and numerical approximation. There is equally a case for thinking that a supercomputing-based paradigm is

also relevant to many areas of GIS, but it should be appreciated that this involves much more than attempts to revamp basic GIS functions using parallel computing. HPC is defined as computer hardware based on vector or parallel processors (or some mixture) that offers at least one order of magnitude increase in computing power over that available from a mid 1990s workstation. In fact, as highly parallel processors take over from the earlier vector machines, the performance gain from using leading edge HPC hardware is more usually at least two orders of magnitude. This whole area is now developing at a rapid rate with most HPCs having a two- to three-year life cycle. For example, in 1996 it was possible to buy for £500 000 (US $800 000) a parallel machine with equivalent performance to one costing about 10 times as much only a few years earlier (see also Longley et al, Chapter 1). The computing world is in the throes of a major technological change; that of highly parallel supercomputing (Hillis 1992). A highly or massively parallel processor (MPP) is a computing system with multiple central processing units (CPUs) that can work concurrently on a single task. This idea is not new but it was only in the mid 1990s that the technology matured sufficiently for multiple CPUs to become the dominant future HPC machine architecture (see also Batty, Chapter 21). A nice feature of MPPs is that both processing capacity and memory are scaleable – if you want more computer power, then simply add more processors. If spatial analysis tools and models are also scaleable then running them on more processors reduces computer wall-clock times in a linear way (Turton and Openshaw 1996).

Openshaw (1994b) has suggested that by 1999 it is quite likely that available HPC hardware will be 10^9 times faster (and bigger in memory) than what was common during the 'Geography Quantitative Revolution' years of the 1960s (when many of the current so-called spatial statistical methods were developed: see Getis, Chapter 16); 10^8 times faster than hardware available during the mathematical modelling revolution of the early 1970s (on which virtually all of the so-called 'intelligent' model based spatial decision support systems employed in today's GISs were based: Birkin et al 1996); 10^6 times since the GIS revolution of the mid 1980s (a time of considerable neglect of quantitative geography), and at least a further 10^2 times faster than what in 1994 was Europe's fastest civilian supercomputer – the Edinburgh Cray T3D.

A widespread problem is that many potential users appear to have failed to appreciate what these developments in HPC mean. For instance, the Edinburgh Cray T3D has 512 processors and has a peak theoretical performance of 76.8 gigaflops – but what does that mean? One way of answering this question is to create a benchmark code that can be run on the widest possible range of computer hardware, ranging from a PC, UNIX workstations, vector supercomputers, and massively parallel machines. The widely used scientific benchmark codes measure machine performance in terms of simple matrix algebra problems, but it is not clear whether this is relevant in a GIS context. Openshaw and Schmidt (1997) have developed a social science benchmark code based on a scaleable spatial interaction model which can be run on virtually any serial and parallel processor. Table 1 provides a preliminary assessment of the performance of some current HPC hardware in terms of processing speed relative to the performance of a 486 66 MHz PC. At the time of writing, the best performance for small problem sizes was the SGI Onyx (a workstation with multiple CPUs), followed by the vector processor (the Fujitsu VPX240) which was about an order of magnitude more expensive. However, once problem sizes increase then soon there is no alternative to the massively parallel Cray T3D with speed gains of about 1335 times for a 10 000 by 10 000 zone matrix (equivalent to the UK ward level journey-to-work or migration data from the 1991 Census). This run took 2.4 seconds of wall-clock time (compared with 18 hours on a workstation), while the even larger 25 000 by 25 000 benchmark required 13 seconds. Note that HPC is not just about speed but also memory. The larger memory sizes required in these latter two runs reflect problems that have previously simply been uncomputable. It is interesting that during the 1990s machine speeds have been doubling every 1.5 to 2 years, and that this is expected to continue for at least another 10 years. One way of explaining what these changes in HPC hardware mean is to ask: how would you approach the spatial analysis challenges presented by GIS if that workstation on your desk were about 5000 times faster and bigger?

The criticism that few GIS end-users will ever be able to afford HPC hardware is irrelevant for two reasons. Firstly, what is possible using, for example, a mid 1990s national HPC research centre machine will within five years be affordable and 'do-able' using many workstations – even earlier if 'workstation farms' are used. Note that five years is probably also the lead time for the research and development cycle of new spatial analysis tools. Second, it is possible that the need for highly specialised but generic analytical functions will be met through the development of *embedded systems*. Embedded systems are unifunctional hardware that typically employ multiple CPUs. They are currently mainly used in signal processing. However, there is no reason why they cannot be programmed to perform specialised spatial analysis functions which may need large amounts of processing power. Such a system could appear to the GIS user as a call to a subroutine, except that the subroutine is in fact located somewhere else on the Internet and is not software but an integrated hardware and software

Table 1 Relative performance of a selection of available HPC hardware on a social science benchmark code in relation to 486 PC.

Hardware	Problem size: numbers of origin and destinations					
	Number of processors	100 by 100	500 by 500	1000 by 1000	10 000 by 10 000	25 000 by 25 000
Massively parallel						
Cray T3D	64	88				
	128		241	258		
	256			545	665	
	512				1335	1598
Parallel						
SGI Onyx	4	218	221	192	np	np
SGI Power Challenge	4	51	66	63	np	np
Vector supercomputer						
VPX240	1	162	195	196	np	np
Cray J90	8	8	35	39	np	np
Workstations						
SGI Indy	1	10	10	9	np	np
HP9000	1	14	12	10	np	np
Sun Ultra 2	1	18	17	16	np	np
Personal computers						
133MHz Pentium	1	3	4	4	np	np

Note: Benchmark problem sizes greater than 1000 by 1000 are not possible on a 486 PC. The times are estimated using linear interpolation which provides a good statistical fit to a range of smaller-sized problems.

system (see Maguire, Chapter 25). This form of heterogeneous distributed GIS is possible now and is one way of handling highly specialist but generic needs for spatial analysis functionality. The problem at present is that there is not yet a single example of such a system in operation, and not many ideas about the nature of the spatial analysis technology that should be run on them. On the other hand, the good news is that the languages and software tools needed to develop portable and future-proofed parallel applications are now quite well developed. Very significant here is the recent international standardisation of both a Highly Parallel FORTRAN (HPF) compiler and of the message passing interface (MPI).

3 A GEOCOMPUTATION PARADIGM FOR GIS

The rise of scaleable parallel hardware dramatically increases the opportunities within GIS for large-scale spatial analysis, using new approaches that seek to solve some of the traditional problems by switching to a more computationally intensive paradigm (see Openshaw 1997 for a review). There are now new ways of approaching spatial analysis using what has been termed a geocomputational paradigm. Geocomputation is itself a relatively new term, defined as the adoption of a large-scale computationally intensive approach to the problems of physical and human geography in particular, and the geosciences in general. Geocomputation is a paradigm that is clearly relevant to GIS, but also goes far beyond it. Spatial data manipulation on parallel supercomputing may well involve a return to flat data held in massive memory spaces, rather than recursive relational and hierarchically structured databases held on disk (see Worboys, Chapter 26). It also involves the development and application of new computational techniques and algorithms that are dependent upon, and can take particular advantage of, supercomputing. The motivating factors are threefold:

1 developments in HPC stimulating the adoption of a computational paradigm to problem solving, analysis, and modelling;

2 the need to create new ways of handling and analysing the increasingly large amounts of spatial information about the world stored in GIS;

3 the increased availability of AI tools and Computational Intelligent methods (Bezdek

1994) that already exist and are readily applicable to many areas of GIS (Openshaw and Openshaw 1997). Geocomputation also involves a fundamental change of style with the replacement of computational minimising technologies that reflect an era of hand calculation by a highly computationally intensive one. It also brings with it some grand ambitions about the potential usefulness that may well result from the fusion of virtually unlimited computing power with smart AI-based technologies that have the potential to open up entirely new perspectives on the ways in which we do geography and perform GIS applications (see Openshaw 1994a, 1995). This new emphasis on geocomputation is an unashamedly applied, problem-solving approach. The challenge is to create new tools which are able to suggest or discover new knowledge and new theories from the increasingly spatial data-rich world in which we live.

4 WHAT SORT OF HPC-POWERED GIS-RELEVANT SPATIAL ANALYSIS TOOLS ARE NEEDED?

A longstanding difficulty with GIS-Relevant Spatial Analysis (GRSA) is the lack of any consensus as to what it means, what its requirements are, what its users want now, and what its users would want if only they knew it were possible (or if the methods existed to stimulate demand). The situation has not improved much over the last five years (see Openshaw 1991). Far too often GRSA is equated with whatever old or new statistical technology a researcher happens to be familiar with, or with what a largely unskilled enduser thinks is required based on knowledge of what a proprietary GIS vendor provides. Yet GIS offers far more than a source of data that can be run with pre-GIS methods! 'Yes', that can be done, and 'yes' it can be useful; but GIS presents a much deeper challenge to spatial analysts. The question 'what kind of spatial analysis do researchers and academics want in GIS?' has to be tempered by the feasibility constraint of 'what kinds of spatial analysis can be implemented in or with GIS?' and the sensibility constraint 'what types of spatial analysis is it sensible to provide for GIS and its user community?'. Another set of general design constraints reflect other very important but hitherto neglected considerations, such as who are the likely users, what it is (in generic terms) that they want, and what sort of analysis technology they can handle given fairly low levels of statistical knowledge and training in the spatial sciences. Table 2 summarises many of the principal design questions. It is noted that the abilities of users are very important and the future viability of whatever technologies are proposed will ultimately depend on the extent to which the methods can be safely packaged for use by non-experts. There is a considerable mismatch between the criteria identified in Table 2 and the capabilities of existing spatial statistical and spatial modelling tools: for example, statistical packages are really of potentially very limited use outside of a research organisation and, in any case, they lack the power to cope with most of the analysis problems created by GIS.

Table 2 Basic design questions.

What kinds of spatial analysis are:
• relevant to GIS data environments?
• sensible given the nature of GIS data?
• reflect likely enduser needs?
• compatible with the GIS style?
• capable of being used by endusers?
• add value to GIS investment?
• can be regarded as an integral part of GIS?
• offer tangible and significant benefits?

A most important challenge is to identify and develop generic spatial analysis tools which are appropriate for use with spatial data in GIS environments. Table 3 shows the ten basic 'GISability' criteria that spatial analysis methods should ideally attempt to meet (Openshaw 1994e). It is important to recognise that GIS creates its own spatial analysis needs and that these needs make it a special and different sub-field of spatial statistics. It is within this context that spatial analysis tools need to be regenerated, rediscovered, or created anew. The present is a good time to tackle these problems.

The debate as to whether these methods should be accessed from within or without a GIS package is irrelevant. There is no reason to insist on only one

Table 3 Openshaw's 10 basic 'GISability' criteria.

1. Can handle large N values
2. Study region invariant
3. Sensitive to the nature of spatial data
4. Mappable results
5. Generic analysis
6. Useful, and valuable
7. Interfacing problems invariant
8. Ease of use and understandable
9. Safe technology
10. Applied rather than research-only technology

Source: Openshaw 1994f

form of integration or interfacing, leaving aside the obvious point that to be a GRSA tool the spatial analysis operation has to be called from, and ultimately end up within, a GIS environment. In an era of heterogeneous distributed computing there is no longer any need for all of the systems to be on the same machine (see Coleman, Chapter 22). Equally, the extent to which methods are perceived as having to be run within a GIS environment is often overplayed, since the special properties that a GIS can offer spatial analysis amount to little more than spatial data and consistently defined contiguity lists. Thus much system complexity can be avoided by the simple expedient of separating out the different components needed by the analysis process and developing a high-level system to call a GIS here, a model or analysis tool there, or a map drawer when one is needed.

These GRSA criteria can be converted into a series of researchable topics that would appear to have considerable relevance in the late 1990s. Participants at a workshop on 'New Tools for Spatial Analysis', held in Lisbon in 1993, were asked to think about the research themes that might be the most useful in the spatial analysis area. A summary of the suggestions that emerged after several hours of discussion and debate spread over a three-day period is given in Table 4. At present there is no funding to develop any of these themes, and this is one of the reasons why spatial analysis relevant to GIS is so backward.

Table 4 A research agenda for spatial analysis in the mid 1990s.

Theme 1: Toolkits for spatial analysis in GIS

Theme 2: Methods and tools for handling uncertainty in spatial data

Theme 3: Methods of automated and exploratory spatial analysis

Theme 4: Data driven modelling and data mining of GIS databases

Theme 5: Statistical aspects of model evaluation and choice

Theme 6: Confidentiality of spatial data and zone design

Theme 7: Impediments to the development and use of spatial analysis methods

Source. Adapted from Openshaw and Fischer 1995

5 A TYPOLOGY OF SPATIAL ANALYSIS TECHNOLOGIES

The developments in HPC environments and the increasing interest in geocomputation as a paradigm relevant to GIS offer a useful perspective on the current state of GRSA methods. A threefold typology is suggested:

1 Type One methods are based on computationally limited technology. Most conventional statistical methods are of this type. It is true that some require supercomputers to invert rank N matrices, where N is the number of spatial observations which can be very large. However, this is still computationally limited technology if, for example, there are only a small number of spatial origins, or if only a few variables can be handled.

2 Type Two methods are computationally intensive, but in a dumb manner. The early uses of supercomputers in spatial analysis resulted in the development of 'brute force'-based exploratory pattern and relationship detectors. For example, they include the Geographical Analysis Machines (GAM) of Openshaw et al (1987), and the GAM/K version of Openshaw and Craft (1991) who used Cray 1, 2, and Cray X-MP and Y-MP vector supercomputers to power a large-scale exploratory search. These methods were originally criticised by some spatial statisticians (Besag and Newell 1991) who suggested simpler variations. Subsequent testing appears to have demonstrated the superiority of the GAM/K

variant, as documented in Openshaw 1997b, although the results were five years late in being published! The Geographical Correlates Exploration Machine (GCEM) of Openshaw et al (1990) is another type of 'brute force' search for localised patterns and geographical associations. Since 1990, both methods can be run on workstations. A few variations on the GAM theme in particular have been suggested by other researchers – for example, Fotheringham and Zhan (1996) describe a procedure almost identical to part of the GAM. Note though that the quality of the results does depend on the purpose of the experiment or investigation. For example, if the objective is to test hypotheses then typically several million Monte Carlo significance tests with repeat replication may be required in order to handle multiple testing problems. The process is highly parallel but probably still needs the next generation of machine. However, if the objective is to use a GAM style of approach as a descriptive tool, indicating areas where to look or perform more detailed work, then this can be avoided. The advantages of this style of approach are essentially those of automating an exploratory spatial analysis search function as well as obviating the need to have prior knowledge about where to look for localised spatial patterning. Type Two methods were used in the first spatial analyses to recognise the importance of searching for localised patterning rather than global patterns.

3 Type Three methods are computationally intensive but also computationally intelligent. The difficulty with the Type Two methods is that as problem sizes increase (e.g. as a consequence of improved data resolution) and as the dimensionality of the data increases from two spatial dimensions to multiple data domains, then this technology breaks down. The answer is to switch to a smart search strategy. Openshaw and Cross (1991) described the use of a genetic algorithm to move hypercircles around a multidimensional map in the search for crime clustering. This technology has been developed further into suggestions for exploratory analysis that can operate in space, in time, and in multivariate data domains. The resulting database exploring creatures (termed Space Time Attribute Creatures – STACs) are described in Openshaw (1994e, 1995). Interpretation of the outcome of analysis can be aided by using computer animation to follow the search behaviour of the STACs as they go on a data pattern hunting safari. This has been developed into a prototype system called **MAPEX** (Map Explorer: Openshaw and Perrée, 1996). The hope is that the inexperienced or unskilled endusers can visualise and discover what is happening in their databases by viewing a library of computer movies illustrating different amounts and types of geographical patterning. The resulting user-centred spatial analysis system uses computationally intensive methods in order to ensure that the results that the users see have been processed to remove or at least reduce the effects of multiple testing and other potential complications. The users are assumed not to be interested in p-values and Type One or Type Two errors, but merely need to know only where the patterns are strongest and whether they can be trusted. The computer animation provides a useful communication tool. The expansion of this technology into the multiple data domains occupied by STACs and its integration into a standard GIS environment is currently underway. The zone design problem of Openshaw (1978) is another example of a Type Three problem which is of practical significance. Its wider application has been delayed until both digital boundaries were routinely available and HPC hardware speeds had increased sufficiently to make zone design a practical proposition.

The challenge for the geocomputational and HPC future is how to evolve more Type Three methods, which can handle rather than ignore the challenges of performing more intelligent spatial analysis, using computational and AI technologies. Spatial analysis needs to become more intelligent and less reliant on the skills of the operator, and this can only be achieved in the long term by a movement towards Type Three methods. The ultimate aim is to develop an intelligent partnership between user and machine, a relationship which currently lacks balance. Many of the statistical and computational components needed to create these systems exist. What has been lacking is a sufficient intensity of understanding of the geography of the problems and of the opportunities provided by GIS in an HPC era. A start has been made but much work still needs to be done.

6 SPATIAL ANALYSIS INVOLVING COMPUTATIONAL ZONE DESIGN

Geographers have been slow to appreciate the importance of spatial representation in their

attempts to describe, analyse and visualise patterns in socioeconomic data. The effects of scale and/or aggregation of zones upon the nature of the mappings that are produced are well known (Openshaw 1984; Fotheringham and Wong 1991). Yet this very significant source of variation is usually ignored because of the absence of methods in existing GIS packages that are able to handle it. Once this mattered much less, inasmuch as users had no real choice since they were constrained to use a small number of fixed zone based aggregations. GIS has removed this restriction and as the provision of digital map data improves, so users are increasingly exposed to the full range of possible zoning systems. Openshaw (1996: 66) explains the dilemma as follows: 'Unfortunately, allowing users to choose their own zonal representations, a task that GIS trivialises, merely emphasises the importance of the MAUP. The user modifiable areal unit problem (UMAUP) has many more degrees of freedom than the classical MAUP and thus an even greater propensity to generate an even wider range of results than before.' The challenge is to discover how to turn this seemingly impossible problem into a useful tool for geographical analysis. What it means is that the same microdata can be given a very large number of broadly equivalent but different spatial representations. As a consequence, it is no longer possible to 'trust' any display or analysis of zone-based spatially aggregated data that just happens to have been generated for a zoning system. Users have to start seriously worrying about the nature of the spatial representations contained in the zoning systems they use. It is argued that this problem mainly affects socioeconomic information where, because of confidentiality constraints and lingering data restrictions, attention is often limited to the display and analysis of data that has been spatially aggregated one or more times.

The only alternative to an 'as is' spatial representation is to develop zone design as a spatial engineering tool. The zone design problem can be formulated as a non-linear constrained integer combinatorial optimisation problem. Openshaw (1977) defines this so-called automated zone design problem (AZP) as optimise F(X) where X is a zoning system containing an aggregation of N original zones into M regions ($M<N$) subject to the members of each region being internally connected and all N zones being assigned to a region. The F(X) can be virtually any function that can be computed from the M region data: for instance, it could be a simple

statistic or a mathematical model that is fitted to the data. Typically, F(X) is non-linear, it could possess multiple suboptima, it need not be globally convex, and it is probably discontinuous (because of the contiguity and coverage constraints on X). Additionally, the user may wish to impose extra constraints either on the regions created by X (e.g. shape or size) and/or on the global data generated by the set of M regions (e.g. normality or spatial autocorrelation properties). This is achieved by converting F(X) into a penalty function. The Powell-Fletcher penalty function method has been found to be very effective (Fletcher 1987). This involves optimising the new function:

$$\Phi(X,\sigma,\theta) = F(X) + 0.5\sum_{i}^{N}\sigma_i(C_i(X) - \theta_i)^2$$

where

$\Phi(X,\sigma,\theta)$ is a penalty function that is dependent on X, σ_i and θ_i;

σ_i and θ_i are a series of parameters that are estimated to ensure gradual satisfaction of the constraints;

$C_i(X)$ is a constraint violation which is some function of the zoning system;

$F(X)$ is defined as previously.

This AZP was solved for small problems in the mid 1970s (Openshaw 1977, 1978). The technology was revived in the early 1990s when the increased availability of digital map data and the GIS revolution highlighted the importance of the problem. Larger datasets required better algorithms for optimising these functions. Openshaw and Rao (1995) compared three different methods and suggested that simulated annealing was the best choice. Unfortunately, they also found that simulated annealing took about 100 times longer to run than the other methods, while the use of additional constraints would have added another factor of 30 or so. Attempts were made to speed-up the simulated annealing approach by switching to parallel supercomputers. The immediate difficulty was the need for a fully parallel simulated annealing algorithm relevant to AZP types of problems. After considerable effort a hybrid simulated annealer with multiple adaptive temperatures controlled by a genetic algorithm was developed and was shown to work very effectively (Openshaw and Schmidt 1996).

It is now possible to use zone design to re-engineer all types of zoning systems with five principal areas of application:

1 to demonstrate the MAUP by seeking minimum and maximum function value zoning systems – this also helps prove the lack of simple minded objectivity in GIS;

2 to design zoning systems with particular properties that are believed to be beneficial for certain applications – for example, electoral re-districting or to minimise data confidentiality risks;

3 as a spatial analysis tool – for example, the zoning system acts as a pattern detector tuned to spot particular patterns;

4 as a visualisation tool – for example, to make visible the interaction between a model and the data it represents;

5 as a planning aid – for example, to define areas of maximum but equal accessibility or regions which are comparable because they share common properties.

These five application areas have been illustrated using the ARC/INFO-based Zone Design System (ZDES: Openshaw and Rao 1995; Alvanides 1995). The system attempts to routinise zone design using a number of generic zone design functions, and has been designed as a portable add-on to ARC/INFO.

7 ZONE DESIGN ANALYSIS OF SPATIAL DISTRIBUTIONS

This section provides an example of a geocomputation approach to zone design that demonstrates some of the potential capabilities of ZDES as a spatial analytical tool. Consider a problem that involves the analysis of non-white population data for England and Wales. This is currently of interest to some telecoms companies seeking to establish retail networks offering cut-price long-distance telephone calls. The 1991 Census data for persons born outside the UK may provide one surrogate indicator of this potential market demand for cheap long-distance calls. Figure 1(a) shows the 54 county zones that cover England and Wales while Figure 1(b) displays a choropleth map of the ethnic population. The key underlying geographical question is whether Figure 1(b) provides a 'meaningful' spatial representation of the Census data. Certainly, the visual patterns appear to identify some areas of apparent concentrations. Yet distortions introduced by differences in county sizes may blur some of the patterns and diminish local concentrations by averaging them out. The map is the outcome of highly complex distortion of the data, caused by its aggregation into counties. Additionally, the use of counties as the object of study introduces an arbitrary geography, since there is no reason to suppose that it has any relevance whatsoever to the factors governing the distribution of ethnic communities in the UK. Indeed the principal attraction of the county zone is the convenience and ease of using a standard geography! Different zoning systems may be expected to offer different levels of data distortion, and some may tell a different story as a consequence. It is interesting, therefore, to explore some of the alternative patterns of the same 1991 Census data aggregated from the underlying 9522 census wards as an exercise in spatial analysis by zone design.

First consider what happens if an attempt is made to create a new set of 54 different 'county' regions that have approximately equal ethnic population counts, and then remap the ethnicity rates (see Figures 2(a) and 2(b)). The constant shading of Figure 2(b) shows the effects of the equal size function – the three small light zones are formed by contiguity islands in the data. One problem is the very intricate zonal boundary patterns formed by the ZDES optimiser as it attempted to create regions of equal ethnic population size in Figure 2(a). In fact, the more efficient the zoning system optimiser becomes so the more intricate are the resulting boundaries because of the imposed constraint of equal population sizes. The simulated annealer used here was very successful in optimising the function, but produced extremely crenulated and irregularly-shaped regions. Perhaps this also says something about the fine-scale spatial distribution of the ethnic population at the ward level – that is, the pattern is concentrated but discontinuous hence the need to link widely separated areas together in order to meet the population size restriction. It is a matter for further research as to whether there is an optimal spatial scale (or level of aggregation) at which the zones suddenly become more regular. This would be an interesting question to try to answer. Another problem here is that there are potentially many different zoning systems that will yield areas of nearly, or approximately, equal population size. It may also be necessary to introduce shape constraints.

A more useful spatial analytic function would be some measure of the spatial dispersion of the population around a set of region centroids. This is broadly equivalent to a type of large location–allocation problem. The objective is to minimise the global sum of the population weighted distances to each of the ward centroids. This function is expressed as follows:

$$\text{minimise} \quad \sum_{j}^{M} \sum_{i \in j}^{N} P_i D_{ij}$$

where

P_i is the population value for ward i;

D_{ij} is the distance from ward i to the population centroid of region j of which i is a member. (Note that this centroid depends on the current membership of region j.)

The restriction on the second summation indicates that the summation only occurs for wards that belong to region j. This is a way of partitioning the zoning system of N Census wards into M regions.

Figure 3 shows the results of minimising this local spatial dispersion function. This map is the outcome of a contest between different parts of the UK, as the ZDES algorithm seeks to trade-off population gains in some parts of the country against losses in others. The resultant zoning system is thus a visualisation of the tension (or interaction) between the objective function, the Census ward zonation and aggregation effects. Those parts of the UK with the largest ethnic populations have relatively small regions. The map in Figure 3(b) seems to offer a more sensitive view of the distribution of ethnic populations compared with Figure 1(b), and it highlights some of the areas where aggregation effects have seemingly removed some of the patterns. This approach is developed further in Figure 4(a), which shows the results for the same objective function but subject to constraints that the total population of all the regions should be at least 75 per cent of the average. This involves solving a penalty function version of ZDES. On this application the simulated annealer needed a Cray T3D to produce the results within a convenient time – that is, one hour. The patterns in Figure 4(b) are considerably more disjointed than in Figure 3(b) with more intricate boundary resolution in the principal areas of ethnic concentration.

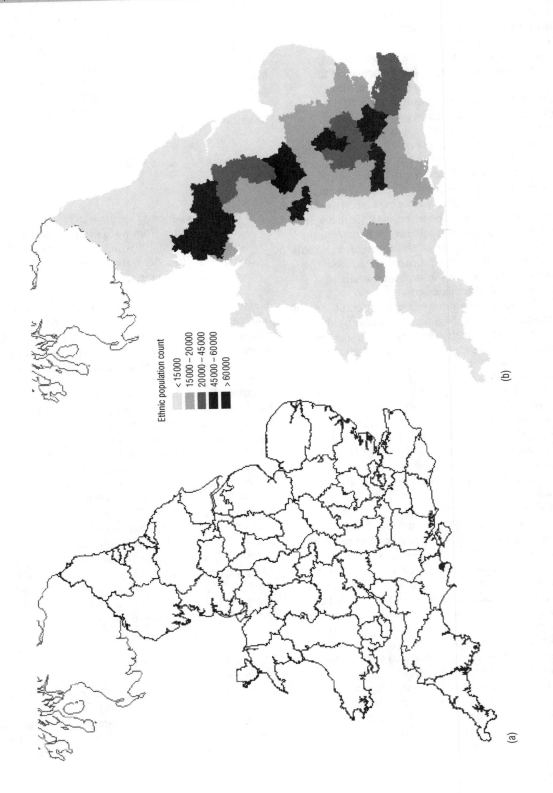

Ethnic population count

< 15000
15000 – 20000
20000 – 45000
45000 – 60000
> 60000

(b)

(a)

Fig 1. (a) The 54 counties in England and Wales; (b) distribution of ethnic population for counties in England and Wales.

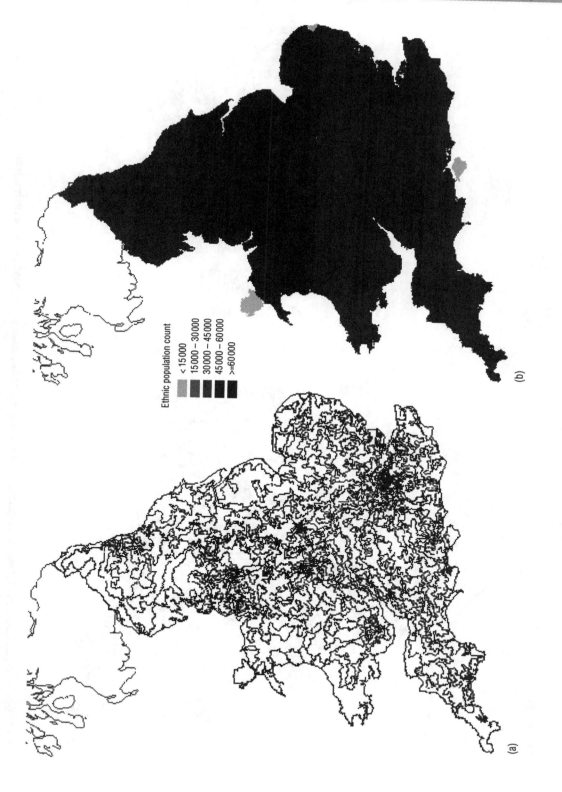

Fig 2. (a) Equal ethnic population regions in England and Wales; (b) distribution of ethnic population in England and Wales for equal population regions.

Ethnic population count

< 15 000
15 000 – 30 000
30 000 – 45 000
45 000 – 60 000
>=60 000

(a)

(b)

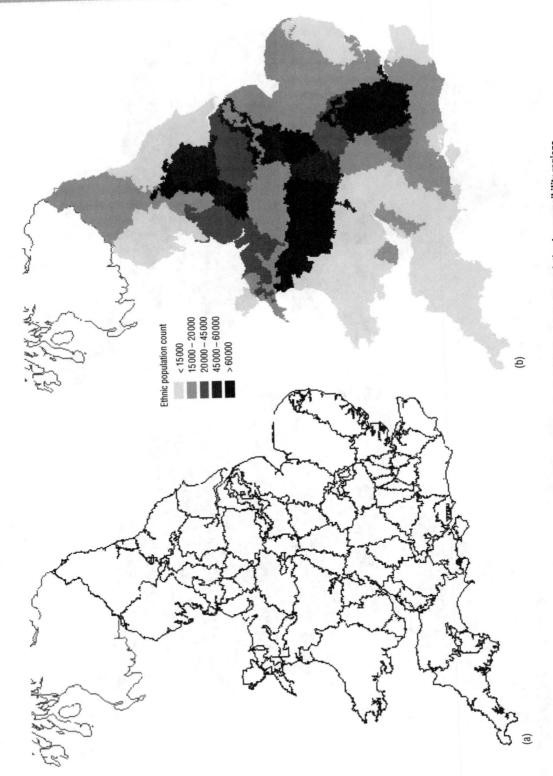

Ethnic population count

< 15000
15000 – 20000
20000 – 45000
45000 – 60000
> 60000

(b)

(a)

Fig 3. (a) Regions in England and Wales that minimise population weighted distances; (b) distribution of ethnic population for accessibility regions.

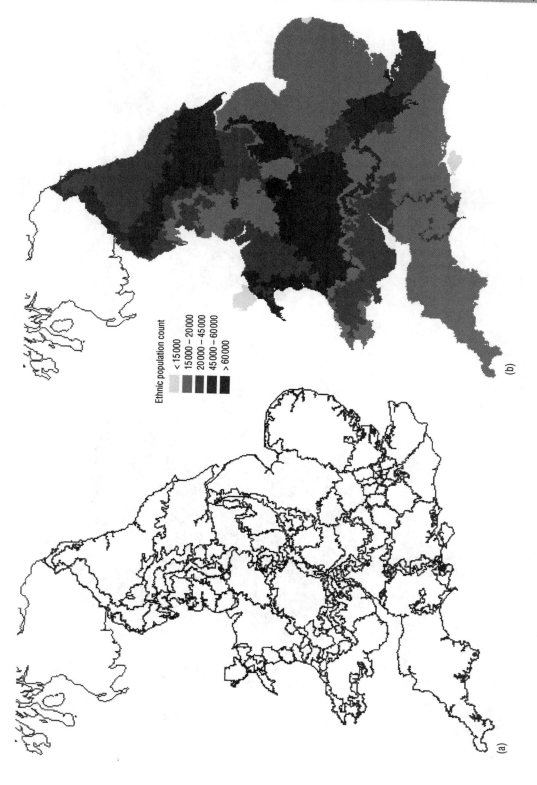

Ethnic population count

- < 15 000
- 15 000 – 20 000
- 20 000 – 45 000
- 45 000 – 60 000
- > 60 000

(b)

(a)

Fig 4. (a) Regions in England and Wales that minimise constrained population weighted distances; (b) distribution of ethnic population for constrained accessibility regions.

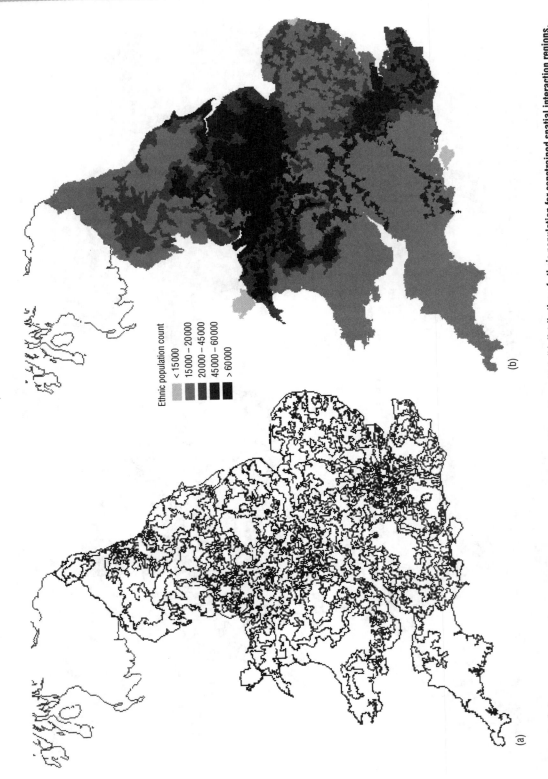

Ethnic population count

< 15000
15000 – 20000
20000 – 45000
45000 – 60000
> 60000

(b)

(a)

Fig 5. (a) Regions in England and Wales that minimise a constrained spatial interaction model; (b) distribution of ethnic population for constrained spatial interaction regions.

A final demonstration involves converting the previous objective function into a measure of population potential as a form of crude spatial interaction model. The function is as follows:

$$\text{minimise} \quad \sum_{j}^{M} \sum_{i \in j}^{N} P_i D_{ij}^{-2}$$

subject to minimum size constraints. The resulting zonation is shown in Figure 5(a). The region boundaries are once again very crenulated, especially in areas of high ethnic populations. The resulting population distribution is mapped in Figure 5(b). Once again the underlying impression is that of local area concentrations surrounded by gaps. This may be a reflection of areas of localised negative spatial autocorrelation. The deficiency of the administrative scheme, represented here by the county geography, is demonstrated by a series of descriptive statistics. Table 5 shows some statistics concerning the ethnic population per zone for each of the different models illustrated earlier. The administrative county geography scores worse than any other scheme with zones extending beyond 700 000 ethnic residents and an extreme standard deviation figure (113 750). The equal ethnic population geography demonstrates the best statistical properties for a representative administration scheme, with a maximum ethnic population very close to the mean ethnic population value (54 650 residents). However, the boundaries of the zones are rather obscure for any policy making exercise and the same problem occurs with a constrained interaction geography, as we saw in Figures 3(a) and 5(a) respectively.

A trade-off between a comprehensible geography and satisfactory statistical properties is the constrained accessibility model illustrated in Figure 4(a). In this case the weighted distance minimisation function works as a shape constrained thus producing relatively compact zones, while the population constraint function restricts the maximum population of zones to

176 200 and retains the standard deviation low enough for the zones to be comparable. Possibly the 'best' zoning systems that would represent this type of socioeconomic data would be very different from the conventional administrative zoning scheme and this forms the subject of ongoing research (Openshaw and Alvanides 1997b).

8 CONCLUSIONS

The case for developing a wide range of GRSA tools is very compelling. GIS is creating an immensely data-rich environment. The technology for spatial data capture, management, and handling has far outstripped the available tools for its analysis. The hope now is that it may be possible to compute our way out of the data swamp by developing new generations of intelligent spatial analysis tools which are better able to cope with the conflicting requirements of large volumes of data, geographical reasonableness, and the endusers. The need is undeniable and there are various ways forward involving the use of zoning systems as data, pattern, and model visualisers; and the development of smart GIS database explorers. The technological basis for these exists: HPC, AI toolkits, computational statistics, large spatial databases, and well-developed GIS. Powerful zone design algorithms have been developed. Artificial life-based 'creatures' or 'agents' can be created that are able to move around space-time-attribute GIS databases under their own control in an endless search for patterns and relationships of possible interest. Computer animation provides the basis for users 'watching' what is happening in highly complex geo-cyberspaces projected on to a 2-dimensional map (Openshaw and Perrée 1996). The basic methods will probably run on a workstation, but as the complexity of the data domain increases or as greater use is made of computational statistics in order to improve performance, so they will need HPC. There is a suggestion that the evolution of new types of spatial analysis technology is about to start.

The challenge then is to solve the principal outstanding spatial analysis problems by developing various geocomputational approaches, to demonstrate they work on a range of generic problems, and then to make them available either within or without current GIS software. This task is becoming increasingly urgent and requires geographers (in particular) to use their understanding of the geography

Table 5 Statistical properties of different geographies for ethnic population.

Geography	Maximum	Mean	Standard deviation
Administrative counties	703 700	46 850	113 750
Equal ethnic population	58 650	54 650	13 200
Population accessibility	320 900	54 650	74 700
Constrained accessibility	176 200	55 700	39 850
Constrained interaction	273 200	55 700	51 500

of the problems to create a new approach to developing geographical analysis methods. A start is being made (here as well as elsewhere in this book) but much still remains to be done.

References

Alvanides S 1995 'The investigation of a Zone Design System for reconstructing census geographies'. Unpublished dissertation 3031 available from School of Geography, University of Leeds

Besag J, Newell J 1991 The detection of clusters in rare diseases. *Journal of the Royal Statistical Society* A 154: 143–55

Bezdek J C 1994 What is computational intelligence? In Zurada J M, Marks R J, Robinson C J (eds) *Computational Intelligence: Imitating Life*. New York, IEEE: 1–12

Birkin M, Clarke G P, Clarke M, Wilson A G (eds) 1996 *Intelligent GIS: location decisions and strategic planning*. Cambridge (UK), GeoInformation International

Fletcher R 1987 *Practical Methods of Optimisation*. Chichester, John Wiley & Sons

Fotheringham A S, Wong D W S 1991 The modifiable areal unit problem in multivariate statistical analysis. *Environment and Planning A 23*: 1025–44

Fotheringham A S, Zhan F B 1996 A comparison of three exploratory methods for cluster detection in spatial point patterns. *Geographical Analysis* 28: 200–18

Hillis W D 1992 What is Massively Parallel Computing and Why is it important? In Metropolis N, Carlo Rota G (eds) *A new era in computation*. Cambridge (USA), MIT Press: 1–15

Openshaw S 1977 A geographical solution to scale and aggregation problems in region-building, partitioning, and spatial modelling. *Transactions of the Institute of British Geographers* 2: 459–72

Openshaw S 1978 An optimal zoning approach to the study of spatially aggregated data. In Masser I, Brown P J B (eds) *Spatial representation and spatial interaction*. Boston (USA), Martinus Nijhoff: 95–113

Openshaw S 1984 *The modifiable areal unit problem*. Concepts and Techniques in Modern Geography 38. Norwich, Geo-Books

Openshaw S 1991c Developing appropriate spatial analysis methods for GIS. In Maguire D, Goodchild M F, Rhind D (eds) *Geographical information systems: principles and applications*. Harlow, Longman/New York, John Wiley & Sons Inc. Vol. 1: 389–402

Openshaw S 1994a A concepts-rich approach to spatial analysis, theory generation, and scientific discovery in GIS using massively parallel computing. In Worboys M F (ed) *Innovations in GIS 1*. London, Taylor and Francis: 123–38

Openshaw S 1994b Computational human geography: towards a research agenda. *Environment and Planning A* 26: 499–505

Openshaw S 1994d GIS crime and spatial analysis. *Proceedings of GIS and Public Policy Conference*. Ulster, Ulster Business School: 22–35

Openshaw S 1994e Two exploratory space-time-attribute pattern analysers relevant to GIS. In Fotheringham A S, Rogerson P (eds) *Spatial Analysis and GIS*. London, Taylor and Francis: 83–104

Openshaw S 1994f What is GISable spatial analysis? In *New Tools for Spatial Analysis*. Luxembourg, Eurostat: 157–62

Openshaw S 1995 Developing automated and smart spatial pattern exploration tools for geographical information systems applications. *The Statistician* 44: 3–16

Openshaw S 1996a Developing GIS relevant zone based spatial analysis methods. In Longley P, Batty M (eds) *Spatial analysis: modelling in a GIS environment*. Cambridge (UK), GeoInformation International: 55–73

Openshaw S 1997 Supercomputing in Geographical Research. *Proceedings of Conference on IT in the Humanities*. London, British Association

Openshaw S, Alvanides S 1997 Designing zoning systems for representation of socioeconomic data. In Frank I, Raper J, Cheylan J (eds) *Time and motion of socioeconomic units*. GISDATA series. London, Taylor and Francis

Openshaw S, Charlton M, Wymer C, Craft A 1987 A mark I Geographical Analysis Machine for the automated analysis of point datasets. *International Journal of Geographical Information Systems* 1: 335–58

Openshaw S, Craft A 1991 Using Geographical Analysis Machines to search for evidence of clusters and clustering in childhood leukaemia and non-Hodgkin Lymphomas in Britain. In Draper G (ed) *The geographical epidemiology of childhood leukaemia and non-Hodgkin lymphomas in Great Britain, 1966–83*. London, Her Majesty's Stationery Office: 89–103

Openshaw S, Cross A 1991 Crime pattern analysis: the development of Arc/Crime. *Proceedings of AGI Annual Conference*. Birmingham (UK)

Openshaw S, Cross A, Charlton M 1990 Building a prototype geographical correlates exploration machine. *International Journal of Geographical Information Systems* 3: 297–312

Openshaw S, Fischer M M 1995 A framework for research on spatial analysis relevant to geostatistical information systems in Europe. *Geographical Systems* 2: 325–37

Openshaw S, Openshaw C A 1997 *Artificial intelligence in geography*. Chichester, John Wiley & Sons

Openshaw S, Perrée T 1996 User-centred intelligent spatial analysis of point data. In Parker D (ed.) *Innovations in GIS 3*. London, Taylor and Francis: 119–34

Openshaw S, Rao L 1995 Algorithms for re-engineering 1991 Census geography. *Environment and Planning A* 27: 425–46

Openshaw S, Schmidt J 1996 Parallel simulated annealing and genetic algorithms for re-engineering zoning systems. *Geographical Systems* 3: 201–20

Openshaw S, Schmidt J 1997 A Social Science Benchmark (SSB/1) Code for serial, vector, and parallel supercomputers. *International Journal of Geographical and Environmental Modelling*

Turton I, Openshaw S 1996 Modelling and optimising flows using parallel spatial interaction models. In Bougé L, Fraigniaud P, Mignotte A, Roberts Y (eds) *Euro-Par 96 Parallel Processing Vol. 2* Lecture Notes in Computer Science (1124). Berlin, Springer: 270–5

Introduction

THE EDITORS

We have already described how the real and absolute costs of computing have continued their precipitous fall of recent years, and how the attendant developments in computer graphics and visualisation have provided necessary (although not in themselves sufficient) conditions for putting GIS principles into practice. Yet this is only part of the story, and here in this Technical Issues part of the book we will explore changes in GIS architecture, issues of data collection and database management, and developments in the ways in which data are transformed and linked together. We begin with an extended discussion of the many ways in which GIS are configured, and the wide range of interactions with computers in general and GIS in particular.

In the opening chapter, Michael Batty provides a succinct yet wide-ranging review of the development of computer technology, in which he emphasises the unpredictability of change and our inability to anticipate it – a point we return to below with respect to the technical predictions made in the first edition of this book. He sees technology as also providing a stimulus to the proliferation of digital data sources, and the consequent increased richness of digital analysis. It seems inevitable that the pace of technical change will continue to accelerate, and that computation will continue to be a fast-changing and diversifying medium. Batty's own general predictions are that data-rich computing will soon be distributed across global networks, that decentralisation of software will bring changes in the way in which it is both licensed and used, that there will be rapid technical advance in voice input and output, and that there will be knock-on consequences for the development of new kinds of virtual realities.

The prospects for GIS are equally dramatic. The vendors of proprietary systems have constantly to adapt to change by packaging software and data into niche market solutions, by facilitating data exchange (particularly across networked environments), and by developing novel approaches to business solutions and consulting.

Software and information exchange is crucial to these technical developments. One of the most far-reaching (yet unanticipated – even in the first edition of this book!) changes in recent years has been the development of data and software transfer across the Internet. David Coleman describes in Chapter 22 how sophisticated network environments and browser systems have developed from the early distributed computing systems of local- and wide-area networks.

An accompanying development has been that as software converges across platforms, so geographical components are becoming more pervasive within general-purpose software – for example, spreadsheets now have GIS capability and vice versa. This is seen most clearly in the development of so-called 'desktop' GIS (Elshaw Thrall and Thrall, Chapter 23) which has come to embrace a range of general-purpose application and consumer mapping products, including digital atlases, digital gazetteers, geographically-enabled spreadsheets, and thematic mapping products. Such 'shrink-wrapped' products lack most, or even all, of the analytical capabilities of 'true GIS', yet are clearly geographical software products. In these circumstances, any distinct identity of GIS inevitably begins to blur. Taken together with the developments in networking, this suggests that GIS is becoming both more specialised in its range of possible applications, and more pervasive in its wider usage as a background technology. The broad picture is that software is both breaking up – fragmenting on the desktop so that users can construct all kinds of tailor-made applications from individual elements – and coming together – in that vendors are providing non-traditional functions within their traditional software, 'hooks', and linkages to other related software and openness to networked environments.

The implication of all of this is that software will emerge which is extremely basic to the computational environment and that programming languages will develop which enable this software to be connected in diverse ways. This is almost a full circle back to the early days of computer cartography and GIS when researchers wrote their own FORTRAN programs, although the elemental building blocks are very different. Now, there is the prospect of GIS users assembling reusable software 'modules' using the highest of high-level languages, in order to develop highly customised solutions to their problems. Already this is possible at a somewhat lower level using the various macro languages which are available within GIS. In the future, graphics, data elements, routing algorithms, modelling methods, and so on are likely to be packaged in whatever manner the user requires, using software which simply exists within appropriate environments – possibly based on networks rather than on desktops. The vendor response to this has been recognition of the need for vendor-neutral computing standards, and so-called 'open systems'. This is the theme of 'GIS interoperability' which is explored by Mark Sondheim, Kenn Gardels, and Kurt Buehler in Chapter 24.

There have been other sea changes in the organisation of the GIS industry. For example, the longest-established GIS companies, such as Intergraph, were initially primarily hardware developers, but are now more concerned with software and consultancy provision; while more recent entrants, such as ESRI, are increasingly involved with the development of desktop systems, network platforms, consulting, and data provision. Overlain onto this is the drive across the industry to develop new niche markets around reusable software modules. This is leading GIS vendors to work closely with their clients to develop customised system specifications, designs, and implementations, with or without the assistance of in-house consultancy services. The means and methods of GIS customisation in a wide range of applications are discussed here by David Maguire (Chapter 25).

This is the second edition of *Geographical Information Systems*, and it is perhaps appropriate to contrast the contemporary technological setting with the first edition's projections for 'GIS 2000' (Rhind et al 1991: 320–2). The first edition did, of course, draw attention to the changes in computer architectures consequent upon the growth of computer power – specifically, the shift away from mainframe and mini computers to PCs and workstations – but did not anticipate the development of the Internet. A second prediction was that the diffusion and wide uptake of GIS would likely lead to dramatic falls in real software

prices. While we might quibble whether the prediction of a 'fully functional GIS for about $500 in the mid 1990s' was strictly speaking realised, each of the contributions to this section demonstrates that GIS has become a more affordable and routine technology. Indeed if our definition of GIS is drawn more loosely than Susan Elshaw Thrall and Grant Thrall's (Chapter 23) 'true GIS' (e.g. to include Microsoft and MapInfo GI products), the number of systems actually in use by the year 2000 now seems set to exceed the first edition prediction of 580 000. The range of application areas has continued to diversify, to the point at which it is perhaps more instructive to comment on areas in which GIS has not been used than to enumerate those in which it has: this is a theme that is picked up in the Applications part to this edition.

Which new technologies were not anticipated in the first edition? As suggested above, it failed entirely to predict the emergence of the Internet and the World Wide Web from its early civilian use in inter-university electronic mail. Second, there was at best only a hazy conception of the many ways in which GIS functionality would be packaged into specific 'shrink-wrapped' applications for the current generation of low-cost, application-specific desktop systems. Third, and related to this, there was still a sense in the first edition of even the most rudimentary GIS applications being the preserve of 'GIS specialists'. The book failed to anticipate the degree that this would be overcome by the development of vastly improved graphical user interfaces. Fourth, there was perhaps an over-emphasis upon the anticipated differentiation of proprietary products at the expense of the drive towards GIS interoperability, a development which has gained impetus with network-based data transfer and the use of the Internet as a GIS platform. Fifth and finally, these technological changes have led to organisational changes in the ways in which GIS is customised to specific applications, and the development of new application tools. These themes are all developed here in this section of the second edition.

Reference

Rhind D W, Goodchild M F, Maguire D J 1991 Epilogue. In Maguire D J, Goodchild, M F, Rhind D W (eds) *Geographical information systems: principles and applications.* Harlow, Longman/New York, John Wiley & Sons Inc. Vol. 2: 313–27

22

Geographical information systems in networked environments

D J COLEMAN

This chapter examines basic approaches to distributed architectures in GIS, with special emphasis on exploitation of local- and wide-area networks and, more recently, the Internet and wireless communications. The first section reviews the design models inherent in host-based systems, distributed networks, and field-based systems. The second section describes selected applications of these design models, with particular emphasis placed on Internet-based geographical information retrieval and computer-supported cooperative work. The final section deals with system performance considerations in networked environments and describes new developments affecting the performance of future systems.

1 INTRODUCTION

Technical design models supporting mainstream GIS technology have evolved from early host-based efforts on mainframe and mini-computers, through stand-alone systems operating on personal computers and workstations, and on to today's distributed computing environments across local- and wide-area networks. Each advance has extended overall flexibility in terms of the relative location of users, processing capabilities, and data storage units (see also Batty, Chapter 21). Recent advances in broadband and wireless communications technologies – as well as the dramatic increase in Internet usage and extension of Internet browsing technology – promise to extend further the reach and range of GIS users working in offices or laboratories, in the field or at home.

This chapter examines the basic approaches to distributed architectures in GIS, with special emphasis on the exploitation of local- and wide-area networks and, more recently, the Internet and wireless communications. The first section of the chapter reviews the design models inherent in host-based systems, distributed networks, and field-based systems. The second section describes specific applications of the latest design models, with particular emphasis placed on emerging issues

associated with Internet-based geographical information retrieval and computer-supported cooperative work. The final section of the chapter deals with the issue of system performance in networked environments under a variety of conditions. After a brief summary of considerations involved in objectively assessing system performance, the author describes new developments affecting the performance of future systems.

2 HOST-BASED SYSTEMS AND EARLY PERSONAL COMPUTING ENVIRONMENTS

2.1 Host-based systems

Early centralised computing environments were characterised by small numbers of large-scale mini- or mainframe computers, with shared storage devices attached via hardware input/output channels and multiple users connected via terminals possessing varying levels of on-board 'intelligence' or processing power (Katz 1991). Such environments defined the architecture for most major data processing and information systems applications until the mid 1980s, including the GIS installations found in major forestry organisations, utilities, municipalities, and land records management programmes.

This architecture implied greater control over data integrity and system security; the database was managed centrally, with responsibility for system and data administration entrusted to experienced data processing specialists. However, performance of such systems would often degrade in unpredictable ways when growing numbers of users demanded computing resources and database access. Conflicts with system administrators over development and maintenance priorities also often resulted in dissatisfaction among endusers in many large organisations.

2.2 Stand-alone, PC-based systems

By 1986, PC-based GIS software packages had begun moving geoprocessing out of the hands of information system managers (Miller 1990). Besides their low cost, these systems offered more predictable response times since the user was the only one on the system. However, the proliferation of stand-alone PC-based systems meant it was much more difficult to share data among several different people in the organisation. Also, the enduser often had to become his or her own system and database administrator. While PC-based systems undoubtedly accounted for the dramatic growth in GIS usage through the late 1980s, they also put greater onus on managers in large organisations to keep effective track of the data being collected and processed by an increasing number of endusers with little experience in routine data management procedures.

3 COMPUTER NETWORKING AND DISTRIBUTED SYSTEMS

Beginning in the mid 1980s, higher-performance workstations connected through local and wide area networks (LANs and WANs) became a viable alternative to host-based and stand-alone configurations. Connectionless LAN and 'LAN-interconnect' services began displacing earlier connection-oriented services which required a dedicated link between the enduser and the host computer (Figure 1).

Based on packet-switching transmission protocols which did not require dedicated connections between user and host, networking permitted users to share access to scarce equipment resources (e.g. printers, plotters, databases etc.), made possible inter-site communication applications like electronic

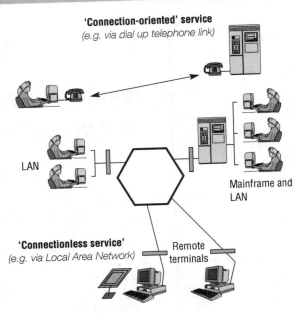

Fig 1. Examples of alternative wide-area networking services: 'connection-oriented' vs 'connectionless' models.

mail and file transfer, and enabled distributed processing. It also allowed users to expand their facilities with a degree of vendor independence and to incorporate special purpose processors, storage units, or input/output devices as required. The term *distributed computing* was coined to describe a situation where processing tasks and data are distributed among separate hardware components connected by a network, with all these various components capable of being accessed in a relatively transparent manner (Champine et al 1980).

Distributed systems today generally offer a combination of: (1) greater access to GIS data stored and managed on a central server; (2) faster response time due to local computing; (3) tighter system security; (4) less complexity; and (5) in many cases, lower-cost computing solutions than more traditional mainframe or minicomputer solutions. At a higher level, the notion of distributed computing often provides a better fit to the complex structures and often multidisciplinary nature of modern organisations and offers greater user involvement in information management activities.

Early investigators of formal network-based approaches to distributed GIS database management systems (DBMS: e.g. Ezigbalike et al 1987; Webster 1988) laid out excellent foundations for future research, but were limited by both the distributed DBMS technology then commercially available at reasonable cost, and the relatively limited use of LANs or WANs by the GIS community at that time. The subject has been revisited more recently by Ingoldsby (1991) and Laurini (1994), among others.

3.1 The client–server model

Most distributed computing today is based on a client–server architecture (Katz 1991). In this model, a collection of workstations (or *clients*) relies on one or more *servers* residing elsewhere on the network for access to data files, application software and, in certain cases, more powerful computing resources. Such servers are really high-volume storage devices with processors which have been optimised to provide high-speed retrieval of large disk-based data or database files. In such an environment, the data retrieval aspects of a database query can be carried out largely independently of the data processing and display tasks.

The time required to execute such a data retrieval operation depends on the individual performance of the storage, processing, and communications components involved in the client–server system. While improvements in all three of these technologies have made key contributions, it has been the introduction of remote file management services which finally provided the transparency required for distributed computing. Remote file system implementations like Sun Microsystems' Network File System (or NFS) on UNIX workstations, PC-NFS and Novell's Netware on PC-compatibles, and Appletalk on Macintoshes are now present in many organisations.

3.2 Network architecture

A computer network architecture can be considered as a set of functions, interfaces, and protocols which enables devices to communicate with one another on-line. The architecture is composed of a layered collection of communication, networking, and application functions implemented such that – while each layer is designed to operate independently – higher-level operations are built on functions provided by the lower layers (Chorafas 1980).

Network protocols are the formal sets of rules or specifications for coding messages exchanged

between two communication processes on a network (Voelcker 1986). Protocols govern data control and format across a network, and a variety of protocols exist to ensure that these communications are conducted effectively.

Two different stacks or suites of layered approaches are in common use today:

- OSI: the seven-layer Open System Interconnect suite of protocols developed by the International Standards Organisation;
- TCP/IP: the four-layer Transmission Control Protocol/Internet Protocol suite originally developed for the ARPAnet research network in the USA.

Originally developed by computer user groups and European telephone companies, the OSI model helped unify world telephony and provided a clear framework and explanation of the functions required for computer communications. However, it was regarded by some as being too cumbersome for high-speed networks (Wittie 1991).

By comparison, the TCP/IP suite of protocols became a *de facto* standard by the early 1980s as a result of its early use in the implementation of the US Defense Department-funded Internet. Its longer-term popularity was secured through subsequent bundling with the 1983 release of the Berkeley UNIX 4.2 operating system. While TCP/IP protocols do not precisely fit into the more general OSI model, the functions performed by each OSI layer correspond to the functions of each part of the TCP/IP protocol suite and provide a good framework for visualising the respective relationships between the various protocols.

3.3 Local area networks

LANs physically or logically connect together multiple workstations, terminals, and peripheral devices via a single cable or shared medium (Pretty 1992). Through the 1980s alone, over 100 000 LANs were set up in offices and laboratories around the world to link workstations to printers, share files and send electronic mail (Wittie 1991). LAN usage in the general computing community grew at a rate of 80 per cent per year between 1985 and 1991 (Pretty 1992), with networks extending into schools, libraries, laboratories, and offices around the developed world using telephone lines, optical fibres, and satellite links.

Several accepted and standardised types of LAN technology now share the market, including IEEE 802.3 (CSMA/CD or Ethernet), IEEE 802.4 (Token Bus), IEEE 802.5 (Token Ring), and ANSI FDDI (fibre distributed data interface). To date, the Ethernet (Figure 2) and Token Ring technologies have dominated the market.

3.4 Metropolitan and wide area networks

Until the early 1990s, most wide-area networking and LAN-interconnect services did not deliver the performance required either to move large GIS data files quickly or to offer real-time access to remote users over very long distances (Craig et al 1991). Where data delivery was an issue, magnetic tapes and diskettes were viewed as the media of choice for the distribution or exchange of data among GIS data producers and users (Newton et al 1992).

Modern metropolitan area networks (MANs) and WANs employ different protocols and technology in order to offer speeds comparable to LANs while operating over greater distances. By 1992, fibre-based packet switching services like SMDS in North America and FASTPAC in Australia began providing high-speed (34–45 Mbit/sec) links between LANs across and between major cities. Today,

dedicated FDDI networks connect users within limited areas at rates up to 100 Mbit/sec, and higher-speed asynchronous transfer mode (ATM) services promise to support a wide range of real-time multimedia applications across long distances.

Results of the performance testing experiments described by Coleman (1994) indicated that future users of these broadband services would be able to access remote disks, processors, and output devices at near-LAN levels of performance. Today, such services have already influenced the nature of data delivery and software support practices to a small but growing group of users in the GIS community (e.g. Annitto and Patterson 1995; Streb 1995).

3.5 The Internet

Especially since 1993, it has been the connectionless services built atop the Internet which have come to define the current paradigms for mass market network usage. The Internet is a collection of interconnected campus, local, provincial, national, and corporate networks in more than 150 countries. Originally designed to connect researchers in university, government, and industrial defence establishments, the Internet is now estimated to

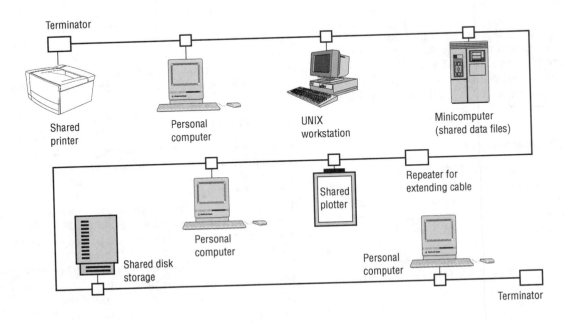

Fig 2. Simplified Ethernet local area network.

service more than 20 million users through a loose collection of more than 5000 registered networks (Reinhardt 1994; Thoen 1995). While electronic mail, file transfer, and remote login services originally accounted for the majority of Internet traffic, it has been ease-of-use and the multimedia capabilities of the World Wide Web (WWW) that have attracted most of the mass-market attention and resultant increases in usage. Over 145 000 Web sites have appeared since its introduction in 1993, and this number is continuing to rise monthly (Webcrawler Survey 1996).

3.6 Wireless, field-based systems

Recent advances in wireless communications, notebook computing, and the integration of GIS and global positioning system (GPS) technology have extended the application of traditional GIS databases into field operations (Lange and Gilbert, Chapter 33). Using normal telephone lines and, more recently, cellular telephone modems, field users tying into enterprise-wide networks may now access information resources previously available only from within the branch office or headquarters. New implementations are extending functionality beyond traditional records management and modelling roles into such areas as property inspection, field updating, facilities maintenance, customer service, and emergency response (Elliot 1994).

At least one industry source predicts the mobile computer/mobile field automation market will grow from $30 billion in 1995 to $80 billion by the year 2001 (FieldWorks Inc. 1996). FieldWorks estimates the 1995 total market for 'ruggedised' portable computers at $500 million with an annual growth rate of 35 per cent.

This predicted growth in the ruggedised portable market is based on current trends within large service firms and government organisations to automate the data collection and communication capabilities of field personnel through greater use of reliable mobile computers. As growth in notebook over desktop sales suggests, purchasers want systems which combine the functionality of desktops with the ability to fulfil specific applications beyond word processing and spreadsheet manipulation. The integration and use of GIS software and GPS technology within mobile computers for field data collection and database updating is one example of such extended functionality.

4 APPLYING THESE NEW MODELS

4.1 Enterprise computing implementations

Large utilities and municipalities were among the first to identify the operational requirements to integrate smaller GIS- and facilities management-related databases on stand-alone systems with larger corporate databases residing on mainframes (Popko 1988). As LANs became more widespread, enterprise computer systems evolved from host-based (or 'single-tier') to two-tier environments with networked PCs and workstations in a client-server architecture replacing connected terminals. This modular approach is attractive to many organisations since it allows them to take quicker advantage of price/performance improvements on modular hardware components and the other advantages of network computing described earlier (Mimno 1996).

While advantageous in many respects, users have found that these two-tier environments may not necessarily retain some of the advantages found in centralised host-based implementations, notably availability, expandability, and reliability of service (Strand 1995). As a result, newer three-tier environments have been developed to facilitate the placement of applications and data in locations in order to optimise these three factors (see Figure 3).

Existing mainframes at Tier 3 – possibly connected over long distances via MANs and WANs – are used to support legacy applications and provide access to large databases. To implement a client/server computing strategy, additional layers of hardware and software are added as a front end to the host computer. These layers consist of shared servers at Tier 2 (interconnected by high-speed LANs), and LAN-based PCs and laptop computers at Tier 1 (Mimno 1996). GIS and desktop mapping applications are included at this first tier, although the data may reside on either the Tier 2 servers or even the Tier 3 hosts.

This enterprise information architecture is highly flexible and can be modified easily to accommodate changes in business requirements. As more large customers adopt this approach, commercial GIS software firms and mainstream DBMS vendors alike are modifying their offerings in response to customer demands. In particular, spatial data management models and processes are becoming a more intrinsic component of an organisation's larger information management architecture. ESRI's Spatial Database Engine and MapInfo's 1996

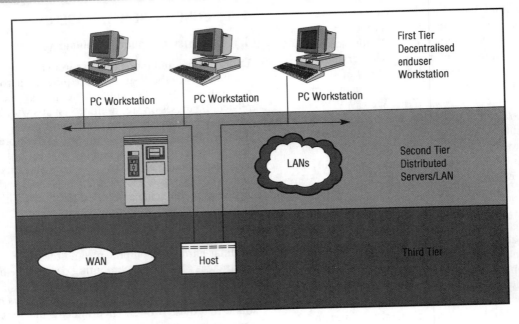

Fig 3. Three-tier enterprise computing environments.
Source: Strand 1995

acquisition of the Spatialware server technology from Unisys Corporation are two examples of how GIS vendors are providing even tighter high-performance links into mainstream relational DBMS packages like ORACLE and others, while based on *de facto* industry standards for object linking and embedding (OLE), structured query language (SQL), third- and fourth-generation languages, and ODBC. Table 1 contains a summary of recent industry offerings in this regard.

At the same time, the international database standards community (through ISO/IEC/JTC1/SC21 – WG3) is extending the proposed SQL MM ('MultiMedia') to accommodate spatial relationships, indexing arrangements, and operators typically

found in GIS packages. Mainstream commercial database vendors are already incorporating some such enhancements into their own RDBMS offerings, with ORACLE's SDO and Informix/Illustra's Spatial DataBlades offerings being two such examples. These developments, combined with the recent formal attention being paid to interoperability between software packages (Sondheim et al, Chapter 24), suggest that future enterprise implementations may rely less on 'full-solution' packages maintained by a single vendor or consortium of vendors. Rather, large institutional users may opt for a series of smaller, lower-cost software components or applications possessing limited functionality on their own, but which seamlessly interact with other database, word processing or advanced modelling applications which may be present in those organisations (see Elshaw Thrall and Thrall, Chapter 23).

To make optimum use of this flexibility, application developers must have the ability to allocate data and processes anywhere in the multi-tier enterprise information architecture – and also reallocate them as operational requirements change. At the time of writing, concerns have been expressed that first-generation client/server application development tools like Visual Basic,

Table 1 Tighter linkages between GIS and RDBMS: 'Spatial Middleware' (from Costello 1996).

Vendor	Client Software product	Middleware Product
Vision Intergraph Corporation	AutoCAD MicroStation	Vision Server Jupiter
ESRI	ArcView MapObjects	Spatial Database Engine (SDE)
MapInfo Corporation	MapInfo	SpatialWare

PowerBuilder, and others may not have the capabilities and performance required to support more complex, enterprise-wide applications adequately (Mimno 1996). While it is likely that newer versions of these and similar tools will address many of these concerns, they must still evolve as the needs and practices of different user groups in these large organisations change over time.

4.2 Data access and delivery

Dozens of major land information and resource inventory programs around the world have demonstrated on-line organisation and electronic distribution of government imagery, mapping, and related land information across proprietary networks. Early examples included Land Information Alberta (McKay 1994) and the Manitoba Land Information Utility (Oswald 1994) projects in Canada, the early commercial ImageNet service in the United States, and the Land Ownership and Tenure System (LOTS) in South Australia (Sedunary 1988). Numerous other early examples may also be found elsewhere in these countries and in Europe.

Originally, such projects offered connection-oriented access to secure databases utilising either direct LAN-interconnect services or high-speed modems. Especially since late 1993, however, the Internet-based WWW has emerged as an alternative means of accessing, viewing, and distributing spatial information. Used in combination with modern Web 'browsing' software packages like Netscape and Microsoft's Internet Explorer, the WWW is emerging as a mainstream tool for

- the distribution of public-domain spatial information;
- online ordering of commercial datasets;
- indexing and cataloguing of related spatial datasets available off-line (Dawe 1996).

An analysis of more than 25 home pages classified WWW usage in the GIS community into four overlapping categories (Coleman and McLaughlin 1997). Setting aside those sites focusing on product or program advertising, the remaining three categories included:

1 *Data distribution.* At such sites, users may search for specific spatial information features or datasets based on either keyword searches or Boolean queries of existing database fields (e.g. Crossley 1994; Nebert 1994; Pleuwe 1994). In some cases, the user may obtain only pointers to

datasets stored off-line (e.g. Marmie 1995). In more sophisticated implementations, the user may retrieve image, map graphics, or attribute information covering a given area of interest (e.g. Conquest and Speer 1996).

2 *Custom map creation and display.* These sites represent the results of special-purpose development projects aimed at the composition, display, and downloading of user-defined custom map products. They may or may not involve the use of a GIS package running in the background. Currently limited in terms of the extent and variety of data coverage, they nevertheless represent an exciting development in providing 'just-in-time' mapping to the general public (Pleuwe 1997).

3 *GIS/WWW integration.* These sites represent the results of integrating front-end query capabilities (supported through standard Internet and WWW interfaces and protocols) with the capabilities of commercial DBMS and GIS software packages residing in the background. User-defined queries are translated into corresponding SQL commands and passed to the 'back-end' GIS database for handling. The resulting response is passed back through the gateway to the user and, in the case of map-based responses, the resulting map is translated into a graphics or bit-map format suitable for fast transmission and viewing across the Internet. User-developed examples of such GIS/WWW integration are documented by Conquest and Speer (1996), Nebert (1996) and Liederkerke et al (1995), among others. Software vendors offering WWW interfaces to spatial data servers or map-based spatial data browsers include Autodesk, ESRI, Genasys, MapInfo, and Universal Systems.

The lines distinguishing these categories are fading quickly. Various electronic telephone and business directories offered in Australia *(http://www.whitepages.com.au)*, Chile *(http://www.chilnet.cl/index.htm)*, and the USA *(http://www.bigbook.com)* now allow users both to obtain contact information and to find the location of selected individuals or businesses on custom-generated electronic street maps (see Elshaw Thrall and Thrall, Chapter 23). Other creative early examples of how these systems may appeal to general users include:

1 The 'Tripquest' service (Figure 4) *(http://www.mapquest.com)*, which enables users to specify a starting point and destination of a

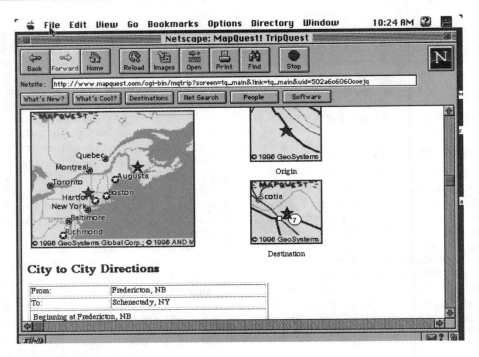

Fig 4. 'Tripquest' service.

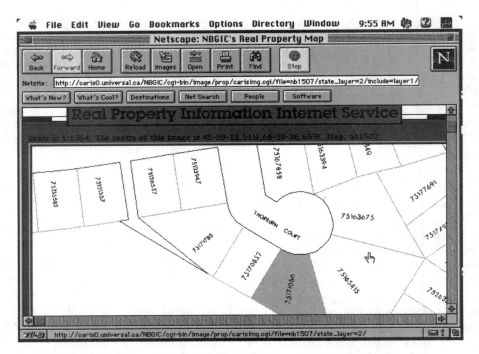

Fig 5. New Brunswick Geographic Information Corporation Property Information Service.

road trip, then receive detailed driving instructions based on a road network database contained at the server site.

2 The LandData BC site *(http://www.landdata. gov.bc.ca)*, which offers on-line access to a wide variety of mapping and land information products from different provincial government departments in British Columbia, Canada.

3 The New Brunswick Geographic Information Corporation Property Information Server (Figure 5), which allows NBGIC subscribers to access and download property ownership, registration, assessment, and parcel-centred mapping information covering any property in New Brunswick, Canada via keyword search, database query or map-based index (Arseneau et al 1996).

The advent of Java – a portable, object-oriented Internet language developed by Sun Microsystems Inc. – promises to remove many of the constraints inherent in early WWW protocols and further extend the capabilities of Web-based data browsing systems (Strand 1995). By moving much of the requisite display, processing, and analysis functionality to the client end of the Internet connection, performance delays due to server load and Internet bandwidth limitations may be greatly reduced (Figures 6 and 7).

Early examples of Java-based applications in GIS have tended to focus on limited tasks, including improved client-based spatial data browsing (Mapguide 1996) and network analysis engines (Fetterer et al 1996). More recent research efforts (Choo 1996) are aimed at using Java to provide an even wider suite of GIS display and modelling capabilities across the Internet. With a number of

vendors and standards organisations (e.g. the Open GIS Consortium) now including it as an important component of an overall network-based GIS architecture, Java-based technology may eventually be one of the keys to developing more open systems of distributed processing of geographical data.

At the time of writing, more extensive and well-maintained lists of WWW sites falling into all these categories may be found at the 'Metadata and WWW Mapping' home page maintained by Katz *(http://www.blm.gov/gis/nsdi.html)* and the 'GeoWeb' site *(http://wings.buffalo.edu/geoweb)* maintained by Pleuwe.

4.3 Collaborative production and group decision-making

Hardware, software, and procedures to support computer-supported cooperative work (CSCW) have been discussed and compared for more than 30 years (Englebart 1963). In the corporate world, shared access to corporate resources and innovative new approaches to collaborative production using groupware tools like Lotus Notes (Coleman 1995) and, more recently, the WWW (Pilon et al 1996) are now being investigated.

When considered in combination with emerging groupware products, the developments in GIS and broadband communications mentioned earlier may also offer a new approach to both collaborative group decision-making and digital map production – enabling some processes to be conducted concurrently rather than sequentially (for examples see Churcher and Churcher 1996; Coleman and Brooks 1995; Faber et al 1995; Karacapilidis et al 1995).

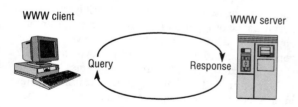

- Query handling and database inquiry
- Interactive point selection
- Zooming/panning across graphics dataset
- Selection/clipping of graphics data of interest
- Raster-to-GIF conversion

Fig 6. Internet spatial data request handling using HTTP.

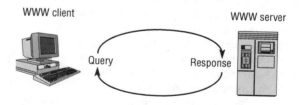

- Downloading of selected datasets
- Interactive point selection
- Zooming and panning
- Selected analytical functions

Java moves processing load off server and increases functionality of client

Fig 7. Internet spatial data request handling using Java language.

Network-based collaborative applications depend on more than just electronic mail. Specifically, they are predicated on the presence of a shared 'database' or collection of files, and rules which permit the definition of group members' roles, task status reporting and tracking, and gateways to electronic mail and other sources of data. Such systems should permit the organisation of correspondence, comments, reports etc. associated with a project or product and should support the management of multiple versions of objects (e.g. images, vector-based charts, video, and sound). Finally, a limited number of applications may require two or more remote users to be able to view the same file simultaneously, modify or add comments to specific entities in the same file where necessary, and communicate via voice, video, or e-mail while making these changes.

A number of organisations are already prototyping the WWW as the medium for collaborative decision-making (Gordon et al 1996), enhanced design, and concurrent engineering applications. Moreover, recent extensions to WWW viewers like Netscape are now introducing much of the communication and threaded-discussion, file transfer, data management and simultaneous data viewing functionality required in these applications (Forrester Research Inc. 1996; Ziegler 1995).

However, while some major corporations have made significant productivity gains through adoption of early groupware packages. The costs and cultural changes involved in such adoptions are likely to blunt the near-term impact of CSCW in the marketplace. Technically, vigorous adoption of collaborative technology will depend on:

- acceptance of the Internet or corporate intranets as the preferred medium for communications and information delivery;
- acceptance of common standards for threaded discussions and calendar management;
- subsequent development of a wider variety of inexpensive and compatible components or 'plug-in' tools capable of operating atop the standards extended from or similar to Netscape's ONE and Microsoft's COM (Forrester Research Inc. 1996).

Culturally, successful groupware implementations are likely to be shaped more by prevailing organisational cultures and constraints than by the technology itself. Since organisations work in different ways, it is best to let them customise their own groupware applications using modular toolkits rather than expecting any single shrink-wrapped package or technology to serve their needs adequately (Schrage 1995).

5 SYSTEM PERFORMANCE IN NETWORKED ENVIRONMENTS

Forthcoming generations of system architecture and broadband telecommunication technology promise to change fundamentally the way many organisations manage, transfer, and utilise their spatial data. However, before adopting such technology, both suppliers and potential customers require a clear understanding of potential network usage and the performance, capacity, and cost trade-offs involved in existing and emerging applications. A study carried out for the Government of Canada (IDON 1990) suggested that designing and building cost-effective GIS networks within and between federal government organisations would require defensible answers to six important groups of questions. These questions concerned the kind of information to be moved, the data volumes involved, response-time requirements of the users, the distance the data were being transferred, the frequency or regularity of data transmissions and, perhaps most important, the available funds which could be devoted to such operations.

Given the ability of today's WWW servers to track the volume and nature of user transactions at and between given sites, it is now easier to address such questions than it was seven years ago. Further, given the ubiquity of the Internet today, the justification of such a network may appear a moot exercise. However, the continuing requirement for 'hard numbers' reinforces the need to determine defensibly the performance of a particular application or group of applications across a network under known conditions, in order to predict the performance of the same application(s) under other conditions.

Numerous authors have suggested the fundamental importance of performance determination within the overall framework of the system life cycle and structured system and database design processes (e.g. Ferrari et al 1983; Jain 1991). However, while performance analysis has formed an important component of many GIS selection processes since the early 1980s, test procedures and results have largely remained unavailable because of the constraints of commercial confidentiality and competitive pressures.

Researchers have been developing more systematic and rigorous approaches to the determination of GIS performance on stand-alone configurations since the mid 1980s (e.g. Goodchild and Rizzo 1986; Hawke 1991; Marble and Sen 1986; Wagner 1991). Investigations at the University of Edinburgh (Gittings et al 1993) and the University of Tasmania (Coleman 1994) have extended these studies into client/server environments across LANs and WANs. The latter two references in particular provide extensive reviews of the literature related to GIS performance testing.

A key limitation of GIS performance testing research to date has been its lack of extendibility. Absolute performance figures are tied closely to the hardware configurations employed. Further, even under controlled network conditions, it is extremely problematic to model end-to-end GIS performance and to predict how response times will be affected by changes in server load and network data traffic.

As an increasing number of applications take advantage of the Internet – which, in itself, can send data packets across randomly changing combinations of high- and low-speed network connections with ever growing traffic loads – it will be increasingly difficult to make any claims (apart from criticisms) concerning application performance in a given setting. Further, if events over the past 15 years are any indication, the storage, memory, and processing requirements of application software and operating systems have a tendency to expand to surpass eventually the capabilities of a given generation of computers and networks alike.

That being said, the following advances in technology and services will – along with many others – help improve GIS application performance over the next five years:

- the increasing optimisation of high-performance disk arrays and parallel processors for use in networked GIS environments (Sloan 1996);
- the proliferation of fibre-based broadband communication services in both the workplace and the home;
- the introduction of higher-speed satellite and cellular data communication links into areas not served by normal services;
- the introduction of higher-speed Internet services (e.g. the proposed 'Internet II') and value-added service offerings which improve performance by 'detouring' subscriber traffic around high-volume links and nodes;

- the accelerating introduction of special plug-ins, Java 'applets' and other modular software components which balance client and server processor loads and further optimise data traffic between workstations on a network;
- (perhaps most importantly) improvements in the collections of underlying algorithms and procedures defining the application software packages themselves, removing many existing bottlenecks and optimising older code to run on newer systems.

References

Annitto R, Patterson B 1994 A new paradigm for GIS data communications. *Journal of the Urban and Regional Information Systems Association* 7: 64–7

Arseneau B, Kearney A, Quek S, Coleman D 1997 Internet access to real property information. *Proceedings GIS 97 Conference, Vancouver.* GIS World Inc.

Champine G A, Coop R D, Heinselman R C 1980 *Distributed computer systems.* Amsterdam, Elsevier Science

Choo Y K 1996 'Interactive distributed geographical information systems (IDGIS)'. Unpublished Master's Research Project, Department of Rangeland Ecology and Management, Texas A&M University. *http://starr-www.tamu.edu/choo/idgis/intro.html*

Chorafas D 1980 *Computer networks for distributed information systems.* New York, Petrocelli Books

Churcher M, Churcher C 1996 GROUPARC: a collaborative approach to GIS. In Pascoe R T (ed.) *Proceedings, Eighth Annual Colloquium of the Spatial Information Research Centre.* Dunedin, University of Otago: 156–63

Coleman D J 1994 'Geographic information systems performance in a broadband communications environment'. Unpublished PhD dissertation, Department of Surveying and Spatial Information Science, University of Tasmania

Coleman D J 1995 An overview of groupware. In Coleman D J, Khanna R (eds) *Groupware technology and applications.* Englewood Cliffs, Prentice-Hall

Coleman D J, Brooks R 1995 Applying collaborative production approaches to GIS data collection and electronic chart production. Santa Barbara, NCGIA and see *http://www.ncgia.ucsb.edu/research/i17/i17papers.html*

Coleman D J, McLaughlin J D 1997a Information access and usage in a spatial information marketplace. *Journal of Urban and Regional Information System*s 9(1)

Conquest J, Speer E 1996 Disseminating ARC/INFO dataset documentation in a distributed computing environment. *Proceedings 1996 ESRI User Conference.* Redlands, ESRI. *http://www.esri.com/resources/userconf/proc96/TO200/PAP16 5/P165.HTM*

Costello B 1996 Trends in client-server architecture in GIS environments. Slide from invited presentation at 1996 Geomatics Atlantic Conference, Fredericton, New Brunswick, Canada

Craig W J, Tessar P, Ali Khan N 1991 Sharing graphic data files in an open system environment. *Journal of the Urban and Regional Information Systems Association* 3: 20–32

Crossley D 1994 WAIS through the Web: discovering environmental information. Paper presented Second International WWW Conference Mosaic and the Web, Chicago, 17–20 October

Dawe P 1996 'An investigation of the Internet for spatial data distribution'. Unpublished Master of Engineering report, Department of Geodesy and Geomatics Engineering, University of New Brunswick, Fredericton, Canada

Elliot W 1994 Moving AM/FM to the field: technology, trends and success stories. *Proceedings of 1994 AM/FM International Conference*: 845–57

Englebart D 1963 A conceptual framework for the augmentation of man's intellect. In Howerton P (ed.) *Vistas in information handling*. Washington DC, Spartan Books 1: 1–29

Ezigbalike I, Cooper R, McLaughlin J 1987 A query management strategy for land information network implementation. *Proceedings of 1987 Annual Meeting of the Urban and Regional Information Systems Association (URISA)* 3: 105–12

Faber B G, Wallace W, Cuthbertson J 1995 Advances in collaborative GIS for land-resource negotiation. *Proceedings of GIS 95 Ninth Annual Symposium on Geographic Information Systems*: 183–9

Ferrari D, Serazzi G, Zeigner A 1983 *Measurement and tuning of computer systems*. Englewood Cliffs, Prentice-Hall

Fetterer A, Goyal B, Agarwal N 1996 Background: Java interactive routing for Minneapolis: explanation of WebRoute project. Spatial Database Laboratory, Department of Computer Science, University of Minnesota. *http://www.ggrweb.com/geojava/*

FieldWorks Inc. 1996 Eden Prairie. *http://www.canaska.com/fw.htm*

Forrester Research Inc 1996 Teams on the Internet. *The Forrester report* 7(6): 2–8. Cambridge (USA), Forrester Research Inc. *http://www.forrester.com*

Gittings B M, Sloan T M, Healey R G, Dowers S, Waugh T C 1993 Meeting expectations: a view of GIS performance issues. In Mather P M (ed) *Geographical information handling*. Chichester and New York, John Wiley & Sons

Goodchild M F, Rizzo B 1986 Performance evaluation and workload estimation for geographic information systems. *Proceedings, Second International Symposium on Spatial Data Handling*: 497–509

Gordon T, Karacapilidis N, Voss H 1996 Zeno: a mediation system for spatial planning. *Proceedings of ERCIM workshop on CSCW and the Web, Sankt Augustin, Germany, 7–9 February. http://orgwis.gmd.de/W4G/proceedings/zeno.html*

Hawke D 1991 Characterising the performance of geographic information systems. *Proceedings of Third Annual Colloquium of the Spatial Information Research Centre, Dunedin*. University of Otago

IDON Corporation 1990 *Final report: a federal geographic information systems (GIS) network requirements study*. Contract report prepared for the Data Communications and Networking Subcommittee, Interagency Committee on Geomatics, Government of Canada

Ingoldsby T R 1991 Transparent access to geographically related data from heterogeneous networked systems. *Proceedings of 1991 Annual Conference of the Urban and Regional Information Systems Association* 3: 14–24

Jain R 1991 *The art of computer systems performance analysis: techniques for experimental design, measurement, simulation, and modeling*. New York, John Wiley & Sons Inc.

Karacapilidis N, Papadias D, Egenhofer M 1995 Collaborative spatial decision making with qualitative constraints. *Proceedings of Third International ACM Workshop on Advances in Geographic Information Systems*: 53–9

Katz R 1991 *High performance network and channel-based storage*. Sequoia 2000 Technical Report 91/2. Computer Science Division, Department of Electrical Engineering and Computer Sciences, University of California, Berkeley

Laurini R 1994 Sharing geographic information in distributed databases. *Proceedings of 1994 Annual Conference of the Urban and Regional Information Systems Association (URISA)*: 441–54

Liederkerke M van, Jones A, Graziani G 1996 The European Tracer Experiment system: where GIS and WWW meet. *Proceedings of 1996 ESRI User Conference*. Redlands, ESRI

MapGuide (1996) Mill Valley, Autodesk Inc. *http://www.mapguide.com*

Marble D F, Sen L 1986 The development of standardized benchmarks for spatial database systems. *Proceedings, Second International Symposium on Spatial Data Handling*: 488–96

Marmie A 1995 Promoting and providing GIS data via the Internet. *Proceedings, 1995 ESRI User Conference*. Redlands, ESRI. *http://www.esri.com/resources/userconf/proc95/to250/p2101.html*

McKay L 1994 Data brokering in land information – Land Information Alberta. *Proceedings, 22nd Annual Conference of AURISA* 1: 171–82

Miller A 1990 GIS in a networking environment. *Proceedings of GIS 90, International Symposium on Geographic Information Systems*: 255–62

Mimno P 1996 *Building enterprise-class client/server applications*. Lisle, Dynasty Technologies, Inc. *http://www.dynasty.com/product/mimno_wp.htm*

Nebert D D 1994 Serving digital information through the WWW and Wide Area Information Server technology. Paper presented at Second International WWW Conference Mosaic and the Web, Chicago

Nebert D D 1996 Supporting search for spatial data on the Internet: what it means to be a clearinghouse node. *Proceedings of 1996 ESRI User Conference, Palm Springs* Redlands, ESRI. *http://www.esri.com/resources/userconf/proc96/TO100/PAP096/P96.htm*

Newton P W, Zwart P R, Cavill M 1992 Inhibitors and facilitators in high-speed networking of spatial information systems. In Newton P W, Zwart P R, Cavill M E (eds) *Networking spatial information systems.* London, Belhaven

Oswald R 1994 Manitoba Land Related Information System land information utility. *Proceedings, 1994 Conference of the Urban and Regional Information Systems Association (URISA)*: 179–92

Pilon D, Whalen T, Palmer C 1996 *Communications toolkit.* Technical Paper. Network Services and Interface Design Laboratory, Communication Research Centre, Industry Canada. *http://debra.dgbt.doc.ca/~daniel/cscw.text.html*

Pleuwe B 1994 The GeoWeb project: using WAIS and the World Wide Web to aid location of distributed datasets. Paper presented at WWW 94 Conference, Chicago, 17–20 October. *http://wings.buffaloedu/~plewe/paperwww.html*

Pleuwe B 1997 *GIS online: information retrieval, mapping, and the Internet.* Santa Fe, Onward Press

Popko E 1988 *An enterprise implementation of AM/FM.* Internal paper. Geo-Facilities Information Systems Application Center, IBM Corporation, Houston, Texas

Pretty R W 1992 LANs, MANs, and WANs: introductory tutorial on computer networking. In Newton P W, Zwart P R, Cavill M E (eds) *Networking spatial information systems.* London, Belhaven Press

Reinhardt A 1994 Building the data highway. *Byte Magazine* 19: 46–74

Schrage M 1995 Groupware requires much more than bandwidth. *Business Communications Review*: 35–8

Sedunary M E 1984 LOTS and the nodal approach to a total land information system. In Hamilton A C, McLaughlin J D (eds) *Proceedings of 'The Decision Maker and Land Information Systems' – FIG Commission III International Symposium*

Sloan T M 1996 'The impact of parallel computing on the performance of geographical information systems'. Unpublished MPhil dissertation, Department of Geography, University of Edinburgh

Strand E J 1995 GIS thrives in three-tier enterprise environments. *GIS World* 8: 38–40

Streb D 1995 Using cable television to distribute GIS access. *Proceedings, 1995 ESRI User Conference.* Redlands, ESRI

Thoen B 1995 Web GIS: toy or tool? *GIS World* (October) *http://www.csn.net/~bthoen/webgis.html*

Voelcker J 1986 Helping computers communicate. *IEEE Spectrum* (March): 61–79

Wagner D F 1991 'Development and proof-of-concept of a comprehensive performance evaluation methodology for geographic information systems'. Unpublished PhD dissertation, Department of Geography, Ohio State University, Columbus

Webcrawler Survey 1996 *World Wide Web size and growth.* Global Network Navigator, Inc. *http://webcrawler.com/WebCrawler/Facts/size.html*

Webster C J 1988 Disaggregated GIS architecture. Lessons from recent developments in multi-site database management systems. *International Journal of Geographical Information Systems* 2: 67–80

Wittie L D 1991 Computer networks and distributed systems. *IEEE Computer*: 67–76

Ziegler B 1995 Internet software poses threat to Notes. *The Wall Street Journal* (7 November)

23

Desktop GIS software

S ELSHAW THRALL AND G I THRALL

This chapter reviews the emergence of desktop GIS, which is defined as products which may have begun as features of workstation GIS but which have been spun off to create new and useful niche market products. As such, desktop GIS is seen to include: digital atlases; interactive street displays and route finding software; mapping on the Internet; spreadsheet and database mapping; clip art and readymade maps; thematic maps; and so-called 'true' desktop GIS. Geographically-enabled programming languages are also considered from the standpoint of add-ons to desktop GIS.

1 INTRODUCTION

GIS keeps getting bigger, encompassing more as technology develops and our knowledge base increases. As GIS grows, so pieces of what had previously been viewed as exclusively the territory of GIS are being sliced away to become market niches in their own right. Those who use new market niche products may even be unaware of the background GIS technology from which it developed and may have no need ever to be directly involved with GIS per se (see also Batty, Chapter 21).

What, then, is 'desktop GIS'? There are a number of ways in which to answer this question. From one viewpoint, desktop GIS is simply that part of the newly emergent desktop computer market that is not developing for any specific market niche. From this perspective, desktop GIS is no more than a software and technology marketing term. And yet in order to understand desktop GIS trends a second, more analytical perspective is required. In this chapter, we choose to define desktop GIS as inclusive of those products that may have begun as features within workstation or desktop GIS software programs and technology, but which have been spun off to create more useful products within new market niches.

How is the market for desktop GIS developing? Thrall, interviewed for the industry magazine *Geo Info Systems* in 1996, made the following five year forecast:

'I expect mainstream [desktop] GIS will be mainstream application software. GIS will lose its distinguishable identity in the market-place. If a map makes sense in an application, then a map will be in that application. If GIS functionality makes sense in an application, then that application will contain GIS functionality. GIS will become application niche software and application niche software will seamlessly include GIS. Customers will expect it. In five years, only us old-timers will remember when GIS was a technology with a culture separate and distinct from mainstream software.' (Thrall and Trudeau 1996: 12)

From the vantage point of the general software user, GIS will be all-pervasive and unrecognisable as a unique technology with a unique history. From the vantage point of the academic, the software developer, and the more sophisticated user, understanding of the development of desktop GIS technology will be one means of distinguishing between casual users and experts.

In this chapter we present an overview of desktop GIS. Our broad definition of desktop GIS will enable us to discern the market trends more readily. Our purpose is to present a categorisation of desktop GIS, and to provide examples of each. This will contribute towards clarification of what desktop GIS is today and how it is distinguished from previous GIS technology. Table 1 presents eight categories of desktop GIS, in increasing order of complexity. As new markets emerge, so new categories will be added: as technology and the markets change, so the ordering of the categories will change.

Table 1 Categories of desktop GIS technology.

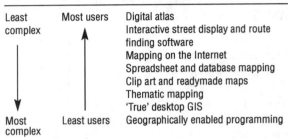

Novice users of desktop GIS might consider any software that fits within any of the categories of Table 1 to be 'GIS'. Conversely, novices may believe that their requirements can only be met by what we refer to as 'true' GIS software; or, because they are producing thematic maps they may believe that they are working within a 'true GIS environment'. Such differences in part reflect the rapidity of development of the field: what may once have been a central component of desktop GIS may have evolved into a niche which is peripheral to 'true' GIS, but which is nevertheless an important emerging new market segment. For example, GIS functionality and geographical data have been key to increasing the productivity of those who need customised maps: however, the new category of readymade maps (which includes no GIS functionality) may now better suit the needs of this user group. The core components of each of the classes of geography software identified in Table 1 will now be discussed in turn.

2 THE DIGITAL ATLAS

The digital atlas has developed from its traditional analogue counterpart using GIS technology (see also Salgé, Chapter 50). Today the digital atlas is typically distributed on CD-ROM. The illustrative products included in Table 2 serve the same purpose as printed atlases. In terms of production costs, the first (master) copy of the digital version may cost as much or more than the first master copy of its printed version counterpart. However, the marginal

cost of producing each additional digital atlas, especially when very large mass-market volumes are considered, becomes insignificant. The digital atlas has further advantages over its printed counterpart by virtue of being compact and lightweight – so that, for example, information can even be accessed with a CD-ROM-equipped laptop computer. The digital atlas may also include a computer-automated 'find' feature: printed atlases require one to go to an index, read cryptic references to the page and row and column that a map feature may be found on, and then – armed with such information – the atlas reader must locate the feature on the map. The digital atlas will find the feature for the user and show the user where the feature is on the appropriate map. Digital street atlases like those discussed below may even allow the user to enter an address and the digital version will then zoom into the likely location of the address. It is traditional for the printed atlas to include some thematic maps and charts. The digital version may allow the user to select from more themes, and to vary the manner in which the theme is displayed.

Table 2 Simple mapping products and digital atlases.

Small Blue Planet	Now What Software
Maps 'n' Facts	Broderbund
Picture Atlas of the World	National Geographic
Expert Maps Gold	Expert Software
World Atlas	Software Toolworks

To sum up, the digital geographical technology offers advantages over traditional printed technology in terms of convenience of use, reduced size and weight, and lower cost. The lower marginal production cost is likely to mean that the product is more frequently updated: by contrast, the traditional atlas in a library may previously only have been replaced when its pages became worn from years of use.

3 INTERACTIVE STREET DISPLAY AND ROUTE FINDING

Interactive street and route finding software is a variation on the digital atlas that is a separate and new emerging technology. Several of the products that are available in the USA, for example, are listed in Table 3. DeLorme has become dominant in the

US market with this type of product because it entered the market early, its street maps are attractively designed and easy to read, it is inexpensive, and (very important for a product of this type) it uses data compression technology to include the software and the geographical data on a single CD-ROM.

Table 3 Interactive street display and route finding.

Global Explorer	DeLorme	multimedia with photographic images
Map 'n' Go	DeLorme	includes route finding
Street Atlas USA	DeLorme	based on US Census TIGER/Line

Beyond the basic functions of street display and route finding, there are also products (e.g. those of Sony Inc.) that combine features from Table 3 with hardware. The hardware is ergonomically designed for ease of use in an automobile. These products may include Global Positioning Systems (GPS) technology (see Lange and Gilbert, Chapter 33) with interactive voice, whereby the device will select the best route based upon current traffic conditions and will audibly inform the driver where to turn (see also Waters, Chapter 59).

4 MAPPING ON THE INTERNET

Simple descriptive maps have also become a feature used to measure user access, also known as 'hits', to World Wide Web (WWW) sites on the Internet. Users may access these 'Where is it?' maps in either of two ways. First, the user utilises a 'find' feature to locate an Internet address, using a similar procedure to that used to interrogate a CD-ROM street atlas. Advertisers will often pay to have their logo and universal reference locator (URL) jump to their site prominently displayed alongside a 'Where is it?' map. Second, the 'Where is it?' map feature may be part of a WWW page that provides the user with a street map so that some desired destination can be found (Figure 1).

Fig 1. 'Where is it?' map feature from a WWW page.

'Where is it?' maps can be downloaded or printed thereby making available to the user custom atlas type features. An example of two kinds of 'Where is it?' map is provided in Table 4. The Vicinity site uses Etak Corporation geographical data and software to create the maps; the maps are used as a draw to their site for the display of advertising. Union Street Links uses on-line maps to inform the user of business locations in order to encourage patronage.

The ability to locate point addresses on a map is considered a standard feature of GIS software; the 'Where is it?' map separates out this feature from the general purpose GIS, allowing the user to display the location of an address or object. As discussed below, to be 'true GIS' the software must be capable of transforming and manipulating the data thereby effectively creating new data. 'Where is it?' maps are highly valuable but are limited to the display of stored geographical information. At the same time, if the program user's needs require only 'Where is it?' capability, then there is no need to use any more complex GIS software. Table 5 includes a list of software vendors that provide the capability to program WWW sites to include GIS features such as the 'Where is it?' capability.

A second type of mapping on the Internet is analogous to placing the geographical and attribute data and GIS software in a client/server configuration. The actual software and the data reside on a server which the user accesses via the Internet. The users (clients) request maps and the information is returned to them. Many of the GIS user tasks, such as opening tables and formatting the

Table 4 Examples of 'Where is it?' maps on the Internet.

Name of	Owner	Description	URL	WWW site
Vector	Vicinity Inc.	Address locator on demand		http:/www.vicinity.com
Union St Links	G&S Thrall	Displays location of business		http://www.afn.org/~links/

Table 5 Vendors of Internet enabled mapping software.

Caliper	http://www.caliper.com
ESRI	http://www.esri.com
MapInfo	http://www.mapinfo.com
Sylvan Ascent	http://www.sylvanmaps.com

Note: see authors' Web page for sample list of URLs using Internet enabled mapping software: http://www.afn.org/~thrall/gitwebs.htm

layers, are performed by the server. Proper software design is imperative, because many of those that access such Internet sites will be novices to geographical technology.

5 GEOGRAPHICALLY-ENABLED SPREADSHEETS

The thematic mapping capability has become mainstream with its introduction as an adjunct to mass consumer market spreadsheets (Thrall 1996d). For example, in the mid 1990s the alliance between Microsoft Corporation and MapInfo Corporation provided users of the Microsoft Excel spreadsheet software with the capability to generate simple thematic maps (Table 6). Similar products are available from Autodesk and Intergraph. While highly limited in terms of numbers of mapping features and options, the Excel/MapInfo product is a good example of how a mapping software program can be designed to be highly user friendly (see also Birkin et al, Chapter 51). The product also includes a limited amount of coarse resolution geographical boundaries, including those of the major land forms, many nations (e.g. the USA, Canada, Mexico, Western Europe, Australia), states and provinces of North America, and US counties. Excel does not include smaller boundary files like census tracts or zip codes.

With Microsoft's Excel, the user selects a range of cells to map. One of the columns of selected cells in the range enables the data to be linked to a geographical object such as a US county. For the attribute data to be geographically enabled, the field with the geographical identifier must follow a rigid standardised convention. The other columns in the

Table 6 Geographically-enabled spreadsheets.

Borland	*Quattro Pro*	Strategic Mapping Inc.
Lotus 1-2-3	*Lotus 1-2-3*	Strategic Mapping Inc.
Microsoft	*Excel*	MapInfo Inc.

Note: In 1996, Strategic Mapping Inc. was purchased by ESRI (Environmental Systems Research Institute)

selected range contain the attribute data to be used to provide values to the mapped themes. All mapping is done by linking attribute data to polygon map objects. Excel cannot map with user-specified latitude and longitude coordinates. Excel does not allow the user to customise polygon boundaries, and polygons may not be joined or buffered. As such, the polygon areas that can be mapped with Excel are highly limited. Yet these very limitations, which make it simple for the novice to use, in combination with its widespread distribution by Microsoft, have created a lucrative market for GIS data vendors. A variety of boundary files and attribute data can be purchased separately and then used within Excel.

Wessex Inc. has released an 'after market' product designed to extend the range of attribute data easily accessible by a mapping program to include the entire US Census of Population and US Census of Housing. Geographical boundary files for census tracts are provided with the Wessex product. Wessex's products include ProFiler (the US Census of Population data), a set of TIGER/Line files (the US Census boundary files in MapInfo format), First Map (for desktop mapping with Excel), and First Street (ArcView Software from ESRI (Environmental Systems Research Institute), Wessex's ProFiler, and Wessex's boundary files in ArcView format).

Lotus Corporation has also linked with a GIS software vendor to provide mapping capability similar to that described above for the Excel/MapInfo product. The choice of Lotus to have aligned with Strategic Mapping Incorporated (SMI) may have been unfortunate since SMI has subsequently ceased to exist and has sold its geographical software division to ESRI – best known for the GIS program ARC/INFO. However, what may appear to have been a mistake for Lotus – to align itself with a troubled GIS software vendor (SMI) – may in the final analysis appear to be fortuitous given the quality and strength of product offered by ESRI. For more discussion on ESRI's takeover of SMI see Thrall (1996a) and see Longley et al (Chapter 1; Chapter 72) for a discussion of rationalisation and change in the GIS software industry.

The mapping capabilities of spreadsheet programs are highly user friendly, but primitive in mapping capability. Prior to the introduction of mapping in spreadsheet software there were perhaps 100 000 users of geographical mapping software in the world, with varying skills and interests. Now with geographical technology included in mass market spreadsheet products the numbers of persons becoming familiar with geographic technology may exceed 20 million. As those high numbers of users proceed through the learning curve they will demand a greater variety of mapping operations, together with the ability to manipulate geographically enabled data. The scene is thus set for GIS to become part of the mainstream mass market computer software industry, in contrast to its past which appears increasingly esoteric and peripheral in comparison. At the time of writing, several database vendors including ORACLE, Informix, Computer Associates, Sybase, and IBM have already announced planned products that will allow the user to work with spatial data (Sonnen 1996).

6 CLIP ART AND READYMADE MAPS

In this section we will discuss clip art maps and a newly emerging product that we refer to here as 'readymade maps'. In Table 1 they could have been listed as two separate product categories. They share a number of common characteristics and are both based upon a newly emergent technology in a highly dynamic market. Indeed, the vendors of what we call readymade maps advertise their products as clip art maps. However, the authors of this chapter believe that once the consumers understand the capabilities of the mapping products, readymade maps will be identified as a new product category (see Thrall and McLean 1997).

Readymade maps combine the technology of clip art imagery with GIS. Clip art has long been a common feature of desktop publishing where the digital artist may begin to construct a computer image by importing an image from a catalogue of images. Clip art usually allows simple manipulation of the image such as changing the colour or deleting a part of the image. Clip art may also stand alone without modification and is used extensively in the design of brochures and other forms of advertising. Clip art maps allow simple editing of a map by

exporting it to a graphics program which is used to manipulate it.

The readymade map is a more-powerful variation of the standard clip art map. Readymade maps differ from clip art maps in the magnitude of the file size and in the complexity of the data file. Clip art is typically one layer of information while readymade maps typically have many layers. Clip art images are typically small while readymade maps can require as many as 64 megabytes of RAM memory just for their editing. Clip art is designed to be brought into a text document – for example as an image in a Microsoft Word text file. Readymade maps, by contrast, generally must be imported into a graphics program to be edited. Most readymade maps are distributed as an ensemble of geographical data and a software program to display them. Readymade map software can also often allow the geographical data to be exported in a variety of formats for subsequent editing.

With readymade maps, the user can do much more than change the colour or insert a map into text. With readymade maps, the user can use advanced features such as masks which allow the addition of boundaries, physical features such as rivers and mountains, graticules, scale bars, shadowing, and so on. Readymade maps can also be edited at a 'sub-image' level using a product like Macromedia's Freehand which allows the map to be edited in layers. Some readymade maps allow the importation of multiple files that automatically register to one another at the same scale and in the correct location without manual manipulation. For example, if a file of a country is opened, then other imported countries will align correctly and at the same scale as the original opened country. Table 7 gives examples of some readymade map vendors.

Usually the quality of readymade maps requires the highest quality printer and a desktop computer specifically configured for high resolution graphics production. These are truly maps ready to publish and bear comparison with the finest maps published anywhere today.

Table 7 Some readymade map vendors and their products.

Mountain High Maps	Digital Wisdom	High resolution relief maps of countries of the world
Cool Maps	Digital Wisdom	Relief maps for desktop publishing produced using visually exciting colours and projections
Globe Shots	Digital Wisdom	Views of the earth from space: many of the earth images are animated for easy inclusion in multimedia

Note: Based upon Thrall and McLean (1997)

Clip art maps and readymade maps provide basic descriptive geography for the publisher and graphics artist/cartographer. The fundamental characteristic that differentiates clip art maps and their more powerful cousin, readymade maps, from, say, a digital atlas is that the digital atlas is targeted towards the end or final consumer whereas clip art maps and readymade maps are intended to be modified using the map program itself or by using a program such as Adobe Photoshop or Macromedia Freelance to prepare the map for the enduser. The finished map may remain in digital form. Macromedia includes the ability to distribute electronic images produced within its software program on diskette. With Macromedia Director comes the ability to produce multimedia productions and to distribute the results over the Internet. By acquiring the desktop GIS capability to edit maps, clipart maps and readymade maps have established a market niche for themselves.

7 THEMATIC MAPPING

Thematic mapping begins one small step beyond the categories of the digital atlas, 'Where is it?' maps, and readymade maps. Thematic mapping software can be quite complex and sophisticated in its data manipulation functionality. Many 'true' desktop GIS software programs include thematic mapping capabilities; however, these capabilities are generally highly limited. 'True' desktop GIS software has greater focus upon the manipulation of spatial data while thematic mapping software has greater focus toward the display of spatial data. Software specialising in thematic mapping generally offers a much greater range of thematic mapping features than is available from desktop GIS software programs. Thus users of GIS who desire more complex thematic mapping capability may process and manipulate their data within the GIS, and then transfer the results of analysis to a thematic mapping program for presentation.

Thematic mapping capability includes the ability to produce shaded choropleth maps where the data ranges are represented by different colours, shading, hatching patterns or dot densities, or by graduated symbols. In dot density maps, dots are randomly scattered within polygons, although the total number of dots represents the number of observations. In graduated symbol maps, the symbol size is proportional to the value of the observation (see Martin 1996; and Kraak, Chapter 11, for a

general discussion of digital symbolisation). Some thematic mapping software also includes the ability to construct isoline and 3-dimensional shaded relief maps, or prism maps in which a polygon can be extended 3-dimensionally above the surface – the height above the surface being proportional to some attribute data value. Thematic mapping software may also allow the mapping of two variables as bivariate maps. However, even though some software allows a virtually unlimited number of themes to be simultaneously displayed, it should be recognised that a map generally loses interpretability when more than two themes are displayed. Each of these types of maps is discussed in fuller detail below in the GIS software section.

Thematic mapping software often requires data input in a very rigid data format and field layout. Thematic mapping software programs also generally have very limited capability for modification of the data. Thus while many GIS software programs include a variety of thematic mapping functionalities, most thematic software programs cannot be classified as true GIS software. Table 8 gives some examples of thematic mapping products.

Table 8 Some thematic mapping products.

Product	Manufacturer	Description
Surfer	Golden Software	Processes the user's data into isolines or contour lines. Provides a robust variety of spatial interpolation features as well as a reasonable set of default options for the novice. For a review and further discussion of Surfer see Thrall (1995a).
Map Viewer	Golden Software	Allows the user's data to be associated with US county and US state boundary files for the display of 3-dimensional prism maps as well as other forms of thematic map that are popular among endusers. Program users may provide their own boundary files. (See Elshaw Thrall 1997.)

8 'TRUE' DESKTOP GIS

'True' desktop GIS software programs include features which allow the user to achieve similar results to all of the above desktop technologies, but also include additional capabilities. True desktop

GIS software programs allow the user to access information using spatial logic, to modify geographically-enabled data, and to visualise the results as a map. This category of desktop software allows the user to query data and map objects using enhanced structured query language (SQL: see Egenhofer and Kuhn, Chapter 28) database operations. The user can also query individual map objects.

It is routine for desktop GIS software to include the capabilities for the input of textual spatial queries, including SQL queries, polygon joins, point-in-polygon operations, and buffering. Each of these uses is discussed in detail below. Table 9 identifies some of the leading desktop GIS software programs.

Who are the users of feature-rich desktop geographical software technology? The user of desktop GIS today is quite a different kind of individual to the user who is reliant upon mainframe and workstation GIS. Large GIS platforms like ARC/INFO or Intergraph are now used primarily by governmental organisations and large research universities. Governmental organisations have not traditionally been highly price sensitive or high productivity-motivated, and therefore they have been able to afford the high expense of large platform software and hardware, as well as the high learning curve required of highly technically trained specialised workers. ESRI, which produces ARC/INFO, has suggested that the learning curve to become fully proficient in ARC/INFO is roughly three years (Thrall 1995b; Thrall et al 1995). Thus GIS such as the mainframe systems of ARC/INFO and Intergraph have traditionally been used by the public sector.

Private businesses are generally unwilling to allocate the high level of expenditure necessary to establish mainframe and workstation GIS. Businesses are also generally unwilling to commit themselves to adding employees who require years of training in a software program which can only indirectly enhance their decision making. In short, businesses are unwilling to become hostage to an exotic technology operated by irreplaceable workers. The negatives of the mainframe and workstation environments make private businesses the primary audience for desktop GIS since, other than the ubiquitous desktop computer, there are no special hardware requirements. Moreover, powerful GIS software of unprecedented user-friendliness can be obtained at prices similar to those of bundled business office software (wordprocessing, spreadsheet, database management, presentation designers, etc.) from major manufacturers. Inexpensive GIS software now comes as standard with the basic geographically-enabled data required for most business decisions (see also Coleman, Chapter 22; Sugarbaker, Chapter 43).

Private businesses will not readily adopt research frontier technology, yet until quite recently GIS has resided within this realm. However, with mapping now included with conventional spreadsheet software, businesses no longer perceive mapping and geographical technology as being the research frontier. Most businesses which do adopt GIS perceive it as an aid in what Thrall has referred to as the first stage of GIS reasoning; namely, the employment of GIS technology to represent spatial phenomena for descriptive purposes (Thrall 1995c). Businesses may understand the need for spatial data visualisation. They may also use GIS for fast manipulation of large volumes of spatial data. For them GIS saves the many weeks or years that might be needed to organise and visualise their own data in association with data that describe their market. Prior to the innovation of desktop GIS, the cost and complexity of such descriptive representations made all but a few businesses stay away from the technology. Desktop GIS has changed all that, so now having this information is considered to be a necessary input for proper business decisions.

Some businesses which better understand geographical concepts may have proceeded to Thrall's second and third stages of GIS reasoning,

Table 9 Leading desktop GIS software programs.

ArcView 3	ESRI (Environmental Systems Research Incorporated)	Redlands, USA
Atlas GIS*	ESRI (Environmental Systems Research Incorporated)	Redlands, USA
Autodesk World	Autodesk Corporation	San Rafael, USA
GeoMedia	Intergraph Corporation	Huntsville, USA
MapInfo	MapInfo Corporation	Troy New York
Maptitude	Capliper Corporation	Newton, USA

* Note that Atlas GIS orginally published by Strategic Mapping Incorporated is now under the ownership of ESRI so the future of Atlas GIS is doubtful.

namely using GIS for explanation and prediction. The ability to visualise large amounts of geographical data has made it easier for businesses to understand correlation and causality. In order to enhance explanation and prediction, mathematical constructs such as gravity models (Haynes and Fotheringham 1984) have been programmed to use data that have been spatially summarised or modified using desktop GIS software (see Birkin et al, Chapter 51; Waters, Chapter 59). Geographical prediction with desktop GIS software is akin to 'what-if' scenarios performed with conventional spreadsheet software. Forecasts of spatial trends and processes, such as the changing market for a business firm, can be created with the desktop GIS software. It is in this context that Thrall (1995c) has written that 'relevance and marketability are key to understanding market-driven GIS today . . .'.

Business utilisation of desktop GIS software has not substantially gone beyond descriptive applications. There has been very little documented use of GIS to enhance business judgement and decision strategy. Business remains largely unaware of the contemporary capabilities of geographical technology and methodology. Today it is essentially the sizzle and pizzazz of a descriptive map that sells GIS. Given that the use to which GIS is put by business has been highly limited, the cost of implementation of GIS in a business environment must be low, in order to overcome the perceived limitations of the technology. As awareness of capabilities increases, business will demand more geographical capability from their GIS software and employees. As a consequence, we anticipate bright prospects for appropriately trained economic geographers. However, knowing the commands of the GIS software does not geographically enable the personnel using the software. By analogy, knowing the software features of a spreadsheet program does not by itself make one an expert in finance or accounting. The need for personnel who are trained in geographical reasoning is a prerequisite for the continued expansion of the desktop GIS industry.

Business requirements for high productivity that accompany inexpensive and user-friendly turnkey GIS solutions translates into a demand for prepackaged GIS data. And the market can respond by offering commercial business data because the volume is high and in the USA at least the production costs are very low (but see Rhind, Chapter 56). Free availability of high quality geographical and geographically enabled attribute data (Smith and Rhind, Chapter 47) has given the USA a competitive edge over those countries where access to similar data is highly restrictive or very costly.

Businesses have not adopted GIS in the same way that governments and related large public institutions have. Still, the chronology of the adoption by business of geographical technology mirrors the earlier process of adoption by government. Outside of military applications (see Swann, Chapter 63), those government agencies and divisions that were the first to adopt GIS were in the fields of natural resources or physical geography (e.g. Hutchinson and Gallant, Chapter 9). Only later did government agencies and divisions concerned with human geography applications adopt GIS. Similarly, those businesses in the natural and earth resources industries, such as forestry, were the first to adopt geographical technology, including desktop GIS. It has been a recent phenomenon that businesses have adopted GIS for use within the fields related to human geography; examples include real estate appraisal and investment, market analysis for retail outlets, and banking (see Birkin et al, Chapter 51; Longley and Clarke 1995).

8.1 True desktop GIS features and functions

Desktop GIS can be used to perform a variety of operations, though many of these operations are not unique to GIS. For example, adding new street line segments to a base map is a feature shared between GIS (computer assisted drawing), and software designed for and limited to digitising. In this section we discuss those features and operations that differentiate desktop GIS software from other software. These features include generating thematic maps, performing spatial queries or selections, joining polygons, performing point-in-polygon operations, and buffering (see Thrall and Marks 1994).

A fundamental application of desktop GIS software is the creation of thematic maps. Although thematic mapping is but one facet to the visualisation capabilities of GIS (see Kraak, Chapter 11), the thematic mapping feature is one that has been carved out from the GIS technology and sold on the market place as a separate product (as described in section 7 above). The advantage of separate smaller products is simplicity of software design, cost of software production and support maintenance, and a quicker learning curve for the program user; the disadvantage is that the product is much more limited in scope.

The GIS thematic map presents an underlying motif of data in spatial form. Data that can be mapped can be numeric or nominal. Nominal data are data with no numeric value that are described by name (Star and Estes 1990: 28): for example, various types of crimes (murder, assault, rape, robbery), or various types of underground cables (electric lines, cable television lines, fibre optic cables). The thematic maps that can be generated on desktop computers fall into several general categories: maps of individual values, choropleth maps, dot density maps, and graduated symbol maps. Another type combines two data themes in the same map and is referred to as a bivariate map. Some thematic mapping programs specialise in isoline, wire frame, and prism maps (see also Beard and Buttenfield, Chapter 15; Kraak, Chapter 11).

Individual value maps show different colours for each unique value of the data. Individual value maps are used primarily for the display of nominal data. A different colour or pattern is used to represent each data value. The attribute data can be associated with any form of geographic object, namely points, lines, or polygons. For example, a crime occurrence has a location and a classification; the colour of a point being red may indicate the crime as murder; likewise, assault could be represented by a brown dot, or a fibre optic cable line can be represented by a blue line and electrical lines represented by black lines. Individual value maps can also present numeric data, though often a map of this type suffers from loss of interpretability. Consider a map of Europe where the colours represent unique population density figures of the countries; each country would then have a different colour since while countries may have similar population density, no two countries will have identical population density values.

Choropleth maps (see Figure 2) are used to present ranges of values by colour or shading or hatching. Appropriate data ranges can usually be

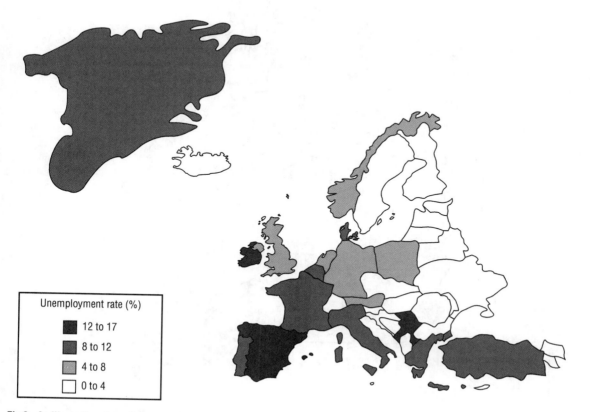

Unemployment rate (%)

■ 12 to 17
▨ 8 to 12
▤ 4 to 8
□ 0 to 4

Fig 2. An illustrative choropleth range map.

identified by the user where: there are the same number of observations in each range; the ranges themselves are of the same size; or the user can customise the ranges to any convenient values. Each value range then is represented by a particular colour, and successive ranges are denoted using lighter or darker shading of the same colour, or by a different hatching pattern.

Dot density maps (see Figure 3) represent raw data values that reside within a polygon. With a dot density map the software assigns one point for each incremental range of data values, such as one point for

every 1000 people in a state or province or county. Desktop GIS software generally displays the dots as being randomly scattered as opposed to clustered where the density of observations might be the greatest.

Finally, the graduated symbol map (see Figure 4) displays a symbol of varying size where size represents the magnitude of the attribute data value. For example, a large aeroplane symbol could represent a major regional airport while a small aeroplane symbol could represent a minor local private airport. The size of symbol could be graded by thousands of passengers served.

Fig 3. Dot density map of Canadian population 1990.

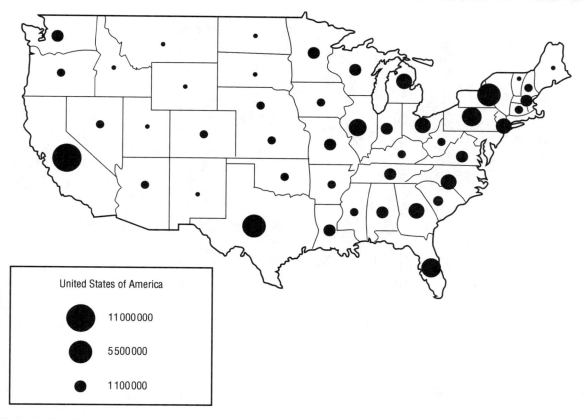

Fig 4. Graduated symbol map of number of housing units per US State 1990.

A bivariate map displays two themes, or two data variables. For example a map may have different symbols representing the type of crop grown in an area (like a corn stalk for corn or wheat shaft for wheat) and then the size of the symbol can represent the amount of yield for the particular crop. Bivariate maps typically use a combination of different symbols with different colours or sizes or the same symbol showing different sizes and colours to represent the two themes. Bar chart maps and pie chart maps are variations on the theme of bivariate maps. With bar chart or pie chart maps the user can have different bars or pie slices represent different fields for each record.

Isoline maps show varying data ranges by means of lines. For example, an isotherm map may show temperature ranges. The lines are labelled with the data value they represent. From these isoline maps, 3-dimensional wire frames can be mathematically interpolated and drawn. Some mapping programs can generate wire frame maps (see Figure 5) also offer features of sun inclination showing on the wire

frame, perhaps with a solid or shaded drape over the frame, to give a more realistic 3-dimensional effect (Thrall 1995a; Dowman, Chapter 31).

Another form of GIS produced thematic map, known as a prism map (see Figure 6), is a variation of a bar chart map. The height of the bar (prism) is proportional to the attribute value being mapped. But the prism itself is more than a plain rectangular 2-dimensional or 3-dimensional bar; it is in the form of the polygon with which the attribute is associated. The effect is to generate a map of varying plateaux or elevations in the shape of the polygons comprising the map.

8.2 True desktop GIS spatial selections

All true desktop GIS software programs include the capability to perform spatial operations. A spatial operation is a database query that is performed on spatial data using spatial criteria (see also Martin, Chapter 6). For example, a regular aspatial query

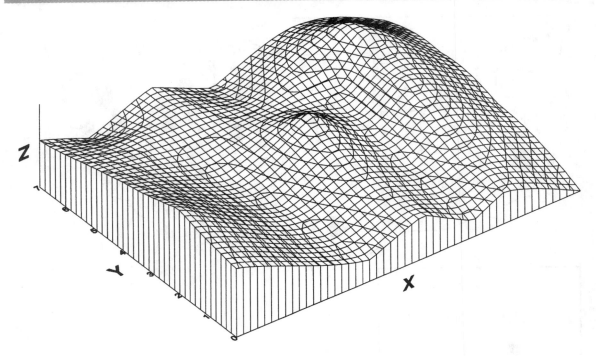

Fig 5. A wire frame map.

might be to select those home sales records for which the value of the sale is above a certain amount. A spatial query might be to select those homes for sale which are within a five kilometre radius of a given school. Again, by itself the capability to perform this operation does not distinguish true GIS from some GIS niche products.

Spatial selections in GIS are usually accomplished using tools selected using a toolbox icon, or via typed queries entered using a dialogue box. Tools for spatial queries using desktop GIS fall into three general categories: selection or pointer-like tools; circular or 'radius select' tools; and polygon (non-circular area) tools. The selection or

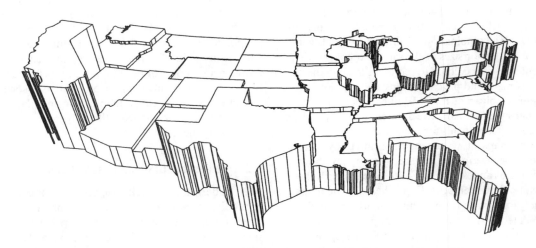

Fig 6. A prism map depicting the relative population sizes of the contiguous United States.

pointer tool is typically in the shape of an arrow head (and is a normal tool found, for example, in windows-based wordprocessors). It is used to point at map objects, and by clicking the main mouse button the user may select items. In a true desktop GIS, the items selected are spatial objects, namely points, lines and polygons, and the associated labels. Generally, the layer that includes the item being selected must be active. Many true desktop GIS software programs allow only one map layer to be active at a time.

The radius selection tool allows the user to draw circular areas around a point such that all records falling within a circle are selected. For example, in order to find all major cities in a database that fall within a 100-km radius of New York City, the software user would click on the central point of the City and drag a radius of 100 km from it. All records falling within the circle defined by the radial distance would be selected (but see Birkin et al, Chapter 51, for a discussion of the limitations of such tools).

The polygon selection tool is similar to the radius selection tool in that it allows the user to draw polygons around areas and to select those records that fall within the polygon. The polygon tool does not limit the user to a circular area, but rather allows the user to construct any regular or irregular closed shape.

Spatial queries can be performed using logic and words as well as pointing and selecting with a tool. Spatial querying using words is known as 'word query'. All regular database queries in true desktop GIS can be performed using word queries, including simple queries using the relational operators (less than, less than or equal to, greater than, etc.) and complex compound queries using the logical operators (and, or, not). In addition to these standard database operators, true desktop GIS software programs include geographic operators such as 'contains', 'falls within', and 'intersects'. 'Contains' is an operator where one object contains another, such as a circle inside of a square. 'Falls within' is an operator where one object falls spatially within another object such as a point residing within a circle. 'Intersects' is an operator where part of one object overlaps or intersects part of another object such as a line that passes through a circle.

Structured query language (SQL) word queries allow the user to perform selection operations which are more complex than simply pointing and clicking

with a mouse (see also Egenhofer and Kuhn, Chapter 28). SQL is a standard database query language. When using a regular word query with the geographical operators, the operators work on fields that are actually in the database. SQL queries can work on fields that are not explicitly in the database, but which can be implicitly calculated from fields and spatial information in the database. For example, a regular word query can be performed on the population of the countries of the world (e.g. 'select countries whose population exceeds 50 million'). A SQL query can be performed on any field so long as that field can be created using a combination of information from other fields. Thus, for example, queries can be made about population density, even though population density is not one of the fields in the database, since population is known and a desktop GIS can calculate the area of a county polygon. The desktop GIS software will create a new data field – population density – which can subsequently be saved to the database. For extended discussions of this and related topics, see Oosterom (Chapter 27) and Worboys (Chapter 26).

8.3 True desktop GIS polygon joins

Another common use of desktop GIS is to join polygons. New polygons in the same or in a new layer are created by joining existing polygons – for example, county polygons might be joined to create sales territories for a sales force, and the newly created layer saved as the 'sales territories' layer. Each map object normally has attribute data assigned to it, such as the population of a county. When joining polygons, the user can choose to discard the attribute data or to save them after modification using a mathematical operator – for example, the user may require the software to add the population of each county that is joined to form a new attribute, the 'population of the sales territory'. For further details see Thrall (1992) and Martin (Chapter 6).

8.4 True desktop GIS point-in-polygon operations

The term 'point-in-polygon search' was originally coined to describe the geometrical operations used to locate points relative to vectorised line boundaries. Point-in-polygon operations also involve performing mathematical operations on attribute data fields, for points that fall within the same polygon. For example, if crimes are identified as points on a map, then the number of crimes within a census tract (polygon) can be measured by commanding the software to count the number of points by polygon. If the crime is burglary and the value loss from the burglary is included as an attribute field, then the software can be used to calculate the total loss from burglary within a census tract. Standard mathematical forms of aggregation include performing counts, averaging, multiplying, calculating standard deviations or variance, and so on.

8.5 True desktop GIS buffering

Buffering is the creation of polygons that surround other points, lines, or polygons. Individual buffers can be created around individual objects, or multiple objects can be buffered to act as one buffer area. The user may wish to create buffers to exclude a certain amount of area around a point, line, or polygon, or to include only the buffer area in a study. For example, in using GIS to help determine the possible sites for a new water well, the location of chemical factories may have a 10-km buffer drawn around them so that these buffered areas are excluded in the list of possible sites. The user of desktop GIS can indicate how many line segments are to be used to make up the boundary of the buffer, thereby controlling the accuracy of the buffer boundary. Buffering of points is performed in a way analogous to the 'radius select' or 'polygon select' tools described above. Buffering of lines and polygons can be accomplished by using the buffer function in the desktop GIS program.

8.6 True desktop GIS programming

GIS is foremost a *spatial* database management tool. One of the uses of database management is application development through the use of the programming language feature of the database management program. General purpose GIS programs are seen by many business managers as being too difficult to learn. They want push-button GIS capability in the software where no special knowledge on how to use a GIS is required. Special purpose GIS or niche GIS is being demanded by those without the knowledge or time to proceed up a steep GIS learning curve. Niche GIS programs are programmed using either true desktop GIS, or written using a geographically-enabled software language.

True desktop GIS software has language capabilities that allow the development of specific or niche applications. As a variation on this theme, the many true desktop GIS software vendors offer the capability to add GIS functionality to programs written in languages such as Visual Basic and C++, or written in database management languages such as PowerBuilder, Visual Fox Pro or Access.

9 GEOGRAPHICALLY-ENABLED PROGRAMMING LANGUAGES

Some desktop GIS have add-on products that allow the user to program with a language embedded within the GIS software. This capability generally requires a fee in addition to the price of the true desktop GIS software. Also, if copies of the resulting program are distributed to other users, additional fees are generally payable.

MapInfo Corporation sells a programming language known as MapBASIC. MapBASIC is similar to the generic BASIC with the addition of geographical operators such as 'is located within' that are used for the distinct geographical operations. Caliper Corporation also offers a stand alone geographically enabled language known as the 'geographic information systems developers kit' (GISDK).

Other true desktop GIS software programs allow the programming of modules using standard programming languages such as Visual C++ or Visual Basic. They permit the use of the generic languages to write modules that can be used in conjunction with the GIS. The modules are called from the GIS as standard dynamic link libraries (DLLs) or using object linking and embedding (OLE). Often, however, communicating with DLLs or using OLE requires knowledge of advanced programming procedures.

If the applications programmer is not starting with a pre-existing GIS program, but is writing a program that will have GIS capabilities, a language such as Visual Basic or C++ will probably be used for the primary program. A commercial off-the-shelf add-on module can then be used to write the GIS component of the program. Standard components called VBXs (16 bit components) or OCXs (32 bit components) exist that allow functions to be added to the Visual Basic or C++ program. Examples of functionality added to software programs include the capabilities of drawing and charting, spreadsheets, wordprocessing, database management, telecommunications, and so on. These add-on modules, called custom controls, save the programmer the time needed to program them from scratch, and are usually moderately priced and very easy to use. The VBX or OCX feature is thus an added tool in the programmer's toolbox. The programmer can add the feature to the program with a click of a mouse on the tool. For further discussion on the role of GIS programming tools and the impact on the GIS industry see Thrall (1996b, 1996c).

10 CONCLUSION

Desktop GIS has changed immeasurably since its inception as a province of desktop computing. It has grown to include many new features and capabilities. The start-up costs of learning 'traditional' GIS software have never been small and with added features and functionality comes a steeper learning curve. Yet users of spatial software have increasingly demanded more user-friendliness. The response to these demands has been that parts of what has always been considered GIS are being spun off into new market niche software. Thus while 'traditional' GIS has been growing in capabilities, its stature is at the same time being eroded by niche software and other generic spreadsheets and databases – as vendors discover that these are symbiotic with certain GIS features. GIS software vendors are finding opportunities to form cooperative ventures with generic software vendors, and those that fail to form such cooperative ventures may in time find their own market lost to niche software vendors.

References

Elshaw Thrall S 1997 MapViewer: first impressions software review. *Geo Info Systems* 7 (November)

Haynes K, Fotheringham A S 1984 *Gravity and spatial interaction models.* Newbury Park, Sage Publications

Longley P, Clarke G 1995 *GIS for business and service planning.* Cambridge (UK), GeoInformation International

Martin D 1996 *Geographic information systems: socioeconomic applications,* 2nd edition. London, Routledge

Sonnen D 1996 Spatial information management: an emerging market. *MapWorld* 1: 24

Star J L, Estes J E 1990 *Geographic information systems.* Englewood Cliffs, Prentice-Hall

Thrall G 1992 Using the JOIN function to compare census tract populations between census years. *Geo Info Systems* 2: 78–81

Thrall G 1995a Surfer: review of 3-dimensional surface modelling software. *Journal of Real Estate Literature* 4: 73–5

Thrall G 1995b New generation of mass-market GIS software: a commentary. *Geo Info Systems* 5: 58–60

Thrall G 1995c The stages of GIS reasoning. *Geo Info Systems* 5: 46–51

Thrall G 1996a Battle builds for business GIS market. *Geo Info Systems* 6: 46–7

Thrall G 1996b SylvanMaps/OCX: first impressions software review. *Geo Info Systems* 6: 47–8

Thrall G 1996c Modular component programming: the foundations of GIS applications. *Geo Info Systems* 6: 45–6

Thrall G 1996d Maps, data and mapplets: first impressions software review. *Geo Info Systems* 6: 48–9

Thrall G, McLean M 1997 Blurring the lines between GIS and desktop publishing: first impressions software review. *Geo Info Systems* 7: 49–53

Thrall G, Marks M 1994 Functional requirements of a geographic information system for performing real estate research and analysis. *Journal of Real Estate Literature* 1: 49–61

Thrall G, Trudeau M 1996 Java and Applets: breakfast of champions. *Geo Info Systems Showcase* 6: 10–15

Thrall G, Valle J del, Elshaw Thrall S 1995 Review of GUI-based GIS software products. *Geo Info Systems* 5: 60–5

25

GIS customisation

D J MAGUIRE

Customisation is the process of adapting a generic system to an individual specification. It is generally considered one of the most expensive non-personnel components of a GIS implementation. Because of the limited size and diversity of the GIS market, many GIS software developers have adopted the approach of developing a generic suite of multi-purpose software routines, together with some type of customisation programming capability. This has allowed core GIS software developers to concentrate effort on engineering robust and reliable generic routines. The task of creating specific-purpose, end-user (or vertical application) customisations is usually seen as the domain of application developers.

In the case of desktop and professional level GIS the process of customisation typically involves modification of a standard graphical user interface and extension of the 'out of the box' tools by writing application programs. More sophisticated users may be allowed access to the underlying core GIS capabilities and database. They may be able to extend the core class libraries or reuse objects within their own programs.

Traditionally, GIS software developers have had to develop their own programming languages. However, with the wider incorporation within GIS of industry standard programming environments – such as Visual Basic, Visual C++, and Java – this is changing.

1 INTRODUCTION

GIS software has been used in an extremely wide range of applications: from archaeological site mapping, to managing land assets, to storm runoff prediction, and global zoological analysis. One of the main reasons why it has been possible to employ GIS software in such a diverse range of applications is because of the customisation capabilities that software developers incorporate into their products. These allow application developers to create specific customisations of generic software systems. This inherent flexibility has been one of the major factors in the success of GIS.

This chapter begins with a short history of GIS customisation and the various approaches adopted to customising GIS software systems. The process of GIS customisation is then described in detail, including some consideration of costs. This is followed by a look at the role of software engineering in GIS customisation. Next, examples of

the main GIS customisation tools are described. Finally, the conclusion draws the main points together and briefly looks towards the future.

In the early days of GIS-relevant software development (the 1960s and 1970s) all of the software systems produced were specific-purpose and highly tailored to each application. These monolithic (sometimes called 'stovepipe') systems were unique islands of information processing functionality incapable of exchanging data with other systems (see also Batty, Chapter 21; Sondheim et al, Chapter 24). This situation arose for several reasons.

- Each system had to be developed from scratch because there were no other systems or common pools of routines to adapt or extend.
- The market was very small and there was no incentive to develop generic systems which could be extended or adapted, and then sold to other users.

- There was limited expertise within the developer community about what constituted a generic application.
- The hardware and software development tool limitations of the day necessitated that each application be highly optimised to give the best possible performance.

As technology developed and the expertise and market size grew, the effects of each of these limitations was ameliorated. In the 1980s GIS software developers began to create generic GIS software systems capable of customisation and then deployment in multiple application areas. The first example of a successful generic GIS software system was ARC/INFO, from Environmental Systems Research Institute Inc. (ESRI), released in 1981. In the first few releases, however, the capabilities for user customisation were limited. It was not until the release of the ARC Macro Language (AML) as part of ARC/INFO 4.0 in 1987 that the software was really capable of being customised by end-users. In the late 1980s and early 1990s virtually all major GIS software vendors adopted this approach of developing a generic suite of multi-purpose software routines, together with some type of customisation programming capability. This has allowed core GIS software developers to concentrate effort on engineering robust and reliable generic routines. The task of creating specific-purpose, end-user customisations is usually seen as the domain of application developers. These application developers may belong to the core software developer's organisation, a user organisation, or some independent third-party organisation. This approach to software and product development has significant implications for the cost of implementing GIS in organisations and also the levels of technical expertise required by users, as the following discussion will demonstrate.

More recently, with the late 1990s developments in technology (Batty, Chapter 21) and the cumulative increase in expertise and market size, the trend has been more towards systems designed for specific endusers with only limited capabilities for applications development (Elshaw Thrall and Thrall, Chapter 23). Currently, this is more evident at the lower end of the market. The typology in Figure 1 shows how mass market end-user products – such as ESRI's *Business*MAP, MapLinx's MapLinx, and AutoRoute from Microsoft – have extremely limited

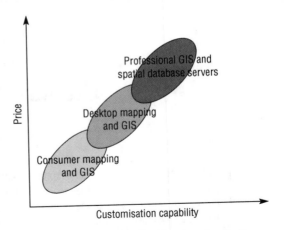

Fig 1. Typology of GIS software based on customisation capability and price. Note that price is approximately inversely proportional to market size.

customisation capabilities. Desktop mapping and GIS software systems, such as ESRI's ArcView and MapInfo's MapInfo, tend to have enduser-orientated customisation capabilities. These typically allow users to change the user interface and add their own macros and programs. Professional, or high end GIS software systems, such as ESRI's ARC/INFO and Spatial Database Engine, Intergraph's MGE and Smallworld's Smallworld GIS, allow customisation to a greater or lesser extent.

Customisation offers both advantages and disadvantages for GIS users and developers. The main advantages for users are that they get systems which incorporate their process-specific business rules and closely match their requirements. As far as developers are concerned, developing a combination of a generic system and a customisation capability is a cost-effective solution for small- to medium-sized markets. It is only in large market sectors that there is a business case to produce a specific-purpose ready-to-run application which does not require customisation capabilities. On the down side, there are some disadvantages of this approach to delivering GIS software solutions. For users, customisation is an expensive and time-consuming exercise (Newell 1993) requiring a considerable degree of input and expertise (e.g. to specify user requirements, perform acceptance testing, and sign off completed customisations). A further problem is that customisations created with high level, user-orientated development languages are inevitably slower than core programs created with a low level, system-orientated language such as C++. Until a given sector (or vertical market) increases to a size sufficient to justify a low level specific-purpose application this will remain the case.

The alternative is bespoke applications development which offers the advantages of an optimised, very specific system with few compromises. Set against this is the fact that these bespoke systems tend to be very expensive (both to build and, more importantly, to maintain), they are very risky projects (because they often start from a low level), and the completed application often cannot be easily adapted, either as the project requirements change or as new projects arise. Furthermore, users with bespoke systems cannot benefit from the continuing general development of commercial-off-the-shelf (COTS) systems for multiple users. These arguments lie behind the decision by most major government and military agencies around the world to move away from proprietary or bespoke systems towards increasing use of COTS solutions: that is, solutions which incorporate as much commercially produced software as possible (see also Bernhardsen, Chapter 41; Meyers, Chapter 57).

2 THE PROCESS OF CUSTOMISATION

Customisation is the process of adapting a system to an individual specification. GIS can be customised in several different ways. In order to explain this it is necessary to describe in general terms the architecture of modern GIS software systems. Figure 2 shows in schematic form a generalised architecture for a desktop or professional GIS software system (see also Elshaw Thrall and Thrall, Chapter 23). Users normally interact with the GIS software via a typically graphical, menu driven, icon-based graphical user interface (GUI). Selections from the GUI make calls to geoprocessing tools (i.e. tools for proximity analysis, overlay processing, or data display). The tools in turn make calls to the data management functions responsible for organising and managing data stored in a database. This three-tier architecture has been widely used (at least conceptually) by many software developers in order to facilitate organisation and management of software development.

It is possible to configure GIS software systems at all three levels. At the GUI level this typically

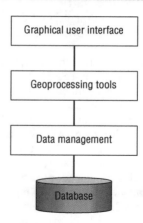

Fig 2. Architecture of a typical desktop or professional GIS software system.

involves configuring the form and appearance of the interface (e.g. adding/removing menu choices and buttons, changing the pattern of icons, and personalising the colour scheme and other characteristics of windows). This is normally carried out using an interactive graphical customisation environment (for an example see Figure 5). At the Tools level customisation involves creating macros to automate frequently required processes and adding new functionality (such as new spatial analysis functions or data translators). This type of customisation can more properly be referred to as programming. More advanced users and software developers are also interested in customising the data management routines within a GIS software system, perhaps to add new datasets to create a new spatial database schema or connect to an external tabular database. These are only a few of the ways in which general-purpose GIS software systems can be customised to create specific-purpose, user-orientated applications.

The above description of the architecture of GIS software systems and the customisation capabilities is useful for explaining the various options available. In practice, however, most of the larger GIS software systems allow users and developers to customise the software at two levels: the application and the core or object code level. In some cases this may be carried out using two programming environments, in others only a single environment is preferred. For example, ESRI's ARC/INFO can be customised at the application (also called 'user' level) by working with the integral fourth generation programming language, AML. Additionally, GIS can be customised at the much lower software development library level using a third generation programming language such as C. In fact the ARC/INFO product is assembled internally by ESRI software development staff who embed the software development library objects in C programs. In contrast, the Smallworld GIS has a single customisation environment, Magik, which is used both internally by Smallworld staff and externally for application customisation by users and developers.

The two level approach adopted by ESRI has the advantages that endusers can use a high level, user-oriented development environment and that they are not exposed to low level programming concepts. The single level approach selected by Smallworld has the advantages that users can gain access to all of the GIS functions and that there is only one development environment for the company to maintain.

3 COSTS OF CUSTOMISATION

It is widely recognised that, along with data capture, customisation is usually the most expensive element of an operational GIS (Antennucci et al 1991; Korte 1996; Smith and Tomlinson 1992).

Table 1 is a generalisation of the approximate breakdown of costs between the various elements of typical GIS implementations based on the author's experience of implementing GIS in over 150 organisations. The 'Low' figures are for a system comprising two server seats (UNIX or Windows NT workstations) and eight clients (desktop PCs). This type of configuration might be found in a small commercial or local government organisation. The 'High' figures are for a corporate implementation comprising 35 UNIX or NT workstations and a UNIX GIS server. This type of configuration might be found in a medium–large local government or utility site. Both configurations run a mixture of desktop and professional GIS software.

Table 1 Breakdown of the percentage costs of typical small desktop (low) and large professional or enterprise (high) operational GIS.

	Low	High
Hardware	22	7
Software	13	12
Data	6	23
Customisation	4	30
Personnel	55	28

Many new users mistakenly believe that hardware and software are the major costs of establishing a GIS. In fact, staff costs are the most expensive. One explanation for this apparent discrepancy is that in many public (and even some private) agencies personnel costs are not included in assessment of GIS implementation costs. It is also often stated by GIS commentators that data are the most expensive component of a GIS (Rhind, Chapter 56). In general data do often comprise a significant proportion of costs. In Table 1 it is assumed that in the Low case the data are purchased and in the High case they are part purchased, part specially captured. If all the data had to be captured for the projects then the figures for data would be several percentage points higher.

4 THE SOFTWARE ENGINEERING APPROACH TO GIS CUSTOMISATION

4.1 GIS software engineering

All GIS implementations, including those involving customisation, have in common the fact that they must meet user requirements and be delivered on time, in budget, and in accordance with quality standards. These goals will be greatly facilitated if a rigorous software engineering approach is adopted and if the process of software development is split up into a series of independent steps which are carried out in sequence. In this so-called software development lifecycle, also referred to as the 'waterfall model', each step is well defined and leads to the creation of a definite product (often a piece of paper), thus allowing the correctness of each step to be checked. More recently, software engineers have questioned the waterfall methodology and have proposed the use of prototyping as an alternative or extension to this approach. These two approaches will be discussed in turn below.

Excellent introductions to software engineering are provided by Bell et al (1992), Flaatten et al (1992) and Gilb (1996). For more critical discussion see Maguire (1994) and Brooks (1995).

4.2 The waterfall model

In the waterfall model (Figure 3), the first stage in the software development lifecycle is to establish user requirements. Essentially, this involves a dialogue between one or more representatives from the user and application developer groups. Initially ideas will be loose and vague in part, but over time they will become clearly defined. This is arguably the most important and sometimes difficult stage of application development.

User requirements need to be formally specified if they are to be of use as a description of the application to be developed. They also form the basis of an acceptance test which will determine if the system meets its requirements. Formal specification is all about describing *what* an application will do rather than *how* it will do it; the latter should be left to the discretion of the programmer.

The design stage involves the application developer creating conceptual, logical, and physical designs of the system. These stages progressively refine the application design from being implementation-independent to being system-specific. In GIS application development the design stage will typically address the user interface, the geoprocessing tools required, and the data management capabilities employed. There are various tools available to assist in this process. These include data flow modelling and various database diagramming techniques such as entity-relation modelling and the object modelling technique (OMT: Date 1995; Rumbaugh et al 1991).

There is now almost universal agreement amongst software developers that applications (indeed all software systems) should be designed and implemented using structured programming techniques (Worboys, Chapter 26). There is also a belief that an application should be created as a series of independent modules. Modular software

development is preferred because it supports software reuse and cooperative development, maintenance, testing, and debugging.

Implementation is all about coding, testing, and debugging the design. In the past this is what people have thought of as 'programming'. The implementation stage usually concludes with a period of acceptance testing and bug fixing before final sign-off by users.

The last stage is the operation and maintenance of an application. This will typically involve a period of user training, system enhancement, bug fixing, and system use. During system use, new user requirements will inevitably arise. This initiates the sequence again (through a 'change order' to the original contract) and so the cycle continues.

In recent years several commentators have questioned the waterfall approach (Flaatten et al 1992). In particular, they have pointed out the length of time it can take to go through the lifecycle using this top-down methodology. Also at issue is the fact that many users do not really understand large specification documents. GIS applications are very visual and until potential users see an application interface they often do not really appreciate what it is and how it will work. They may 'sign off' the document but still be surprised when they see the first release. A further issue is that over long periods user requirements and technology can change, particularly in a fast moving area like GIS. This problem is compounded by the fact that GIS is still relatively new for many users. It is also the case that as new users understand the technology better their ideas and aspirations change, often leading to new and enhanced requirements. Changing the requirements and incorporating new elements into the design has proven to be very expensive for systems based on the waterfall approach (typically the later they are incorporated, the more expensive they become).

4.3 The prototyping approach

Prototyping involves creating working designs of a system rather than designs on paper. Functional requirements documents are still required, however, since they form the basis of a contract document and a series of milestones defining acceptance criteria and a payment schedule. These prototypes are demonstrated and evaluated, and form the basis of future prototypes. The prototyping approach allows closer user involvement in system creation, focusing effort on producing user-oriented systems

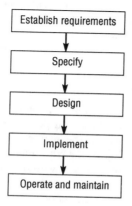

Fig 3. The waterfall approach to software development.

153

and catering for users' evolving knowledge. As Figure 4 shows, prototyping is an incremental process. Some suggest that it can be represented in the form of a spiral with major issues and decisions taken in the early stages and progressively more detail covered in subsequent iterations.

An important issue associated with prototyping is whether prototypes should be discarded or used as the starting point for the next iteration. Unless evolving systems are built on firm foundations which do not compromise system integrity, there is a danger that the final system will be weak. The increasing use of object-oriented approaches to software development, with support for extensibility and modular replacement/development, has helped to minimise this problem.

Many of the modern application development tools are particularly suited to the prototyping approach. Development environments based on the object-oriented paradigm and those which support interactive graphical development are useful for rapid application development based on prototyping (e.g. Visual Basic, described below, and Delphi). Other advantages of prototyping include the availability of early results and deliverables, as well as reduced costs because of the greater likelihood of developing the application that users actually want.

On the downside, prototyping is less helpful for providing fixed prices for development, nor is it appropriate for some small jobs (although some organisations still prefer this approach). In situations where the application is well understood and relatively simple, prototyping is generally regarded as inefficient. Experience suggests that it is very difficult to persuade users to discard prototypes even though their underlying architectures may not be suitable for supporting the large mission-critical developments which must be accommodated over the long term.

4.4 Discussion

On balance most technical GIS people would agree that all GIS application development projects require a clear implementation plan which identifies a sequence of well-defined tasks. Whether the classical waterfall approach is adopted, or whether prototyping is used, is open to discussion by the user and developer. Prototyping seems most useful in the areas of user interface design, performance estimation, and functional requirements analysis.

5 GIS APPLICATION DEVELOPMENT TOOLS

The earlier sections of this chapter have examined various approaches to GIS application development. This section looks at the practicalities of implementing an application and the tools and techniques which are available for GIS customisation. The application customisation capabilities of three software systems are discussed. Unfortunately, given the author's background these are all ESRI software systems. It must be emphasised, however, that the principles are common to virtually all GIS software products. First, the customisation capabilities of ARC/INFO, a high end professional GIS, will be discussed. Next, ArcView, a general purpose desktop GIS software product, will be described. Finally, the customisation capabilities of MapObjects, a highly customisable object-based developer product, will be addressed.

5.1 ARC/INFO

ARC/INFO is an example of a high end or professional level, general-purpose GIS software system. Figure 5 shows the interactive graphical application development environment of ARC/INFO. This example shows the software running in a Microsoft Windows environment on a Windows NT workstation. The functionality is similar but the look and feel of the development environment is slightly

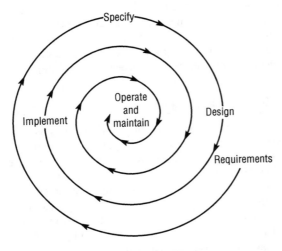

Fig 4. The prototyping approach to software development.

different on UNIX systems (reflecting the different windowing standards of the two operating systems). The display shows a series of windows depicting some of the productivity tools available to help application developers create custom applications. The 'LINE Theme Properties' menu is a complete menu which will be part of the final application (it is a generic menu which will be used to set the symbolisation characteristics of themes or data layers). The 'Untitled – FormEdit' menu and the 'Slider Properties' menu are parts of ARC/INFO's interactive graphical menu-builder called FormEdit. The 'Untitled – FormEdit' window is a widget palette showing the types of widgets which users can place (drag and drop) on menus. The right-hand window of the two shows the property sheet for a slider bar; the second widget down on the left in the FormEdit window is a slider bar (it contains the numbers 750, 0 and 25 000). Changes made using FormEdit can be run immediately in the application to determine their impact and to assist debugging. This greatly facilitates the application development process. In the bottom left-hand corner of Figure 5 is a text dialogue window in which messages and text dialogue will appear (in this instance the person doing the customisation has opted for white text on a black background). In the bottom right of the screen there is a Text Editor window which is used to edit programs and menu descriptions in ASCII format. These programs are typically attached to menu buttons and executed when a user presses the button in the interface. In this case Microsoft Word is being employed as an editor but the developer could choose any text editor. In the bottom centre there is a file browser window (labelled 'Coverage Manager'). This is used to navigate through the file system in order to locate workspaces and specific files.

Figure 5 has been designed to show the main types of tools available to developers. The figure is not a direct example of how an application developer would actually customise a GIS. Usually only a selection of the windows are open at any one time. Most application developers prefer instead to close or iconise windows until they are needed.

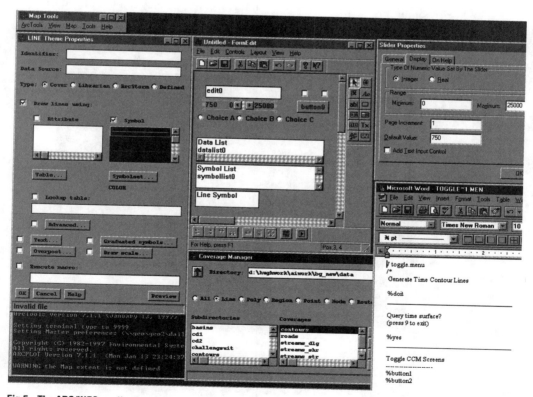

Fig 5. The ARC/INFO application development environment.

155

Normally menus would be developed using FormEdit and programs would be created with the Text Editor. Both of these are really part of the same run time environment in which the final applications will operate. This allows menus and programs to be tested and debugged interactively before they are released as part of a complete application (perhaps comprising 20–100 menus and 50–500 programs, along with additional elements like 'help' files, documentation, and an install script).

Application developers use this type of environment to create a menu-driven interface for an application. Business rules and process tasks are included as AML programs which are called from the interface. Once an application has been created the menus and code are saved in persistent computer files. Typically, a master program will be called at application start-up. This will then call other programs and menus depending on user selections from the interface and answers to prompts issued by the application. AML is an interpreted language and programs do not need to be compiled into machine object code prior to execution.

5.2 ArcView

A second example of a GIS development environment is provided by ESRI's ArcView desktop GIS. ArcView GIS has been designed and developed to provide stand alone and corporate-wide (using client-server network connectivity) integration of spatial data. ArcView is an object-oriented GIS with an object-oriented customisation language (Avenue – itself based on C++).

Within ArcView's application customisation environment there are various programmer resources which support application development and customisation (Figure 6). This ArcView screen shows six PC windows: the ArcView application itself (ArcView GIS Version 3.0); the Project window (Untitled); a View (View1) containing a World Country Theme (World94.shp) and a Degree Lines Theme (Deg30.shp); the Document Designer (Customize: Untitled); the Script Editor (Script 1); and the Script Manager. These have been designed to be easy to use so that endusers can develop their own applications. Scripts can be created using the integral editor or they can be entered from text files.

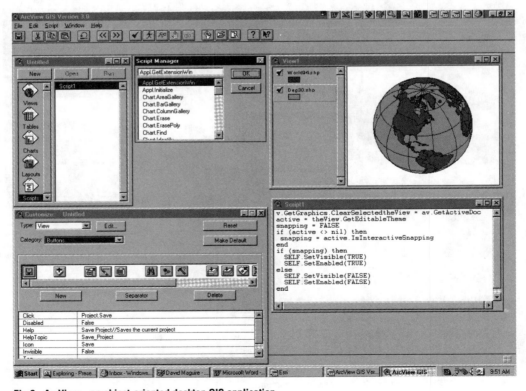

Fig 6. ArcView – an object-oriented desktop GIS application.

They are organised using the Script Manager. Users can attach the scripts to controls in the GUI of each document using the Document Designer (the window labelled 'Customize: Untitled'). This is also used to control the appearance and functionality available within the GUI. Scripts contain requests (messages) to other objects which return an object. Typically scripts are made accessible to users by being placed behind a menu choice or button in the GUI. In event-driven systems such as ArcView, scripts can be run whenever an event occurs (such as resizing of a window, arrival of new data from an external source, or when a user clicks on a button). ArcView additionally has a form interface builder (not shown) used to create multi-widget screen forms.

Like ARC/INFO, Avenue is an interpreted development language and any applications created can be immediately run by the user. Stand-alone applications can be developed and delivered to other users by wrapping the scripts and user interface amendments together as an ArcView Extension (an application which operates within and extends the standard COTS ArcView application).

5.3 MapObjects

ESRI's MapObjects is a collection of GIS objects which conform to Microsoft's Object Linking and Embedding (OLE) / Component Object Model (COM) or ActiveX software specification. Software developers can create applications which embed MapObjects (and other non-GIS objects which conform to the COM specification) within programs written with any industry standard OLE/COM compliant software development environment (e.g. Visual Basic, Visual C++, Visual J++ or Delphi).

Figure 7 shows the Visual Basic application development environment. All elements of the display are part of Visual Basic with the exception of the additional map icon on the bottom right of the left-hand tool palette (this is used to create a MapObjects custom control [an instance of a map window] on the form – the large window in the centre of the screen called Form1). The white area in the centre of Form1 is the map custom control window. When the application is executed this would normally display some geographical data. The form can be thought of as the application user interface. Visual Basic developers create applications by placing controls (maps, buttons, scrolling lists, etc.)

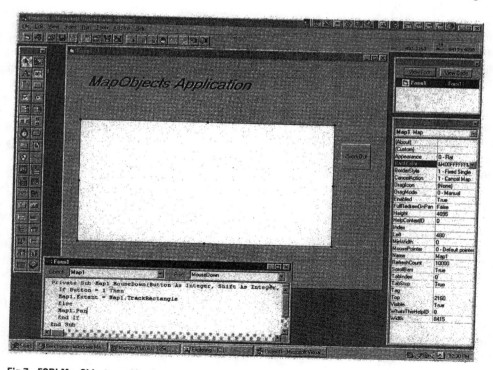

Fig 7. ESRI MapObjects working in a Visual Basic development environment.

on a form. Code is then attached to the controls. The lower, Form1, code window shows the Visual Basic code attached to the map control. This particular program will execute a zoom in or pan operation when a user clicks on the map. The right-hand window, called Properties Form1, is used to set the properties of controls (e.g. font size, type, and colour) on buttons such as the 'Zoom Out' button placed on the form. The top right-hand window, called Project1, is used to organise and register all the windows, controls, and code for the application.

Once a Visual Basic programmer has created an application embedding MapObjects the application can be compiled into a stand-alone executable program which users may run from a menu, icon, or other user interface object.

6 PRACTICAL ISSUES ASSOCIATED WITH GIS CUSTOMISATION

There is clearly more to application development than just obtaining some software and writing code. Any GIS project which adopts a pure technology focus is doomed to failure and customisation projects are no exception.

Because customisation can be a time consuming and expensive activity it is very important to involve senior management and sponsorship at the earliest opportunity. Educating management will help to get their support for the process, as well as further assistance should there be any problems with the implementation.

It is generally recommended that all large customisation projects employ the basic concepts of software engineering as described earlier. While this will not ensure success it will certainly ease the process and provide a framework for assessing progress. A key part of software engineering is specifying the precise scope and content of the system as early as possible. A timetable for deliverables should then be established and agreed upon by the developer(s) and the user(s). If the project is large then staged delivery with 'sign off' should be considered. If either party is inexperienced in using GIS then it is a good idea to adopt a prototyping approach. This will ensure that users learn what is possible with GIS and developers learn what users want to do with the system.

A further important practical recommendation is that the GIS implementation accounting model should budget for requirements analysis, specification, training, documentation, and acceptance testing. It is also as well to remember that coding will typically take only about 15 per cent of the time and that testing and documentation can take up to 30 per cent of the time. Even after a customisation project has been completed there may well be a period of 'institutionalisation' during which the application becomes incorporated into the business practices of an organisation.

Finally, experience suggests that it usually takes about twice as long as inexperienced users first estimate to do customisation. However, if the basic rules and suggestions outlined above are borne in mind then it should be possible to produce reasonably reliable cost and timescale estimates.

7 CONCLUSIONS

This chapter has defined the process of customisation and has shown why GIS customisation is necessary. The formal, well-established steps in the software development lifecycle have been described in outline terms. This included a comparison of the waterfall and prototyping methods. Information has been presented about the main development tools available to application developers. The integrated GIS-specific development environments of ARC/INFO and ArcView have been contrasted with the contemporary industry standard development environment offered by Microsoft's Visual Basic. In essence it seems that users and developers want a GIS customisation environment that is standard across GIS and non-GIS applications, easy to use and highly productive, yet offers access to powerful and sophisticated tools.

As the GIS market grows it is to be expected that more end-user products will be created by core software vendors and third party developers. When this happens, the need for end-user customisation will decrease. There will, however, always be a need for GIS customisation for the simple reasons that GIS is a very diverse field and all users and projects are different in some way.

The utilisation of object technology, such as Microsoft's OLE/COM, has already been cited as a key development in GIS customisation. As the Open Geodata Consortium releases implementation details of their Open Geodata Interoperability Specification (OGIS) and software vendors releases products which conform to it, it is expected that many more developers will be able to develop specific purpose GIS applications and customisations (Sondheim et al, Chapter 24). End-users will benefit enormously from this activity because they will be able to work with ready to run domain specific applications rather than generic GIS products.

References

Antenucci J, Brown K, Croswell, Kevany M 1991 *Geographic information systems – a guide to the technology*. New York, Van Nostrand Reinhold

Bell D, Morrey I, Pugh J 1992 *Software engineering: a programming approach*, 2nd edition. New York, Prentice-Hall

Brooks F 1995 *The mythical man month. Essays on software engineering*. Reading (USA), Addison-Wesley

Date C J 1995 *Introduction to database systems*, 6th edition. Reading, Addison-Wesley

Flaatten P O, McCubbrey D J, O'Riordan P D, Burgess K 1992 *Foundations of business systems*, 2nd edition. Orlando, The Dryden Press

Gilb T 1996 *Principles of software engineering management*, 2nd edition. Reading, Addison-Wesley

Korte G B 1996 Weighing GIS benefits with financial analysis. *GIS World* 9(7): 48–52

Maguire S A 1994 *Debugging the development process*. Redmond, Microsoft Press

Newell R G 1993 Customising a GIS. *GIS Europe* 2(1): 20–1

Rumbaugh J, Blaha M, Premerlani W, Eddy F, Lorensen W 1991 *Object-oriented modelling and design*. Englewood Cliffs, Prentice-Hall

Smith D A, Tomlinson R F 1992 Assessing costs and benefits of geographical information systems: methodological and implementation issues. *International Journal of Geographical Information Systems* 6: 247–56

Introduction

THE EDITORS

Like many other technologies, digital computers have evolved to provide ever more sophisticated environments for their users. Early programmers worked with very simple languages that in principle could do anything, but in practice were limited by the complexity of the necessary programming. Today's programming languages, and application programmer interfaces, allow far more to be achieved with much less effort. While it may have taken a million lines of code to write an early GIS in the 1960s or 1970s, the same could probably have been achieved with at least two orders of magnitude less, had it been possible to take advantage of the sophisticated programming environments available today. Languages like Tcl/Tk, for example, allow easy-to-use graphic interfaces to be constructed quickly that would have taken vastly more programmer effort 20 years ago.

Such progress relies on a simple principle: that if enough commonality can be identified between the needs of a sufficiently large number of users, then it makes sense to embed those common needs in the computing environment. Like the human mind, the digital computer is capable of supporting ever more complex concepts provided they can be constructed from simpler ones (and ultimately a 'hard-wired' base) in well-defined ways. Besides obvious gains in efficiency and productivity, such approaches provide consistency and rigour, offer simplicity by hiding the complex workings of operations from the programmer or user, and allow for uniform approaches to such issues as integrity.

By the mid 1960s, the computer industry had begun to see how this principle might be applied to the datasets processed by digital computers. Computer applications had been growing rapidly in various areas of industry and commerce, and were requiring and producing increasingly complex masses of data. Rather than treat each application as unique, and program its operations from scratch, there appeared to be sufficient commonality in the ways these applications interacted with data to justify the development of generic structures and approaches. Thus the database industry was born, in the form of special software applications to manage the interactions between programs and data. By assigning standard data management operations to generic systems, these so-called database management systems (DBMS) relieved the programmer of much inherently repetitive programming. They also encouraged a more disciplined approach to data management, which was perceived to have its own benefits in terms of increased efficiency and control.

While the database industry is by definition generic, and the characteristics of geographical data and GIS widely acknowledged to be special in many respects, nevertheless by the late 1970s significant efforts were under way to take advantage of database technology in GIS applications. Instead of a monolithic, stand-alone software application, GIS was increasingly perceived as layered, with specialised software working in conjunction with, or conceptually on top of, a standard DBMS. ESRI's ARC/INFO was one of the first of these, released in 1981 and incorporating an existing DBMS into a specialised GIS environment. Today, more and more of the functionality of GIS is assigned to increasingly sophisticated but still generic database products, many of which now include the capability to store and process explicitly spatial data.

These moves towards reliance on underlying DBMS reflect several important priorities and concerns in the GIS industry. First, if GIS and underlying DBMS are at least partially independent, then one DBMS can be easily replaced with another. This is attractive to many GIS customers, who may be able to share the DBMS among many computing applications within the organisation, and value the freedom to update the DBMS independently of the GIS. Second, the DBMS may be perceived as more reliable than less generic approaches to data management, because of the relative size of the DBMS industry – an industry more sophisticated in its

161

approach to data management, with better ways of ensuring data integrity; offering greater interoperability between software environments; and with greater adherence to general standards.

Michael Worboys begins this section with a discussion of database models (Chapter 26). The first generations of database systems, appearing in the 1960s, were regarded as too general for effective use in GIS, and it was not until the emergence of the relational model, with its greater sophistication, that GIS began to adopt database solutions in earnest. The term 'georelational' is often used to describe the particular implementation of the relational model for geographical data, in which geographical relationships between entities become the basis for many of the common keys or linkages between relational tables. Nevertheless, this idea took some time to emerge, and early uses of relational databases in GIS were driven largely by the more general advantages of database systems listed earlier.

Worboys takes the reader beyond the relational model into more recent research and thinking in database systems for GIS, notably the concepts broadly known as 'object-orientation'. Just as the relational model gave GIS users a natural way to represent geographical relationships, object-oriented models provide a natural way to manipulate the various entities found on the geographical landscape, and to describe their behaviours. As Worboys notes, object-oriented databases are in their infancy, and although several successful object-oriented GIS have appeared in recent years, there is still much work to be done in identifying the exact limits of the application of object-oriented thinking in GIS.

The designer of a generic solution to management of data must make decisions based on expectations about usage that will inevitably reflect the needs of the largest segment of users. As a specialised application and a relatively small part of the DBMS market, GIS has its own particular needs that are often difficult to promote in the wider arena of DBMS design. GIS databases tend to be large (a single remotely-sensed image or topographic map can easily require 100 million bytes of storage); and searching for geographical objects based on their locations is inherently multidimensional. DBMS solutions for GIS have often encountered disastrously poor performance, even though it is often possible to 'tune' a modern DBMS for the particular characteristics of a given application. Early users of relational DBMS for GIS found it necessary to develop complex implementation guidelines to ensure minimally

acceptable performance. Unfortunately, it is almost always true that the benefits of generic solutions must be balanced against the inability to optimise a generic design for the specific needs of a complex application.

Spatial indexing offers one of the most powerful tools to affect and improve performance in a GIS application, just as indexing in publishing or library cataloguing affects the usefulness of those fields. In the second chapter of this section, Peter van Oosterom reviews the state of the indexing art in spatial databases. Many indexing schemes have been devised, and it seems unlikely that any one is optimal over any significant domain of GIS applications. Many different schemes have been implemented, but although spatial indexing is often invisible to the user, it seems likely that in those applications where performance is critical, some degree of involvement of the user in the implementation of indexing will always be necessary.

Early DBMS followed one or other of the standard models for databases, but used proprietary languages for interaction with the user. Even though the underlying structure was essentially the same, a user wanting to move from one DBMS product to another often had to learn an entirely new language. The introduction of standard query languages, notably SQL, across entire sections of the DBMS industry led to much greater interoperability between systems, and greatly reduced the complications of training users. Recent efforts to extend SQL to the needs of GIS are reviewed by Worboys, while Max Egenhofer and Werner Kuhn in Chapter 28 give an overview of user interaction in general, comparing the query language approach to other, newer, and more powerful methods of user interface design. As an inherently visual technology, GIS stands to benefit enormously from graphic user interfaces, which offer the potential to make GIS much easier to use, and much easier to learn. Egenhofer and Kuhn review the various metaphors that are guiding contemporary user interface design for GIS, and that make use of an increasing number of distinct media.

In the final chapter in this section, Yvan Bédard adds a distinctly practical flavour to the topics discussed in the previous three. While databases provide the broad framework for describing the geographical world, the specific details of implementation can be critical, in determining performance, and essential to the success of any given application. Generic tools have been developed for database design, and much effort has gone into adapting these to the special needs of GIS.

26

Relational databases and beyond

M F WORBOYS

This chapter introduces the database perspective on geospatial information handling. It begins by summarising the major challenges for database technology. In particular, it notes the need for data models of sufficient complexity, appropriate and flexible human-database interfaces, and satisfactory response times. The most prevalent current database paradigm, the relational model, is introduced and its ability to handle spatial data is considered. The object-oriented approach is described, along with the fusion of relational and object-oriented ideas. The applications of object-oriented constructs to GIS are considered. The chapter concludes with two recent challenges for database technology in this field: uncertainty and spatio-temporal data handling.

1 INTRODUCTION TO DATABASE SYSTEMS

1.1 The database approach

Database systems provide the engines for GIS. In the database approach, the computer acts as a facilitator of data storage and sharing. It also allows the data to be modified and analysed while in the store. For a computer system to be an effective data store, it must have the confidence of its users. Data owners and depositors must have confidence that the data will not be used in unauthorised ways, and that the system has fail-safe mechanisms to cope with unforeseen events. Both data depositors and data users must be assured that, as far as possible, the data are correct. There should be sufficient flexibility to give different classes of users different types of access to the store. Most users will not be concerned with how the database works and should not be exposed to low-level database mechanisms. Data retrievers need flexible methods for establishing what is in the store and for retrieving data according to their requirements and skills. Users may have different conceptions of the organisation of the data in the store. The database interface should be sufficiently flexible to respond equally well to both single-time users with unpredictable and varied requirements, and to regular users with little variation in their requirements. Data should be retrieved as effectively as possible. It should be possible for users to link pieces of information together in the database to get the benefit of the added value from making the connections. Many users may wish to use the store, maybe even the same data, at the same time and this needs to be controlled. Data stores may need to be linked to other stores for access to pieces of information not in their local holdings.

1.2 Database history

Database management systems have grown out of file management systems that perform basic file handling operations such as sorting, merging, and report generation. During the 1950s, as files grew to have increasingly complex structures, an assortment of data definition products came into use. These became standardised by the Conference on Data Systems and Languages (CODASYL) in 1960 into the Common Business-Oriented Language, that is the COBOL programming language, which separates the definition on file structure from file manipulation. In 1969, the DataBase Task Group (DBTG) of CODASYL gave definitions for data description and definition languages, thus paving the way for hierarchal and network database management systems. The underlying model for these systems is navigational, that is connections between records are made by navigating explicit relationships between them. These relationships were 'hard-wired' into the database, thus limiting the degree to which such databases could be extended or distributed to other groups of users.

The acknowledged founder of relational database technology is Ted Codd, who in a pioneering paper (Codd 1970) set out the framework of the relational model. The 1970s saw the advent of relatively easy-to-use relational database languages such as the Structured Query Language, SQL, originally called the Structured English Query Language, SEQUEL (Chamberlin and Boyce 1974) and Query Language, QUEL (Held et al 1975), as well as prototype relational systems such as IBM's System R (Astrahan et al 1976) and University of California at Berkeley's Interactive Graphics and Retrieval Systems, INGRES (Stonebraker et al 1976).

From the latter part of the 1970s, shortcomings of the relational model began to become apparent for particular applications, including GIS. Codd himself provided extensions to incorporate more semantics (Codd 1979). Object-oriented notions were introduced from programming languages into databases, culminating in prototype object-oriented database systems, such as O_2 (Deux 1990) and ORION (Kim et al 1990). Today, object-oriented systems are well established in the marketplace, as are object-oriented extensions of relational systems, which may be where the future really is. Early developments in object-relational systems are described in Haas et al (1990) and Stonebraker (1986). SQL has developed into the international standard SQL-92, and SQL3 is being developed.

1.3 Data models

The data model provides a collection of constructs for describing and structuring applications in the database. Its purpose is to provide a common computationally meaningful medium for use by system developers and users. For developers, the data model provides a means to represent the application domain in terms that may be translated into a design and implementation of the system. For the users, it provides a description of the structure of the system, independent of specific items of data or details of the particular implementation.

A clear distinction should be made between data models upon which database systems are built, for example the relational model, and data models whose primary roles are to represent the meaning of the application domains as closely as possible (so-called *semantic data models*, of which entity-relationship modelling is an example: see also Martin, Chapter 6; Raper, Chapter 5). It might be that a semantic data model is used to develop applications for a database system designed around another model: the prototypical example of this is the use of the entity-relationship model to develop relational database applications. The three currently most important data modelling approaches are *record-based*, *object-based* and *object-relational*.

1.4 Human database interaction

Humans need to interact with database systems to perform the following broad types of task:

1 *Data definition*: description of the conceptual and logical organisation of the database, the database schema;
2 *Storage definition*: description of the physical structure of the database, for example file location and indexing methods;
3 *Database administration*: daily operation of the database;
4 *Data manipulation*: insertion, modification, retrieval, and deletion of data from the database.

The first three of these tasks are most likely to be performed by the database professional, while the fourth will be required by a variety of user types possessing a range of skills and experience as well as variable needs requirements in terms of frequency and flexibility of access.

User interfaces are designed to be flexible enough to handle this variety of usage. Standard methods for making interfaces more natural to users include menus, forms, and graphics (windows, icons, mice: see Egenhofer and Kuhn, Chapter 28; Martin 1996). Natural language would be an appropriate means of communication between human and database, but successful interfaces based on natural language have not yet been achieved. For spatial data, the graphical user interface (GUI) is of course highly appropriate. Specialised query languages for database interaction have been devised.

1.5 Database management

The software system driving a database is called the *database management system* (DBMS). Figure 1 shows schematically the place of some of these components in the processing of an interactive query, or an application program that contains within the host general-purpose programming language some database access commands. The DBMS has a query compiler that will parse and analyse a query and, if all is correct, generate execution code that is passed to the runtime database processor. Along the way, the compiler may call the query optimiser to optimise the code so that performance on the retrieval is improved. If the database language expression had been embedded in a general-purpose computer language such as C++, then an earlier precompiler stage would be needed. To retrieve the required data from the database, mappings must be made between the high-level objects in the query language statement and the physical location of the data on the storage device. These mappings are made using the system catalogue. Access to DBMS data is handled by the stored data manager, which calls the operating system for control of physical access to storage devices.

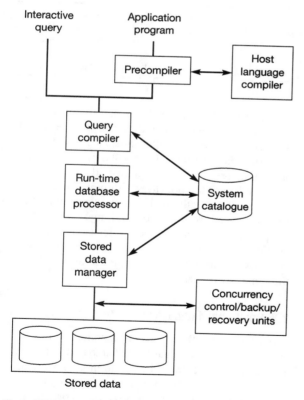

Fig 1. DBMS components used to process user queries.

Labels in figure: Interactive query; Application program; Precompiler; Host language compiler; Query compiler; Run-time database processor; System catalogue; Stored data manager; Concurrency control/backup/recovery units; Stored data.

The logical atom of interaction with a database is the *transaction*, broadly classified as *create*, *modify* (*update*), and *delete*. Transactions are either executed in their entirety (committed) or not at all (rollback to previous commit). The sequence of operations contained in transactions is maintained in a system log or journal, hence the ability of the DBMS to roll back. When a 'commit' is reached, all changes since the last commit point are then made permanent in the database. Thus, a transaction may be thought of as a unit of recovery. The DBMS seeks to maintain the so-called ACID properties of transactions: Atomicity (all-or-nothing), Consistency (of the database), Isolation (having no side-effects and unforeseen effects on other concurrent transactions), and Durability (ability to survive even after system crash).

2 RECORD-BASED DATA MODELS: RELATIONAL DATABASES

2.1 Introduction to the relational model

A record-based model structures the database as a collection of files of fixed-format records. The records in a file are all of the same record type, containing a fixed set of fields (attributes). The early network and hierarchical database systems, mentioned earlier, conform to the record-based data model. However, they proved to be too closely linked to physical implementation details, and they have been largely superseded by the relational model.

A relational database is a collection of tabular relations, each having a set of attributes. The data in a relation are structured as a set of rows. A row, or *tuple*, consists of a list of values, one for each attribute. An attribute has associated with it a domain, from which its values are drawn. Most current systems require that values are atomic – for example they cannot be decomposed as lists of further values – so a single cell in a relation cannot contain a set, list or array of values. This limits the possibilities of the pure relational model for GIS.

A distinction is made between a *relation schema*, which does not include the data but gives the structure of the relation (its attributes, their corresponding domains, and any constraints on the data) and a *relation*, which includes the data. The relation schema is usually declared when the database is set up and then remains relatively

Table 1 Tuples from the Country relation.

Name	Population (millions)	Land area (thousand sq. miles)	Capital
Austria	8	32	Vienna
Germany	81	138	Berlin
Italy	58	116	Rome
France	58	210	Paris
Switzerland	7	16	Bern

Table 2 Tuples from the City relation.

Name	Country	Population (thousands)
Vienna	Austria	1500
Berlin	Germany	3400
Hamburg	Germany	1600
Rome	Italy	2800
Milan	Italy	1400
Paris	France	2100
Zurich	Switzerland	300
Bern	Switzerland	100

Table 3 Tuples from the Country relation after a project operation.

Name	Population (millions)
Austria	8
Germany	81
Italy	58
France	58
Switzerland	7

Table 4 Tuples from the City relation after a restrict operation.

Name	Country	Population (thousands)
Berlin	Germany	3400
Rome	Italy	2800
Paris	France	2100

Table 5 Tuples from the joined Country and City relations.

Name	Country population (millions)	Land area (thousand sq. miles)	Capital	Country city	City population (thousands)
Austria	8	32	Vienna	Austria	1500
Germany	81	138	Berlin	Germany	3400
Italy	58	116	Rome	Italy	2800
France	58	210	Paris	France	2100
Switzerland	7	16	Bern	Switzerland	100

unaltered during the lifespan of the system. A relation, however, will typically be changing frequently as data are inserted, modified and deleted. A *database schema* is a set of relation schemata and a *relational database* is a set of relations, possibly with some constraints. An example of a database schema, used throughout this chapter, comprises two relations Country and City, along with their attributes, as shown:

Country (Name, Population, Land Area, Capital)
City (Name, Country, Population).

Tables 1 and 2 show part of an example database according to this schema. Each row of a relation in a relational database is sometimes called a tuple.

The primitive operations that can be supported by a relational database are the traditional set operations of union, intersection, and difference, along with the characteristically relational operations of project, restrict, join, and divide. The structure of these operations and the way that they can be combined is provided by relational algebra, essentially as defined by Codd (1970). The set operations union, intersection, and difference work on the relations as sets of tuples. The project operation applies to a single relation and returns a new relation that has a subset of attributes of the original. For example, Table 3 shows the Country relation projected onto its Name and Population attributes. The restrict operation acts on a relation to return only those tuples that satisfy a given condition. For example, Table 4 shows a restriction of the City relation, retrieving from the City relation those tuples containing cities with populations greater than two million. The join operation makes connections between relations, taking two relations as operands, and returns a single relation. The relation shown in Table 5 is a join of the Country and City relations, matching tuples when they have the same city names.

2.2 Relational database interaction and SQL

From the outset, there has been a collection of specialised query languages for database interaction. For relational databases, the Structured or Standard Query Language (SQL) is a *de facto* and *de jure* standard. SQL may either be used on its own as a means of direct interaction with the database, or may be embedded in a general-purpose programming language. The most recent SQL standard is SQL-92 (also called SQL2: ISO 1992). There is a large effort to move forward to SQL3.

2.2.1 Schema definition using SQL

The data definition language component of SQL allows the creation, alteration, and deletion of relation schemata. It is usual that a relation schema is altered only rarely once the database is operational. A relation schema provides a set of attributes, each with its associated data domain. SQL allows the definition of a domain by means of a CREATE DOMAIN expression.

A relation schema is created by a CREATE TABLE command as a set of attributes, each associated with a domain, with additional properties relating to keys and integrity constraints. For example, the relation schema City may be created by the command:

```
CREATE TABLE     City
(Name            PlaceName,
Country          PlaceName,
Population       Population,
PRIMARY KEY      (Name)
```

This statement begins by naming the relation schema (called a table in SQL) as City. The attributes are then defined by giving each its name and associated domain (assuming that we have already created domains PlaceName and Population). The primary key, which serves to identify a tuple uniquely, is next given as the attribute Name. There are also SQL commands to alter a relation schema by changing attributes or integrity constraints and to delete a relation schema.

2.2.2 Data manipulation using SQL

Having defined the schemata and inserted data into the relations, the next step is to retrieve data. A simple example of SQL data retrieval resulting in the relation in Table 4 is:

```
SELECT *
FROM City
WHERE Population > 2000000
```

The SELECT clause indicates the attribute to be retrieved from the City relation (* indicates all attributes), while the WHERE clause provides the restrict condition. Relational joins are effected by allowing more than one relation (or even the same relation called twice with different names) in the FROM clause. For example, to find names of countries whose capitals have a population less than two million people, use the expression:

```
SELECT Country.Name
FROM Country, City
WHERE Country.Capital = City. Name
AND City. Population < 2000000
```

In this case, the first part of the WHERE clause provides the join condition by specifying that tuples from the two tables are to be combined only when the values of the attributes Capital in Country and Name in City are equal. Attributes are qualified by prefixing the relation name in case of any ambiguity.

Most of the features of SQL have been omitted from this very brief summary. The documentation on the SQL2 standard is about 600 pages in length. The reader is referred to Date (1995) for a good survey of the relational model and SQL2.

2.3 Relational technology for geographical information

There are essentially two ways of managing spatial data with relational technology: putting all the data (spatial and non-spatial) in the relational database (integrated approach), or separating the spatial from the non-spatial data (hybrid approach). The benefits of using an integrated architecture are considerable, allowing a uniform treatment of all data by the DBMS, and thus not consigning the spatial data to a less sheltered existence outside the database, where integrity, concurrency, and security may not be so rigorously enforced. In theory, the integrated approach is perfectly possible: for example, Roessel (1987) provides a relational model of configurations of nodes, arcs, and polygons. However, in practice the pure relational geospatial model has not up to now been widely adopted because of unacceptable performance (Healey 1991). Essentially, problems arise because of:

1 slow retrieval due to multiple joins required of spatial data in relations;
2 inappropriate indexes and access methods, which are provided primarily for 1-dimensional data types by general-purpose relational systems;

3 lack of expressive power of SQL for spatial queries.

The first problem arises because spatial data are fundamentally complex – polygons being sequences of chains, which are themselves sequences of points. The object-oriented and extended relational models are much better able to handle such data types. With regard to the second problem, extended relational models allow much more flexibility in declaring indexes for different types of data. For the third problem, the limitations of SQL have been apparent for some time in a number of fields (for example, CAD/CAM, GIS, multimedia databases, office information systems, and text databases). SQL3, currently being developed as a standard, promises much in this respect.

3 OBJECT-BASED DATA MODELS

3.1 Introduction and the entity–relationship–attribute approach

The primary components of an object-based model are its *objects* or *entities*. The entity–relationship–attribute (ERA) model and the object-oriented (OO) models are the two main object-based modelling approaches. The ERA approach is attributed to Chen (1976) and has been a major modelling tool for relational database systems for about 20 years. In the ERA approach, an entity is a semantic data modelling construct and is something (such as a country) that has an independent and uniquely identifiable existence in the application domain. Entities are describable by means of their attributes (for example, the name, boundary, and population of a country). Entities have explicit relationships with other entities. Entities are grouped into entity types, where entities of the same type have the same attribute and relationship structure. The structure of data in a database may be represented visually using an ERA diagram. Figure 2 shows an ERA diagram representing the structure of these data in the example database schema in Tables 1 and 2. Entity types are represented by rectangles with offshoot attributes and connecting edges showing relationships. The ERA approach is fully discussed by Bédard (Chapter 29), and so is not considered

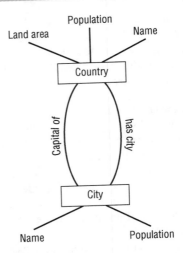

Fig 2. Example of an ERA diagram.

further here.

3.2 The object-oriented approach

3.2.1 Objects, classes, encapsulation, and identity

For many application domains, including GIS, ERA modelling has proved too limited and is being superseded by the OO approach. The OO approach is in use both as a method of semantic data modelling and as a model of data handled by object-oriented programming and database management systems. From the database systems viewpoint, the OO model adapts some of the constructs of object-oriented programming languages to database systems. The fundamental idea is that of encapsulation which places a wrapper around an identifiable collection of data and the code that operates upon it to produce an object. The state of an object at any time is determined by the value of the data items within its wrapper. These data items are referred to as *instance variables*, and the values held within them are themselves objects. This is an important distinction between objects (in the OO sense) and entities (in the ERA sense) which have a two-tier structure of entity and attribute.

The American National Standards Institute (ANSI 1991) Object-Oriented Database Task Group Final Technical Report describes an object as something 'which plays a role with respect to a request for an operation. The request invokes the operation that defines some service to be performed.' The code associated with a collection of data in an object provides a set of methods that can be performed upon it. As well as executing methods on its own data, an object may as part of one of its methods send a message to another object, causing that object to execute a method in response. This highly active environment is another feature that distinguishes between OO and ERA, which is essentially a collection of passive data. An object has both state, being the values of the instance variables within it, and behaviour, being the potential for acting upon objects (including itself). Objects with the same types of instance variables and methods are said to be in the same object class. Figure 3 shows some instance variables and methods associated with classes Country and Polygon and the manner in which the class Polygon is referenced as an instance variable by the class Country. Figure 4 shows in schematic form an object encapsulating state and methods, receiving a message from another object

```
Country

    Population: Integer
    Name: String
    Capital city: City
    Extent: Polygon

    Update City: City → City
    Insert City: → City
    Delete City: City →
```

```
Polygon

    Boundary: Set (Segment)

    Area: Polygon → Real
```

Fig 3. Part of the class descriptions for *Country* and *Polygon*.

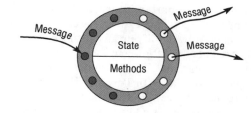

Fig 4. State, methods, and messages of an object.

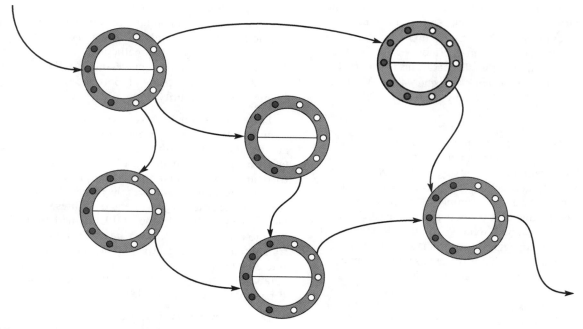

Fig 5. Messages between objects.

169

and executing methods which result in two messages output. Figure 5 shows the interaction of several objects in response to a message to one of them.

With encapsulation, the internal workings of an object are transparent to users and other objects, which can communicate with it only through a set of predefined message types that the object can understand and handle. To take an example from the real world, I usually do not care about the state of my car under the bonnet (internal state of object class Car) provided that when I put my foot on the accelerator (send message) the car's speed increases (change in the internal state leading to a change in the observable properties of the object). From the viewpoint external to the object, it is only its observable properties that are usually of interest.

3.2.2 Inheritance and composition of objects

Inheritance is an important system and semantic modelling construct, and involves the creation of a new object class by modifying an existing class. Inheritance in an object-oriented setting allows inheritance of methods. Thus, Triangle and Rectangle are subclasses of Polygon. The subclasses inherit all the instance variables and methods from the superclass as well as adding their own. In this example, Triangle and Rectangle may have specialised methods, for example the algorithm implementing the operation Area may be different for Rectangle and Triangle, and each will be different from an Area algorithm for Polygon. This phenomenon, where an operator with the same name has different implementations in different classes, is called *operator polymorphism*. An example of an inheritance hierarchy of spatial object classes is given below (see Figure 8).

Object *composition* allows the modelling of objects with complex internal structures. There are several ways in which a collection of objects might be composed into a new object. *Aggregation* composes a collection of object classes into an aggregate class. For example, an object class Property might be an aggregate of object classes Land Parcel and Dwelling. To quote Rumbaugh et al (1991), 'an aggregate object is semantically an extended object that is treated as a unit in many operations, although physically is made up of several lesser objects'. *Association* groups objects all from the same class into an associated class. For example, an object class Districts might be an association of individual district object classes.

As an illustration of some of these constructs, Figure 6 shows the object class Country (as an abstract object class, represented as a triangle) with three of its instance variables Name, Population, and Area. Variables Name and Population reference printable object class Character String (represented as an oval) and Area references abstract class Polygon. The class Polygon has instance variable Boundary referencing an association of class Segment (the association class shown in the figure as a star and circle). Each segment has a Begin and End Point, and each Point has a Position which is an aggregation (shown as a cross and circle) of printable classes X-coordinate and Y-coordinate.

3.3 Object-oriented database management systems

3.3.1 Making OO persistent

Object-Oriented Programming Languages (OOPLs) such as C++ and Smalltalk provide the capabilities to support the OO approach described above, including the creation, maintenance, and deletion of objects, object classes, and inheritance hierarchies. Object-Oriented Database Management Systems (OODBMS) supplement these capabilities with database functionality, including the ability to support:

1 persistent objects, object classes, and inheritance hierarchies;
2 non-procedural query languages for object class definition, object manipulation, and retrieval;
3 efficient query handling, including query optimisation and access methods;
4 appropriate transaction processing (ACID properties), concurrency support, recovery, integrity, and security.

There are essentially two choices for the developer of an OODBM system: extend a relational system to handle OO, or build a database system around an OO programming language. Both choices have been tried, and section 3.4 on object-extensions to relational technology explores the former. With regard to the latter, object-oriented features will already be supported by the OOPL, so there is the need to add persistency, query handling, and transaction processing. An approach to persistency is to add a new class Persistent Object and allow all database classes to inherit from this class. The class Persistent Object will include methods to:

1 create a new persistent object;
2 delete a persistent object;

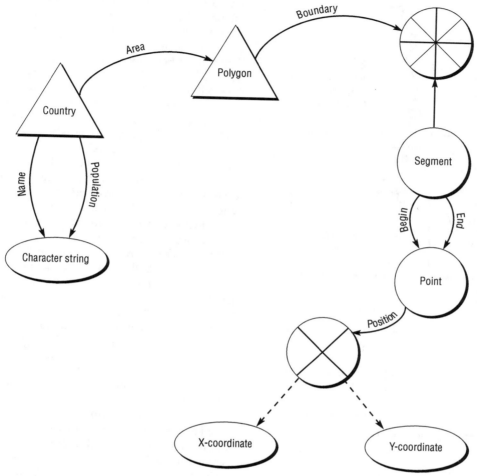

Fig 6. Complex objects.

3 retrieve the state of a persistent object;
4 provide concurrency control;
5 modify a persistent object.

A key benefit of an **OODBMS** is the support it provides for a unified programming and database environment. However, most current **OODBMS** treat persistent and non-persistent data differently. A fundamental distinction between RDBs and OODBs is that between call-by-value and call-by-reference. In an RDB, relationships are established by value matching. In our example, to retrieve the population of the capital of Germany, a join between Country and City is made using the value of the Capital field, Berlin. In an OODB (see Figure 7), the connection is made by navigation using the object identifiers (OIDs). The Capital instance variable of Country points to the appropriate City object.

3.3.2 Standardisation of OO systems

The OO approach is more complex than the relational model and has not yet crystallised into a set of universally agreed constructs; even basic constructs like inheritance have been given several different interpretations. Nevertheless, there has been considerable work to arrive at some common definitions.

The Object Management Group (OMG) is a consortium of hardware and software vendors, founded in 1990 with the aim of fostering standards for interoperability of applications within the OO approach. To this end, it has defined the OMG Object Model (see, for example, Kim 1995) many of the concepts of which have been discussed above. An important component of the OMG work is the

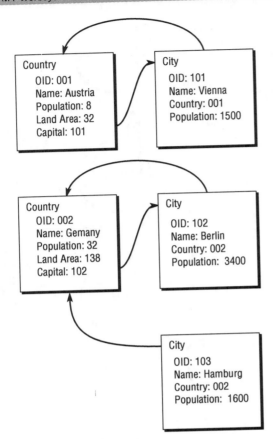

Fig 7. Navigation using object identifiers.

Common Object Request Broker Architecture (CORBA) standard, which specifies an Interface Definition Language for distributed access to objects (see Sondheim et al, Chapter 24). The Object Database Management Group (ODMG) is a consortium of OODBMS vendors, founded in 1990 with the aim of arriving at a commonly agreed OO database interface. The ODMG has defined object definition, manipulation, and query languages, corresponding to data definition and manipulation languages in relational systems.

3.4 Object extensions to relational technology

Object-relational models combine features of object-based and record-based models. They enhance the standard relational model with some object-oriented features, as opposed to OODBMS, that build

database functionality around an OO programming language. Enhancements include complex, possibly user-defined data types, inheritance, aggregation, and object identity. Early work at the University of California at Berkeley on the inclusion of new data types in relational database systems (Stonebraker 1986) led to the POSTGRES DBMS (Stonebraker and Rowe 1986). Parallel developments at the IBM Research Laboratories at San Jose, California resulted in the STARBURST project (reported in Haas et al 1990). These developments have led to contemporary proprietary object-relational systems as well as to the addition of object features to new releases of widely used proprietary relational systems. With regard to query languages that support OO extensions to the relational model, SQL3 is currently under development as an international standard. SQL3 is upwardly compatible with SQL-92, and adds support for objects, including multiple inheritance and operators. The goal for object-relational systems is to provide the wide range of object-oriented functionality that has proved so useful for semantic data modelling and programming systems, while at the same time giving the efficient performance associated with the relational model.

Objects are structured by the relational model as tuples of atomic values such as integers, floats, Booleans, or character strings. This provides only a limited means to define complex data types. Object-relational systems allow non-atomic types. A common extension of the relational model to provide for complex data types is to allow *nested relations*. In a nested relation, values of attributes need not be atomic but may themselves be relations.

Relational database systems provide hashing and B-tree indexes for access to standard, system-provided data types. A major extension that an object-relational system allows is the provision of more appropriate indexes for user-defined types. Object-relational systems provide for the definition of a range of indexes appropriate to a heterogeneous collection of object classes.

3.5 OOGIS

A basic requirement for any OO approach to GIS is a collection of spatial object classes. Figure 8 uses the notation of Rumbaugh et al (1991) to represent an inheritance hierarchy of some basic classes. Class Spatial is the most general class, which is specialised into classes Point and Extent (sets of points). Spatial extents may be classified according to dimension, and examples of classes in one dimension (Polyline) and two dimensions (Polygon) are given. Class Polyline is further specialised into classes Open Polyline and Closed Polyline, the former having two distinct end-points while the latter is joined and has no end-points. Of course, this is just an example of some basic spatial object classes. Table 6 shows some sample methods that will act upon these classes. The name of the method is given, along with the classes upon which it acts and the class to which the result belongs.

The OO approach to geospatial data management is now well established. For some time there have been innovatory proprietary GIS that provide OO programming language support, including spatial object classes, overlaying flat-file, or relational databases. There now exist proprietary GIS that incorporate a full OODBMS. Papers that survey the

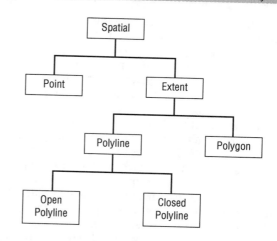

Fig 8. Spatial object class inheritance hierarchy.

Table 6 Inheritance hierarchy of spatial object classes.

Method	Operand	Operand	Result
Equals?	Spatial	Spatial	Boolean
Belongs?	Point	Extent	Boolean
Subset?	Extent	Extent	Boolean
Intersection	Extent	Extent	Extent
Union	Extent	Extent	Extent
Difference	Extent	Extent	Extent
Boundary	Polygon		ClosedPolyline
Connected?	Extent		Boolean
Extremes	OpenPolyline		Set(Point)
Within?	Point	ClosedPolyline	Boolean
Distance	Point	Point	Real
Bearing	Point	Point	Real
Length	Polyline		Real
Area	Polygon		Real
Centroid	Polygon		Point

application of OO to GIS include Egenhofer and Frank (1992); Worboys (1994); Worboys et al (1990).

4 CONCLUSIONS AND CHALLENGES

The purpose of a database is to serve a user community as a data store for a particular range of applications. Regarding geospatial applications, relational databases have fallen short of effectively achieving that purpose for two main reasons:

1 the relational model has not provided a sufficiently rich set of semantic constructs to allow users to model naturally geospatial application domains;
2 relational technology has not delivered the necessary performance levels for geospatial data management.

This chapter has argued that the OO approach provides part of the solution to these difficulties. It might be that the rapprochement between OO and relational technologies offers the best possible way forward.

There are still significant challenges for the database community in this area, and the chapter concludes by mentioning two of them. First, handling uncertain information has always played a major part in GIS, because many phenomena in the geographical world cannot be represented with total precision and accuracy (Fisher, Chapter 13). Reasoning with uncertain information and managing the associated data in a database remains an important research topic. Deductive databases incorporate logical formalisms, usually subsets of first-order logic, into databases, thereby increasing the expressive power of the query languages and allowing richer semantics for the data models (see, for example, Ceri et al 1990). There has also been work on the fusion of deductive and object technologies for GIS (Paton et al 1996).

Second, the world is in a continual state of change. Classical database technology provides only the capability to manage a single, static snapshot of the application domain. There are two ways in which this can be extended: *temporal* databases manage multiple snapshots (history) of the application domain as it evolves; and *dynamic* databases where a single snapshot changes in step with a rapidly and continuously changing application domain. There are many geospatial applications for both types of extension. Temporal GIS are required to handle such diverse applications as spatio-temporal patterns of land ownership and use, navigation, and global environmental modelling (see Peuquet, Chapter 8). Dynamic systems are required to model such rapidly changing contexts as transportation networks. Problems with development of such systems include the enormous volumes of data required for temporal databases and real-time transaction processing requirements in dynamic systems. These matters are covered in more detail by Bédard (Chapter 29).

References

ANSI 1991 *Object-Oriented Database Task Group final report*. X3/SPARC/DBSSG OODBTG. American National Standards Institute

Astrahan M, Blasgen M, Chamberlin D, Eswaran K, Gray J, Griffiths P, King W, Lorie R, McJones P, Mehl J, Putzolu G, Traiger I, Wade B, Watson V 1976 System R: a relational approach to database management. *ACM Transactions on Database Systems* 1: 97–137

Ceri S, Gottlob G, Tanca L 1990 *Logic programming and databases*. Berlin, Springer

Chamberlin D, Boyce R 1974 SEQUEL: a structured English query language. *Proceedings ACM SIGFIDET Workshop Conference, New York*. ACM Press: 249–64

Chen P P-S 1976 The entity–relationship model – toward a unified view of data. *ACM Transactions on Database Systems* 1: 9–36

Codd E 1970 A relational model for large shared data banks. *Communications of the ACM* 13: 377–87

Codd E 1979 Extending the relational database model to capture more meaning. *ACM Transactions on Database Systems* 4: 397–434

Date C J 1995 *An introduction to database systems*, 6th edition. Reading (USA), Addison-Wesley

Deux O 1990 The story of O_2. *Institute of Electrical and Electronics Engineers Transactions on Knowledge and Data Engineering* 2: 91–108

Egenhofer M J, Frank A 1992 Object-oriented modeling for GIS. *Journal of the Urban and Regional Information Systems Association* 4: 3–19

Haas L M, Chang W, Lohman G M, McPherson J, Wilms P F, Lapis G, Lindsay B, Piransesh H, Carey M J, Shekita E 1990 Starburst mid-flight: as the dust clears. *IEEE Transactions on Knowledge and Data Engineering* 2: 143–60

Healey R G 1991 Database management systems. In Maguire D J, Goodchild M F, Rhind D W (eds) *Geographical information systems: principles and applications*. Harlow, Longman/New York, John Wiley & Sons Inc. Vol. 1: 251–67

Held G D, Stonebraker M R, Wong E 1975 INGRES: a relational database system. *Proceedings AFIPS 44, Montvale*. AFIPS Press: 409–16

ISO 1992 *Database language SQL*. Document ISO/IEC 9075. International Organisation for Standardisation

Kim W (ed) 1995 *Modern database systems: the object model, interoperability, and beyond*. New York, ACM Press

Kim W, Garza J, Ballou N, Woelk D 1990 Architecture of the ORION next-generation database system. *IEEE Transactions on Knowledge and Data Engineering* 2: 109–25

Martin D J 1996 *Geographic information systems: socioeconomic applications*, 2nd edition. London, Routledge

Paton N, Abdelmoty A, Williams M 1996 Programming spatial databases: A deductive object-oriented approach. In Parker D (ed.) *Innovations in GIS 3*. London, Taylor and Francis: 69–78

Roessel J W van 1987 Design of a spatial data structure using the relational normal forms. *International Journal of Geographic Information Systems* 1: 33–50

Rumbaugh J, Blaha M, Premerlani W, Eddy F, Lorensen W 1991 *Object-oriented modeling and design*. Englewood Cliffs, Prentice-Hall

Stonebraker M 1986 Inclusion of abstract data types and abstract indexes in a database system. *Proceedings 1986 IEEE Data Engineering Conference, Los Alamitos*. IEEE Computer Society: 262–9

Stonebraker M, Rowe L 1986 The design of POSTGRES. *ACM SIGMOD International Conference on Management of Data, New York*. ACM Press: 340–55

Stonebraker M, Wong E, Kreps P 1976 The design and implementation of INGRES. *ACM Transactions on Database Systems* 1: 189–222

Worboys M F 1994b Object-oriented approaches to georeferenced information. *International Journal of*

27

Spatial access methods

P VAN OOSTEROM

This chapter first summarises why spatial access methods are needed. It is important to note that spatial access methods are not only useful for spatial data. Some early main memory spatial access methods are described (section 2), followed by an overview of space-filling curves (section 3). As it is impossible to present all of the spatial access methods described in the recent literature, only the following characteristic families are presented: quadtree, grid-based methods, and R-tree (sections 4–6). Special attention is paid to spatial access methods taking multiple scales into account (section 7). Besides the theory of spatial access methods, the issue of using them in a database in practice is treated in the conclusion of this chapter (section 8).

1 WHY ARE SPATIAL ACCESS METHODS NEEDED?

The main purpose of spatial access methods is to support efficient selection of objects based on spatial properties. For example, a range query selects objects lying within specified ranges of coordinates; a nearest neighbour query finds the object lying closest to a specified object (see Worboys, Chapter 26). Further, spatial access methods are also used to implement efficiently such spatial analyses as map overlay, and other types of spatial joins. Two characteristics of spatial datasets are that they are frequently large and that the data are quite often distributed in an irregular manner. A spatial access method needs to take into account both spatial indexing and clustering techniques. Without a spatial index, every object in the database has to be checked to see whether it meets the spatial selection criterion; a 'full table scan' in a relational database. As spatial datasets are usually very large, such checking is unacceptable in practice for interactive use and most other applications. Therefore, a spatial index is required, in order to find the required objects efficiently without looking at every object. In cases when the whole spatial dataset resides in main memory it is sufficient to know the addresses of the requested objects, as main memory storage allows

random access and does not introduce significant delays. However, most spatial datasets are so large that they cannot reside in the main memory of the computer and must be stored in secondary memory, such as its hard disk. Clustering is needed to group those objects which are often requested together. Otherwise, many different disk pages will have to be fetched, resulting in slow response. In a spatial context, clustering implies that objects which are close together in reality are also stored close together in memory. Many strategies for clustering objects in spatial databases adopt some form of 'space-filling curve' by ordering objects according to their sequence along a path that traverses all parts of the space.

In traditional database systems, sorting (or ordering) of the data forms the basis for efficient searching, as in the B-tree approach (Bayer and Creight 1973). Although there are obvious bases for sorting text strings, numbers, or dates (1-dimensional data), there are no such simple solutions for sorting higher-dimensional spatial data. Computer memory is 1-dimensional but spatial data is 2-dimensional or 3-dimensional (or even higher dimensioned), and must be organised somehow in memory. An intuitive solution is to use a regular grid just as on a paper map. Each grid cell has a unique name, e.g. 'A3', 'C6', or 'D5'. The cells are stored in some order in memory and can each contain a (fixed) number of object references. In a grid cell, a reference is stored to an object whenever the object (partially) overlaps the cell. However, this will not be very efficient because of the irregular data distribution of spatial data: many cells will be empty (e.g. in the ocean), while many other cells will be over-full (e.g. in the city centre). Therefore, more advanced techniques have been developed.

2 MAIN MEMORY ACCESS METHODS

Though originally not designed for handling very large datasets, main memory data structures show several interesting techniques with respect to handling spatial data. In this section the KD-tree (adaptive, bintree) and the BSP-tree will be illustrated.

2.1 The KD-tree

The basic form of the KD-tree stores K-dimensional points (Bentley 1975). This section concentrates on the 2-dimensional case. Each internal node of the KD-tree contains one point and also corresponds to a rectangular region. The root of the tree corresponds to the whole region of interest. The rectangular region is divided into two parts by the x-coordinate of the stored point on the odd levels and by the y-coordinate on the even levels in the tree; see Figure 1. A new point is inserted by descending the tree until a leaf node is reached. At each internal node the value of the proper coordinate of the stored point is compared with the corresponding coordinate of the new point and the proper path is chosen. This continues until a leaf node is reached. This leaf also represents a rectangular region, which in turn will be divided into two parts by the new point. The insertion of a new point results in one new internal node. Range searching in the KD-tree starts at the root, checks

whether the stored node (split point) is included in the search range and whether there is overlap with the left or right subtree. For each subtree which overlaps the search region, the procedure is repeated until the leaf level is reached.

A disadvantage of the KD-tree is that the shape of the tree depends on the order in which the points are inserted. In the worst case, a KD-tree of n points has n levels. The adaptive KD-tree (Bentley and Friedman 1979) solves this problem by choosing a splitting point (which is not an element of the input set of data points), which divides the set of points into two sets of (nearly) equal size. This process is repeated until each set contains one point at most; see Figure 2. The adaptive KD-tree is not dynamic: it is hard to insert or delete points while keeping the tree balanced. The adaptive KD-tree for n points can be built in $O(n \log n)$ time and takes $O(n)$ space for $K=2$. A range query takes $O(\mathrm{sqrt}(n)+t)$ time in two dimensions where t is the number of points found. Another variant of the KD-tree is the bintree (Tamminen 1984). Here the space is divided into two equal-sized rectangles instead of two rectangles with equal numbers of points. This is repeated until each leaf contains one point at the most.

A modification that makes the KD-tree suitable for secondary memory is described by Robinson (1981) and is called the KDB-tree. For practical use, it is more convenient to use leaf nodes containing more than one data point. The maximum number of points that a leaf may contain is called the 'bucket size'. The bucket size is chosen in such a way that it fits within one disk page. Moreover, internal nodes are grouped and each group is stored on one page in order to minimise the number of disk accesses. Robinson describes algorithms for deletions and insertions under which the KDB-tree remains balanced. Unfortunately, no reasonable upper bound for memory usage can be guaranteed.

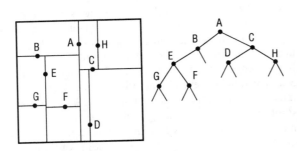

Fig 1. The KD-tree.

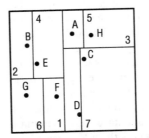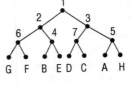

Fig 2. The adaptive KD-tree.

Matsuyama et al (1984) show how the geometric primitives polyline and polygon may be incorporated using the centroids of bounding boxes in the 2-D-tree. Rosenberg (1985) uses a 4-D-tree to store a bounding box by putting the minimum and maximum points together in one 4-dimensional point. This technique can be used to generalise other geometric data structures that are originally suited only for storing and retrieving points. The technique works well for exact-match queries, but is often more complicated in the case of range queries. In general, geometrically close 2-dimensional rectangles do not map into geometrically close 4-dimensional points (Hutflesz et al 1990). The ranges are transformed into complex search regions in the higher-dimensional space, which in turn results in slow query responses.

2.2 The BSP-tree

We begin here by describing the binary space partitioning (BSP)-tree, before presenting a variant suitable for GIS applications: the multi-object BSP-tree for storing polylines and polygons. The original use of the BSP-tree was in 3-dimensional computer graphics (Fuchs et al 1980; Fuchs et al 1983). The BSP-tree was used by Fuchs to produce a hidden surface image of a static 3-dimensional scene. After a preprocessing phase it is possible to produce an image from any view angle in $O(n)$ time, with n the number of polygons in the BSP-tree.

In this chapter the 2-dimensional BSP-tree is used for the structured storage of geometric data. It is a data structure that is not based on a rectangular division of space. Rather, it uses the line segments of the polylines and the edges of the polygons to divide the space in a recursive manner. The BSP-tree reflects this recursive division of space. Each time a (sub)space is divided into two subspaces by a so-called splitting primitive, a corresponding node is added to the tree. The BSP-tree represents an organisation of space by a set of convex subspaces in a binary tree. This tree is useful during spatial search and other spatial operations. Figure 3(a) shows a scene with some directed line segments. The 'left' side of the line segment is marked with an arrow. From this scene, line segment A is selected and space is split into two parts by the supporting line of A. This process is repeated for each of the two sub-spaces with the other line segments. The splitting of space continues until there are no line segments left. Note that sometimes the splitting of a

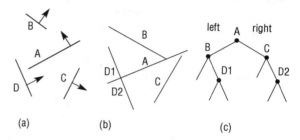

Fig 3. The building of a BSP-tree: (a) 2-dimensional scene; (b) convex sub-spaces; and (c) BSP-tree.

space implies that a line segment (which has not yet been used for splitting itself) is split into two parts. Line D, for example, is split into D1 and D2. Figure 3(b) shows the resulting organisation of the space, as a set of (possibly open) convex subspaces. The corresponding BSP-tree is drawn in Figure 3(c).

The BSP-tree, as discussed so far, is suitable only for storing a collection of (unrelated) line segments. In GIS it must be possible to represent objects, such as polygons. The multi-object BSP-tree (Oosterom 1990) is an extension of the BSP-tree which caters for object representation. It stores the line segments that together make up the boundary of the polygon. The multi-object BSP-tree has explicit leaf nodes which correspond to the convex subspaces created by the BSP-tree. Figure 4(a) presents a 2-dimensional scene with two objects, triangle T with sides abc, and rectangle R with sides defg. The method divides the space in the convex subspaces of Figure 4(b). The BSP-tree of Figure 4(c) is extended with explicit leaf nodes, each representing a convex part of the space. If a convex subspace corresponds to the 'outside' region, no label is drawn in the figure. If no more than one identification tag per leaf is allowed, only mutually exclusive objects can be stored in the multi-object BSP-tree, otherwise it would be possible also to deal with objects that overlap. A disadvantage of this

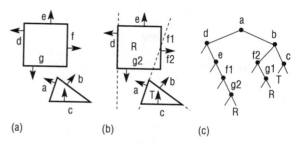

Fig 4. The building of a multi-object BSP-tree: (a) object scenes; (b) convex sub-spaces; and (c) multi-object BSP-tree.

177

BSP-tree is that the representation of one object is scattered over several leaves, as illustrated by rectangle R in Figure 4. The (multi-object) BSP-tree allows efficient implementation of spatial operations, such as pick and rectangle search.

The choice of which line segment to use for dividing the space very much influences the building of the tree. It is preferable to have a balanced BSP-tree with as few nodes as possible. This is a very difficult requirement to fulfil, because balancing the tree requires that line segments from the middle of the dataset be used to split the space. These line segments will probably split other line segments. Each split of a line segment introduces an extra node in the BSP-tree. However, Paterson and Yao (1989) prove that, if the original line segments are disjoint, then it is possible to build a BSP-tree with $O(n \log n)$ nodes and depth $O(\log n)$ using an algorithm requiring only $O(n \log n)$ time.

3 SPACE-FILLING CURVES

This section presents an overview and some properties of space-filling curves. Space-filling curves order the points in a discrete 2-dimensional space. This technique is also called tile indexing. It transforms a 2-dimensional problem into a 1-dimensional one, so it can be used in combination with a well known data structure for 1-dimensional storage and retrieval, such as the B-tree (Bayer and McCreight 1973). This presentation is based on several papers (Abel and Mark 1990; Goodchild and Grandfield 1983; Jagadish 1990; Nulty and Barholdi 1994) and the book by Samet (1989).

Row ordering simply numbers the cells row by row, and within each row the points are numbered from left to right; see Figure 5(a). Row-prime (or snake like, or boustrophedon) ordering is a variant in which alternate rows are traversed in opposite directions; see Figure 5(b). Obvious variations are column and column-prime orderings in which the roles of row and column are transposed. Bitwise interleaving of the two coordinates results in a 1-dimensional key, called the Morton key (Orenstein and Manola 1988). The Morton key is also known as the Peano key, or N-order, or Z-order. For example, row 2 = 10_{bin} column 3 = 11_{bin} has Morton key 14 = 1110_{bin}; see Figure 5(c). Hilbert ordering is based on the classic Hilbert-Peano curve, as drawn in Figure 5(d). Gray ordering is obtained by bitwise interleaving the Gray codes of the x and y

coordinates. As Gray codes have the property that successive codes differ in exactly one bit position, a 4-neighbour cell differs only in one bit; see Figure 5(e) (Faloutsos 1988). In Figure 5(f) the Cantor-diagonal ordering is shown. Note that the numbering of the points is adapted to the fact that we are dealing with a space that is bounded in all directions; for example, row 3 column 1 has order number 10 instead of 11 and row 3 column 3 has order number 15 instead of 24. Spiral ordering is depicted in Figure 5(g). Finally, Figure 5(h) shows the Sierpinski curve, which is based on a recursive triangle subdivision.

Figure 6 shows the geometric construction of the Peano curve: at each step of the refinement each vertex of the basic curve is replaced by the previous order curve. A similar method, but now also including reflection, is used to construct the reflected binary Gray curve; see Figure 7. The Hilbert curve is constructed by rotating the previous order curves at vertex 0 by –90 degrees and at vertex 3 by 90 degrees; see Figure 8. The Sierpinski curve starts with two triangles; each triangle is split into two new triangles and this is repeated until the required resolution is obtained. Note that the orientation and ordering of the triangles is important; see Figure 9.

Abel and Mark (1990) have identified the following desirable qualitative properties of spatial orderings:

- An ordering is continuous if, and only if, the cells in every pair with consecutive keys are 4-neighbours.
- An ordering is quadrant-recursive if the cells in any valid sub-quadrant of the matrix are assigned a set of consecutive integers as keys.
- An ordering is monotonic if, and only if, for every fixed x, the keys vary monotonically with y in some particular way, and vice versa.
- An ordering is stable if the relative order of points is maintained when the resolution is doubled.

Ordering techniques are very efficient for exact-match queries for points, but there is quite a difference in their efficiency for other types of geometric queries, for example range queries. Abel and Mark (1990) conclude from their practical comparative analysis of five orderings (they do not consider the Cantor-diagonal, the spiral and the Sierpinski orderings) that the Morton ordering and the Hilbert ordering are, in general, the best options. Some quantitative properties of curves are: the total length of the curve, the variability in unit lengths (path between two cells next in order), the average of

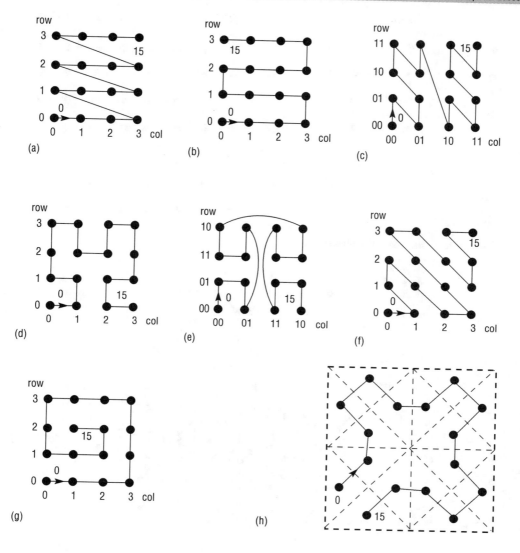

Fig 5. Eight different orderings: (a) row; (b) row prime; (c) Peano; (d) Hilbert; (e) Gray; (f) Cantor-original; (g) spiral; and (h) Sierpinski (triangle).

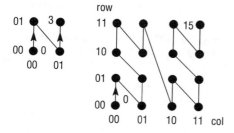

Fig 6. Geometric construction of the Peano curve.

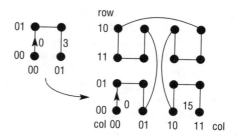

Fig 7. Geometric construction of the Gray curve.

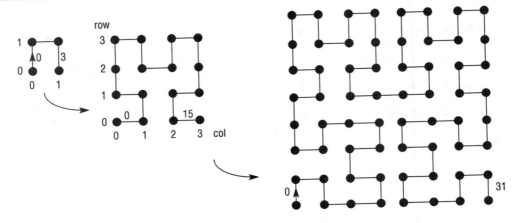

Fig 8. Geometric construction of the Hilbert curve.

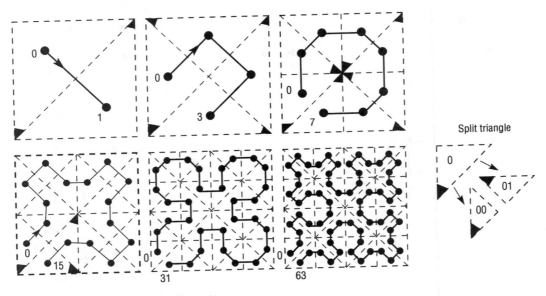

Fig 9. Geometric construction of the Sierpinski curve.

the average distance between 4-neighbours, and the average of the maximum distance between 4-neighbours. Goodchild (1989) proved that the expected difference of 4-neighbour keys of an n by n matrix is $(n+1)/2$ for Peano, Hilbert, row and row-prime orderings, indicating that this is not a very discriminating property. Therefore, Faloutsos and Roseman (1989) suggest a better measure for spatial clustering: the average (Manhattan) maximum distance of all cells within $N/2$ key value of a given cell on a N by N grid; see Table 1. Another measure for clustering is the average

number of clusters for all possible range queries. Note that a cluster is defined as a group of cells with consecutive key value; see Table 2 and Figure 10.

Table 1 Average Manhatten maximum distance of cell within N/2 key value (after Faloutsos and Roseman 1989).

grid	Hilbert	Gray	Peano
2*2	1.00	1.00	1.50
4*4	2.00	2.75	2.75
8*8	3.28	5.00	4.84
16*16	4.89	8.52	7.91

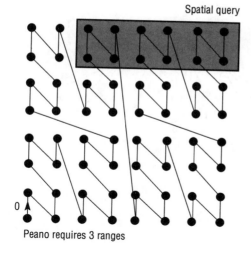

Spatial query

Peano requires 3 ranges

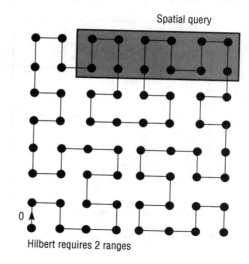

Spatial query

Hilbert requires 2 ranges

Fig 10. Number of clusters for a given range query.

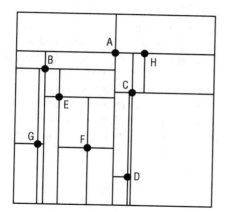

Fig 11. The point quadtree.

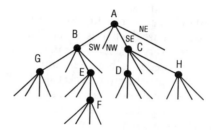

Table 2 Average number of clusters for all possible range queries. (After Faloutsos and Roseman 1989.)

grid	Hilbert	Gray	Peano
2*2	1.11	1.11	1.22
4*4	1.64	1.92	2.16
8*8	2.93	4.02	4.41
16*16	5.60	8.71	9.29

4 THE QUADTREE FAMILY

The quadtree is a generic name for all kinds of trees that are built by recursive division of space into four quadrants. Several different variants have been described in the literature, of which the following will be presented here: point quadtree, PR (point region) quadtree, region quadtree, and PM (polygonal map) quadtree. Samet (1984, 1989) gives an excellent overview.

4.1 Point quadtree and PR quadtree

The point quadtree resembles the KD-tree described in section 2. The difference is that the space is divided into four rectangles instead of two; see Figure 11. The input points are stored in the internal nodes of the tree. The four different rectangles are typically referred to as SW (southwest), NW (northwest), SE (southeast), and NE (northeast).

181

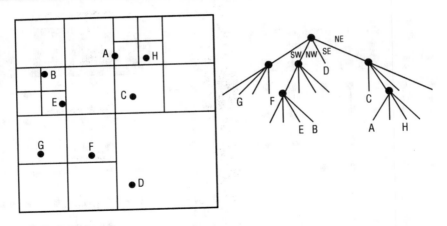

Fig 12. The PR quadtree.

Searching in the quadtree is very similar to the KD-tree: whenever a point is included in the search range it is reported and whenever a subtree overlaps with the search range it is traversed.

A minor variant of the point quadtree is the **PR** quadtree, which does not use the points of the data set to divide the space. Every time it divides the space, a square, into four equal subsquares, until each contains no more than the given bucket size (e.g. one object). Note that dense data regions require more partitions and therefore the quadtree will not be balanced in this situation; see Figure 12.

4.2 Region quadtree

A very well known quadtree is the region quadtree, which is used to store a rasterised approximation of a polygon. First, the area of interest is enclosed by a square. A square is repeatedly divided into four squares of equal size until it is completely inside (a black leaf) or outside (a white leaf) the polygon or until the maximum depth of the tree is reached (dominant colour is assigned to the leaf); see Figure 13. The main drawback is that it does not contain an exact representation of the polygon. The same applies if the

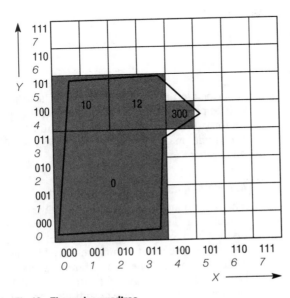

Fig 13. The region quadtree.

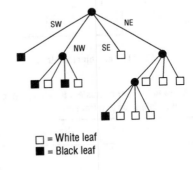

☐ = White leaf
■ = Black leaf

(note that SW=0, NW=1, SE=2, NE=3)

Quadcode 0 has Morton range: 0–15
Quadcode 10 has Morton range: 16–19
Quadcode 12 has Morton range: 24–27
Quadcode 300 has Morton range: 48–48

region quadtree is used to store points and polylines. This kind of quadtree is useful for storing raster data.

4.3 PM quadtree

A polygonal map, a collection of polygons, can be represented by the PM quadtree. The vertices are stored in the tree in the same way as in the PR quadtree. The edges are segmented into q-edges which completely fall within the squares of the leaves. There are seven classes of q-edges. The first are those that intersect one boundary of the square and meet at a vertex within that square. The other six classes intersect two boundaries and are named after the boundaries they intersect: NW, NS, NE, EW, SW, and SE. For each non-empty class, the q-edges are stored in a balanced binary tree. The first class is ordered by an angular measure and the other six classes are ordered by their intercepts along the perimeter. Figure 14 shows a polygonal map and the corresponding PM quadtree. The PM quadtree provides a reasonably efficient data structure for performing various operations: inserting an edge, point-in-polygon testing, overlaying two maps, range searching and windowing.

The extensive research efforts on quadtrees in the last decade resulted in more variants and algorithms to manipulate these quadtrees efficiently. For example, the CIF quadtree is particularly suited for rectangles. Another interesting example is the linear quadtree: in this representation there is no explicit quadtree, but only an enumeration of the quadcodes belonging to the object; e.g. 0, 10, 12, and 300 in Figure 13. Rosenberg (1985) and Samet (1984) describe these variants.

5 GRID-BASED METHODS

The intuitively attractive approach of organising space by imposing a regular grid has been refined in several different ways, in order to avoid the problems of dealing with irregular distributed data. In this section two different approaches are described: the grid file and the field-tree.

5.1 The grid file

The principle of a grid file is the division of a space into rectangles (regular tiles, grids, squares, cells) that can be identified by two indices, one for the x-direction and the other for the y-direction. The grid file is a non-hierarchical structure. The geometric primitives are stored in the grids, which are not necessarily of equal size. There are several variants of this technique. In this subsection the file structure as defined by Nievergelt et al (1984) is described.

The advantage of the grid file as defined by Nievergelt et al is that, unlike most other grid files, it adjusts itself to the density of the data: however, it is also more complicated. The cell division lines need not be equidistant; for each of x and y there is a 1-dimensional array (in main memory) giving the actual sizes of the cells. Neighbouring cells may be joined into one bucket if the resulting area is a rectangle. The buckets have a fixed size and are stored on a disk page. The grid directory is a 2-dimensional array, with a pointer for each cell to the correct bucket. Figure 15 shows a grid file with linear scales and grid directory. The grid file has good dynamic properties. If a bucket is too full to store a new primitive and it is used for more than one cell, then the bucket may be divided into two buckets. This is a

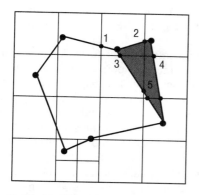

Tree: same as PR Quadtree
2 nodes with their balanced binary trees

node with q-edges

1,2,3	4,5	
vertex tree	NS-tree	SW-tree
1 2 3	4	5
other trees empty	other trees empty	

Fig 14. The PM quadtree.

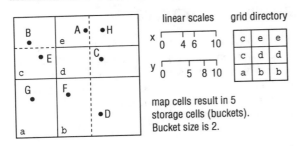

Fig 15. The grid file.

minor operation. If the bucket is used for only one cell, then a division line is added to one of the linear scales. This is a little more complex but is still a minor operation. In the case of a deletion of primitives, the merging process is performed in a manner analogous to the splitting process for insertion.

5.2 The field-tree

The field-tree is suited to store points, polylines, and polygons in a non-fragmented manner. Several variants of the field-tree have been published (Frank 1983; Frank and Barrera 1989; Kleiner and Brassel 1986). In this subsection attention will be focused on the partition field-tree. Conceptually, the field-tree consists of several levels of grids, each with a different resolution and a different displacement/origin; see Figure 16. A grid cell is called a field. The field-tree is not, in fact, a hierarchical tree, but a directed acyclic graph, as each field can have one, two, or four ancestors. At one level the fields form a partition and

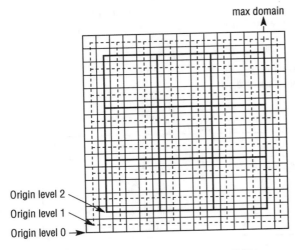

Fig 16. The positioning of geometric objects in the field-tree.

therefore never overlap. In another variant, the cover field-tree (Frank and Barrera 1989), the fields may overlap. It is not necessary at each level that the entire grid be explicitly present as fields.

A newly inserted object is stored in the smallest field in which it completely fits (unless its importance requires it to be stored at a higher level). As a result of the different displacements and grid resolutions, an object never has to be stored more than three levels above the field size that corresponds to the object size. Note that this is not the case in a quad-tree-like structure, because here the edges at different levels are collinear. The insertion of a new object may cause a field to become too full. In this case an attempt is made to create one or more new descendants and to reorganise the field by moving objects down. This is not always possible. A drawback of the field-tree is that an overflow page is sometimes required, as it is not possible to move relatively large or important objects from an over-full field to a lower level field.

6 THE R-TREE FAMILY

Instead of dividing space in some manner, it is also possible to group the objects in some hierarchical organisation based on (a rectangular approximation of) their location. This is the approach of the R-tree and in this section several variants are also described: the R+-tree, R*-tree, Hilbert R-tree and sphere-tree.

6.1 The R-tree

The R-tree is an index structure that was defined by Guttman in 1984. The leaf nodes of this multiway tree contain entries of the form (I, object-id), where object-id is a pointer to a data object and I is a bounding box (or an axes-parallel minimal bounding rectangle, MBR). The data object can be of any type: point, polyline, or polygon. The internal nodes contain entries of the form (I, child-pointer), where child-pointer is a pointer to a child and I is the MBR of that child. The maximum number of entries in each node is called the branching factor M and is chosen to suit paging and disk I/O buffering. The Insert and Delete algorithms assure that the number of entries in each node remains between m and M, where $m \leq \lceil M/2 \rceil$ is the minimum number of entries per node. An advantage of the R-tree is that pointers to complete objects (e.g. polygons) are stored, so the objects are never fragmented.

When inserting a new object, the tree is traversed from the root to a leaf choosing each time the child which needs the least enlargement to enclose the object. If there is still space, then the object is stored in that leaf. Otherwise, the leaf is split into two leaves. The entries are distributed among the two leaves in order to try to minimise the total area of the two leaves. A new leaf may cause the parent to become over-full, so it has to be split also. This process may be repeated up to the root. During the reverse operation, delete, a node may become under-full. In this situation the node is removed (all other entries are saved and reinserted at the proper level later on). Again, this may cause the parent to become under-full, and the same technique is applied at the next level. This process may have to be repeated up to the root. Of course, the MBRs of all affected nodes have to be updated during an insert or a delete operation.

Figure 17 shows an R-tree with two levels and $M = 4$. The lowest level contains three leaf nodes and the highest level contains one node with pointers and MBRs of the leaf nodes. Coverage is defined as the total area of all the MBRs of all leaf R-tree nodes, and overlap is the total area contained within two or more leaf MBRs (Faloutsos et al 1987). In Figure 17 the coverage is A $\cup$ B $\cup$ C and the overlap is A $\cap$ B. It is clear that efficient searching demands both low coverage and low overlap.

6.2 Some R-tree variants

Roussopoulos and Leifker (1985) describe the Pack algorithm which creates an initial R-tree that is more efficient than the R-tree created by the Insert algorithm. The Pack algorithm requires all data to be known *a priori*. The R+-tree (Faloutsos et al 1987),

a modification of the R-tree, avoids overlap at the expense of more nodes and multiple references to some objects; see Figure 18. Therefore, point queries always correspond to a single-path tree traversal. A drawback of the R+-tree is that no minimum space utilisation per node can be given. Analytical results indicate that R+-trees allow more efficient searching, in the case of relatively large objects.

The R*-tree (Beckmann et al 1990) is based on the same structure as the R-tree, but it applies a different Insert algorithm. When a node overflows, it is not split right away, but first an attempt is made to remove p entries and reinsert these in the tree. The parameter p can vary. In the original paper it is suggested that p be set to be 30 per cent of the maximum number of entries per node. In some cases this will solve the node overflow problem without splitting the node. In general this will result in a fuller tree. However, this reinsert technique will not always solve the problem and sometimes a real node split is required. Instead of only minimising the total area, an attempt is also made to minimise overlap between the nodes, and to make the nodes as square as possible.

The Hilbert R-tree uses the centre point Hilbert value of the MBR to organise the objects (Kamel and Faloutsos 1994). When grouping objects (based on their Hilbert value), they form an entry in their parent node which contains both the union of all MBRs of the objects and the largest Hilbert value of the objects. Again, this is repeated on the higher levels until a single root is obtained. Inserting and deleting is basically done using the (largest) Hilbert value and applying B-tree (Bayer and McCreight 1973) techniques. Searching is done using the MBR and applying R-tree techniques. The B-tree technique makes it possible to get fuller nodes, because '2-to-3' or '3-to-4' (or higher) split policies can be used. This results in a more compact tree,

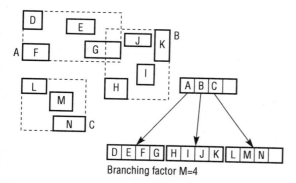

Fig 17. The R-tree.

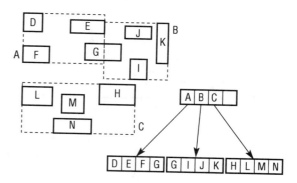

Fig 18. The R+-tree.

which is again beneficial for performance. The drawback of the Hilbert R-tree is that the real spatial aspects of the objects are not used to organise them, but instead their Hilbert values. It is possible that two objects which are very close in reality have Hilbert values which are very different. Therefore, these two objects will not end up in the same node in spite of the fact that they are very close; instead each of them may be grouped with other objects (further away in reality, but with closer Hilbert value). This will result in larger MBRs and therefore reduced performance.

The sphere-tree (Oosterom and Claassen 1990) is very similar to the R-tree (Guttman 1984), with the exception that it uses minimal bounding circles (or spheres in higher dimensions, MBSs) instead of MBRs; see Figure 19. Besides being orientation-insensitive, the sphere-tree has the advantage over the R-tree in that it requires less storage space. The operations on the sphere-tree are very similar to those on the R-tree, with the exception of the computation of the minimal bounding circle, which is more difficult (Elzinga and Hearn 1972; Megiddo 1983; Sylvester 1857).

7 MULTISCALE SPATIAL ACCESS METHODS

Interactive GIS applications can be supported even better if importance (resolution, scale) is taken into account in addition to spatial location when designing access methods (see also Weibel and Dutton, Chapter 10). Think of a user who is panning and zooming in a certain dataset. Just enlarging the objects when the user zooms in will result in a poor map. Not only must the objects be enlarged, but they must be displayed with more

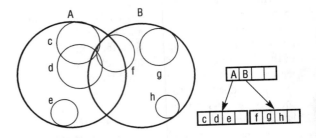

Fig 19. The sphere-tree.

detail (because of the higher resolution), and less significant objects must also be displayed. A simple solution is to store the map at different scales (or levels of detail). This would introduce redundancy with all the related drawbacks of possible inconsistency and increased memory usage. Therefore, geographical data should be stored in an integrated manner without redundancy and, if required, be supported by a special data structure. Detail levels are closely related to cartographic map generalisation techniques. Besides being suited for map generalisation, these multiscale data structures must also provide spatial properties; e.g. it must be possible to find all objects within a specified region efficiently. The name of these types of data structures is reactive data structures (Oosterom 1989, 1991, 1994).

The simplification part of the generalisation process is supported by the binary line generalisation tree (Oosterom and Bos 1989) based on the Douglas–Peucker algorithm (Douglas and Peucker 1973); see Figure 20. The reactive-tree (Oosterom 1991) is a spatial index structure that also takes care of the selection part of generalisation. The reactive-

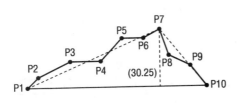

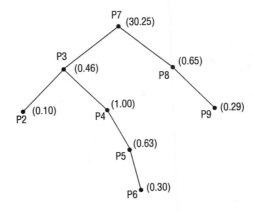

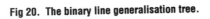

Fig 20. The binary line generalisation tree.

tree is based on the R-tree (Guttman 1984) with the difference that important objects are not stored at leaf level, but are stored at higher levels according to their importance; see Figures 21 and 22. The further one zooms in, the more tree levels must be addressed. Roughly stated, during map generation based on a selection from the reactive-tree, one should try to choose the required importance value such that a constant number of objects will be selected. This means that if the required region is large only the more important objects should be selected, and if the required region is small then the less important objects must also be selected. When using the reactive-tree and the binary line generalisation (BLG)-tree for the generalisation of an area partitioning, some problems are

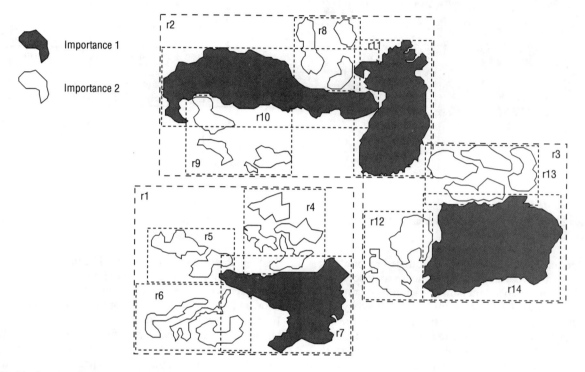

Fig 21. An example of reactive-tree rectangles.

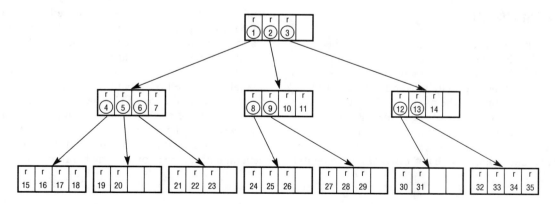

Fig 22. The reactive-tree.

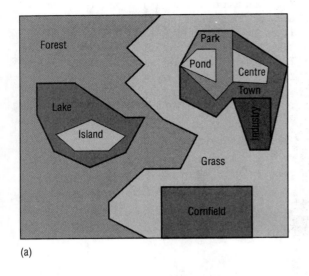

(a)

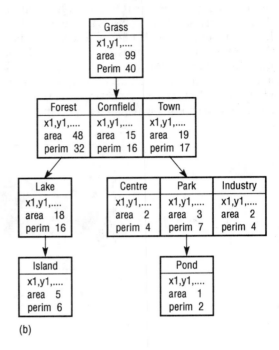

(b)

Fig 23. (a) The scene and (b) the associated GAP-tree.

encountered: gaps may be introduced by omitting small area features and mismatches may occur as a result of independent simplification of common boundaries. These problems can be solved by additionally using the generalised area partitioning (GAP)-tree. Using the reactive-tree, BLG-tree, and the GAP-tree, it is possible to browse interactively through large geographical data sets at very different scales (Oosterom and Schenkelaars 1995).

8 CONCLUSION

Though many spatial access methods have been described in the literature, the sad situation is that only a few have been implemented within the kernel of any (commercial) database and are ready to be used: exceptions are the use of the R-tree in Illustra (Informix) and the use of the Hilbert R-tree in CA-OpenIngres. At the moment several layered 'middleware' solutions are provided; for example, the spatial data engine (SDE) of ESRI (Environmental

Systems Research Corporation). The drawback of the layered approaches is that the database query optimiser does not know anything about the spatial data, so it cannot generate an optimal query plan. Further, all access should be through the layer, yet already many database applications do exist which have direct access to the database. In this situation consistency may become a serious problem. Thus, in practice, possibilities are quite limited and users have to develop their own solutions. An approach for this is the Spatial Location Code (SLC: Oosterom and Vijlbrief 1996), which has been designed to enable efficient storage and retrieval of spatial data in a standard (relational) DBMS. It is used for indexing and clustering geographical objects in a database and it combines the strong aspects of several known spatial access methods (quadtree, field-tree, and Morton code) into one SLC value per object. The unique aspect of the SLC is that both the location and the extent of possibly non-zero-sized objects are approximated by this single value. The SLC is quite general and can be applied in higher dimensions.

188

References

Abel D J, Mark D M 1990 A comparative analysis of some 2-dimensional orderings. *International Journal of Geographical Information Systems* 4: 21–31

Bayer R, McCreight E 1973 Organization and maintenance of large ordered indexes. *Acta Informatica* 1: 173–89

Beckmann N, Kriegel H-P, Schneider R, Seeger B 1990 The R-tree: an efficient and robust access method for points and rectangles. *Proceedings ACM/SIGMOD, Atlantic City*. New York, ACM: 322-31

Bentley J L 1975 Multi-dimensional binary search trees used for associative searching. *Communications of the ACM* 18: 509–17

Bentley J L, Friedman J H 1979 Data structures for range searching. *Computing Surveys* 11: 397–409

Douglas D H, Peucker T K 1973 Algorithms for the reduction of points required to represent a digitized line or its caricature. *Canadian Cartographer* 10: 112–22

Elzinga J, Hearn D W 1972 Geometrical solutions for some minimax location problems. *Transportation Science* 6: 379–94

Faloutsos C 1988 Gray codes for partial match and range queries. *IEEE Transactions on Software Engineering* SE-14: 1381–93

Faloutsos C, Roseman S 1989 Fractals for secondary key retrieval. *Eighth ACM SIGACT-SIGMOD-SIGART Symposium on Principles of Database Systems (PODS)*: 247–52

Faloutsos C, Sellis T, Roussopoulos N 1987 Analysis of object oriented spatial access methods. *ACM SIGMOD* 16: 426–39

Frank A U 1983 Storage methods for space-related data: the field-tree. Tech. rep. Bericht no. 71. Zurich, Eidgenössische Technische Hochschule

Frank A U, Barrera R 1989 The Field-tree: a data structure for geographic information systems. *Symposium on the Design and Implementation of Large Spatial Databases, Santa Barbara, California*. Berlin, Springer: 29–44

Fuchs H, Abram G D, Grant E D 1983 Near real-time shaded display of rigid objects. *ACM Computer Graphics* 17: 65–72

Fuchs H, Kedem Z M, Naylor B F 1980 On visible surface generation by a priori tree structures. *ACM Computer Graphics* 14: 124–33

Goodchild M F 1989 Tiling large geographical databases. *Symposium on the Design and Implementation of Large Spatial Databases, Santa Barbara, California*. Berlin, Springer: 137–46

Goodchild M F, Grandfield A W 1983 Optimizing raster storage: an examination of four alternatives. *Proceedings AutoCarto 6*: 400–7

Guttman A 1984 R-trees: a dynamic index structure for spatial searching. *ACM SIGMOD* 13: 47–57

Hutflesz A, Six H-W, Widmayer P 1990 The R-file: an efficient access structure for proximity queries. *Proceedings IEEE Sixth International Conference on Data Engineering, Los Angeles, California*. Los Alamitos, IEEE Computer Society Press: 372–9

Jagadish H V 1990 Linear clustering of objects with multiple attributes. *ACM/SIGMOD, Atlantic City*. New York, ACM: 332–42

Kamel I, Faloutsos C 1994 Hilbert R-tree: an improved R-tree using fractals. *VLDB Conference*

Kleiner A, Brassel K E 1986 Hierarchical grid structures for static geographic databases. In Blakemore M (ed.) *AutoCarto London*. London, The Royal Institution of Chartered Surveyors: 485–96

Matsuyama T, Hao L V, Nagao M 1984 A file organization for geographic information systems based on spatial proximity. *Computer Vision, Graphics, and Image Processing* 26: 303–18

Megiddo N 1983 Linear-time algorithms for linear programming in R3 and related problems. *SIAM Journal on Computing* 12: 759–76

Nievergelt J, Hinterberger H, Sevcik K C 1984 The grid file: an adaptable, symmetric multikey file structure. *ACM Transactions on Database Systems* 9: 38–71

Nulty W G, Barholdi J J III 1994 Robust multi-dimensional searching with space-filling curves. In Waugh T C, Healey R G (eds) *Proceedings, Sixth International Symposium on Spatial Data Handling, Edinburgh, Scotland*. London, Taylor and Francis: 805–18

Oosterom P van 1989 A reactive data structure for geographic information systems. *Auto-Carto 9*, April 1989. American Society for Photogrammetry and Remote Sensing: 665–74

Oosterom P van 1990 A modified binary space partitioning tree for geographic information systems. *International Journal of Geographical Information Systems* 4: 133–46

Oosterom P van 1991 The Reactive-tree: a storage structure for a seamless, scaleless geographic database. In Mark D M, White D (eds) *AutoCarto 10*. American Congress on Surveying and Mapping: 393–407

Oosterom P van 1994 *Reactive data structures for geographic information systems*. Oxford, Oxford University Press

Oosterom P van, Claassen E 1990 Orientation insensitive indexing methods for geometric objects. *Fourth International Symposium on Spatial Data Handling, Zürich, Switzerland*. Colombus, International Geographic Union: 1016-29

Oosterom P van, Schenkelaars V 1995 The development of an interactive multi-scale GIS. *International Journal of Geographical Information Systems* 9: 489–507

Oosterom P van, Bos J van den 1989 An object-oriented approach to the design of geographic information systems. *Computers & Graphics* 13: 409–18

Oosterom P van, Vijlbrief T 1996 The spatial location code. In Kraak M-J, Molenaar M (eds) *Proceedings, Seventh International Symposium on Spatial Data Handling, Delft, The Netherlands*. London, Taylor and Francis

Orenstein J A, Manola F A 1988 PROBE spatial data modeling and query processing in an image database application. *IEEE Transactions on Software Engineering* 14: 611–29

Paterson M S, Yao F F 1989 Binary partitions with applications to hidden-surface removal and solid modeling. *Proceedings, Fifth ACM Symposium on Computational Geometry*. New York, ACM: 23–32

Robinson J T 1981 The K-D-B-tree: a search structure for large multidimensional dynamic indexes. *ACM SIGMOD* 10: 10–18

Rosenberg J B 1985 Geographical data structures compared: a study of data structures supporting region queries. *IEEE Transactions on Computer Aided Design* CAD-4: 53–67

Roussopoulos N, Leifker D 1985 Direct spatial search on pictorial databases using packed R-trees. *ACM SIGMOD* 14: 17–31

Samet H 1984 The quadtree and related hierarchical data structures. *Computing Surveys* 16: 187–260

Samet H 1989 *The design and analysis of spatial data structures*. Reading (USA), Addison-Wesley

Sylvester J J 1857 A question in the geometry of situation. *Quarterly Journal of Mathematics* 1: 79

Tamminen M 1984 Comment on quad- and octrees. *Communications of the ACM* 27: 248–9

28

Interacting with GIS

M J EGENHOFER AND W KUHN

The user interface of a GIS determines to a large extent how usable and useful that system is for a given task for a user. The usability of GIS has significantly improved in recent years, through changes in our understanding of what system use should achieve, which communication channels it should exploit, and in what form the interaction should occur. This chapter analyses these characteristics of human interaction with GIS, looking back at what has been accomplished in practice, and forward to what can be expected from current research and development efforts. The emphasis lies on an overview of interaction paradigms, modalities, and styles in GIS, rather than on an exemplary discussion of particular systems. While other chapters discuss cognitive, social, institutional, and economic factors affecting GIS usability, the emphasis here is upon the technical aspects, as they are investigated in the field of human–computer interaction.

1 INTRODUCTION

Over the past decade, human interaction with GIS has received increasing attention among researchers, developers, and users. In the 1970s and early 1980s, GIS–user interfaces were dominated by command-style query languages accessible only to expert users. In the second half of the 1980s, graphical user interfaces (GUIs) used the same syntactic structures, but hid them behind icons and menus. This development has primarily improved the familiarity of systems by reducing the need to memorise commands and by providing visual feedback to users.

In the late 1990s, the state-of-the-art in commercial GIS–user interfaces is characterised by the use of windows, icons, menus, and pointing devices (WIMP-style interfaces). These group related operations into understandable chunks and represent them in menus or visually through icons with elementary direct-manipulation operations, like selection by pointing and clicking, or moving by dragging. Their prevailing paradigm of interaction is that of querying a geographical database and presenting the results in maps and tables (Egenhofer and Herring 1993). The map is used as a presentation medium for query results in their spatial context and as a referencing mechanism to indicate location by pointing (Frank 1993). This paradigm has recently been carried over from GIS to digital map libraries where users browse through datasets in a similar style, providing an abstraction from the physical location of geographical datasets when querying or searching them (Smith 1996). The World Wide Web (WWW) offers easy and wide access to these kinds of geographical data (Coleman, Chapter 22).

The functionality offered by current GIS–user interfaces primarily includes the selection of data layers; the identification of objects by location, name, and elementary spatial relations; and the modification of graphical output parameters such as colours and patterns. Most GIS designers have attempted to provide users with a wide range of functions, allowing them to ask as many as possible of the following generic spatial and temporal questions (adapted from Rhind and Openshaw 1988):

where is … ?
what is at location …?
what is the spatial relation between …?
what is in a particular spatial relation to …?

what is similar to …?
where has … occurred?
what has changed since …?
what will change if …?
what spatial pattern(s) exist(s) and where are anomalies?

While current interaction styles represent a significant improvement over the state-of-the-art a decade ago (Frank and Mark 1991), they generally fall short of achieving the usability necessary to solve spatial problems without being a GIS specialist. The user interface itself too often remains an impediment to effective system use in problem-solving or decision-making (Medyckyj-Scott and Hearnshaw 1993).

The proliferation of so-called GIS viewers with reduced functionality, tailored to the inspection of datasets, has reduced the complexity of system use at the inevitable expense of losing more powerful operations (see Elshaw Thrall and Thrall, Chapter 23). Recent developments toward more flexible and adaptable interfaces have eased customisation, but have not resolved this dilemma. In the absence of GIS-specific interface design guidelines, a growing variety of layouts and arrangements have appeared, limiting the possibility for knowledge transfer from one system to the other and improving usability at best marginally. At the same time, modern application programming tools for interface and system design have become widely available and are being used to implement and customise GIS applications. These range from simple macro languages to sophisticated programming languages including such mechanisms as inheritance and polymorphism (Maguire, Chapter 25).

Several ongoing developments are now pointing the way towards interfaces that offer substantial improvements in the usability of GIS. Among them are the focus on specific tasks and on the operations needed to accomplish them (Davies and Medyckyj-Scott 1995). Examples of such tasks include map digitising, where productivity is dramatically influenced by usability (Haunold and Kuhn 1994), and the interaction with car navigation systems where drivers have specific and very limited needs for spatial information, such as distance to and direction of the next turn (Waters, Chapter 59). Virtual reality systems have demonstrated that the concurrent use of multiple modes of interaction dramatically increases the engagement of the user by coming closer to natural ways of interacting with the world itself rather than with maps or other static models of it (Neves and Câmara, Chapter 39; Jacobson 1992). The primary interaction modes beyond keyboard input and visual presentation are sketching and gesturing in the visual channel; speech and other sound input and output; tactile input and feedback. Novel interaction devices come with displays in a wide variety of sizes and resolutions, enabling interaction that is tailored to special user needs like field portability or group work (Florence et al 1996). The current emphasis on multimedia technologies, in particular on video, CD-ROMs and the WWW, increases both the need and the possibilities for spatial representations and interaction. Scientific visualisation is creating and manipulating worlds of its own, in which motion, perspectives, and multiple representations convey information otherwise hidden in datasets (see Anselin, Chapter 17). The traditional textual query languages assume a new role as 'intergalactic data speak' (Stonebraker et al 1990), supporting interoperability by allowing heterogeneous systems to exchange data with each other (Sondheim et al, Chapter 24). Most of these developments are happening outside of the GIS field and come with their own theories of interaction and collaboration. They are, however, rapidly becoming the determining factors shaping human interaction with GIS.

This chapter reviews the state-of-the-art in GIS query and manipulation languages, measures it against the current understanding of usability requirements, and proceeds to an outlook of how current scientific and technological developments will shape future GIS–user interfaces. A survey of existing query languages or of tools for data capture, data manipulation, and application programming lies beyond its scope. The chapter compiles and integrates the various ways of comparing GIS–user interfaces from a user perspective, offering an introduction to the commonly-used terminology and criteria. It is structured in the following way for classifying interaction languages:

- the interaction paradigms (e.g. querying, browsing, interviewing, analysing, updating, experiencing);
- the interaction modalities (e.g. text, speech and sound, graphics, animation and video, sketching);
- the interaction styles (e.g. command-line, direct manipulation, filtering, delegation).

2 INTERACTION PARADIGMS

People want to do different things with a GIS: some use it as a substitute for a collection of paper maps; others consider it a repository of geographical data that they want to feed into a simulation model; and others think of a GIS as a model of reality in which they want to find interesting places, configurations, and relationships. These differences in the use of a GIS reflect different understandings or paradigms of use (Kuhn 1992). Some of these have traditionally been dominant (querying, browsing, analysing) and are, therefore, better supported by current commercial GIS software than others (interviewing, updating, experiencing).

2.1 Querying

Querying refers to the retrieval of information from a database using a language with well-defined syntax and semantics. The concept is adopted from database systems, where the query language provides a uniform way to access stored data. The idea of a database query is that the user specifies properties of the desired result. For example, if one wanted to know the population of all capitals in the European Union (EU), the query should ask for the names and populations of those cities in Europe that are capitals of a country that is an EU member.

Query languages allow users to retrieve data not only as they were stored, but also in combinations through which information can be obtained that is not directly stored. The combinations are based on logical operators such as conjunction, disjunction, and negation, plus some simple arithmetics. Though different types of query languages exist, the most prominent kind of languages is based on the Structured Query Language (SQL: Worboys, Chapter 26; Melton 1996). Written in SQL, the above example could read as follows:

SELECT name, pop FROM cities, countries
WHERE name = capital AND eumember = 'true'.

The advantage of using a query language rather than low-level programming is that a query language works independently of the contents of a database and, consequently, is useful for different database schemata. This data independence guarantees access without having to worry about storage locations and formats. Furthermore, logical and numerical computations need not be programmed. The result is a very powerful data retrieval mechanism based on a simple syntax and clean semantics (Frank 1982).

However, these benefits have their limitations. First of all, the above example shows that the user needs to know how the data are structured in the relational tables (what attributes exist in what table and what they are called). Second, most users find it difficult to phrase their requests in expressions involving logical connectors (e.g. AND, OR) whose semantics clash with those of their natural language counterparts (Mark, Chapter 7; Reisner 1981), and systems perform poorly in constructing and comparing different nested queries (Luk and Kloster 1986). Third, different data models may expose different properties that cannot be addressed by one and the same query language. For example, the use of a relational query language for a system based on an object-oriented data model would not support particular object-oriented characteristics, such as identity, encapsulation and inheritance (Atkinson et al 1989). In a similar vein, spatial characteristics have required extensions to query languages, primarily for the inclusion of spatial relations to allow users to make selections based on high-level spatial properties (Egenhofer 1994). Such spatially-extended query languages are based on an extension of the relational algebra to include spatial data types and operators (Güting 1988).

For human interaction with a GIS as a whole, database query languages are obviously biased toward the retrieval of data (Egenhofer and Kuhn 1991). Although the term 'query language' has taken on a broader meaning in the context of GIS (Egenhofer and Herring 1993), most GIS query languages treat geographical data as sets of attributes that can be logically combined in multiple ways. A query is usually considered a separate, standalone interaction and the next query does not build on the context established by the answers to previous queries (Egenhofer 1992). The narrow focus of a query language makes it difficult to support other tasks, such as asserting whether a certain set of facts is true or exploring whether an interesting configuration can be found in a dataset.

2.2 Browsing

With the proliferation of spatial datasets, finding those datasets that are of interest to a user has become an essential task. Consider digital libraries or the whole WWW acting as a huge but unstructured repository for spatial datasets. In such contexts, users commonly have little knowledge about the contents of the available datasets and cannot specify precisely what they want to retrieve. The paradigm of querying does not help here, as it would require exact knowledge of the configurations sought in order to specify appropriate constraints in a query. Rather, users want to be able to browse the data collections, enabling them to recognise rather than having to describe the desired data. The issue is not to find a needle in the haystack, but to look in a haystack for a straw of a shape and texture that the user is unable to specify prior to seeing it. Even if a user were to apply a query language to start a browsing activity, the result set of such a query would usually be very large and require a tedious case-by-case examination.

Browsing, like querying, is content-oriented, as users are looking for datasets that contain specific configurations; however, browsing generally requires additional information at a higher level of abstraction. Traditionally, browsing in GIS has been limited to searching by file names, providing very little and arbitrary information about contents. The current approach is to supply aggregate information and descriptions of dataset properties as metadata. It attempts to ease the user's task of selecting a dataset without having to examine all choices in detail. Emerging metadata standards, however, focus more on what the data producers have to say than on what the data users need to know (Timpf and Raubal 1996). Important descriptors are often unstructured text fields whose content is subject to chance and interpretation. While users can query and read such metadata prior to purchasing, downloading, or examining a dataset, they cannot yet browse them the same way as they would skim newspaper headlines and perceive relevant articles at a glance.

Browsable meta-information tailored to user needs will require search engines incorporating intelligent agents. These can learn about a user's task, mine through geographical datasets, and extract representative subsets, effectively bridging the gap between user needs and data sources.

2.3 Interviewing

The interaction with traditional information systems can be seen as based on the metaphor of an interview with a single person. The user asks questions and the information system responds.

If the interviewed system contradicts itself, the user starts to lose trust in its statements. As long as users perceive each information system as the integrated coherent collection of a single person's knowledge, this paradigm is not challenged.

With a new generation of information systems, however, the understanding of what it means to use an information system may change. These information systems exploit multiple distributed datasets, using them to generate query answers or suggestions supporting the user's decisions. Much like the text-based search engines available today on the WWW, such distributed GIS will face a huge number of dispersed and often isolated data collections. The paradigm for using such information systems changes from the interview with a single person to the interview with multiple and diverse informants or agents, some of them digital, some of them human. These interviewees all hear the same question and they answer when they believe they have something to contribute; however, they are located at different sites such that they generally do not hear each others' answers. Some use this opportunity sparsely and provide input only when they believe that they have something significant to contribute. Others may answer each and every question asked. Some in the audience may be experts in certain areas, whereas others may have a less sophisticated background, but still voice their views. Some may have had recent insights that may affect the answer to a question, while others may contribute their outdated views. Since interviewers hear the views of a group of people, they are expected to make better-informed decisions.

2.4 Analysing

The mere retrieval of stored data is often insufficient since users want to relate and combine data to 'see' patterns and connections among different data elements. Such connections may be visible in reality, but different data collection methods may have hidden them. Or they may be truly discovered through the process of spatial analysis (see Getis, Chapter 16 and Openshaw and Alvanides, Chapter 18 for overviews of competing spatial analysis discovery paradigms). Traditional database query languages offer only modest support for combining data in the form of joining tables over common attribute values. Finding other relationships requires further mathematical analysis, often using statistical or computational methods.

The most common analysis tool for combining data in GIS is the overlay operation. It integrates two or more thematic layers through various analytical operations in order to generate a new layer. Map algebra (Tomlin 1990) has been the common framework for such operations, allowing users to specify (1) what layers to combine, (2) with what operations, and (3) what to do with the result. For example, in order to find the probability of pieces of land near a river being flooded, a layer representing flood risk zones can be overlain with a cadastral layer.

While the source domain of the map overlay metaphor in GIS – stacking transparent sheets on top of each other on a light table (Steinitz et al 1976) – implies an arithmetic multiplication of the layers, the digital environment enables a much larger set of operations. Tomlin's semi-formal model is strictly raster-oriented, but corresponding overlay operations have also been implemented for vector data. Map algebra remains a core functionality in GIS–user interfaces, served by a variety of user interface styles from command line to iconic, direct-manipulation languages (Bruns and Egenhofer 1997; Kirby and Pazner 1990).

More advanced analysis operations rely on sophisticated spatial analysis techniques such as those explored in the spatial analysis chapters of the Principles part of this volume and by Eastman (Chapter 35). Their integration into spatial query languages (Svensson and Zhexue 1991) occurs in highly dispersed ways that have so far shown no common thread toward generic GIS analysis (Yuan and Albrecht 1995). The state-of-the-art solution is to adopt a 'toolbox' metaphor, allowing for an open collection of analysis tools, though often at the expense of limited usability.

2.5 Updating

Few geographical datasets are static and keep their currency and validity over extended periods of time. Updates, corrections, and the generation of 'what if?' scenarios are important operations through which users want to change geographical data. The changes include adding new data and modifying or removing existing data. Adding new data to a dataset means to embed the new data within the setting of existing data (see Peuquet, Chapter 8, for the case of temporal updating; and Goodchild and Longley, Chapter 40, for a discussion of the 'life of a dataset').

Most query languages offer some constructs for elementary updates, though not always in ways that readily support user tasks or are easy to learn and use. Special languages for data manipulation, on the other hand, are much less developed and standardised than query languages. The reason for this is their tight coupling to specific application requirements. For example, a language to design and modify the geometry of a cadastral database requires entirely different operations from one designed to maintain the accuracy of a statistical database (Kuhn 1990).

Operations to update geographical data can be classified according to their scope. At the elementary level, all manipulation involves changing a particular value or adding (deleting) an object. Both of these kinds of operations can apply to data themselves or to the schema of a database, although the latter is not always supported at the end-user interface of a GIS. At a higher level, these elementary operations can be aggregated to changes of value or object collections. Various constraints on these collective changes propagate down from the user's task (e.g. to split a parcel) to the elementary value and object level. In order to maintain the consistency of a database, the commitment of these changes has to be coordinated by transactions (Bédard, Chapter 29). Their granularity can vary from single value changes to modifications sweeping through entire databases. A key problem with current query and manipulation languages is that they operate primarily at lower levels, leaving the management of transactions to the discretion of organisational, rather than purely technical, considerations.

2.6 Experiencing

An entirely different paradigm of interaction with GIS is that of operating within the modelled world itself rather than asking questions about it or writing instructions for changes. Virtual reality, as an immersive variant of direct manipulation, offers this possibility (Neves and Câmara, Chapter 39). In the context of GIS, the idea of 'living' in the model rather than just looking at it has an immediate appeal. Since the phenomena modelled by GIS exist at human or larger scales, it appears natural to provide digital equivalents for human ways of interacting with the world, such as turning one's head, walking, driving, flying, gesturing, and manipulating objects. Systems providing these have been in use for years in applications like flight simulation or games. Their greatest potential for GIS lies in the possibility of overlaying sensory input from the real world with that from one or more models. These ideas have, however, not yet significantly influenced the traditional, map-oriented architectures of GIS interfaces (Kuhn 1991). For some years to come, the interface between our bodies and GIS models is likely to remain the tiny area of our fingertips (on a keyboard or mouse) and overlays will be limited to digital map layers.

3 INTERACTION MODALITIES

Communication between people exploits multiple modalities (spoken, written, gestured, graphical) which map onto different communication channels (visual, auditory, tactile). Each of the modalities has its own strengths and the unavailability of one or the other may seriously impede how people are able to interact with each other and with their environment. A key to successful communication is the appropriate combination and redundant use of different modalities: people gesture while talking, or annotate drawings with spoken or written text. Similar to communication among people, the choice of modalities plays a key role in the success or failure of the interaction between one or more users and a GIS. GIS have often been referred to as early examples of multimedia systems, integrating alphanumerical with map-like and other graphical representations. Ample opportunities for the use of modalities beyond static text and graphics exist and GIS query languages are increasingly making use of them. Almost no theory exists, however, on how to combine multiple modalities appropriately (Egenhofer 1996a).

3.1 Text

Text has been the principal mode for providing instructions to a GIS – be they typed or selected from menus or forms. Unlike in conventional information systems, however, the presentation of query results to a user in textual form has always been secondary to graphical presentations. This is partially so because most GIS replaced map use or production systems in instances where there was little need to convey text to users. Only with the advent of the requirement for data exchange among GIS did the emphasis on

textual information strengthen, and ASCII text files have become a standard way to move data from one system to another. Text is then used to transmit attribute data as well as encoded geometry, but it is another system, not an interactive user, that requests the data.

Text input or output can come in more-or-less-structured form, ranging from free-form natural language texts, to structured tables, to expressions coded in a formal language. Independently of the form, however, textual descriptions of spatial situations are frequently ambiguous and may lead to misinterpretations (cf. the difficulties when using textual directions in way-finding). The use of traditional textual spatial query languages has serious limitations, because geographical concepts are often vague, imprecise, little understood, or not standardised (Fisher, Chapter 13). The dilemma is most apparent in the semantics of spatial terms (Mark, Chapter 7). What does it take to refer to something as a 'mountain' and when would it be more appropriate to call it a 'hill'? Likewise, what paths would qualify to 'cross' the Rocky Mountains and when is a restaurant really 'in town'? These difficulties make most current textual spatial query languages error-prone and difficult to use. They leave the interpretation of terms up to the designers, while users have to comply with their (often hidden) judgements.

On the other hand, the ambiguity inherent in textual descriptions of spatial situations can be exploited to express just that: situations that are not fully determined. Linguists and designers of visual languages have found that icons or sketches (see section 3.5) often over-specify spatial relations (Haarslev and Wessel 1997). For example, one can ask verbally for an object located 'outside' a region, while a corresponding icon always implies a certain distance and especially a direction from the region. In such cases, text is actually more appropriate than graphics or sketches to represent spatial configurations.

3.2 Speech and sound

Some situations of GIS use make it impractical to enter text by typing or selecting; for instance, when working with a mobile GIS in the field. Similarly, it can be preferable to receive output in spoken or other auditory rather than visual form, for instance when driving, assisted by a vehicle navigation system (Waters, Chapter 59). Speech recognition and synthesis techniques are today sufficiently advanced to allow effective communication in situations with a limited vocabulary and few users. An even more limited interaction language allows for a broader range of users (e.g. call-in information systems for public transportation schedules), and vice versa (e.g. text entry systems on a PC).

There are, so far, few examples of using sound in human interaction with GIS (see Shiffer, Chapter 52; Shiffer 1995; Weber 1997). Apart from the intrusiveness of sound in professional as well as private environments, lack of understanding of the role sound plays in cognition (e.g. to identify and locate objects) seems to hinder a broader use of this medium. Considering that sound plays a crucial, though often unconscious, role in our interaction with the world, it can be expected to become more important in GIS–user interfaces. An important reason for further development of auditory interfaces is that they represent the only practical modality supporting visually handicapped users in highly interactive settings.

3.3 Graphics

The strong traditional link between geographical information and graphical communication has led to a higher emphasis on graphics in GIS–user interfaces than in those of most other information systems. These graphics are predominantly maps, which have evolved to become the most sophisticated means of communication about geography during the past 3500 years. Most GIS offer at least some map output to present query results or support data entry. These screen maps are slowly becoming more versatile than simple digital versions of their paper ancestors, offering ways of interacting to refine a query or ask further queries. Users can often point to features on maps to obtain or enter information about them, select map features as input for operations, outline a zoom window to get a map with different contents at a larger scale, or select different layers of information.

In addition to maps, business graphics supply a graphic modality to represent attribute data by graphs and diagrams (e.g. see Elshaw Thrall and Thrall, Chapter 23). Where they occur in isolation, there is nothing that distinguishes them in a GIS from other software that visualises tabular data. Where they are combined with base maps, however, they achieve one of the most refined modes of representing information visually: thematic maps.

The key interaction issue posed by them is how users can choose among symbolisation options and how system designers can provide reasonable default symbolisation (see Kraak, Chapter 11).

3.4 Animation and video

Moving pictures have yet to find their role in communicating geographical information in a GIS (but see Raper, Chapter 5). Attempts at using them have so far concentrated on video clips, offering more intuitive perspectives on buildings and landscapes than maps or textual descriptions can provide (Shiffer 1995; but see Batty, Chapter 21). While such animated views can be very useful (for instance, in a system assisting home buyers), they also consume considerable system and user resources and are difficult to integrate with other parts of a GIS–user interface. Nevertheless, video sequences represent highly valuable information sources within a GIS, and the real issue is to develop appropriate indexing and retrieval systems that can support users in monitoring changes. Time has been noted as a domain of considerable research and development efforts regarding GIS languages. These efforts focus on modelling aspects, but query languages will look very different depending on what kinds of models they support.

3.5 Sketching

Sketching has been used primarily for design tasks as they occur in computer-aided design (CAD) systems. SketchPad (Sutherland 1963) and ThingLab (Borning 1986) were initial approaches to formulate spatial constraints graphically, a principle that was later introduced to GIS interfaces (Egenhofer 1996b; Kuhn 1990). Sketching has also been explored to describe consistency constraints in spatial databases through the construction of situations that would establish unacceptable database states (Pizano et al 1989) or the definition of spatial relations by examples (Petersen and Kuhn 1991). These approaches have confirmed that sketches, like all graphic representations, are good at describing single configurations, but fail in scenarios that require multiple geometric specifications ('this or that', 'this without that') or topological information only ('across', 'outside'). Despite these shortcomings, sketching offers great potential for interaction with GIS when it is given its appropriate role (describing unique, but not exactly determined situations) and combined with other modalities (primarily speech; Egenhofer 1996a, 1996b). It also shows the way to a much more prominent role for gestures in interaction. The very limited understanding of how people assign meaning between speech, gestures, and other means of communication, however, still hinders the development of broadly usable interaction techniques along these lines.

4 INTERACTION STYLES

Independently of the chosen interaction paradigms and modalities, query and manipulation languages can be classified according to their interaction styles. The style of an interaction captures how users express queries or updates and how they receive results: by written commands, direct manipulation, dynamic queries, or delegation.

4.1 Command-line input

Command-line systems represented the state-of-the-art before GUIs became available in the early 1980s. Their syntax can be formally defined, making command interpreters and compilers relatively easy to create. For all their learnability and other usability problems, command-line systems have some definite advantages, particularly for experienced users. Macro commands are easy to write by grouping commands in a text file. Adding programming constructs, such as variables, functions, branching, and looping, can make the power and flexibility of a complete programming language available to the user. However, command-line interfaces have poor cognitive characteristics. The users are interacting through text only and a screen full of text has too high a density of information. Textual objects are more difficult to identify and locate on a screen than graphical objects. Also, typing is physically tiring and error-prone.

The major problems with command-line interfaces are: (1) determining the appropriate command for a task; (2) remembering its name and the names of variables; and (3) entering commands in the correct syntax. The last two problems are significantly reduced in form-based interfaces, which often come as parts of direct manipulation interfaces, but do not constitute a dominating GIS interaction style by themselves.

4.2 GUIs, WIMPs, and direct manipulation

Contemporary GUIs are essentially extensions of window-based operating systems. Such window environments depend on high-quality bit-mapped raster displays and some kind of pointing device (mouse, track ball, joystick etc.). This dependency explains why the terms WIMP (windows, icons, menus, and pointing devices) and GUI are often used interchangeably. GUIs have a much stronger visual component than command-line interfaces and have therefore also been called visual interfaces. The windows, icons, and menus determine the visual characteristics of the user interface.

Icons are symbolised pictorial representations for objects or operations in a user interface. The small space allocated to their pictorial part necessitates careful design and testing. Many icons found in current GIS–user interfaces are more important as place holders for commands (exploiting the user's spatial memory) than as symbols whose meaning can be understood on first sight. Icons can be enhanced through the use of animation or sound.

While most menus are textual, menus with graphics or icons, for instance in the form of tool bars, have become more popular. Dynamic or contextual menus constrain a user to allowable actions at any given time, significantly reducing the chance for errors. Windows allow users to switch rapidly and coherently between multiple tasks or multiple parts of a single task. Various strategies exist for managing the organisation of several screen windows.

Pointing and typing characterise the physical aspects of GUIs. Pointing is used to select objects and operations, typically within a window. For example, a map algebra or SQL expression may be created in a form by pointing to various icons and objects on a map. The users are still composing a command line, but instead of typing, they recognise and select tokens and the system does the typing.

The interaction style supported by GUIs or WIMPs has been introduced in practice by the Xerox Star development (Smith at al 1982) and in theory by Ben Shneiderman (Shneiderman 1983). Although the principles of direct manipulation are by now commonplace, they are far from always being satisfied in GIS interfaces and warrant careful consideration in every design process of the need for:

- visual presentation of objects and operations;
- visual presentation of results;
- rapid, incremental, and reversible operations;

- selection by pointing rather than typing;
- immediate and continuous feedback.

All variations of direct manipulation share a few key qualities: the appearance of and interaction with a system are based on metaphors and multiple metaphors need to be combined (Kuhn 1995; Mark 1992). When moving text in a wordprocessor or dragging a document icon to a folder, users are engaged in a multi-modal activity. What they see and what they do are closely coupled, both physically and conceptually.

Direct manipulation is an appropriate interaction style for the primarily visual operations in a GIS, like zooming, panning or map overlays. Historically, the process of map overlay has been a visual and tactile operation, presenting a rich source domain for direct manipulation metaphors. By enforcing a visual representation of data and operations, direct manipulation also fosters exploratory data analysis (Anselin, Chapter 17). Exploration requires a dynamic, absorbing, and engaging task environment. Users need to become less aware of the existence of the user interface and more immersed in their analytical tasks. Such genuinely empowering environments are still rare. The use of metaphors and direct manipulation alone does not automatically lead to them. The metaphors must draw on the visual and physical characteristics inherent in the user's understanding of a task.

4.3 Filtering

The consequent application of direct manipulation principles to querying led to a variety of interaction techniques that can best be characterised as interactive filters for spatial data. While traditional GIS query languages apply direct manipulation to the composition of queries which are then sent as commands to the database, interactive filters are directly evaluated while the user sets some parameters, with update rates in the order of 100 milliseconds. The effect is that users have a much greater sense of control over the database and an opportunity to 'play' with the data. This supports to a large extent the paradigm of exploration, and supports browsing as well as exploratory data analysis. Examples for such filtering techniques are dynamic queries, dynamic filters, and magic lenses.

Dynamic queries give users interactive control over the setting of query parameters, usually in the form of sliders or buttons (Shneiderman 1994).

They were invented with GIS as one of the key applications in mind. An example is a dynamic 'home finder' with sliders for distances to two places, number of bedrooms, and cost, plus buttons for home type and home features. While the user manipulates these, a map displays the location of homes satisfying the criteria. Clicking at these points reveals detailed descriptions of homes.

Dynamic filters continuously control the density of information shown on screen and provide panning and zooming techniques to focus on portions of the displayed contents (Ahlberg and Shneiderman 1994). Magic lenses (Stone et al 1994) are yet another filtering tool that allows users to change the presentation of objects over which the lens is laid. For example, a portion of a topographic map can change into a weather map or a population map when a special lens is applied to it.

The exploratory nature of the filtering interaction style allows for discovering patterns in the data, forming and testing hypotheses about correlations, and identifying outliers. As such, filtering is a practical approach to interactive data mining (Fayyad et al 1996). Technical problems are posed by the bottleneck of accessing databases and displaying the data rapidly. A more fundamental issue is the application-specific nature of each interface: a home finder looks quite different from, say, a cancer rate visualisation tool. This shows again the persistent trade-off between generality and usability of interaction languages.

4.4 Delegation

Delegation is a style of interaction founded on a special, terse form of communication, where the system takes on the role of an agent or assistant to the user. It has the advantage of establishing a restricted, fairly simple, and familiar communication protocol. Delegation is often seen as an antithesis to direct manipulation (Negroponte 1989) and gets naturally associated with speech-based interaction. It has the potential to compensate for the negligence of the auditory channel in visual interfaces. By its terseness, delegation suits current speech recognition technology quite well. On the other hand, the difficulties of knowing what to ask for and of being able to express it, as well as the decoupling of input and output in the interaction process, have so far prevented successful delegation interfaces in GIS. A more fundamental obstacle is the lack of formal theories for talking about space (Mark and Gould 1991).

5 LOOKING AHEAD

The overview of various kinds of GIS interaction languages in this chapter has been written from a human–computer interaction perspective. It focused on paradigms, modalities, and styles of interaction, as developed in theory and practice over the past two decades. Today, user interface design and interaction research are increasingly driven by broader cognitive, social and economic concerns (Mark and Frank 1991; Nyerges et al 1995). Cognitive sciences investigate how people think and communicate about their applications or about space and how this affects the usability of technology (Mark, Chapter 7). Social studies focus on the role of technology in society, which is strongly influenced by system usability. Economic approaches such as business re-engineering study the core processes of organisations and how they can better be supported by interactive systems and shared databases (Campbell, Chapter 44). The impact of these developments goes beyond traditional GIS applications to areas where geographical information plays a role in decision-making, but GIS cannot yet be applied due to their complexity.

A conclusion emerging from our analysis of interaction issues is that GIS will probably never offer a common query language, let alone common ways of manipulating data. Such generic approaches attempt too much for too many different settings. Indeed, today's market shows a departure altogether from the idea of a GIS platform common to the large and rapidly expanding spectrum of applications. A wider perspective on geoprocessing is emerging, considering it an integral part of enterprise computing. This idea has forged the Open GIS Consortium (OGC, see *http://www.opengis.org*), a cooperative effort between researchers, developers, and users to standardise object-oriented software interfaces for open, interoperable GIS (Sondheim et al, Chapter 24). It is based on the consensus that standardising software interfaces (in the sense of an 'intergalactic data speak', i.e. relatively low-level data retrieval languages along the lines of extended SQL) while diversifying user interfaces will significantly improve overall usability.

In this scenario, integration will occur with other tools at the user sites (databases, spreadsheets, groupware, workflow management, etc.) rather than with GIS in other application domains. Application programming interfaces will support specific classes of GIS tasks and users much better than today. This change in thinking about GIS architectures is almost certain to affect GIS user interfaces at least as much as the introduction of visual interfaces.

With the integration of geoprocessing into mainstream computing comes also a chance to turn things around and use spatial forms of interaction in non-spatial applications (Kuhn 1996). Using spatial metaphors to structure interaction with non-spatial information is at least as old as the desktop metaphor (Smith et al 1982). More recent examples include the use of a landscape metaphor to visualise, explore, and query non-spatial data (Wise and Thomas 1995). With novel interaction paradigms like that of experiencing, and modalities like speech and gestures further developing, entirely new interaction styles along the lines of filtering will develop within and outside the GIS field. Space and spatial interaction, as fundamental categories of human cognition, will be one of their key characteristics.

References

Ahlberg C, Shneiderman B 1994 Visual information seeking: tight coupling of dynamic query filters with starfield displays. In Adelson B, Dumais S, Olson J (eds) *Human factors in computing systems, CHI 94 conference proceedings*. Boston, ACM Press: 313–17

Atkinson M, Bancilhon F, DeWitt D, Dittrick K, Maier D, Zdonik S 1989 The object-oriented database system manifesto. *Proceedings, First International Conference on Deductive and Object-Oriented Databases, Kyoto, Japan*. Amsterdam, Elsevier Science

Borning A 1986 Defining constraints graphically. In Mantei M, Orbeton P (eds) *Human factors in computing systems, CHI 86 conference proceedings*. Boston, ACM Press: 137–43

Bruns T, Egenhofer M 1997 User interfaces for map algebra. *Journal of the Urban and Regional Information Systems Association* 9: 44–54

Davies C, Medyckyj-Scott D 1995 Feet on the ground: studying user-GIS interaction in the workplace. In Nyerges T L, Mark D M, Laurini R, Egenhofer M J (eds) *Cognitive aspects of human–computer interaction for GIS*. Dordrecht, Kluwer: 123–41

Egenhofer M J 1992 Why not SQL! *International Journal of Geographical Information Systems* 6: 71–85

Egenhofer M J 1994 Spatial SQL: a query and presentation language. *IEEE Transactions on Knowledge and Data Engineering* 6: 86–95

Egenhofer M J 1996a Multi-modal spatial querying. In Kraak M-J, Molenaar M (eds) *Advances in GIS research II (Proceedings, Seventh International Symposium on Spatial Data Handling)*. London, Taylor and Francis: 785–99

Egenhofer M J 1996b Spatial query-by-sketch. In *VL 96: IEEE Symposium on Visual Languages*. IEEE Computer Society: 60–7

Egenhofer M J, Herring J 1993 Querying a geographical information system. In Medyckyj-Scott D, Hearnshaw H (eds) *Human factors in geographical information systems*. London, Belhaven Press: 124–36

Egenhofer M J, Kuhn W 1991 Visualising spatial query results: the limitations of SQL. In Knuth E, Wegner L (eds) *IFIP Transactions A-7: Visual Database Systems, II, Proceedings of IFIP TC2/WG2.6 Second Working Conference on Visual Database Systems, Budapest, Hungary*. Amsterdam, North-Holland: 5–18

Fayyad U, Haussler D, Stolorz P 1996 Mining scientific data. *Communications of the ACM* 39: 51–7

Florence J, Hornsby K, Egenhofer M 1996 The GIS WallBoard: interactions with spatial information on large-scale displays. In Kraak M-J, Molenaar M (eds) *Advances in GIS research II (Proceedings of Seventh International Symposium on Spatial Data Handling)*. London, Taylor and Francis: 449–63

Frank A U 1982 MAPQUERY: database query language for retrieval of geometric data and its graphical representation. *ACM SIGGRAPH* 16: 199–207

Frank A U 1993 The user interface is the GIS. In Medyckyj-Scott D, Hearnshaw H (eds) *Human computer interaction and geographic information systems*. London, Belhaven Press: 3–14

Frank A U, Mark D M 1991 Language issues for GIS. In Maguire D J, Goodchild M F, Rhind D W (eds) *Geographic information systems: principles and applications*. Harlow, Longman/New York, John Wiley & Sons Inc. Vol. 1: 147–63

Güting R 1988 Geo-relational algebra: a model and query language for geometric database systems. In Schmidt J, Ceri S, Missikoff M (eds) *Advances in database technology – EDBT 88 International Conference on Extending Database Technology, Venice, Italy*. New York, Springer 303: 506–27

Haarslev V, Wessel M (1997) Querying GIS with animated spatial sketches. In Tortona J (ed.) *Thirteenth IEEE Symposium on Visual Languages, Capri, Italy*. IEEE Society

Haunold P, Kuhn W 1994 A keystroke level analysis of a graphics application: manual map digitising. In Adelson B, Dumais S, Olson J (eds) *Human factors in computing systems, CHI 94 conference proceedings*. Boston, ACM Press: 337–43

Jacobson B 1992 The ultimate user interface. *Byte* 17: 175–82

Kirby K C, Pazner M 1990 Graphic map algebra. In Kishimato H, Brassel K (eds) *Proceedings, Fourth International Symposium on Spatial Data Handling, Zurich, Switzerland*. Columbus, International Geographical Union 1: 413–22

Kuhn W 1990 Editing spatial relations. In Kishimoto H, Brassel K (eds) *Proceedings, Fourth International Symposium on Spatial Data Handling, Zurich, Switzerland*. Columbus, International Geographical Union 1: 423–32

Kuhn W 1991 Are displays maps or views? In Mark D, White D (eds) *Proceedings, Tenth International Symposium on Computer-Assisted Cartography (AutoCarto 10)*. American Congress of Surveying and Mapping: 261–74

Kuhn W 1992 Paradigms of GIS use. *Proceedings, Fifth International Symposium on Spatial Data Handling*. Columbus, International Geographical Union 1: 91–103

Kuhn W 1995 7±2 questions and answers on metaphors for GIS user interfaces. In Nyerges T L, Mark D M, Laurini R, Egenhofer M J (eds) *Cognitive aspects of human–computer interaction for geographic information systems*. Dordrecht, Kluwer: 113–22

Kuhn W 1996 Handling data spatially: spatialising user interfaces. In Kraak M-J, Molenaar M (eds) *Advances in GIS research II (Proceedings Seventh International Symposium on Spatial Data Handling)*. London, Taylor and Francis: 877–93

Luk W S, Kloster S 1986 ELFS: English language from SQL. *ACM Transactions on Database Systems* 11: 447–72

Mark D M 1992 Spatial metaphors for human–computer interaction. In Cowen D (ed.) *Proceedings, Fifth International Symposium on Spatial Data Handling*. International Geographical Union 1: 104–12

Mark D M, Frank A U (eds) 1991 *Cognitive and linguistic aspects of geographic space*. NATO ASI Series D: Behavioural and Social Sciences, Vol. 63. Dordrecht, Kluwer

Mark D M, Gould M 1991 Interaction with geographic information: a commentary. *Photogrammetric Engineering and Remote Sensing* 57: 1427–30

Medyckyj-Scott D, Hearnshaw H M (eds) 1993 *Human factors in geographical information systems*. London, Belhaven Press

Melton J 1996 SQL language summary. *ACM Computing Surveys* 28: 141–3

Negroponte N 1989 An iconoclastic view beyond the desktop metaphor. *International Journal of Human–Computer Interaction* 1: 109–13

Nyerges T L, Mark D M, Laurini R, Egenhofer M J (eds) 1995 *Cognitive aspects of human-computer interaction for geographic information systems*. NATO ASI Series – Series D: Behavioural and Social Sciences. Dordrecht, Kluwer

Petersen J K, Kuhn W 1991 Defining GIS data structures by sketching examples. *ACSM/ASPRS Annual Convention*. American Congress on Surveying and Mapping 2: 261–9

Pizano A, Klinger A, Cardenas A 1989 Specification of spatial integrity constraints in pictorial databases. *IEEE Computer* 22: 59–71

Reisner P 1981 Human factors studies of database query languages: a survey and assessment. *ACM Computing Surveys* 13: 13–31

Rhind D W, Openshaw S 1988 The analysis of geographical data: data rich, technology adequate, theory poor. *Proceedings, Fourth International Working Conference on Statistical and Scientific Database Management*. Berlin, Springer: 427–54

Shiffer M J 1995b Geographic interaction in the city planning context: beyond the multimedia prototype. In Nyerges T L, Mark D M, Laurini R, Egenhofer M (eds) *Cognitive aspects of human–computer interaction for geographic information systems*. Dordrecht, Kluwer: 295–310

Shneiderman B 1983 Direct manipulation: a step beyond programming languages. *IEEE Computer* 16: 57–69

Shneiderman B 1994 Dynamic queries for visual information seeking. *IEEE Software* 11: 70–7

Smith D C, Irby C, Kimball R, Verplank B, Harslem E 1982 Designing the Star user interface. *Byte* 7: 242–82

Smith T 1996 Alexandria Digital Library. *Communications of the ACM* 4: 61–2

Steinitz C, Parker P, Jordan L 1976 Hand-drawn overlays: their history and prospective uses. *Landscape Architecture* 66: 444–55

Stone M, Fishkin K, Bier E 1994 The movable filter as a user interface tool. In Adelson B, Dumais S, Olson J (eds) *Human factors in computing systems, CHI 94 conference proceedings*. Boston, ACM Press: 306–12

Stonebraker M, Rowe L A, Lindsay B, Gray J, Carey M, Beech D 1990 Third-generation database system manifesto. *SIGMOD Record* 19: 31–44

Sutherland I 1963 SketchPad: a man-machine graphical communication system. *Proceedings of AFIPS Spring Joint Computer Conference*: 329–46

Svensson P, Zhexue H 1991 Geo-SAL: a query language for spatial data analysis. In Günther O, Schek H-J (eds) *Advances in spatial databases – second symposium, SSD 91*. New York, Springer 525: 119–40

Timpf S, Raubal M 1996 Experiences with metadata. In Kraak M-J, Molenaar M (eds) *Advances in GIS research II (Proceedings Seventh International Symposium on Spatial Data Handling)*. London, Taylor and Francis: 815–27

Tomlin C D 1990 *Geographic information systems and cartographic modeling*. Englewood Cliffs, Prentice-Hall

Weber C 1997 The representation of spatio-temporal variation in GIS and cartographic displays: the case for sonification and auditory data representation. In Egenhofer M, Golledge R (eds) *Spatial and temporal reasoning in GIS*. New York, Oxford University Press: 74–85

Wise J A, Thomas J 1995 Visualising the non-visual: spatial analysis and interaction with information from text documents. *Proceedings of IEEE Visualisation 95*: 51–8

Yuan M, Albrecht J 1995 Structural analysis of geographic information and GIS operations: from a user's perspective. In Frank A U, Kuhn W (eds) *Spatial information theory – a theoretical basis for GIS*. Lecture notes in computer science 988. Berlin, Springer: 107–22

Introduction

THE EDITORS

Prior to the widespread development of digital data capture, data were captured by manual recording of attribute measurements and then reconciling them to some kind of georeferenced frame. This section begins with a 'back-to-basics' rendition of the impact of the science of geodesy upon the way in which we position ourselves on the Earth's surface. In Chapter 30 Hermann Seeger's clear message is that in an age in which surveying techniques have been apparently de-skilled and reference frameworks globalised, GIS users will nevertheless ignore the scientific basis to global positioning at their peril. After a description of the principles used to locate points on the Earth's surface, Seeger reviews the techniques available for conversion of coordinate data between different positioning systems. With respect to GIS usage, he goes on to suggest that a range of potential complexities and gross errors may arise in geographical positioning and that, in particular, conflation of conventional map information with that derived from global positioning systems (GPS: see Lange and Gilbert, Chapter 33) may create considerable practical problems. The message is thus a need for a good understanding of the science of geodesy coupled with adequate knowledge of the genealogy of the maps from which other GIS-based data have been derived. Goodchild and Longley (Chapter 40) return to this issue in the context of the quest for 'perfect positioning' at the end of the Technical Issues Part.

The early years of GIS were characterised more by rapid developments in computer hardware and software, and as such GIS was technology – rather than data – led. Many of the data for early GIS applications were captured from existing analogue sources which had already been reconciled with spatial referencing systems. Even though many of the chapters elsewhere in this book (especially those discussing spatial analysis in Section 1c) have made much of the subsequent creation of huge digital datasets and the emergence of a digital information economy, it is important to remember that the costs of data are still very often the most significant component of a GIS (see also Bernhardsen, Chapter 41; Rhind, Chapter 56). Moreover, a recurring theme through this book has been the interdependencies between the conception, measurement, representation, and analysis of geographical phenomena, with the implication that involvement (or at least acquaintance) with each of these successive stages is likely to lead to more sensitive and intelligent geographical information handling (e.g. Goodchild and Longley, Chapter 40). Founding a GIS upon dubious or poor quality data is simply not good enough – and such data will have obvious repercussions of the 'garbage in – garbage out' variety when used in analysis or decision-making. If source data were originally captured in a different medium, and were intended for use in different ways, then we need to understand the characteristics of the original data – including their quality, the processes used to produce the document to be encoded, and the geometric characteristics of the data (Dowman, Chapter 31). We also need to understand and anticipate the additional problems that can arise in the transfer of analogue information to digital media.

In the second contribution to this section, Ian Dowman presents a review of the simplification, codification, and generalisation that takes place in capturing data from hard copies, together with the checks and balances that can be used to minimise the additional error introduced by the data transfer process. The pitfalls associated with digitising from paper maps are widely known, and the checks that may be used to ensure adequate spatial referencing are straightforward. This is not so clearly the case with regard to aerial photographs, since digital mapping requires specification of the geometry of the image forming system. Moreover, and especially at the scales characteristic of satellite images, relief causes displacement together with changes in scale across the image: these can only be removed by developing a stereoscopic 3-dimensional model or by using ancillary information from a digital elevation model (see Hutchinson and Gallant, Chapter 9). Even digitally corrected orthoimages may still be distorted by tilting of the sensor platform (and will require reconciliation with ground control points), and a range of other transformations may be necessary to reconcile images with spatially referenced systems.

The 'data bottleneck' experienced in early GIS applications stood in stark contrast to the experience of the remote-sensing community, in which the handling and processing of spatially extensive digital images remains a major activity. Satellite imagery has become both more detailed, with developments in satellite technology, and more widely available, particularly from satellites belonging to the former Soviet Union. As a consequence, there has been some convergence of GIS and remote sensing and an opening up of new application areas. Mike Barnsley (Chapter 32) provides an overview of remote sensing as an enabling technology for exploiting the fuller potential of GIS. The availability of more data and the improved positional accuracy associated with them is part of this story. Of more wide-ranging import (at least in applications terms), Barnsley is at pains to emphasise the roles of inference and estimation in moving from surrogate (e.g. land cover) to target (e.g. land use) variables. This provides a reminder that in remote sensing, as elsewhere in GIS, our understanding of the process of inference is far from complete (cf. Getis, Chapter 16) and that irrespective of technique used (cf. Fischer, Chapter 19) we only ever end up with estimates of our chosen

classification. That said, there are a number of important developments in image classification using artificial neural network techniques and contextual information about the apparent configuration of land use: Barnsley sees these as offering the prospect of freeing the classification process from conventional assumptions about statistical distributions and opening up new domains of application, such as urban remote sensing. From a purely technical standpoint, Barnsley also examines the relationship between the design of new sensors and their abilities to estimate the intrinsic properties of the Earth's surface.

The final contribution to this section, by Lange and Gilbert, describes the development of global positioning systems (GPS) as a GIS-enabling technology. They review a range of civilian applications and describe the use of base stations to improve levels of geographical accuracy. The development of GPS vividly illustrates how far technical aspects of GIS data collection have come: Dowman's contribution to this section provided a review of the ways in which digital abstractions can be captured from paper maps which themselves are highly abstract and selective descriptions of geographical reality; by analogy, the GPS user creates an abstraction of the surface of the Earth through the act of moving across it (and perhaps recording additional attribute layers at the same time). This provides a vivid illustration of how much richer GIS data models have become, and how much more control the informed GIS user now has over some aspects of data collection.

32

Digital remotely-sensed data and their characteristics

M BARNSLEY

This chapter explores the nature and properties of digital remotely-sensed data. Rather than simply summarising the ever-growing range of airborne and satellite sensor systems, together with their technical characteristics, the chapter is divided into three distinct parts, namely: (a) the interaction of electromagnetic radiation with Earth surface materials, focusing on the physical, chemical, and biological properties that control their reflectance, emittance, and scattering characteristics; (b) the impact of sensor and platform design on the ability to record the surface-leaving radiation and the nature of the data that are produced; and (c) the production of data-processing algorithms to translate the recorded signals into estimates of the intrinsic properties of the observed surfaces.

1 PHYSICAL PRINCIPLES

1.1 Remote sensing: inference and estimation

Broadly speaking, the subject matter of terrestrial remote sensing encompasses the set of instruments (sensors), platforms, and data-processing techniques that are used to derive information about the physical, chemical, and biological properties of the Earth's surface (i.e. the land, atmosphere, and oceans) without recourse to direct physical contact. Information is derived from measurements of the amount of electromagnetic radiation reflected, emitted, or scattered from the Earth surface, and its variation as a function of wavelength, angle (direction), wave polarisation, phase, location, and time. A variety of sensors is commonly employed in this context – both passive (i.e. those reliant on reflected solar radiation or emitted terrestrial radiation) and active (i.e. those generating their own source of electromagnetic radiation) – operating throughout the electromagnetic spectrum from visible to microwave wavelengths (see also Dowman, Chapter 31). The platforms on which these instruments are mounted are similarly diverse: although Earth-orbiting satellites and fixed-wing

aircraft are by far the most common, helicopters, balloons, masts, and booms are also used. Finally, a wide range of data-processing techniques has been developed, often in response to advances in sensor technology, but increasingly to meet the demands of a growing set of applications.

The problem with the definition of remote sensing outlined above is that it focuses on the *technology,* as opposed to the *science,* of remote sensing. In doing so, it obscures two fundamental aspects of the remote sensing process, namely *inference* and *estimation.* The role of inference becomes clear when it is understood that very few properties of interest to the environmental scientist can be measured directly by remote sensing. Instead, they must be inferred from measurements of reflected, emitted, or scattered radiation using some form of mathematical model, or via their relationship with a surrogate variable (e.g. land cover) that can be derived more readily from the remotely-sensed data (see also Bibby and Shepherd, Chapter 68; Fisher, (Chapter 13). The accuracy with which a given property can be inferred is therefore dependent on the quality (generality, applicability, reliability, etc.) of the model and algorithms used, or on the degree of correlation between the surrogate and target variables, together with the accuracy and suitability of any land-cover classification scheme involved and the classes that this defines. Unfortunately, our understanding of these relationships is, in many instances, still quite poor (see also Dowman, Chapter 31). Even where our knowledge is well developed, we may be forced to employ models involving a number of approximations or simplifications, perhaps to reduce the computational load in time-critical applications. As a consequence, the values inferred from remotely-sensed data are almost always estimates of the actual quantities of interest.

1.2 Sources of information

There are five main sources of information that can be exploited by remote sensing systems. These relate to variations in the recorded signal as a function of:

- wavelength ('spectral');
- angle ('directional');
- wave polarisation;
- location ('spatial');
- time ('temporal').

1.2.1 Inference and estimation from spectral variations

Most remote sensing studies attempt to exploit spectral (i.e. wavelength dependent) variations in the radiation emanating from the Earth's surface: these are controlled by the physical and chemical properties of Earth surface materials. In the case of healthy green leaves, for example, the principal controlling factors are plant pigments (e.g. chlorophyll, xanthophyl, and the carotenoids), lignin, cellulose, protein, and leaf-water content (Asrar 1989; Jacquemoud and Baret 1990). In the case of soils, the most important factors are the content of moisture, iron oxides, and organic matter, together with surface structure (Price 1990; Jacquemoud et al 1992).

There are two main ways in which the relationship between surface properties and spectral response can be exploited. At one level, the aim may be simply to distinguish different types of surface material. In this case, the objective is to identify those wavelengths at which the contrast between their reflectance, emittance, or scattering characteristics is maximised. Since not all surface materials can be distinguished at a given wavelength, it is common to record data in several parts of the electromagnetic spectrum (i.e. multispectral remote

sensing). A subsequent aim may be to identify the nature of the surface materials; that is, to assign each a label from a set of pre-defined classes, typically expressed in terms of land cover.

The second major use of multispectral data is to estimate values for selected properties of the observed surface materials. For example, many studies have attempted to derive information on the above-ground biomass, leaf area index (LAI), and levels of photosynthetic activity of vegetation canopies. This is commonly based on linear combinations of data recorded in two or more spectral wavebands, generally centred on the visible red and near-infrared wavelengths (Myneni and Williams 1994). Use of this type of empirical model – referred to generically as vegetation indices – is widespread, despite their well-known limitations (Baret and Guyot 1991; Myneni et al 1995); indeed, new indices are continually being developed. The enduring attraction of vegetation indices lies in their conceptual and computational simplicity. This goes some way to explain the enduring popularity of the normalised difference vegetation index (NDVI), most recently for mapping and monitoring vegetation at regional and global scales (Townshend et al 1995).

Recent advances in sensor technology, specifically those relating to improvements in spectral resolution (Vane and Goetz 1993), have prompted more detailed studies of the relationships between spectral response and surface biochemical properties (Wessman et al 1988; Hunt 1991). Many of these studies continue to make use of simple empirical transformations (such as ratios) of data measured in a number of spectral wavebands (Danson et al 1992). Attention has also been focused on locating the so-called 'red edge' (the wavelength of maximum slope in the spectral response of vegetation between 690μm and 740μm; Figure 1), using this as an indicator of photosynthetic activity and leaf biochemistry (Boochs et al 1990; Filella and Peñuelas 1994; Curran et al 1995). More importantly, attempts have also been made to develop physically-based models to account for the optical properties of individual leaves in terms of their chemical and physical characteristics (Jacquemoud and Baret 1990). In principle, these models should be less data-dependent and site-specific than their empirical counterparts. It may also be possible to invert them, so that estimates of their parameters – and, hence, the surface biophysical properties to which they relate – can be derived from multispectral measurements made by remote sensing systems.

Whichever methods are employed, it should be noted that vegetation canopies do not behave simply as 'big leaves', so that problems arise in attempting to apply the techniques described above directly to remotely-sensed images. More specifically, relationships determined *in vitro*, or *in vivo* at the scale of a single leaf, are complicated by differences in, among other things, the spatial and geometric structure of vegetation canopies and variations in the soil substrate (Goel 1989; Asrar 1989). For this reason, attempts to estimate biophysical or biochemical properties at the canopy scale require the use of coupled models of canopy and leaf reflectance (Jacquemoud 1993; Jacquemoud et al 1995).

1.2.2 Inference and estimation from directional variations
The detected reflectance of most Earth surface materials varies, sometimes considerably, as a

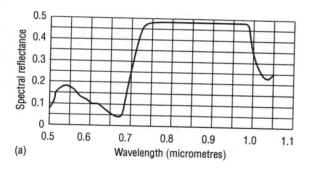

(a)

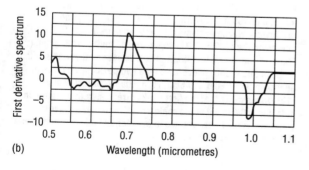

(b)

Fig 1. (a) Example leaf reflectance spectrum at visible and near-infrared wavelengths; (b) first derivative spectrum produced from Figure 1(a), showing position of the 'red-edge' (peak in derivative spectrum) at 0.693µm.

function of the angles at which they are illuminated by the Sun and viewed by the sensor. The form and magnitude of this effect are controlled by:

- the optical properties of the component elements of the surface material (e.g. the spectral reflectance and transmittance of plant leaves, stems, tree crowns, and soil facets);
- the spatial and geometric arrangement of these elements;
- the spectral and angular distribution of the incident solar radiation (Goel 1989; Barnsley 1994).

The angular distribution of the reflected radiation is described by the bidirectional reflectance distribution function (BRDF). Research in this area has focused on the development and implementation of various mathematical models (Myneni et al 1990), ranging from simple empirical (e.g. Walthall et al 1985) and semi-empirical functions (e.g. Roujean et al 1992), to models with a more direct foundation in physical principles (e.g. Ahmad and Deering 1992; Kuusk 1994). Interest in these models arises from their potential to derive quantitative estimates of certain biophysical properties of the Earth surface (e.g. the LAI). This can be achieved by inverting the model against measurements of reflected radiation made at a number of different sensor view angles and solar illumination angles with respect to a fixed point on the Earth surface (Plate 21; Goel 1989; Barnsley et al 1994). Estimates of the surface albedo can also be obtained through numerical or analytical integration of the modelled BRDF (Kimes et al 1987; Barnsley et al 1997a).

1.2.3 Inference and estimation from wave polarisation
Electromagnetic radiation considered in wave form has two fields (electric and magnetic) which are perpendicular both to one another and to the direction of propagation (Rees 1990). The orientation of these two fields, known as the wave polarisation of the radiation, has been observed to change as a result of scattering and reflection within the atmosphere and at the Earth surface. The majority of studies in this area have employed data from the microwave region of the electromagnetic spectrum. For example, polarimetric radar data have been used to distinguish different stands in coniferous forests (Grandi et al 1994) and to assess their biophysical characteristics (Baker et al 1994). A smaller number of studies has explored

polarisation characteristics of Earth surface materials at visible and infrared wavelengths (e.g. Vanderbilt et al 1991; Ghosh et al 1993). This is partly attributable to the paucity of appropriate airborne and satellite sensors, and partly because the polarisation signal is dominated by the atmosphere at these wavelengths.

1.2.4 Inference and estimation from spatial variations

The amount of radiation reflected, emitted, or scattered from the Earth's surface varies spatially in response to changes in the nature (type) and properties of the surface materials. These variations may be continuous, discrete, linear, or localised, depending on the controlling environmental processes (Davis and Simonett 1991). They may also be manifest at a variety of different spatial scales (Townshend and Justice 1990; Barnsley et al 1997b; Figure 2). The relationships between surface type, surface properties, and spatial variability in land-leaving radiance has been exploited using measures of:

- texture – the statistical variability of the detected signal, typically based on the grey-level co-occurrence matrix, measured at the level of individual pixels (Richards 1993);
- pattern – including the size and shape of discrete spatial entities (regions), typically land-cover parcels, identified within the scene, as well as the spatial relations between them (LaGro 1990; Lam 1990);
- context – referring to the structural and semantic relations between discrete spatial entities identified within the scene (Barr and Barnsley 1997).

1.2.5 Inference and estimation from temporal variations

The reflectance, emittance, and scattering properties of most Earth surface materials vary with respect to time. This may be in response to diurnal effects (e.g. changes in the leaf-angle distribution of vegetation canopies attributable to moisture deficiency or heliotropism), seasonal effects (e.g. phenology), episodic events (e.g. rainfall and fire), anthropogenic influences (e.g. deforestation), or long-term climate change. There are several ways in which these temporal variations can be exploited, namely:

- to assist in distinguishing surface materials, by selecting the time of day or year at which the contrast between their reflectance, emittance, or scattering properties is greatest;
- to detect a change in the dominant land-cover type or biophysical property of an area by measuring

(a)

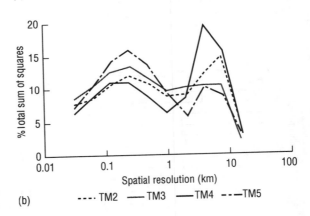

(b) ---- TM2 —— TM3 —— TM4 --- TM5

Fig 2. (a) LANDSAT-TM sub-scene (2048 by 2048 pixels; near-infrared waveband) covering part of southeast England; (b) Scale variance analysis (Townshend and Justice 1990) applied to the LANDSAT-TM sub-scene of southeast England (a). The diagram shows the different scales of spatial variability that occur in this scene, indicated by the two peaks in variance at approximately 250m and 5km, respectively. Barnsley et al (1977b) suggests that the first peak corresponds to variation in detected reflectance at the scale of individual field parcels, while the second peak relates to broader edaphic and geological differences across the scene.

variations in the amount of surface-leaving radiation over time (known as change detection);
- to determine the physical, chemical, and biological properties of Earth surface materials.

For example, the third approach has been used to produce land cover maps at regional and global scales from coarse spatial resolution satellite sensor images (Lloyd 1990). Lloyd's approach is based on an analysis of the date-of-onset, duration, and amplitude of the 'greening-up' curve, derived from a time-series of NDVI values (Figure 3).

2 MEASURING THE SIGNAL

This section considers the impact of sensor and platform design on the ability to record surface-leaving radiation. While some space is dedicated to specific sensor systems and the characteristics of the data that they produce, the intention is not to provide a summary of current and future remote sensing devices. Rather, the aim is to examine the way in which their general design affects the ability to translate the recorded signals into estimates of the intrinsic properties of the Earth surface. Thus, the sensor system is viewed both as a measurement device and as 'filter' to the surface-leaving signal.

2.1 Spectral resolution and spectral coverage

Since most remote sensing studies – particularly those concerned with the use of optical instruments – exploit spectral variations in surface-leaving radiation, it seems appropriate to begin with a consideration of the spectral characteristics of remote sensing systems, namely:

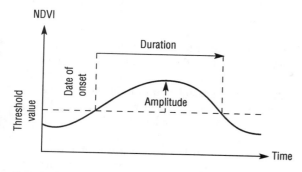

Fig 3. Diagrammatic representation of the variation in the normalised difference vegetation index (NDVI) for a vegetated surface over the growing season. The figure illustrates the concepts of the date-of-onset, the duration, and the amplitude of the 'greening-up' curve used by Lloyd (1990) to map land cover at the regional scale using coarse spatial resolution satellite sensor images.

- the number of spectral wavebands in which the sensor operates (see also Estes and Loveland, Chapter 48);
- the position of these spectral wavebands within the electromagnetic spectrum (spectral coverage);
- the range of wavelengths covered by each waveband (spectral bandwidth or spectral resolution) (Davis and Simonett 1991).

Clearly, the specific configuration adopted for a given sensor is determined by the scientific objectives of the mission, but it is also conditioned by a number of fundamental technical constraints. The latter include the need to locate the wavebands within 'atmospheric windows', the total volume of data that must be handled (including telemetry to Earth, in the case of spaceborne sensors), and the need to achieve an acceptable signal-to-noise ratio (SNR).

2.1.1 Atmospheric windows

The atmosphere scatters and absorbs radiation during its passage from the Sun to the Earth's surface and from the Earth's surface to the sensor. In doing so, it attenuates the amount of radiation reaching the ground and, subsequently, the sensor. It also alters the spectral composition and the angular distribution of this radiation (Diner and Martonchik 1985; Kaufman 1988). The magnitude of these effects varies strongly with wavelength. Sensors designed to study either the land surface or the oceans operate in regions of the electromagnetic spectrum in which the transmission of radiation through the atmosphere is high – known as 'atmospheric windows'. Even so, solar radiation may be scattered within the atmosphere into the path of the sensor without interacting with the Earth surface. Among other things, this component of the signal detected by the sensor, known as path radiance, reduces the apparent contrast between surface materials within the resultant image (Kaufman 1993).

2.1.2 Data volumes and spectral redundancy

Over the last twenty years or so, there has been a continuing trend towards sensors that are able to record data in a greater number of (typically narrower) spectral wavebands, resulting in an increase in the total volume of data acquired. In general, however, the amount of useful information that can be extracted from these data does not increase linearly with the number of available spectral wavebands: there is often a strong statistical correlation between the data recorded in different parts of the electromagnetic spectrum, particularly those from adjacent spectral wavebands. As a result, the intrinsic dimensionality of a multispectral dataset may be very considerably smaller than the number of available wavebands. This is sometimes referred to as 'spectral redundancy'.

2.1.3 Imaging spectrometers and signal-to-noise ratio

Despite the observations made above, one of the major developments in optical sensor technology in recent years has been the advent of imaging spectrometers and imaging spectroradiometers – instruments capable of recording data in tens, or even hundreds, of very narrow (typically contiguous) spectral channels (Vane and Goetz 1993). The manner in which data from these sensors are generally employed differs from that of conventional multispectral scanners. Instead of focusing on the data simply as a set of 2-dimensional images, emphasis is placed on an analysis of the detailed spectral response recorded for each pixel. These can be compared against spectra for a range of different surface materials, drawn either from an on-line library or from representative pixels sampled within the image itself. In addition to the overall shape of the spectra, comparisons can be made in terms of the presence, depth, and width of absorption features associated with specific biochemical constituents. This may allow the analyst to derive detailed information on the nature, properties, and proportions of the different surface materials present in the corresponding area on the ground.

The Advanced Visible and Infrared Imaging Spectrometer (AVIRIS) instrument operated by NASA is one example of an imaging spectrometer (Vane et al 1993). This airborne sensor records data in approximately 200 narrow spectral channels in the region 0.4μm to 2.5μm. One of the penalties commonly associated with the use of narrow spectral wavebands is a reduction in the SNR of the sensor, because of the smaller number of photons admitted to the detector. This can be compensated for by increasing the sensor dwell-time (at the expense of a reduction in the effective sensor spatial resolution) or by combining images from several successive flights over the same target.

A spaceborne imaging spectrometer, known as HIRIS (High Resolution Imaging Spectrometer), was originally scheduled for launch at the end of the decade as part of NASA's 'Mission to Planet Earth'

programme (Goetz and Herring 1989). The instrument was, however, de-selected at an early stage because of budgetary constraints. A second imaging spectrometer, known as MODIS (Moderate Resolution Imaging Spectrometer) is due to be launched in 1998/9, although this sensor has a much smaller number of spectral wavebands (30, cf. ~200 for HIRIS) and a considerably coarser spatial resolution (250m to 1 km, cf. 30m for HIRIS; Ardanuy et al 1991).

2.2 Radiometric resolution and radiometric calibration

2.2.1 Radiometric resolution

The radiometric resolution of a sensor can be thought of as its ability to distinguish different levels of reflected, emitted, or scattered radiation. Expressed more precisely, radiometric resolution involves three key concepts, namely:

- quantisation;
- signal-to-noise ratio;
- dynamic range.

A digital remote sensing device converts the radiation incident on its detectors initially into an analogue signal (i.e. an electrical voltage) and subsequently into a digital signal. After the analogue-to-digital (A-to-D) conversion, the detected signal is represented as a numerical value, referred to (somewhat tautologously) as a digital number (DN). The set of possible values for the DN is determined by the quantisation level: thus, if a sensor has 8-bit quantisation, it will record values in the range 0 to 255 (i.e. 2^8 or 256 different levels of incident radiation), where a value of 0 indicates the lowest level of detectable radiance and 255 the highest. The sensor designer must also decide how the range of DN are to be used to record incident radiation. It is possible, for example, to design an instrument that is capable of recording the full range of radiances expected under normal illumination conditions from surfaces with reflectances varying between 0 and 1. Alternatively, if the intention is primarily to observe relatively dark targets, such as the oceans, the dynamic range of the instrument might be limited accordingly. Thus, the set of available DNs would be optimised to distinguish surfaces in the desired range of reflectances, although the instrument response would saturate over brighter targets.

2.2.2 Radiometric calibration

One characteristic of some of the most recent and many of the proposed future satellite sensors is the greater attention that is being given to their absolute radiometric calibration. This ensures that the recorded DN can be related accurately to known levels of surface reflectance (or emittance) (Price 1987). This assists in the retrieval of datasets expressed in terms of standard geophysical units and ensures greater consistency between datasets generated by different sensors, or by the same sensor over a prolonged period of time (Hall et al 1991). The lack of accurate radiometric calibration in early satellite sensors has been one of the major hindrances to the use of these data for long-term regional and global-scale environmental monitoring (Hall et al 1995).

2.3 Spatial resolution

In simple terms, the spatial resolution of a sensor determines the level of spatial detail that it provides about features on the Earth's surface. Beyond this, spatial resolution can be defined in a number of different ways (Forshaw et al 1983). For example, the instantaneous field-of-view (IFOV) defines the (nominal) angle, subtended at the sensor, over which the instrument records radiation emanating from the Earth's surface at a given instant in time. The area on the Earth's surface to which this corresponds, known as the ground resolution element (GRE), is therefore controlled by the IFOV and the height of the sensor above the ground. The actual area of ground from which radiation is incident on the detector is, however, larger than this and is determined by the sensor's point spread function (PSF). Finally, 'pixel size' denotes the area of ground covered by a single pixel in the resultant image. This may differ from the GRE because of the effects of over-sampling, variations in the height of the terrain below the sensor, and variations in the attitude and altitude of the platform on which the sensor is mounted (Forshaw et al 1983).

2.3.1 Impacts of sensor spatial resolution

Images produced by digital sensors can be thought of as 2-dimensional grids or arrays of data cells ('picture elements' or pixels). The spatial resolution of the sensor defines the size of these cells, in terms of the area that they represent on the ground (Plate 22). Thus, a remotely-sensed image represents a spatial regularisation of the observed scene (Jupp et

al 1988). One effect of this process is that two or more surface materials may fall within a single pixel, producing a 'mixed pixel' (or 'mixel'). The extent to which this occurs is, of course, dependent on the spatial resolution of the sensor and the spatial variability of the observed surface. The mixed pixel effect has a number of implications for information retrieval from remotely-sensed images. First, the detected spectral response of a mixel will be some composite of the individual spectral signals from the constituent surface materials (Smith et al 1985). Second, the size, shape, and spatial arrangement (pattern) of the major spatial entities present within the scene will be to some extent obscured (Woodcock and Strahler 1987). The first of these two problems has been addressed through the development of a number of techniques designed to 'un-mix' the component spectral responses contained in each pixel (Ichoku and Karnieli 1996). The most widely used of these is linear mixture modelling, in which the composite signal is assumed to be a linear summation of the spectral curves for the component land-cover types, weighted by their relative abundance (i.e. proportion of ground covered) within the pixel. The second problem has received rather less attention in the field of remote sensing, although it is the subject of detailed investigation by landscape ecologists (Barnsley et al 1997b).

2.3.2 Current and future directions

In recent years, there has been an intriguing bifurcation in the spatial resolution of spaceborne optical sensors. One element of this has been the widespread development of 'moderate' (or 'medium') resolution (~1km) devices (as is shown in Table 1 and Figure 4; see also Ardanuy et al 1991; Diner et al 1991; Prata et al 1990). The lineage of these sensors is simple to trace – deriving from the outstanding and, to a certain extent, unanticipated success of the current generation of NOAA's Advanced Very High Resolution (AVHRR) sensors in monitoring the land surface at continental and global scales.

Hand-in-hand with this, there is a continuing trend – initiated with LANDSAT-TM and SPOT-HRV during the mid 1980s – towards sensors with an increasingly fine spatial resolution. This trend has been extended through the availability of data from a range of Russian satellite sensors (e.g. KFA-1000 and KFA-3000), as well as the Panchromatic sensor on-board the Indian satellite IRS-1C, and is set to

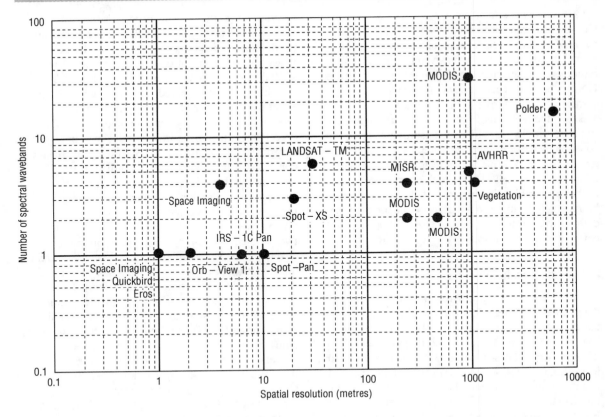

Fig 4. Diagrammatic representation of the range of current and proposed satellite sensors, together with their main spectral and spatial characteristics.

Table 1 Characteristics of a number of 'medium' or 'moderate' spatial resolution satellite sensors currently in operation or scheduled for launch in the next few years. Some other 'standard' remote sensing image sources are set out in Estes and Loveland (Chapter 48) and Dowman (Chapter 31 Table 1).

Sensor	Satellite	Spatial resolution (at nadir)	Number of spectral bands	Year of launch (actual or projected)
ATSR-1	ERS-1	1km	4	1994
ATSR-2	ERS-2	1km	6	1995
POLDER	ADEOS-1	6km by 7km	16	1996
VEGETATION	SPOT-4	1.15km	4	1997/8
MODIS	EOS-1 AM	250m, 500m and 1km	30	1998/9
MERIS	ENVISAT-1	250m (land) and 1km (oceans)	15 (programmable in position and width)	1988
MISR	EOS-1 AM	250m and 1km	4	1998/9

continue with the launch of a number of new, commercially-operated satellite devices (Table 2). Each of these instruments will be capable of producing digital image data with a spatial resolution of between 1 and 5 metres (McDonald 1995; Fritz 1996).

2.4 Angular sampling

Interest in the directional reflectance properties of Earth surface materials has grown considerably in recent years, partly in response to the increasing availability of satellite sensors that can record data at several different angles with respect to the Earth surface. This can be achieved in a number of ways (Barnsley 1994), namely:

- by means of a very wide across-track field-of-view (e.g. NOAA's AVHRR sensors, and the proposed SPOT-4 VEGETATION, MODIS, and MERIS [Medium Resolution Imaging Spectrometer] instruments);
- through the use of a very wide field-of-view in both the along-track and across-track directions (e.g. the POLDER [polarisation and directionality of the Earth's reflectances] sensor on board the ADEOS-1 satellite);
- by pointing the sensor off-nadir in the across-track direction (e.g. the HRV [high resolution visible] instruments on the SPOT-series of satellites), the along-track direction, or both;
- through the use of multiple sensors pointed forward, nadir and aft of the platform (e.g. the multi-angle imaging Spectroradiometer [MISR] scheduled for launch as part of NASA's Earth Observing System); or
- through the use of a conical scanning motion (e.g. the Along-Track Scanning Radiometer ATSR on the European remote sensing [ERS] satellites).

The range of view angles and solar illumination angles over which a given instrument can acquire data is controlled not only by the geometry of the sensor, but also by the orbital characteristics of the satellite on which it is mounted (Barnsley et al 1994). In most cases, the actual number of angles at which

Table 2 Characteristics of a number of very high spatial resolution satellite sensors currently in operation or planned for launch in the near future (see also Dowman, Chapter 31 Table 1).

Sensor	Year of launch (actual or projected)	Spatial resolution Pan	XS	Swath width Pan	XS	Stereoscopic viewing capability
KFA-1000	1994	6m	—	80km	—	No
KFA-3000	1994	3m	—	27.5km	—	No
IRS-1C	1995	5.8m	—	70km	—	Yes
KVR-1000	1996*	2m	—	40km	—	No
Earlybird	1997	3m	15m	6km	30km	Yes
AVNIR	1997	8m	16m	80km	80km	Yes
EROS	1997	1m	1.5m	15km	15km	Yes
Quickbird	1997	1m	4m	6km	36km	Yes
Space Imaging	1997	1m	4m	11km	11km	Yes
Orbview-1	1998	1–2m	4m	8km	8km	Yes
GDE	1998	1m	—	15km	—	Yes

* digital format

reflectance data can be sampled is quite limited (Figure 5). Appropriate mathematical models are therefore required to interpolate between, and to extrapolate beyond, these sample measurements to describe and account for the full BRDF. If the models are also invertible, it may be possible to retrieve estimates of certain properties of the surface (e.g. LAI) from the sample directional reflectance data. Various BRDF models have been developed for this purpose, ranging from simple empirical formulations through to more complex, physically-based models. While the latter offer significant advantages in principle, inversion of such models typically demands the use of computationally-intensive numerical procedures. For this reason, attention is currently being focused on the use of so-called 'semi-empirical, kernel-driven' BRDF models, which can be inverted analytically (Wanner et al 1995; Barnsley et al 1997c). These models, however, tend not to be specified in terms of measurable biophysical properties of the land surface, so that further work is required to establish the relationships between such properties and the model parameters.

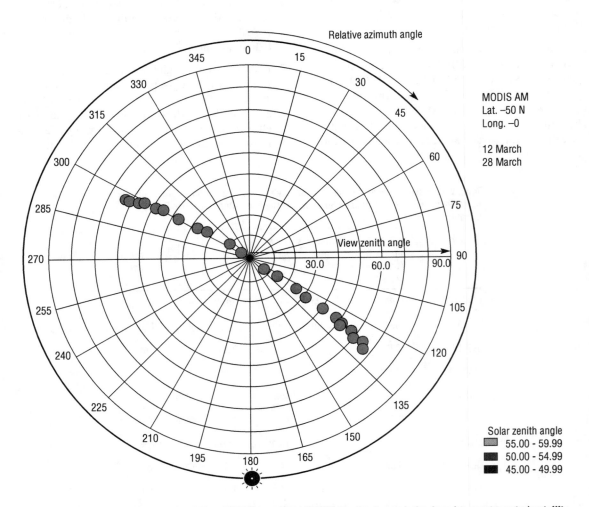

Fig 5. Angular (directional) sampling capability of NASA's proposed MODIS (moderate resolution imaging spectrometer) satellite sensor (in the form of a polar plot) for a fixed site at 50°N over a 16-day period around the vernal equinox (Barnsley et al 1994). Each dot indicates a single occasion on which the sensor is able to observe the target; the position of the dot in the plot indicates the angles at which this was achieved. The figure illustrates the comparatively sparse sample of directional reflectance data that can be acquired using this and other, similar sensors.

2.5 Wave polarisation

The majority of sensors able to measure the polarisation properties of Earth surface materials operate in the microwave region of the electromagnetic spectrum. Sensors such as these can transmit and receive (record) radiation in a given plane polarisation. Where the transmitted and received radiation have the same plane polarisation, the signal is referred to as being like-polarised; where they are of different polarisations, the signal is said to be cross-polarised (Rees 1990). Imaging radar polarimeters can derive a measure of backscattered radiation for any configuration of transmitted and received radiation using a process known as polarisation synthesis (Zyl et al 1987).

The POLDER instrument, launched on board the ADEOS-1 satellite in 1996, offers the capability to measure polarisation properties at visible and near-infrared wavelengths (Deschamps et al 1990). The primary use envisaged for these data is, however, to derive information on atmospheric aerosol properties, rather than the biophysical characteristics of land surface materials.

2.6 Temporal resolution

In simple terms, temporal resolution refers to the frequency with which repeat data can be acquired for a given area on the Earth's surface. This is controlled both by the geometry of the sensor and by the orbital characteristics of the satellite on which it is mounted. In terms of the latter, two main types of orbit are used by Earth-observing sensors, namely: (a) Sun-synchronous, near-polar; and (b) geo-stationary (or geo-synchronous).

Satellites in the first of these two orbits progress in a near-circular path at an altitude of between 500km and 1000km above the Earth's surface. The orbital plane is inclined, so that the satellite passes close to, but not over, the poles (hence 'near-polar'). By taking advantage of precession in the satellite's orbit, it is possible to ensure that the satellite crosses the equatorial plane at approximately the same local solar time on each orbit (hence 'Sun-synchronous'). The rotation of the Earth beneath the satellite means that successive orbits pass over different regions of the surface. Eventually the satellite will complete its sequence of orbits and begin to trace the path of the first orbit again. For a point on the equatorial plane, the period between two such orbits is known as the repeat cycle; LANDSAT-5, for example, has a repeat

cycle of 16 days. It is possible to observe a given point on the Earth's surface more frequently than this, depending on the latitude of the site and the configuration of the sensor. Since the satellite's orbital paths converge towards the poles, there is increasing overlap between images acquired on different orbits at higher latitudes; such sites can therefore be imaged more frequently than those at lower latitudes. This effect is also controlled by the field-of-view and, hence, the swath width of the sensor: the wider the swath width, the greater the number of occasions on which a given point can be imaged during the nominal repeat cycle. Even so, key episodic and seasonal events may still be missed because of cloud cover or simply because the event took place while the satellite was tracing another orbit. The former is, of course, less of a problem for active microwave systems, since these can penetrate cloud. The latter can be offset, to a certain extent, by tilting (or pointing) the sensor away from the sub-satellite point (i.e. off-nadir). This allows the sensor to target an area for repeated imaging, even though the satellite is not directly overhead (Barnsley et al 1994).

The second major type of satellite orbit referred to above is the geo-stationary or geo-synchronous orbit. Here, the satellite maintains a fixed position above the Earth's surface, usually at an altitude of around 36 000 km. This orbit is generally reserved for operational meteorological satellites which require frequent coverage (i.e. once every 20 to 30 minutes) of very large areas at a comparatively low spatial resolution (1–5 km) (Kramer 1994).

3 SELECTED DEVELOPMENTS IN DATA PROCESSING

It is not possible, within the scope of this chapter, to provide a comprehensive review of the full range of techniques and algorithms that are used to derive useful information from digital remotely-sensed images; although some have already been mentioned briefly in the preceding sections. Rather, this section attempts to highlight just a few of the most important, recent developments in image data processing.

3.1 Modelling surface-radiation interactions and data assimilation

Perhaps the most significant developments in the processing of remotely-sensed data over recent years have been the increasing focus on converting the detected signals into estimates of key geophysical units and the assimilation of these data into numerical models of various environmental processes (Hall et al 1995; Townshend et al 1995). The first of these two elements is being achieved through the use of increasingly sophisticated, deterministic, radiative-transfer and energy-balance models, some of which have been alluded to in the preceding sections. This general approach is, for example, central to NASA's 'Mission to Planet Earth' programme, embodied in the Earth Observing System (EOS) and its constituent satellite sensors (Running et al 1994). The significance of this development cannot be overstated: it marks the continuing transition of digital remote sensing from being principally an instrument for large-scale (land cover) mapping, to a more comprehensive, robust, and effective scientific tool for environmental monitoring.

The second element – assimilation of remotely-sensed data products into models of, for example, the global carbon cycle, the surface energy balance, and the net primary productivity of the land surface and oceans – is also receiving widespread attention (Running et al 1994; Townshend et al 1995). It reflects the recognition among much of the remote sensing community that, in addition to developing the science and technology to underpin remote sensing *sensu stricto*, there is a need to generate data products that are both appropriate to, and immediately usable by, the broader community of environmental scientists: that the rationale for remote sensing lies not simply in the development of sensors and algorithms, but more importantly in addressing real environmental problems. The scientific challenges that this creates include the requirement (a) to handle very large volumes (i.e. Tera-bytes) of data, often acquired by more than one sensor and/or satellite; (b) to process these data using robust, computationally-efficient, and validated algorithms, based on methods acceptable to most, if not all, of the target community; and (c) to generate usable products at the appropriate spatial and temporal scales, often in near-real time. Ultimately, remote sensing will be measured against how successful it is in meeting these stringent challenges.

3.2 Image classification and segmentation

At a somewhat different level, the production of thematic maps from digital, remotely-sensed images –

commonly referred to as image classification – remains an area of considerable research interest. Attention is, however, shifting from the use of standard, statistical classification algorithms to the wider application of artificial neural network (ANN), fuzzy-set, knowledge-based and evidential reasoning techniques (Fischer, Chapter 19; Bezdek et al 1984; Mesev et al 1995; Schalkoff 1992; Wilkinson 1996). The attraction of ANNs, for instance, lies in their ability to 'learn' by example, as well as their relative freedom from assumptions about the statistical distributions of the candidate classes (cf. conventional statistical classifiers, such as the maximum likelihood algorithm; Foody 1992). Fuzzy-set techniques, on the other hand, move away from the notion that each pixel must be assigned a single label drawn from a set of discrete, mutually exclusive classes. In doing so, they provide another way to account for the mixed pixel ('mixel') effect in remotely-sensed images (Foody 1992; see also Fisher, Chapter 13). Finally, both knowledge-based and evidential reasoning approaches offer ways to incorporate ancillary data (e.g. digital map data exported from a GIS), heuristics, and facts or evidence into the classification process (Wilkinson 1996).

Despite these developments, the overwhelming majority of studies continue to use image-classification algorithms that operate at the level of the individual pixel; that is, algorithms in which each pixel is assigned a label solely on the basis of its multispectral response, without reference to those of neighbouring pixels or the context of that pixel within the scene as a whole. By comparison, relatively limited use has been made of syntactic (or structural) pattern-recognition techniques, which operate on discrete, multi-pixel regions (i.e. meaningful spatial entities or 'objects') to infer further, higher-level information about the scene (Schalkoff 1992). Notable exceptions include the studies by Moller-Jenson (1990) and Nichol (1990) – on the spatial generalisation of thematic maps derived from remotely-sensed data – and, more recently, by Barr and Barnsley (1997) – to infer information on land use in urban areas from satellite sensor images. The comparative lack of attention given to syntactic pattern-recognition techniques in remote sensing to date is probably because of the relatively coarse spatial resolution of the images acquired by the current generation of satellite sensors. This results in uncertainty, not only about the nature (i.e. land cover type) of the principal spatial entities present within the scene, but also their morphological properties (e.g. size, shape, and boundaries) and the spatial (e.g. adjacency, containment, distance, and direction) and structural (e.g. 'forms part of') relations between them. This situation is, however, likely to change with the advent of the new generation of very high spatial resolution (<5m), commercial satellite sensors (but see Smith and Rhind, Chapter 47, for a discussion of some residual limitations). Indeed, data acquired by these new sensors demand alternatives to the conventional, per-pixel classification algorithms, if we are to derive information other than simple land cover about the observed scenes. Syntactic pattern-recognition techniques offer considerable potential in this context.

3.3 Integration of GIS and remote sensing technologies

The relationship between remote sensing and GIS has received considerable attention in the literature and, indeed, remains the subject of continuing discussion (Hinton 1996; Wilkinson 1996). Much of this discussion revolves around the scientific and technical issues relating to 'integration' of the two technologies (Ehlers et al 1989, 1991), so that remotely-sensed images can be used both as a source of spatial data within GIS and to exploit the functionality of GIS in processing remotely-sensed data. Despite this, the actual progress towards the goal of full integration is surprisingly slow. While this is undoubtedly due to the very considerable technical challenge of accessing, manipulating, and visualising vector, raster, and tabular data simultaneously, it seems unlikely that technical constraints have been the sole barrier to achieving full integration. It might be argued that competing imperatives in both remote sensing and GIS have tended to draw attention away from the issue of integration. For instance, from a remote sensing perspective, the recent focus on monitoring global environmental change using coarse (~1km) spatial resolution sensors – and the assimilation of the data that they produce into various environmental simulation models – has deflected some of the attention away from the traditional issues of large-scale mapping, which are more closely allied to the concerns and use of GIS. Similarly, one can see a number of other developments – such as the emergence of GIS functionality (albeit fairly limited) within standard office software, the potential for wider access to GIS software via network/Web

platforms, and the role and application of multimedia technology within GIS – that have consumed much of the research and development effort in the field of GIS.

Nevertheless, there are at least two reasons why the issue of integration is likely to receive fresh impetus in the near future. The first is the increasing availability of data from the very high spatial resolution, commercial satellite sensors that are scheduled for launch over the next few years. These will produce data appropriate to many of the large-scale mapping projects in which GIS have often been used, and are likely to compete directly with the traditional aerial photography market. The second is that these high resolution images require the development of new data processing algorithms, such as syntactic (or structural) pattern-recognition techniques, to extract the maximum amount of information about the observed scene. There is a considerable overlap between the objectives and functionality of these techniques and those used in mainstream GIS, at least in terms of their potential for spatial analysis, and this may also bring the two communities closer together. Estes and Loveland (Chapter 48) provide a more detailed overview of the management of the data products of new remote sensing technologies.

4 CONCLUSIONS

This chapter has attempted to provide a broad overview of the nature of digital remote sensing, including: (a) the physical, chemical, and biological properties that control the interaction of electromagnetic radiation with Earth surface materials; (b) the impact of sensor and platform design on the ability to record these signals and the nature of the data that are produced; and (c) the derivation of useful information from these data. The coverage has necessarily been brief and somewhat partial. It is impossible, within the scope of this chapter, to do justice to all aspects of the subject. For example, little has been mentioned of the development of remote sensing as it relates to the study of the Earth's oceans and atmosphere, or to the exciting advances that have been made in the application of interferometric synthetic aperture radar (SAR) to measure the morphology and deformation of the Earth's crust. Perhaps some of these aspects are of less relevance to the wider GIS

217

community. What should be apparent, however, is the rapid developments taking place in – and the increasing breadth of – digital remote sensing at the present time. Thus, while remote sensing will continue to be an important source of spatial data that can be used within GIS, the nature of these data is set to change in terms of an increase in their diversity and an improvement in their utility, accuracy, and reliability.

References

Ahmad S P, Deering D W 1992 A simple analytical function for bidirectional reflectance. *Journal of Geophysical Research* 97: 18 867–86

Ardanuy A, Han D, Salomonson V V 1991 The Moderate Resolution Imaging Spectrometer (MODIS): science and data system requirements. *IEEE Transactions on Geoscience and Remote Sensing* 29: 75–88

Asrar G (ed.) 1989 *The theory and applications of optical remote sensing*. New York, John Wiley & Sons Inc.

Baker J R, Mitchell P L, Cordey R A, Groom G B, Settle J J, Stileman M R 1994 Relationships between physical characteristics and polarimetric radar backscatter for Corsican pine stands in Thetford Forest, UK. *International Journal of Remote Sensing* 15: 2827–50

Baret F, Guyot G 1991 Potentials and limits of vegetation indices for LAI and APAR assessment. *Remote Sensing of Environment* 35: 161–73

Barnsley M J 1994 Environmental monitoring using multiple view angle (MVA) remotely-sensed images. In Foody G, Curran P J (eds) *Environmental remote sensing from regional to global scales*. London, Taylor and Francis: 181–201

Barnsley M J, Allison D, Lewis P 1997a On the information content of multiple view angle (MVA) images. *International Journal of Remote Sensing* 18: 1937–60

Barnsley M J, Barr S L, Tsang T 1997b Scaling and generalisation issues in land cover mapping, In Gardingen J van (ed.) *Scaling up*. Cambridge (UK), Cambridge University Press

Barnsley M J, Lewis P, Sutherland M, Muller J-P 1997c Estimating land surface albedo through inversion of two BRDF models against ASAS image data for HAPEX-Sahel. *Journal of Hydrology* 188–9; 749–78

Barnsley M J, Strahler A N, Morris K P, Muller J-P 1994 Sampling the surface BRDF: 1. Evaluation of current and future satellite sensors. *Remote Sensing Reviews* 8: 271–311

Barr S L, Barnsley M J 1997 A region-based, graph-theoretic data model for the inference of second-order thematic information from remotely-sensed images. *International Journal of Geographical Information Science*, 11: 555–16

Bezdek J C, Ehrlich R, Full W 1984 FCM: the fuzzy *c*-means clustering algorithm. *Computers and Geosciences* 10: 191–203

Boochs F, Kupfer G, Dockter K, Kuhbauch W 1990 Shape of the red-edge as a vitality indicator for plants. *International Journal of Remote Sensing* 11: 1741–53

Curran P J, Windham W R, Gholz H L 1995 Exploring the relationship between reflectance red, edge and chlorophyll concentration in slash pine leaves. *Tree Physiology* 15: 203–6

Danson F M, Steven M D, Malthus T J, Clark J A 1992 High-spectral resolution data for determining leaf water content. *International Journal of Remote Sensing* 13: 461–70

Davis F W, Simonett D S 1991 GIS and remote sensing. In Maguire D J, Goodchild M F, Rhind D W (eds) *Geographical information systems: principles and applications*. Harlow, Longman/New York, John Wiley & Sons Inc. Vol. 1: 191–213

Deschamps P Y, Herman M, Podaire A, Leroy M, Laporte M, Vermande P 1990 A spatial instrument for the observation of polarisation and directionality of Earth reflectances: POLDER. *Proceedings of International Geoscience and Remote Sensing Symposium (IGARSS90)*. Washington DC, IEEE Publications: 1769–74

Diner D J, Bruegge C J, Martonchik J V, Bothwell G W, Danielson E D, Ford V G, Hovland L E, Jones K LK, White M L 1991 A multi-angle image spectroradiometer for terrestrial remote sensing with the Earth Observing System. *International Journal of Imaging Systems and Technology* 3: 92–107

Diner D J, Martonchik J V 1985 Atmospheric transmittance from spacecraft using multiple view angle imagery. *Applied Optics* 24: 3503–11

Ehlers M, Edwards G, Bédard Y 1989 Integration of remote sensing with geographical information systems: a necessary evolution. *Photogrammetric Engineering and Remote Sensing* 55: 1619–27

Ehlers M, Greenlee D, Smith T, Star T 1991 Integration of remote sensing and GIS: data and data access. *Photogrammetric Engineering and Remote Sensing* 57: 669–75

Filella I, Peñuelas J 1994 The red-edge position and shape as indicators of plant chlorophyll content, biomass, and hydric status. *International Journal of Remote Sensing* 15: 1459–70

Foody G M 1992 A fuzzy sets approach to the representation of vegetation continua from remotely sensed data: an example from lowland heath. *Photogrammetric Engineering and Remote Sensing* 58: 221–5

Forshaw M R B, Haskell A, Miller P F, Stanley D J, Townshend J R G 1983 Spatial resolution of remotely-sensed imagery: a review paper. *International Journal of Remote Sensing* 4: 497–520

Fritz L W 1996 The era of commercial earth observation satellites. *Photogrammetric Engineering and Remote Sensing* 62: 39–45

Ghosh R, Sridhar V, Venkatesh H, Mehta A, Patel K 1993 Linear polarisation measurements of a wheat canopy. *International Journal of Remote Sensing* 14: 2501–8

Goel N S 1989 Inversion of canopy reflectance models for estimation of biophysical parameters from reflectance data. In Asrar G (ed.) *Theory and applications of optical remote sensing*. New York, John Wiley & Sons Inc.: 205–51

Goetz A F H, Herring M 1989 The High Resolution Imaging Spectrometer (HIRIS) for EOS. *IEEE Transactions on Geoscience and Remote Sensing* 27: 136–44

Grandi G de, Lemoine G G, Groof H de, Lavalle C, Siebert A J 1994 Fully polarimetric classification of the Black Forest MAESTRO 1 AIRSAR data. *International Journal of Remote Sensing* 15: 2755–76

Hall F G, Strebel D E, Nickeson J E, Goetz S J 1991 Radiometric rectification: towards a common radiometric response among multidate, multisensor images. *Remote Sensing of Environment* 35: 11–27

Hall F G, Townshend J R G, Engman E T 1995 Status of remote sensing algorithms for estimation of land surface state parameters. *Remote Sensing of Environment* 51: 138–56

Hinton J C 1996 GIS and remote sensing integration for environmental applications. *International Journal of Geographical Information Systems* 10: 877–90

Hunt E R Jr 1991 Airborne remote sensing of canopy water thickness scaled from leaf spectral data. *International Journal of Remote Sensing* 12: 643–9

Ichoku C, Karnieli A 1996 A review of mixture modelling techniques for sub-pixel land cover estimation. *Remote Sensing Reviews* 13: 161–86

Jacquemoud S 1993 Inversion of the PROSPECT + SAIL canopy reflectance model from AVIRIS equivalent spectra: theoretical study. *Remote Sensing of Environment* 44: 281–92

Jacquemoud S, Baret F 1990 PROSPECT: a model of leaf optical properties spectra. *Remote Sensing of Environment* 34: 75–92

Jacquemoud S, Baret F, Andrieu B, Danson F M, Jaggard K 1995 Extraction of vegetation biophysical parameters by inversion of the PROSPECT + SAIL models on sugar beet canopy reflectance data application to TM and AVIRIS sensors. *Remote Sensing of Environment* 52: 163–72

Jacquemoud S, Baret F, Hanocq J F 1992 Modelling spectral and bidirectional soil reflectance. *Remote Sensing of Environment* 41: 123–32

Jupp D L B, Strahler A H, Woodcock C E 1988 Autocorrelation and regularisation in digital images: 1. Basic theory. *IEEE Transactions on Geoscience and Remote Sensing* 26: 463–73

Kaufman Y J 1988 Atmospheric effect on spectral signature – measurement and corrections. *IEEE Transactions on Geoscience and Remote Sensing* 26: 441–50

Kaufman Y J 1993 Aerosol optical thickness and atmospheric path radiance. *Journal of Geophysical Research* 98: 2677–92

Kimes D S, Sellers P J, Diner D J 1987 Extraction of spectral hemispherical reflectance (albedo) of surfaces from nadir and directional reflectance data. *International Journal of Remote Sensing* 8: 1727–46

Kramer H J 1994 *Observation of the Earth and its environment: survey of missions and sensors*. Berlin, Springer

Kuusk A 1994 A multispectral canopy reflectance model. *Remote Sensing of Environment* 50: 75–82

LaGro J 1990 Assessing patch shape in landscape mosaics. *Photogrammetric Engineering and Remote Sensing* 57: 285–93

Lam N S-M 1990 Description and measurement of LANDSAT-TM images using fractals. *Photogrammetric Engineering and Remote Sensing* 56: 187–95

Lloyd D 1990 A phenological classification of terrestrial vegetation cover using shortwave vegetation index imagery. *International Journal of Remote Sensing* 11: 2269–79

McDonald R A 1995 Opening the Cold War sky to the public: declassifying satellite reconnaissance imagery. *Photogrammetric Engineering and Remote Sensing* 61: 385–90

Mesev T V, Longley P A, Batty M, Xie Y 1995 Morphology from imagery – detecting and measuring the density of urban land-use. *Environment and Planning A* 27: 759–80

Moller-Jenson L 1990 Knowledge-based classification of an urban area using texture and context information in LANDSAT-TM imagery. *Photogrammetric Engineering and Remote Sensing* 56: 899–904

Myneni R B, Hall F G, Sellers P J, Marshak A L 1995 The interpretation of spectral vegetation indices. *IEEE Transactions on Geoscience and Remote Sensing* 33: 481–6

Myneni R B, Ross J, Asrar G 1990 A review of the theory of photon transport in leaf canopies. *Agricultural Forestry Meteorology* 45: 1–153

Myneni R B, Williams D L 1994 On the relationship between FAPAR and NDVI. *Remote Sensing of Environment* 49: 200–11

Nichol D G 1990 Region adjacency analysis of remotely-sensed imagery. *International Journal of Remote Sensing* 11: 2089–101

Prata A J, Cechet R P, Barton I J, Llewellyn-Jones D T 1990 The Along-Track Scanning Radiometer for ERS-1: scan geometry and data simulation. *IEEE Transactions on Geoscience and Remote Sensing* 28: 3–13

Price J C 1987 Calibration of satellite radiometers and the comparison of vegetation indices. *Remote Sensing of Environment* 21: 15–27

Price J C 1990 On the information content of soil reflectance spectra. *Remote Sensing of Environment* 33: 113–21

Rees W G 1990 *Physical principles of remote sensing*. Cambridge (UK), Cambridge University Press

Richards J A 1993 *Remote-sensing digital image analysis: an introduction*. Berlin, Springer

Roujean J-L, Leroy M, Deschamps P Y 1992 A bidirectional reflectance model of the Earth's surface for the correction of remote sensing data. *Journal of Geophysical Research* 97: 20 455–68

Running S W, Justice C O, Salomonson V, Hall D, Barker J, Kaufmann Y J, Strahler A H, Huete A R, Muller J-P, Vanderbilt V, Wan Z M, Teillet P, Carneggie D 1994 Terrestrial remote sensing science and algorithms planned for EOS/MODIS. *International Journal of Remote Sensing* 15: 3587–620

Schalkoff R J 1992 *Pattern recognition: statistical, structural and neural approaches.* New York, John Wiley & Sons Inc.

Smith M O, Johnson P E, Adams J B 1985 Quantitative determination of mineral types and abundances from reflectance spectra using principal components analysis. *Journal of Geophysical Research* 90: 792–804

Townshend J R G, Justice C O 1990 The spatial variation of vegetation changes at very coarse scales. *International Journal of Remote Sensing* 11: 149–57

Townshend J R G, Justice C O, Skole D, Malingreau J-P, Chilar J, Teillet P, Sadowski F, Ruttenberg S 1995 The 1-km resolution global dataset: needs of the International Geosphere Biosphere Programme. *International Journal of Remote Sensing* 15: 3417–42

Vanderbilt V C, Grant L, Ustin S L 1991 Polarisation of light by vegetation. In Myneni R, Ross J (eds) *Photon-vegetation interactions: applications in optical remote sensing and plant ecology.* Berlin, Springer: 191–228

Vane G, Goetz A F H 1993 Terrestrial imaging spectrometry: current status, future trends. *Remote Sensing of Environment* 44: 117–26

Vane G, Green R O, Chrien T G, Enmark H T, Hansen E G, Porter W M 1993 The Airborne Visible/Infrared Imaging Spectrometer (AVIRIS). *Remote Sensing of Environment* 44: 127–44

Walthall C L, Norman J M, Welles J M, Campbell G, Blad B L 1985 Simple equation to approximate the bi-directional reflectance from vegetative canopies and bare soil surfaces. *Applied Optics* 24: 383–7

Wanner W, Li X, Strahler A H 1995 On the derivation of kernels for kernel-driven models of bi-directional reflectance. *Journal of Geophysical Research* 100: 21 077–89

Wessman C A, Aber J D, Peterson D L, Melillo J M 1988 Remote sensing of canopy chemistry and nitrogen cycling in temperate forest ecosystems. *Nature* 335: 154–6

Wilkinson G G 1996 A review of current issues in the integration of GIS and remote sensing data. *International Journal of Geographical Information Systems* 10: 85–101

Woodcock C E, Strahler A H 1987 The factor of scale in remote sensing. *Remote Sensing of Environment* 21: 311–32

Zyl J J van, Zebker H A, Elachi C 1987 Imaging radar polarisation signatures: theory and observation. *Radio Science* 22: 529–43

SECTION 2(d): DATA TRANSFORMATION AND LINKAGE

Introduction

THE EDITORS

In this final Technical Issues section, attention turns to the algorithms that allow GIS to transform and link data. Section 2(a) dealt with trends in the computing industry, in computer architectures, programming environments, and communications networks. All of these are likely to influence GIS profoundly over the coming decade. The next section looked at spatial databases, where the general solutions offered by the database industry intersect with the special needs of spatial databases for high performance, multi-dimensional search, and complex structures. The third section dealt with selected technical aspects of spatial data collection. Having addressed architectures, programming environments and data, this final set of chapters examines technical issues in the processing of data in GIS, and focuses on areas where algorithms are particularly challenging and complex, and the subject of ongoing research.

Several disciplines are contributing actively to research in advanced spatial data processing algorithms. The field of computational geometry addresses specifically the computer processing of geometric objects, and has widespread applications in GIS. For example, a fundamental algorithm in computational geometry detects whether two line segments intersect, and if so, where; it finds applications in such GIS processes as polygon overlay. GIS adds a particular context to computational geometry – one of its most distinguishing features is that position can never be exact, and must always be subject to a prescribed tolerance. GIS is an important application field for 'finite precision geometry', the field that attempts to formalise and theorise about such fuzzy spatial problems.

Another distinguishing feature of many GIS applications is the importance of performance, and the ability of algorithms to scale over a very wide range of problem sizes. The field of algorithmic complexity deals formally with performance, and the effects of problem size, and is widely applied in research on advanced GIS algorithms (see De Floriani and Magillo, Chapter 38, for example).

Image processing is yet another cognate field, with many applications to geographical data and GIS. As we saw in the last section, GIS colours these applications with particular characteristics. The surface of the Earth is not flat, unlike a photographic plate, and algorithms are needed: to register images to it accurately; to register images of the same feature to each other; to 'rubber sheet' images for geometric correction; or to change from one projection to another (see Dowman, Chapter 31). Another distinguishing characteristic of image processing when applied to geographical data is the widespread absence of 'truth' (see also Barnsley, Chapter 32) – when imaging the human body, for example, it is possible to regard the label 'liver' as 'true' of a certain part of the image; but suitably precise geographical analogies are much harder to come by – the label 'lake' is confused by many problems of definition (when is a lake a swamp, or a reservoir? See Mark, Chapter 7).

The ability to link data is often cited as the distinguishing feature of GIS. Location on the Earth's surface forms a convenient common key between otherwise disparate datasets and forms of information, allowing data to be linked across the boundaries of disciplines, departments, and agencies. When events occur in the same place, or near to each other, it is easy to believe that they also influence each other, and that both need to be taken into account in making decisions. Contemporary thinking on environmental management urges us to think of all things on the Earth's surface as connected and inter-related – in Tobler's 'first law of geography', 'all things are related, but nearby things are more related than distant things' (Tobler 1970).

Several different forms of linkage can be identified in GIS. Consider two datasets, A and B, covering the same geographical area, and imported to a GIS from different sources. At one extreme, the information provided by B is entirely distinct ('independent' or 'orthogonal' in a statistical sense), and both A and B are necessary in some application. The application will likely require that A and B be overlaid, and the ability to do so will depend on the formats of the two datasets – most GIS will require that they be both raster, or both vector; if both raster, that the rasters be congruent. In another case, the common key between A and B may be a feature identifier, rather than a geographical location. Such cases occur when B provides tabular information to be added to the geographical features in A as additional attributes. Yet another case occurs when B is a source of selective updates for the information in A – for example, B might contain more accurate coordinates for the features shared between it and A. When no common feature identifier is available, the features in B must be matched geometrically to those in A, a process that has been termed 'conflation', and is itself the subject of intensive current research.

Some forms of data transformation and linkage in GIS are straightforward, and do not justify particular attention. Others are made sufficiently special by their geographical context to have emerged as strong subfields for research and development within GIS. In this section the editors have selected several of these, while recognising that the set is by no means exhaustive, and may not survive the test of time – five or ten years from now research and development attention may have turned to quite different problems.

The section contains seven chapters. In the first, Lubos Mitas and Helena Mitasova review the state of the art in spatial interpolation, a vital component in the GIS arsenal because it provides estimates of the value of a variable z at locations (x, y) where it has not been measured. Spatial interpolation is essential in resampling, when data must be shifted from one raster to another; in transformation between representations, such as from a grid to contours; or in dealing with the problems caused by incompatible reporting zones. Mitas and Mitasova review the methods currently available, discuss the bases on which they can be evaluated, and review the applications of the methods in GIS.

Data linkage across GIS layers provides the theme for Ronald Eastman's contribution. Multi-criteria evaluation is concerned with the allocation of land to suit a specific objective on the basis of a variety of attributes that the selected areas should possess. This implies an apparently straightforward GIS-based overlay exercise, yet this process is complicated on the one hand by differences in data structure (raster versus vector) and, on the other, by ambiguities in the ways in which criteria should be standardised and aggregated into a single summary coverage. Eastman reviews these problems and suggests the use of fuzzy measures as a means of reconciling and developing current practice. In this way, the harshness of using Boolean operators to identify intersection and union of data layers is replaced with an approach which also provides improved standardisation of criteria and better evaluation of decision risk (the likelihood that the decision made will be incorrect).

One aspect that distinguishes GIS from other forms of spatial data processing, notably computer-assisted design (CAD), is its emphasis on representing fields, or variables having a single value at every location on the Earth's surface. Examples of fields are elevation, mean rainfall, or soil type. Because a field is continuous by definition, it must somehow be rendered discrete in order to be represented in the finite space of a digital store. Methods for discretising fields have been discussed at many points in this volume. Among them, two achieve their objective by dividing a plane surface into regular or irregular tiles, forming a 'tessellation'. The mathematics and statistics of tessellations, and their representation and processing in GIS, are important topics for research. The third chapter in this section, by Barry Boots, reviews the state of the art. Some of this research goes well beyond the current state of GIS implementation, particularly in the area of weighted tessellations, but it is easy to see how such methods could be implemented and applied.

GIS has a long history of successful implementation of digital models of the Earth's terrain, collectively known as digital terrain models (DTMs). They include triangulated irregular networks (TINs); and the commonest form of DTM, the digital elevation model (DEM), a rectangular array of spot elevations. DTMs are available in one form or another for much of the Earth's terrestrial surface and for the ocean floor, although the sampling density and conditions of availability vary enormously. One major factor driving the development of DTM technology is its importance in military applications, particularly missile guidance systems ('cruise' missiles navigate largely by recognising the geometric form of the terrain under them).

Many sciences, hydrology and geomorphology in particular, have an interest in the form of the Earth's surface, and its influence on the environment. The availability of DTMs, and the facilities in GIS for processing them, have led to an explosion of research on DTM analysis techniques. GIS is now a very significant tool in these fields, and DTMs are also useful in such practical applications as transmission tower location. The fourth and fifth chapters in this section discuss recent DTM research from two perspectives. Lawrence Band reviews the importance of DTMs and related datasets in hydrography and the analysis of landforms; while Leila De Floriani and Paola Magillo discuss representations, and associated algorithms, for the transformation of DTM data into useful information on intervisibility.

The last two chapters in the section move somewhat away from this intensive discussion of theory and algorithms. In Chapter 39, Jorge Nelson Neves and Antonio Câmara look at the role of GIS in the expanding field of virtual reality, or Virtual Environments (VEs). VEs are clearly an important area of application for GIS, particularly if the environment being simulated is in any sense related to the real, geographical world; and even totally artificial environments must be constrained by certain characteristics of the real world if they are to be convincing. VEs require many standard GIS techniques, as well as more generic techniques of image processing and visualisation; and also require much higher performance than traditional GIS applications.

In the final chapter, Michael Goodchild and Paul Longley discuss the issues encountered in using GIS as a data linkage technology. These range from the technical issues of accuracy, compatible data formats and data models, and rules for conflation, to the capabilities of contemporary communication networks for supporting data search and sharing, to institutional issues that are beyond the scope of this section.

Reference

Tobler W R 1970 A computer movie simulating urban growth in the Detroit region. *Economic Geography* 46: 234–40

35

Multi-criteria evaluation and GIS

J R EASTMAN

Multi-criteria evaluation in GIS is concerned with the allocation of land to suit a specific objective on the basis of a variety of attributes that the selected areas should possess. Although commonly undertaken in GIS, it is shown that the approaches commonly used in vector and raster systems typically lead to different solutions. In addition, there are ambiguities in the manner in which criteria should be standardised and aggregated to yield a final decision for the land allocation process. These problems are reviewed and the theoretical structure of fuzzy measures is offered as an approach to the reconciliation and extension of the procedures currently in use. Specifically, by considering criteria as expressions of membership in fuzzy sets (a specific instance of fuzzy measures) the weighted linear combination aggregation process common to raster systems is seen to lie along a continuum of operators mid-way between the hard intersection and union operators typically associated with Boolean overlay in vector systems. A procedure for implementing this continuum is reviewed, along with its implications for varying the degrees of 'ANDORness' and trade-off between criteria. In addition, the theoretical structure of fuzzy measures provides a strong logic for the standardisation of criteria and the evaluation of decision risk (the likelihood that the decision made will be incorrect).

1 INTRODUCTION

One of the most important applications of GIS is the display and analysis of data to support the process of environmental decision-making. A decision can be defined as a choice between alternatives, where the alternatives may be different actions, locations, objects, and the like. For example, one might need to choose which is the best location for a hazardous waste facility, or perhaps identify which areas will be best suited for a new development.

Broadly, decisions can be classified into two extensive categories – policy decisions and resource allocation decisions. Resource allocation decisions, as the name suggests, are concerned with control over the direct use of resources to achieve a particular goal. Ultimately, policy decisions have a similar aim. However, they do so by establishing legislative instruments that are intended to influence the resource allocation decisions of others. Thus, for example, a government body might reduce taxes on land allocated to a particular crop as an incentive to its introduction. This is clearly a policy decision; but it is the farmer who makes the decision about whether to allocate land to that crop or not.

To be rational, decisions will be necessarily based on one or more criteria – measurable attributes of the alternatives being considered, that can be combined and evaluated in the form of a decision rule. In some circumstances, allocation decisions can be made on the basis of a single criterion. However, more frequently, a variety of criteria is required. For example, the choice between a set of waste disposal sites might be based upon criteria such as proximity to access roads, distance from residential and protected lands, current land use, and so on.

This chapter focuses on the very specific problems of spatial resource allocation decisions in the context of multiple criteria – a process most commonly known as *multi-criteria evaluation*

(MCE) (Voogd 1983). In some instances, this term has also been used to subsume the concept of *multi-objective* decision-making (e.g. Carver 1991; Janssen and Rietveld 1990). However, it is used here in a more specific sense. An *objective* is understood here to imply a perspective, philosophy, or motive that guides the construction of a specific multi-criteria decision rule. Thus in siting a hazardous waste facility, the objective of a developer might be profit maximisation while that of a community action group might be environmental protection. The criteria they each consider and the weights they assign to them are likely to be quite different. Each is likely to develop a multi-criteria solution – but a different multi-criteria decision. The resolution of these differing perspectives into a single solution is known as multi-objective decision-making – a topic which will not be covered in this chapter (see Campbell et al 1992 and Eastman et al 1995 for two prominent approaches to this problem in GIS).

Almost all of the case study examples in this chapter are based on an analysis of suitability for industrial development for the region of Nakuru, Kenya. Nakuru is a region of strong agricultural potential that has experienced rapid urban development in recent years. It is also the location of one of the more important wildlife parks in Kenya (the large area of restricted development to the south of Plate 32) – one of Kenya's soda lakes in the Great Rift Valley, it is the home of over two million flamingoes as well as a wide variety of other species.

2 TRADITIONAL APPROACHES TO MCE IN GIS

In GIS, multi-criteria evaluation has most typically been approached in one of two ways. In the first, all criteria are converted to Boolean (i.e. logical true/false) statements of suitability for the decision under consideration. (The term *Boolean* is derived from the name of the English mathematician, George Boole, who first abstracted the basic laws of set theory in the mid 1800s. It is used here to denote any crisp spatial mapping in which areas are designated by a simple binary number system as either belonging or not belonging to the designated set.) In many respects, these Boolean variables can be usefully thought of as constraints, since they serve to delineate areas that are not suitable for consideration. These constraints are then combined by some combination of intersection (logical AND) or union (logical OR)

operators. This procedure dominates MCE with vector software systems, but is also commonly used with raster systems. For example, Figure 1 shows how Boolean images, along with their intersection achieved through the characteristic overlay operation of a GIS, may be used here to find all areas suitable for industrial development, subject to the following criteria: suitable areas will be near to a road (within 1 km – upper left), near to a labour force (within 7.5 km of a town – middle left), on low slopes (less than 5 per cent – upper right), and greater that 2.5 km from designated wildlife reserves (middle right). In addition, development is not permitted in wildlife reserves (lower left). These criteria are aggregated by means of an intersection (logical AND) operator, yielding the result on the lower right. Note that the distance to labour force was calculated from a cost distance surface that accounted for road and off-road frictions.

In the second most common procedure for MCE, quantitative criteria are evaluated as fully continuous variables rather than collapsing them to

Fig 1. An example of multi-criteria evaluation using Boolean analysis.

Boolean constraints. Such criteria are typically called *factors*, and express varying degrees of suitability for the decision under consideration. Thus, for example, proximity to roads would be treated not as an all-or-none buffer zone of suitable locations, but rather, as a continuous expression of suitability according to a special numeric scale (e.g. 0–1, 0–100, 0–255, etc.). The process of converting data to such numeric scales is most commonly called standardisation (Voogd 1983).

Traditionally, standardised factors are combined by means of weighted linear combination – that is, each factor is multiplied by a weight, with results being summed to arrive at a multi-criteria solution. In addition, the result may be multiplied (i.e. intersected) by the product of any Boolean constraints that may apply (Eastman et al 1995). For example:

$$\text{suitability} = \sum w_i X_i * \prod C_j$$

where w_i = weight assigned to factor i
X_i = criterion score of factor i
C_j = constraint j

Figure 2 illustrates this approach where a comparable example is developed to that in Figure 1. Again, the intention is to find areas suitable for industrial development, subject to the following criteria: suitable areas will be near to a road (as near as possible – upper left), near to a labour force (as near as possible – middle left), on low slopes (as low as possible – upper right) and as far from the wildlife reserve as possible (middle right). As in Figure 1, development is not permitted in wildlife reserves (lower left) through use of a Boolean constraint. These criteria are aggregated by means of a weighted average of the criterion scores. In this case, all criteria were standardised before weighting to a common numeric range using the most commonly used (but not necessarily recommended) technique – linear scaling between the minimum and maximum values of that criterion. The linear rescaling is to a consistent range (0–255) as follows:

$$X_i = (x_i - \min_i) / (\max_i - \min_i)$$

where X_i = criterion score of factor i
x_i = original value of factor i
$\min_i$ = minimum of factor i
$\max_i$ = maximum of factor i

In addition, to provide the most direct comparison to the results of Figure 1, equal weight (0.25) was assigned to each criterion with the wildlife reserve constraint acting as an absolute barrier to

Fig 2. An example of multi-criteria evaluation using weighted linear combination.

development. The result of the averaging process is shown on the lower left. The image on the lower right shows the result of selecting the best areas from this suitability map in order to match the total area of that selected by Boolean analysis in Figure 1. Note that as in Figure 1, the distance to labour force was calculated from a cost distance surface that accounted for road and off-road frictions.

The continuous suitability map shown in Figure 2 has the same numeric range as the standardised factors if the weights that are applied sum to 1.0. A specific decision can then be reached by rank ordering the alternatives (in this case, pixels) and selecting as many of the best ranked areas as is required to meet the objective of the analysis in question. In Figure 2, this has been done in order to select as many of the best areas as were selected by the Boolean analysis in Figure 1.

This procedure of weighted linear combination dominates multi-criteria approaches with raster-based GIS software systems. However, there are a number of problems with both approaches to multi-criteria evaluation.

First, despite a casual expectation that the two procedures should yield similar results, they very often do not. For example, the results of the decision portrayed in the lower right of Figures 1 and 2 are in agreement only by 53 per cent. The reason clearly has to do with the logic of the aggregation operation. For example, Boolean intersection results in a very hard AND – a region will be excluded from the result if any single criterion fails to exceed its threshold. Conversely, the Boolean union operator implements a very liberal mode of aggregation – a region will be included in the result even if only a single criterion meets its threshold. Weighted linear combination is quite unlike these options. Here a low score on one criterion can be compensated by a high score on another – a feature known as *trade-off* or *substitutability*. While human experience is replete with examples of both trade-off and non-substitutability in decision making, the tools for flexibly incorporating this concept are poorly developed in GIS. Furthermore, a theoretical framework that can link the aggregation operators of Boolean overlay and weighted linear combination has, until recently (Eastman and Jiang 1996), been lacking.

The second problem with MCE has to do with the standardisation of factors in weighted linear combination. The most common approach is to rescale the range to a common numerical basis by simple linear transformation (Voogd 1983), as was applied in Figure 2. However, the rationale for doing so is unclear. Indeed, there are many instances where it would seem logical to rescale values within a more limited range. Furthermore, there are cases where a non-linear scaling may seem appropriate. Since the recast criteria really express suitability, there are many cases where it would seem appropriate that criterion scores asymptotically approach the maximum or minimum suitability level.

The third issue concerns the weights that are applied. Clearly they can have a strong effect on the outcome produced. However, not much attention has been focused in GIS on how they should be developed. Commonly they represent the subjective (but no less valid) opinions of one or more experts or local informants. How can consistency and overt validity be established for these weights? Furthermore, how should they be applied in the context of varying trade-off between factors?

A fourth problem concerns decision risk. Decision risk may be considered as the likelihood that the decision made will be wrong. For both procedures (Boolean analysis and weighted linear combination) it is a fairly simple matter to propagate measurement error through the decision rule and subsequently to determine the risk that a given location will be assigned to the wrong set (i.e. the set of selected alternatives or the set of those not to be included). However, the continuous criteria of weighted linear combination would appear to express a further uncertainty that is not so easily accommodated (see Fisher, Chapter 13). The standardised factors of weighted linear combination each express a perspective of suitability – the higher the score, the more suitable. However, there is no real threshold that can definitively allocate locations to one of the two sets involved (areas to be chosen and areas to be excluded). How are these uncertainties to be accommodated in expressions of decision risk? If these criteria really express uncertainties, why are they combined through an averaging process?

The surprising feature of multi-criteria evaluation is that, despite its ubiquity in environmental management, so little is understood of its character in GIS. In the following sections we survey the issues involved, and offer a perspective on a resolution through the concept of *fuzzy measures*.

3 FUZZY MEASURES

This discussion of fuzzy measures is adapted from Eastman and Jiang (1996). The term fuzzy measure refers to any set function which is monotonic with respect to set membership (Dubois and Prade 1982; see also Fisher, Chapter 13). Notable examples of fuzzy measures include *probabilities*, the *beliefs*, and *plausibilities* of Dempster-Shafer theory, and the *possibilities* of fuzzy sets. Interestingly, if we consider the process of standardisation in MCE to be one of transforming criterion scores into set membership statements (i.e. the set of suitable choices), then standardised criteria *are* fuzzy measures.

A common trait of fuzzy measures is that they follow DeMorgan's Law in the construction of the intersection and union operators (Bonissone and Decker 1986). DeMorgan's Law establishes a triangular relationship between the intersection, union, and negation operators such that:

$T(a,b) = \sim S(\sim a, \sim b)$

where T = intersection (AND) = T-Norm
and S = union (OR) = T-CoNorm
and $\sim$ = negation (NOT)

The intersection operators in this context are known as *triangular norms*, or simply T-*Norms*, while the union operators are known as *triangular co-norms*, or T-*CoNorms*.

4 FUZZY MEASURES AND AGGREGATION OPERATORS

A T-Norm can be defined as (Yager 1988):

a mapping $T: [0,1] * [0,1] \rightarrow [0,1]$ such that:

$T(a,b) = T(b,a)$ (commutative)
$T(a,b) \geq T(c,d)$ if $a \geq c$ and $b \geq d$ (monotonic)
$T(a,T(b,c)) = T(T(a,b),c)$ (associative)
$T(1,a) = a$

Some examples of T-norms include:

$\min(a,b)$ (the intersection operator of fuzzy sets)
$a * b$ (the intersection operator of classical sets)
$1 - \min(1,((1-a)^p + (1-b)^p)^{(1/p)})$ (for $p \geq 1$)
$\max(0,a + b - 1)$

Conversely, a T-CoNorm is defined as:

a mapping $S: [0,1] * [0,1] \rightarrow [0,1]$ such that:

$S(a,b) = S(b,a)$ (commutative)
$S(a,b) >= S(c,d)$ if $a \geq c$ and $b \geq d$ (monotonic)
$S(a,S(b,c)) = S(S(a,b),c)$ (associative)
$S(0,a) = a$

Some examples of T-CoNorms include:

$\max(a,b)$ (the union operator of fuzzy sets)
$a + b - a*b$ (the union operator of classical sets)
$\min(1,(a^p + b^p)^{1/p})$ (for $p \geq 1$)
$\min(1,a + b)$

Interestingly, while the intersection ($a*b$) and union ($(a+b) - (a*b)$) operators of Boolean overlay represent a T-Norm/T-CoNorm pair, the averaging operator of weighted linear combination is neither, because it lacks the property of associativity (Bonissone and Decker 1986). Rather, it has been determined (Bonissone and Decker 1986) that the averaging operator falls midway between the extreme cases of the T-Norm (AND) of fuzzy sets (the minimum operator) and its corresponding T-CoNorm (OR – the maximum operator) – in essence, a perfect ANDOR operator. In fact, Yager (1988) has proposed that weighted linear combination is one of a continuum of aggregation operators that lies between these two extremes. Further, he has proposed the concept of an *ordered weighted average* that can produce the entire continuum. Recently, Eastman and Jiang (1996) have implemented this operator, with modifications, in a raster GIS. In doing so, the traditional aggregation operators of vector and raster GIS have been united into a single theoretical framework.

4.1 The ordered weighted average

With the ordered weighted average, criteria are weighted on the basis of their rank order rather than their inherent qualities. Thus, for example, we might decide to apply order weights of 0.5, 0.3, 0.2 to weight a set of factors A, B, and C based on their rank order. Thus if at one location the criteria are ranked BAC (from lowest to highest), the weighted combination would be $0.5*B + 0.3*A + 0.2*C$. However, if at another location the factors are ranked CBA, the weighted combination would be $0.5*C + 0.3*B + 0.2*A$. In the implementation of Yager's concept by Eastman and Jiang (1996), the concept of weights that apply to specific factors has also been incorporated, yielding two sets of weights – *criterion weights* that apply to specific criteria and *order weights* that apply to the ranked criteria, after application of the criterion weights.

The interesting feature of the ordered weighted average is that it is possible to control the degree of ANDORness and trade-off between factors within limits. For example, using order weights of [1 0 0] yields the minimum operator of fuzzy sets, with full ANDness and no trade-off. Using order weights of [0 0 1] yields the maximum operator of fuzzy sets with full ORness and no trade-off. Using weights of [0.33 0.33 0.33] yields the traditional averaging operator of MCE with intermediate ANDness and ORness, and full trade-off. Trade-off is thus controlled by the degree of dispersion in the weights while ANDness or ORness is governed by the amount of skew. For example, order weights of [0 1 0] would yield an operator with intermediate ANDness and ORness, but no trade-off, while the original example with order weights of [0.5 0.3 0.2] would yield an operator with substantial trade-off and a moderate degree of ANDness.

This quality of variable ANDORness has interested some in the decision science field (e.g. Yager 1988) because of the recognition that in human perception of decision logics, it is not uncommon to wish to combine criteria with something less extreme than the hard operations of union or intersection. In the context of GIS, however, it is the property of trade-off that is of special interest. The minimum operator is occasionally used in GIS applications, and represents a form of limiting factor analysis. Here the intent is one of risk aversion, by characterising the suitability of a location in terms of its worst quality. The maximum operator is the opposite, and can thus be thought of as a very optimistic aggregation operator – an area will be suitable to the extent of its best quality. Both of these operations permit no trade-off between the qualities of the criteria considered. Furthermore, in cases where set membership approaches certainty, the results from fuzzy sets will be identical to those of Boolean overlay. However, with weighted linear combination, trade-off is clearly and fully present.

Figure 3 illustrates various degrees of trade-off (and ANDORness) between the same four factors considered in Figures 1 and 2. The same criteria were used as for Figure 2, except that the scaling was changed to facilitate comparison to the results in Figure 1 – a sigmoidal fuzzy membership function was used such that the thresholds used to create the Boolean images in Figure 1 correspond to membership values of 0.5 for the fuzzy criteria in this example (i.e. scaling was asymptotic to membership values of 1.0 and 0.0 at values of 0–2 km for proximity to roads, 0–15 km for proximity to the labour force, 0–10 per cent for slope gradients, and 5–0 km for distance from designated wildlife reserves). In addition, to facilitate comparison, equal criterion weights were applied as in Figure 2. Thus the differences between these aggregations arise solely from the effects of different order weights. The v-shaped sequence, from top to bottom, used order weights of [1 0 0 0], [.60 .20 .15 .05], [.4 .3 .2 .1], [.25 .25 .25 .25], [.1 .2 .3 .4], [.05 .15 .20 .60], and [0 0 0 1]. This sequence progresses from full ANDness and no trade-off for the first (the minimum function), to intermediate ANDORness and full trade-off for the middlemost (equivalent to a standard weighted linear combination), to full ORness and no trade-off with the last (corresponding to the maximum operator). The

image on the middle left is a median operator produced with order weights of [0 .5 .5 0], producing an aggregation with intermediate ANDORness (like weighted linear combination) but almost no trade-off.

The similarity of the result with full ANDness (and thus no trade-off) to the Boolean result in Figure 1 is striking. In fact, when these suitability values are rank ordered and enough of the best pixels are selected to equal the area of the Boolean result, the solution is identical. Thus the reason for the difference in the Boolean and weighted linear combination results is clear – the characteristic Boolean overlay operation of vector GIS produces an aggregation of criteria with full ANDness and no trade-off while the typical weighted linear combination operation of raster GIS produces intermediate ANDness and full trade-off. The results are different because the aggregation operators are different.

Recognising that a full spectrum of aggregation operators exists opens up a much richer set of possibilities for implementing decision rules in GIS.

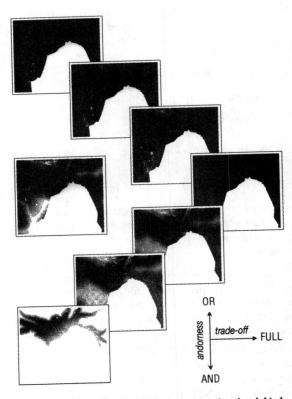

Fig 3. A spectrum of aggregations using the ordered weighted average procedure.

For example, Plate 32 illustrates the effects of combining three of the factors with trade-off and one without. In this case, proximity to roads is given a criterion weight of 0.45, proximity to the labour force is given a weight of 0.12, and the slope factor is given a weight of 0.43. These are combined using a standard weighted linear combination. This result is then combined with the distance from wildlife reserve factor using a minimum operator. The absence of trade-off in this last step is clear – the distance from wildlife reserve factor dominates the result until it no longer represents the limiting factor. The effect is clearly similar to that of a constraint, but lacks the crispness of a traditional constraint. In effect, the minimum operator with a fuzzy measure represents a form of soft constraint. Soft constraints are particularly useful where a specific boundary cannot be reasonably established. Indeed, it might be argued that this is more commonly appropriate than the artificial boundaries of traditional constraints.

5 FUZZY MEASURES AND STANDARDISATION

Clearly, this consideration of fuzzy measures has implications beyond those of the aggregation process alone. It also provides a very strong logic for the process of standardisation. In this context, the process of standardising a criterion can be seen as one of recasting values into a statement of set membership – the degree of membership in the final decision set. Indeed, Eastman and Jiang (1996) argue that such statements of set membership in fact constitute fuzzy sets (a particular form of fuzzy measure), while those of Boolean constraints represent classical sets. This clearly opens the way for a broader family of set membership functions than that of linear rescaling alone. For example, the commonly used sigmoidal (s-shaped) function of fuzzy sets provides a simple logic for cases where a function is required that is asymptotic to 0 and 1. It also suggests that the minimum and maximum raw factor values should not blindly be used as the anchor points for such a function. Rather, anchor points that are consistent with the logic of set membership are clearly superior. For example, in Figure 4, sigmoidal membership functions were created for each factor, with anchor points set at the points where the factor begins to have an effect and where the effect is no longer relevant. The distance

to wildlife reserve factor, for instance, starts to rise above 0.0 immediately at the park boundary, but approaches 1.0 at a distance of 5 kilometres. Further distance does not lead to an increase in the factor score since the distance is far enough.

6 DETERMINATION OF WEIGHTS

Given the consideration of factors as fuzzy sets and the nature of the aggregation process, the criterion weights of weighted linear combination clearly represent trade-off weights – that is, expressions of the manner in which they will trade with other factors when aggregated in multi-criteria evaluation. Rao et al (1991) have suggested that a logical process for the development of such weights is the procedure of pairwise comparisons developed by Saaty (1977). In this process each factor is rated for its importance relative to every other factor using a 9-point reciprocal scale (i.e. if 7 represents substantially more important, 1/7 would indicate substantially less important). This leads to a $n \times n$ matrix of ratings (where n is the number of factors being considered). Saaty (1977) has shown that the principal eigenvector of this matrix represents a best fit set of weights. Figure 5, for example, illustrates this rating scale along with a completed comparison matrix and the best fit weights produced. Eastman et al (1993) have implemented this procedure in a raster GIS with a modification that also allows the degree of consistency to be evaluated as well as the location of inconsistencies to allow for an orderly re-evaluation. The process is thus an iterative one that converges on a consistent set of consensus weights.

A problem still exists, however, in how these weights should be applied in the context of the ordered weighted average discussed above. It seems clear that these weights will have full effect with the weighted linear combination operator (where full trade-off exists), and that they should have no effect when no trade-off is in effect (i.e. with the minimum and maximum operators). It seems logical, therefore, that their effect should be graded between these extremes as the degree of trade-off is manipulated with the ordered weighted average process. However, the logic for this gradation has not been established. In their implementation of the ordered weighted average for GIS, Eastman and Jiang (1996) have used a measure of relative

	1/9	1/7	1/5	1/3	1	3	5	7	9
	extremely	very	strongly	moderately	equally	moderately	strongly	very	extremely

less important ⟶ more important

(a)

	Proximity to roads	Proximity to labour force	Slope gradient	Distance from wildlife reserves
Proximity to roads	1			
Proximity to labour force	1/3	1		
Slope gradient	1	4	1	
Distance from wildlife reserves	1/3	2	1/2	1

(b)

Factor	Derived weight
Proximity to roads	0.3770
Proximity to labour force	0.0979
Slope gradient	0.3605
Distance from wildlife reserves	0.1647

(c)

Fig 4. Saaty's pairwise comparison procedure for the derivation of factor weights. Using a 9-point rating scale (a) each factor is compared to each other factor for its relative importance in developing the final solution (b). The principal eigenvector of this matrix is then calculated to derive the best-fit set of weights (c).

dispersion (a measure closely related to the entropy measure of information theory) as the basis for this gradation. However, further research is needed on this important aspect of the ordered weighted average procedure.

7 DECISION RISK

Uncertainty in the decision rule, and in the criteria that are considered, implies some risk that the decision made will be wrong. In the case of measurement error, the effects of uncertainty can fairly easily be propagated to the suitability map that is produced in MCE (see Heuvelink, Chapter 14; Heuvelink 1993). Furthermore, Eastman (1993) has developed a simple operator that can convert such an evaluation into a mapping of the probability

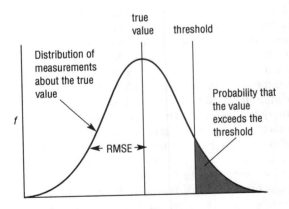

Fig 5. A procedure for calculating decision risk. Assuming a normal distribution of errors, the probability that a data value exceeds or is exceeded by a threshold can be calculated as the area of the normal curve subtended by that threshold. [RMSE = root mean square error]

that locations belong to the decision set (the PCLASS operator of the Idrisi software system). The operator assumes a normal distribution of errors and calculates the area under the normal curve subtended by a threshold that can distinguish the cases that belong in the decision set from those that do not (Figure 5). The result is an expression of decision risk that is directly analogous to the concept of a Type II error in statistical hypothesis testing – that is, the likelihood that the alternative does not belong to the decision set if we assume that it does (Plate 33). The resulting probability map can subsequently be thresholded to see the nature of the decision set at any specified risk level (again see Plate 33).

To the extent that measurement error can be quantified and propagated through an analysis, an expression of decision risk is thus not very difficult to achieve. However, the recognition of factors in MCE as fuzzy sets implies a very different form of uncertainty from that of measurement error. The suitability map that results from weighted linear combination is a clear expression of uncertainty about the suitability of any particular piece of land for the objective under consideration. However, as an expression of uncertainty, it has no relationship to the frequentist notion of probability that underlies the treatment of decision risk in the context of measurement error. Thus a traditional treatment of decision risk as the probability that the decision made will be wrong cannot be developed. Eastman (1996) has therefore suggested that decision risk for such cases be expressed by the concept of *relative risk*.

A mapping of relative risk can be very simply achieved by rank ordering the alternatives and dividing the result by the maximum (i.e. worst) rank that occurs. The outcome is a proportional ranking that can directly be interpreted as relative risk. Then in cases where no specific area requirement for the decision set is being sought (e.g. the best 10 hectares), the final decision set can be established by selecting the alternatives where the relative risk does not exceed a specific threshold (e.g. the best 5 per cent of the areas under consideration). Figure 6 illustrates such a mapping of relative risk for the result of Plate 32 along with a mapping of the best (least risky) 10 per cent of cases outside the wildlife reserve.

Such an expression of relative risk is quite familiar in human experience. By rank ordering the alternatives (on the basis of suitability) and choosing the best ones, we use a procedure that

strives to pick the least risky alternatives (i.e. the ones that are least likely to be poor choices). For example, in screening applicants for employment we may make use of a variety of criteria (e.g. test scores, reference evaluations, years of experience, etc.) that will allow the candidates to be ranked. Then by choosing only the highest ranked candidates we minimise our risk of choosing someone who will perform poorly. However, we do not know the actual degree of risk we are taking; only that the candidates we have chosen are the least risky of the alternatives considered.

From the perspective of considering the criteria of MCE as fuzzy measures, then, it would appear that the expression of decision risk needs to be different from that which arises from a consideration of measurement error. However, in most cases, both

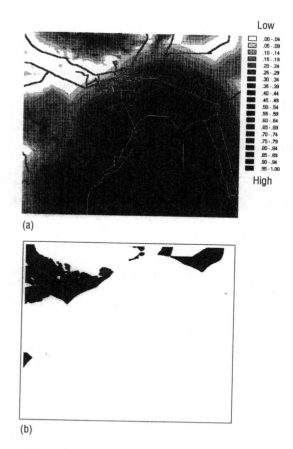

(a)

(b)

Fig 6. (a) A mapping of relative risk for the result in Plate 32. The procedure involves a simple rank ordering of alternatives followed by division by the maximum rank. (b) Areas outside the wildlife reserve with a relative risk of 10 per cent or less.

233

forms of uncertainty exist. Thus one might anticipate the problem of having to express both forms of risk. For example, given the presence of measurement error in the development of a multi-criteria suitability mapping, and a propagation of those errors to the final suitability map, one realises that this mapping is only one of a large number of possible outcomes that might be produced by randomly introducing the uncertainties in measurement that exist. Thus by Monte Carlo simulation (a capability that unfortunately exists in only a small number of GIS software programs) one could thus tabulate the proportion of simulations in which each location falls within a specific threshold of relative risk, or a specific areal requirement. This then restores the frequentist notion of probability and the usual expression of decision risk.

8 CONCLUSION

In this chapter an attempt has been made to reconcile the differences between the typical approaches to MCE used in vector and raster GIS. By using the theoretical structure of fuzzy measures, both approaches can be seen as special cases of a single family of aggregation operators. In the case of Boolean overlay as very typically used in vector GIS, the decision problem is treated as one of classical set membership, with the intersection and union operations resulting in strict ANDness or ORness with no trade-off. However, it has been shown here that these hard constraints represent no more than the crisp extremes of an underlying logic of fuzzy sets. By considering the more general class of fuzzy measures (of which fuzzy sets are a member) it has been shown that similar operations exist for the continuous factors more commonly associated with raster systems (the minimum and maximum operators). In addition, it has been shown that the weighted linear combination operator commonly used with such factors lies on a continuum with these operators, where it represents the case of intermediate ANDness and ORness, and full trade-off between the factors considered. Furthermore, it has been shown that a more general operator (the ordered weighted average) can produce all of these results along with a continuum of other operators with varying degrees of trade-off and ANDORness. This not only acts as a strong theoretical framework for consideration of the aggregation operator, but also provides a logic for the standardisation of factors, a rationale for the expression of decision risk, and a high degree of flexibility in the land allocation decision process.

References

Bonissone P P, Decker K 1986 Selecting uncertainty calculi and granularity: an experiment in trading-off precision and complexity. In Kanal L N, Lemmer J F (eds) *Uncertainty in artificial intelligence.* Amsterdam, Elsevier Science

Campbell J C, Radke J, Gless J T, Wirtshafter R M 1992 An application of linear programming and geographic information systems: cropland allocation in Antigua. *Environment and Planning A* 24: 535–49

Carver S J 1991 Integrating multi-criteria evaluation with geographical information systems. *International Journal of Geographical Information Systems* 5: 321–39

Dubois D, Prade H 1982 A class of fuzzy measures based on triangular norms. *International Journal of General Systems* 8: 43–61

Eastman J R 1993 *Idrisi Version 4.1.* Worcester (USA), Clark University

Eastman J R 1996 Uncertainty and decision risk in multi-criteria evaluation: implications for GIS software design. *Proceedings, International Institute for Software Technology Expert Group Workshop on Software Technology for Agenda 21: Decision Support Systems* Section 8

Eastman J R, Jiang H 1996 Fuzzy measures in multi-criteria evaluation. *Proceedings, Second International Symposium on Spatial Accuracy Assessment in Natural Resources and Environmental Studies.* Fort Collins, GIS World Inc.: 527–34

Eastman J R, Jin W, Kyem P A K, Toledano J 1995 Raster procedures for multi-criteria/multi-objective decisions. *Photogrammetric Engineering and Remote Sensing* 61: 539–47

Heuvelink G B M 1993 'Error propagation in quantitative spatial modelling: applications in geographical information systems'. PhD thesis. Utrecht, Netherlands Geographical Studies 163

Janssen R, Rietveld P 1990 Multi-criteria analysis and geographical information systems: an application to agricultural landuse in the Netherlands. In Scholten H J, Stillwell J C H (eds) *Geographical information systems for urban and regional planning.* Amsterdam, Kluwer: 129–39

Rao M, Sastry S V C, Yadar P D, Kharod K, Pathan S K, Dhinwa P S, Majumdar K L, Sampat Kumar D, Patkar V N, Phatak V K 1991 *A weighted index model for urban suitability assessment – a GIS approach.* Bombay, Bombay Metropolitan Regional Development Authority

Voogd H 1983 *Multi-criteria evaluation for urban and regional planning.* London, Pion

Yager R 1988 On ordered weighted averaging aggregation operators in multicriteria decision-making. *IEEE Transactions on Systems, Man, and Cybernetics* 8: 183–90

40

The future of GIS and spatial analysis

M F GOODCHILD AND P A LONGLEY

The chapter explores factors affecting spatial analysis, in theory and practice, and their likely impacts. A model is presented of the traditional role of spatial analysis, and is examined from the perspectives of increasing costs of data, the increased sharing of data between investigators across a wide range of disciplines, the emergence of new techniques for analysis, and new computer architectures. Practical problems are identified that continue to face investigators using GIS to support spatial analysis, including accuracy, the technical problems of integration, and the averaging of different feature geometries. The chapter ends with a speculation on spatial analysis of the future.

1 INTRODUCTION

GIS and spatial analysis have enjoyed a long and productive relationship over the past decades (for reviews see Fotheringham and Rogerson 1994; Goodchild 1988; Goodchild et al 1992). GIS has been seen as the key to implementing methods of spatial analysis, making them more accessible to a broader range of users, and hopefully more widely used in making effective decisions and in supporting scientific research. It has been argued (e.g. Goodchild 1988) that in this sense the relationship between spatial analysis and GIS is analogous to that between statistics and the statistical packages. Much has been written about the need to extend the range of spatial analytic functions available in GIS, and about the competition for the attention of GIS developers between spatial analysis and other GIS uses, many of which are more powerful and better able to command funding. Specialised GIS packages directed specifically at spatial analysis have emerged (e.g. Idrisi; see also Bailey and Gatrell 1995). Openshaw and Alvanides (Chapter 18) have set out some of the ways in which developments in computation may feed through to enhanced GIS-based spatial analysis. Finally, Anselin (Chapter 17), Getis (Chapter 16) and others have discussed the ways in which implementation of spatial analysis

methods in GIS is leading to a new, exploratory emphasis.

The purpose of this chapter is to explore new directions that have emerged recently, or are currently emerging, in the general area of GIS and spatial analysis, and to take a broad perspective on their practical implications for GIS-based spatial analysis. In the next section, we argue that in the past the interaction between GIS and spatial analysis has followed a very clearly and narrowly defined path, one that has more to do with the world of spatial analysis prior to the advent of GIS than with making the most of both fields – the path is, in other words, a legacy of prior conditions and an earlier era (see also Openshaw and Alvanides, Chapter 18). The following section expands on some of the themes of the introduction to this volume by identifying a number of trends, some related to GIS but some more broadly based, that have changed the context of GIS and spatial analysis over the past few years, and continue to do so at an increasing rate. The third section identifies some of the consequences of these trends, and the problems that are arising in the development of a new approach to spatial analysis. The chapter concludes with some comments about the complexity of the interactions between analysis, data and tools, and speculation on what the future may hold, and what forms of spatial analysis it is likely to favour.

2 TRADITIONS IN SPATIAL ANALYSIS

2.1 The linear project design

In the best of all possible worlds, a scientific research project (the term 'research' will be interpreted broadly to include both scientific and decision-making activities) begins with clearly stated objectives. Some decision must be made, some question of scientific or social concern must be resolved by resorting to experiment or real-world evidence. A research design is developed to resolve the problem, data are collected, analyses are performed, and the results are interpreted and reported. Although this implies a strictly linear sequence of events, the most robust research designs also include feedbacks and checks in order to ensure that the principles of good scientific research are not overly compromised in practical implementation. This simple, essentially linear, structure with recursive feedbacks underlies generations of student dissertations, government reports, and research papers. It is exemplified by the classic social survey research design illustrated in Figure 1. The sequential events in this design together constitute a holistic research project, and the feedbacks are all internal to the research design. Thus once the project has been initiated, the availability of existing data has no further influence upon problem definition; methods of analysis that are consistent with the type, quality, and amount of data to be collected are identified at the design stage (and the data collection method changed if no suitable analytical method exists); the sample design is not guided by considerations and priorities that lie outside the remit of the research; and so on.

In this simple, sequential world, the selection of methods of analysis can be reduced to a few simple rules (in the context of statistical analysis, see for example Levine 1981: chapter 17; Marascuilo and Levin 1983: inside cover; Siegel 1956: inside cover). Choice of analytic method depends on the type of inference to be drawn (e.g. whether two samples are drawn from the same, unknown population, or whether two variables are correlated), and on the characteristics of the available data (e.g. scale of measurement – nominal, ordinal, interval, or ratio: Wrigley 1985). Inference about, and exploration of, the research problem will take place in what is loosely described as the confirmatory (hypothesis testing and

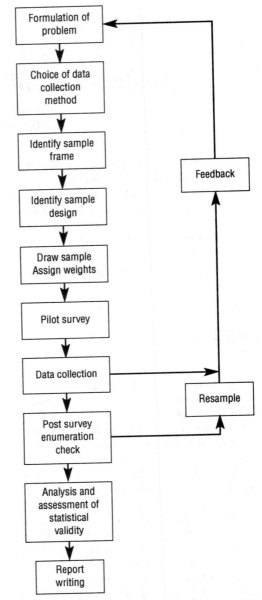

Fig 1. Sequential stages of a typical social survey research design.

inference seeking) and exploratory (pattern or anomaly seeking) stages of the research.

In contrast to this coherent research design, the terms 'data-driven' and 'technique-driven' are highly pejorative in research generally, as are such phrases as 'a technique in search of a problem' – in this ideal world, the statement of the problem strictly precedes the collection of data and the performance of analysis.

2.2 Spatial analysis

Spatial analysis, or spatial data analysis, is a well-defined subset of the methods of analysis available to a project. One might define spatial analysis as a set of methods useful when the data are spatial, in other words when the data are referenced to a 2-dimensional frame. More narrowly, the Earth's surface provides a particular instance of such a frame, the geographical frame, with its peculiar properties of curvature. This definition of spatial analysis is arguably too broad, because in basing the definition on the properties of data it does not address the question of whether the 2-dimensional frame actually matters – could the same results have been obtained if the frame were distorted in some way, or if objects were repositioned in the frame? More precisely, then, spatial analysis can be defined as that subset of analytic techniques whose results depend on the frame, or will change if the frame changes, or if objects are repositioned within it. To distinguish analytic methods from more mundane operations, they might be defined as methods for processing data with the objective of solving some scientific or decision-making problem.

Methods of spatial analysis have accumulated in a literature that spans many decades, indeed centuries (see Getis, Chapter 16). They have been invented in many disciplines, including mathematics, particularly geometry; statistics, particularly spatial statistics and statistical geometry; and in geography and other Earth sciences. Compendia have been published (among others, see Bailey and Gatrell 1995; Berry and Marble 1968; Haining 1990; Taylor 1977; Unwin 1981), and various approaches proposed for structuring this body of technique. Spatial analytic techniques may also be classified into those which are confirmatory and those which are exploratory. Choice of analytical method also relates to data characteristics – documented since Chorley and Haggett's (1965) analogies between (respectively) nominal, ordinal, interval and ratio data and point, line, area, and surface objects (see also Chrisman 1997; Martin 1996).

2.3 The well-informed analyst

Traditionally, the responsibilities of the inventor of a technique ended when the technique had been tested and described. Even the testing of a technique can be suspect in an academic world that often values theory over empiricism, and is suspicious of empirical results that cannot be demonstrated to be generally true. The advent of the digital computer changed this world fundamentally because it became possible for a scientist to perform a method of analysis automatically, without taking personal responsibility for every aspect of the performance. It was now possible using the 'black box' of the computer to perform an analysis that one did not know everything about – that one could not perform by hand. Methods emerged, beginning in the 1970s and particularly in the area of multivariate statistics, that would be impossibly impractical to perform by hand. Pedagogically, a fundamental shift became possible in how analysis was taught – that one might learn about a technique by studying the nature of its response to particular inputs, rather than by studying the procedure which generated the response. But there is a fundamental difference between these two positions: between whether one understands the results of a principal components analysis, for example, as the extraction of eigenvalues from a specific matrix, or the generation of statistics that broadly indicate some concept of 'relative importance' without presuming any understanding of what eigenvalues are and *how* they formalise the structure in data.

Exactly where this change occurred is open to debate, of course. It may have occurred when students were no longer required to perform statistical analyses by hand before being let loose on computer packages; or when FORTRAN appeared, making it necessary to understand less about how instructions were actually carried out; or when the growth of the scientific enterprise had reached such a level that potential replication of every result was a practical impossibility.

Of course the digital computers that were introduced to the scientific community beginning in the late 1950s produced rapid change in the labour demands of many statistical methods. The intricate calculations of factor analysis (Harman 1976) could be performed by a fully automatic machine, provided the researcher could command sufficient computer time, and provided labour was available to punch the necessary cards. Computers and the brains of young humans are in some ways similar: both begin 'hard wired' with the primitive elements of reasoning (e.g. binary processing in computers, linguistic abilities in infants) and both can build enormously complex structures out of simpler ones,

237

apparently ad infinitum (see Fischer, Chapter 19, for a broader discussion of computer 'reasoning'). What began in the 1960s as a set of uncoordinated efforts by individual scientists writing their own programs had developed by the 1990s into a complex of enormously sophisticated tools, each integrating a large number of methods into an easy-to-use whole.

If software packages and user-friendly computer environments have made aspiring spatial analysts less aware of the computational and statistical context to inference, then the opposite is true to some extent of exploratory analysis, where the innovation of computer graphics and windows, icons, mice, and pointers (WIMPs) has created a more intuitive context to the interrogation of spatial data (see Anselin, Chapter 17; Kraak, Chapter 11). Indeed one of the criticisms of GIS-based graphics developments from the spatial analysis community has been that the computer graphics medium has been allowed to dominate the spatial analysis message, by analogy to 'data-led' thinking as described in section 2.1.

2.4 Extending the functions of analytic software

Although they show clear evidence of their roots, the packages used by the scientists of the 1990s are different in fundamental respects from the programs of the 1960s. Besides implementing large numbers of statistical methods, today's packages also provide support for the maintenance of data and the creation of information. There will be tools for documenting datasets, and describing their properties, such as accuracy and history. Other tools will support the sharing of data, in the form of format converters or interfaces to the Internet. In short, the functions of today's digital computers in supporting research go far beyond those of a simple calculating machine, carrying out well-defined methods of analysis. The same digital computer may now be involved in:

- the selection and formulation of a problem, by providing access to automated library catalogues and on-line literature;
- the collection of data through support for real-time data acquisition;
- management of data, performance of analysis, visualisation of results, writing of conclusions;
- even publication through access to the Internet and the World Wide Web.

The computer is no longer part of the research environment – we are rapidly approaching a world in which the computer *is* the research environment.

These trends are all echoed strongly in GIS. Although a particular scientist might use a GIS in ways that are more analogous to the early days of statistical computing, by performing a single buffering operation, for example, scientific applications are much more likely to include integration of many GIS functions. Today's scientist or decision-maker is likely to see a GIS as an environment for research, rather than as a means of automating analysis. The GIS is likely to be involved in the project from beginning to end, and to be integrated with other tools and environments when these are needed. GIS will be used for collecting, assembling, verifying, and editing the data; performing some of the analyses required by the project; and presenting and interpreting the results. Moreover, much GIS use may not be tied to a specific project – GIS finds extensive use in the collection of data for purposes that may be generic, or not well-defined, or may be justified in anticipation of future demand. Even though these may not be 'spatial analysis' in the sense of the earlier discussion, analysis may still be necessary as part of the data production process – for example, when a soil scientist must analyse data to produce a soil map.

2.5 When to choose GIS

A related issue is the extent to which GIS remains a separately identifiable technology, and in what senses the 'GIS environment' is distinctive. The general drift of many of the chapters in this 'Technical issues' Part of this book is that GIS is increasingly becoming both a background technology (more akin to wordprocessing than, say, spatial interaction modelling), and a technology that can be broken up and packaged as niche products (Elshaw Thrall and Thrall, Chapter 23). And yet the various discussions in the 'Principles' Part of this book document the important agenda for spatial analysis set in the environment of GIS, and why GIS-based spatial analysis is likely to remain a distinctive area of activity for the foreseeable future.

If GIS has multiple roles in support of science and problem-solving, then one might not be surprised to find that the choice between GIS alternatives is complex and often daunting. The many GIS packages offer a wide range of combinations of analysis functions, housekeeping support, different ways of representing the same phenomena, variable levels of sophistication in visual display, and performance. In addition, choice is often driven by: the available hardware, since not all GIS run on all platforms; the format in which the necessary data have been supplied; the personal preferences and background of the user; and so forth. Even the extensive and frequently updated comparative surveys published by groups such as GIS World Inc. can be of little help to the uninitiated user.

The existence of other classes of analytic software complicates the scene still further. Under what circumstances is a problem better solved using a package that identifies itself as a GIS, or using a statistical package, or a mathematical package, or a scientific visualisation package? Under what circumstances is it better to fit the square peg of a real problem into the round GIS hole? GIS are distinguished by their ability to handle data referenced to a 2-dimensional frame, but such capabilities also exist to a more limited extent in many other types of software environment. For example, it is possible to store a map in a spreadsheet array, and with a little ingenuity to produce a passable 'map' output; and many statistical packages support data in the form of images.

Under what circumstances, then, is an analyst likely to choose a GIS? The following conditions are suggested, although the list is certainly not complete, and the items are not intended to be mutually exclusive:

- when the data are geographically referenced;
- when geographical references are essential to the analysis;
- when the data include a range of vector data types (support for vector analysis among non-GIS packages appears to be much less common than support for raster analysis);
- when topology – representation of the connections between objects – is important to the analysis;
- when the curvature of the Earth's surface is important to the analysis, requiring support for projections and for methods of spatial analysis on curved surfaces;
- when the volume of data is large, since alternatives like spreadsheets tend to work only for small datasets;

- when data must be integrated from a variety of sources, requiring extensive support for reformatting, resampling, and other forms of format change;
- when geographical objects under analysis have large numbers of attributes, requiring support from integrated database management systems, since many alternatives lack such integration;
- when the background of the investigator is in geography, or a discipline with strong interest in geographical data;
- when the project involves several disciplines, and must therefore transcend the software traditions and preferences of each;
- when visual display is important, and when the results must be presented to varied audiences;
- when the results of the analysis are likely to be used as input by other projects, or when the data are being extensively shared.

3 ELEMENTS OF A NEW PERSPECTIVE

This section reviews some of the changes that are altering the context and face of spatial analysis using GIS. Some are driven by technological change, and others by larger trends affecting society at the turn of the millennium.

3.1 The costs of data creation

The collection of geographical data can be extremely labour-intensive. Early topographic mapping required the map-maker to walk large parts of the ground being mapped; soil mapping requires the exhausting work of digging soil pits, followed often by laborious chemical analysis; census data collection requires personal visits to a substantial proportion of (sometimes all) household respondents; and forest mapping requires 'operational cruise', the intensive observation of conditions along transects. Although many new methods of geographical data creation have replaced the human observer on the ground with various forms of automated sensing, there is no alternative in those areas that require the presence of expert interpreters in the field.

Many of the remaining stages of geographical data creation are also highly labour-intensive. There is still no alternative to manual digitising in cases where the source document is complex, compromised, or difficult to interpret. The processes of error detection and correction are difficult if not impossible to automate, and the methods of cartographic generalisation used by expert cartographers have proven very difficult to formalise and replace. In short, despite much technical progress over the past few decades, geographical data creation remains an expensive process that is far from fully automated.

Labour costs continue to rise at a time when the resources available to government, the traditional source of geographical data, continue to shrink (see Elshaw Thrall and Thrall, Chapter 23). Many geographical datasets are collected for purposes which may be far from immediate, and it is difficult therefore to convince taxpayers that they represent an essential investment of public funds, especially in peacetime. Governments in financial straits call for evidence of need: for example, census organisations are under continual pressure to demonstrate that their costly operations do not replicate information that is available elsewhere; and many governments have moved their mapping operations onto a semi-commercial basis in order to allow demand to be expressed through willingness to pay (see Rhind, Chapter 56). To date, the US Federal mapping agencies have resisted the trend, but internationally there is more and more evidence of the emergence of a market in geographical information.

Within the domain of geographical data the pressures of increased labour costs favour data that can be collected and processed automatically. Given a choice between the labour-intensive production of vector topographic data, and the semi-automated generation of such raster products as digital elevation models and digital orthophotos, economic pressures can lead only in one direction. It is easy to imagine a user trading off the ability to identify features by name against the order of magnitude lower cost, and thus greater potential update frequency, of raster data.

The broader context to these changes is that we now live in a digital world, in which far more data are collected about us, in computer readable form, than ever before. This is what has been termed the 'information economy', in which government no longer has a monopoly in the supply of geographical information, and in which information has become both a tradeable commodity and a strategic organisational resource. Global trends such as deregulation and privatisation, allied to the increasing competitive edge of consumer-led markets, are multiplying the potential number of sources of information, yet at the expense of system-wide standardisation (Rhind, Chapter 56). Such data are not ideally suited to the linear research design set out in section 2.1, yet (in socioeconomic research at least) they frequently are far richer in detail than anything that has been collected hitherto. A clear challenge to spatial analysis is therefore to reconcile diverse datasets with different data structures or spatial referencing systems, and to gauge how representative they are with reference to existing (more limited or less frequently updated) public sources. A good example of this concerns the development of geodemographic indicators which have traditionally been derived from census data. These are typically updated only every ten years and are frequently reliant upon very indirect indicators of likely consumer behaviours (Longley and Clarke 1995). 'Lifestyles' approaches based upon questionnaire returns from a range of self-selecting respondents (Birkin 1995) offer the prospect of 'freshening up' and in time replacing conventional census-based geodemographics, although thorny issues of representativeness and bias must be grappled with before credible 'data fusion' can be deemed to have taken place.

Of course, the principle of information commerce is alien to the scientific community, which is likely to resist strongly any attempt to charge for data that is of interest to science, even peripherally. But here too there are pressures to make better use of the resources invested in scientific data collection. Research funding agencies increasingly require evidence that data collected for a project have been disseminated, or made accessible to others, while recognising the need to protect the interests of the collector.

Trends such as these, while they may be eminently rational to dispensers of public funds, nevertheless fly directly in the face of the traditional model of science presented earlier. For example, the best-known definition of the discipline of geography is that it is 'concerned to provide accurate, orderly, and rational description and interpretation of the variable character of the Earth surface' (Hartshorne 1959). As a general rule, commercial datasets are not accurate (they provide little indication of the sources of unknown errors in data collection or the ways in which they are likely to operate in analysis); they are orderly only in a minimalist sense (for example, satellites provide frequently-updated *coverage* information yet cannot comprehensively measure land *use*; 'lifestyles' data do not provide information about all groups in society); and they are not rational in that they separate still further the analyst from the context to the research problem and lead to data- or machine-led thinking. How can projects fail to be driven by data, if data are forced to obey the economic laws of supply and demand? Where in traditional science are the rules and standards that allow scientists to trade off economic cost against scientific truth? It seems that economic necessity has forced the practice of science to move well beyond the traditions that are reflected in accepted scientific methodologies and philosophies of science.

3.2 The life of a dataset

In the traditional model presented earlier data were collected or created to solve a particular problem, and had no use afterwards except perhaps to historians of science. But many types of geographical data are collected and maintained for generic purposes, and may be used many times by completely unrelated projects. For other types, the creation of data is itself a form of science, involving the field skills of a soil scientist, for example, or a biologist. Thus a dataset can be simultaneously the output of one person's science, and the input to another's. This is to conceive of spatial analysis within GIS as the process of building 'models of models' – whereby the outcome of a 'higher level' spatial analysis is dependent upon data inputs which are themselves a previously modelled version of reality. These relationships have become further complicated by the rise of multidisciplinary science, which combines the strengths and expertise of many different sciences, and partitions the work among them. Once again, the linear model of science is found wanting, unable to reflect the complex relationships between projects, datasets, and analytic techniques that exist in modern science. The notion that data are somehow subsidiary to problems, methods and results is challenged, and traditional dicta about not including technical detail in scientific reports may be counterproductive.

In truth, of course, this is nothing new in the sense that most spatial analysis in the socioeconomic realm has been based upon crude surrogate data, obtained for inappropriate areal units in obsolete time periods. Thirty-five years ago we were all 'information poor' and the limited data-handling capabilities of early

spatial analysis methods reflected this. Arguably, it was this impoverishment that was the root cause of the failure of many such methods to generate detailed insights into the functioning of social systems (see Openshaw and Alvanides, Chapter 18). The potential to build detailed data-rich depictions of reality within GIS will make some problems more transparent, yet others will likely be further obscured. From a pessimist's standpoint, data-rich modelling within GIS represents a return to the shifting sands of naive empiricism. For the optimist, sensitive honing of such data to context allows data-rich models to shed light upon a wider range of social and economic research problems.

In this new world, a given set of data is likely to fall into many different hands during its life. It may be assembled from a mixture of field and remote sensing sources, interpreted by a specialist, catalogued by an archivist or librarian, used by scientists and problem-solvers, and passed between its custodians using a range of technologies (Figure 2). It is quite possible in today's world that the various creators and users of data share little in the way of common disciplinary background, leaving the dataset open to misunderstanding and misinterpretation. Recent interest in metadata, or ways of describing the contents of datasets, is directed at reducing some of these problems, but the easy access to data provided by the Internet and various geographical data archives has tended to make the problem of inappropriate use or application worse.

These issues are particularly prominent in the case of data quality, and the ability of the user of a dataset to understand its limitations, and the uncertainty that exists about the real phenomena the data are intended to represent. To take a simple example, suppose information on the geodetic datum underlying a particular dataset – potentially a very significant component of its metadata – were lost in transmission between source and user; alternatively, suppose that the user simply assumed the wrong datum, or was unaware of its significance. This loss of metadata, or specification of the data content, is equivalent in every respect to an actual loss of accuracy equal to the difference between the true datum and the datum assumed by the user, which can be several hundreds of metres. This is perhaps an obvious example, but what, say, is the magnitude of error associated with soil profile delineation? What is the magnitude of likely ecological fallacy associated

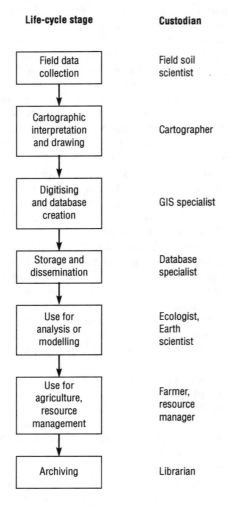

Life-cycle stage	Custodian
Field data collection	Field soil scientist
Cartographic interpretation and drawing	Cartographer
Digitising and database creation	GIS specialist
Storage and dissemination	Database specialist
Use for analysis or modelling	Ecologist, Earth scientist
Use for agriculture, resource management	Farmer, resource manager
Archiving	Librarian

Fig 2. The life-cycle of a soil database: an example of the complex patterns of custodianship and transfer now common for many types of spatial data.

with comparison of a geodemographic classifier with the results of a survey? In short, the quality of a dataset to a user is a function of the difference between its contents and the user's understanding of its meaning, not the creator's.

3.3 Data sharing

In this new world of shared data, the term metadata has come to function as the equivalent of documentation, cataloguing, handling instructions, and production control. The US Federal Geographic Data Committee's Content Standard for Digital

Geospatial Metadata (FGDC 1994) has been very influential in providing a standard, which has been emulated frequently (see Salgé, Chapter 50). If the custodian of a large collection of geographical datasets provides metadata in this form, it is possible for others to search its records for those that match their needs. The FGDC's National Geospatial Data Clearinghouse (*http://www.fgdc.gov*) is one such directory (see also the Alexandria Digital Library project to provide distributed library services for geographically referenced datasets: Smith et al 1996; and see *http://alexandria.sdc.ucsb.edu*).

The user of a traditional library will rarely know the exact subject of a search – instead, library search has an essential fuzziness, which is supported by the traditional library in several essential ways. By assigning similar call numbers to books on similar subjects, and shelving by call number, the traditional library is able to provide an environment that allows the user to browse the collection in a chosen area. But this support is missing when the records of a metadata file are searched using simple Boolean methods. It would make better sense to model the search process as one of finding the best fit between a metadata record representing the user's ideal, and metadata records representing the datasets available. It is very unlikely, after all, that data exist that perfectly match the needs of a given problem, especially in the ideal world of problem-solving represented earlier. This is especially true when the object of search is to find data covering a particular location, as it often is in the GIS context. In such cases, it seems very unlikely that there will be an exact fit between the area requested by the user, and the area covered by a data set in an archive.

3.4 New techniques for analysis

Several chapters in this section have focused on new methods of spatial analysis, particularly new methods that have emerged in the data-rich computational environment now available to scientists. These include neural nets (Fischer, Chapter 19), new methods of optimisation such as simulated annealing and genetic techniques, and computationally intensive simulation. The term *geocomputation* (Openshaw and Alvanides, Chapter 18) has been suggested. Anselin (Chapter 17) and others have extended the principles of exploratory data analysis (Tukey 1970) to spatial data.

In science generally, the combination of vast new sources of data and high-speed computation have led to an interest in methods of *data mining*, which implies the ability to dredge data at very high speed in a search for patterns of scientific interest. In a geographical context, the vague notion of 'scientific interest' might suggest the need for methods to detect features or measurements that are inconsistent with their surroundings, in apparent violation of Tobler's 'first law of geography' (Tobler 1970; see also Johnston, Chapter 3). Linearities in images are of potential interest in geological prospecting; and one can imagine circumstances in which atmospheric scientists might want to search large numbers of images for patterns consistent with weather events. Such techniques of pattern recognition were pioneered many years ago in particle physics, to search vast numbers of bubble-chamber photographs for the tracks characteristic of rare new particles.

One might argue that such techniques represent a renewal of interest in inductive science – the search for regularities or patterns in the world that would then stimulate new explanatory theories (see Fischer, Chapter 19). Inductivism has fallen out of fashion in recent decades, at least in disciplines that focus on geographical data, leading one to ask whether a renewal of interest represents a fundamental shift in science, or merely a response to the opportunities for data-led thinking offered by more powerful technology. On this issue the jury is clearly still 'out' – geocomputation has not yet provided the kinds of new insights that might support a broad shift to inductivism.

3.5 New computer architectures

The communication technologies that have emerged in the past decade have allowed a fundamental change in the architecture of computing systems. Instead of the early mainframes and later stand-alone desktop systems, today's computers are linked with high-speed networks that allow data, software, and storage capacity located in widely scattered systems to be integrated into functioning wholes. Data can now be 'served' from central sites on demand, avoiding the need to disseminate many copies, with subsequent confusion when updates are needed. Coleman (Chapter 22) reviews the architectural alternatives now common in computing systems, and their technical impacts on GIS.

The new approaches to computing that are possible in this interconnected environment are having a profound effect on spatial analysis. Because it is no longer possible to assume a lifetime association between a user and a particular system design, there are mounting pressures for standards and interoperability between systems to counter the high costs of retraining of staff and reformatting of data.

The proprietary GIS that once dominated the industry attempted to provide a full range of GIS services in one homogeneous environment. Data were stored in proprietary formats, often kept secret by vendors to maintain market position, but making it difficult for others to expand the capabilities of the system by programming extra modules. The 'open GIS' movement (Buehler and McKee 1996 and see *http://www.ogis.org*) mirrors efforts in other areas of the electronic data processing world to promote interoperability, open standards and formats, and easy exchange from one system to another. While such ideas were often regarded as counter to the commercial interests of vendors, there is now widespread acceptance in the industry that they represent the way of the future.

The implications of open systems for spatial analysis are likely to be profound. First, they offer the potential of a uniform working environment, in which knowledge of one system is readily transferable to another. To make this work, however, it will be necessary to achieve a uniform view, and its acceptance across a heterogeneous user community. There is no prospect of interoperability and open systems without agreement on the fundamental data models, terminology, and objectives of GIS-based analysis. Thus much effort will be needed on the part of the inventors and implementors of spatial analysis to develop this uniform view.

Second, the possibility of easy sharing of data across systems gives even greater momentum to efforts to make geographical information more shareable, and even greater demands on the existence and effectiveness of metadata.

Third, interoperability is likely to create an environment in which it is much easier to implement methods of spatial analysis in GIS. Traditionally, vendors of monolithic systems have added functions when market demand appears to justify the development costs. It has been impossible, in a world of proprietary systems, for third parties to add significant functionality. Thus expansion of spatial analytic capabilities has been slow, and has tended to reflect the needs of the commercial market, rather than those of science and problem-solving, when these diverge. In a world of open systems it will be much easier to add functions, and the new environment will encourage the emergence of small companies offering specialised functionality in niche markets.

Finally, new interoperable approaches to software will encourage the modularisation of code (Sondheim et al, Chapter 24). It is already possible in some mainstream software environments to launch one specialised application within another – for example, to apply spreadsheet functions to information in a word processing package. This 'plug and play' environment offers enormous scope to GIS, since it will lead ultimately to a greater integration of GIS functions, and map and imagery data in general, into mainstream electronic data processing applications.

The scientific world has grown used to a more or less complete separation between data, and the functions that operate on and manipulate data. Functions are part of 'analysis', which plays a role in the traditional approach to problem-solving outlined earlier that is clearly distinct from that of data. But it has already been argued that in a world of extensive data sharing and interaction between disciplines it is impossible to think of data in isolation from its description, or metadata, which allows the meaning of information to be shared.

In the abstract world of object-oriented methods, it is argued that the meaning of data lies ultimately in the operations that can be performed. If datasets exist in two systems, and pairs of functions exist in both systems that produce the same answers, then the two datasets are the same in information content, irrespective of their specific formats and arrangements of bits. It makes sense, then, to *encapsulate* methods with data. When more than one method is available to perform a given function, it makes sense for the choice to be made by the person best able to do so, and for the method thereafter to travel with the data. For example, a climatologist might encapsulate an appropriate method for spatial interpolation with a set of point weather records, because the climatologist is arguably better able to select the best method of spatial interpolation, given his or her knowledge of atmospheric processes.

In future, and especially given the current trend in computing to object-oriented methods, it is likely that the distinction between data and methods will become increasingly blurred. Commonly used techniques of spatial analysis, such as spatial interpolation, may become encapsulated with data in an extension of the concept of metadata to include methods. Of course this assumes that methods are capable of running in a wide variety of host systems, which takes the discussion back to the issue of interoperability introduced earlier.

4 SPATIAL ANALYSIS IN PRACTICE

At this stage, it seems useful to introduce a discussion of the practical problems which face the users of today's GIS. While it is now possible to undertake a wide range of forms of spatial analysis, and to integrate data from a range of sources that would have seemed inconceivable as little as five years ago, there continue to be abundant limitations that impede the complete fulfilment of the technology's promise. The following subsections discuss several of these enduring impediments.

4.1 Absolute and relative position

First, and perhaps foremost, are problems of varying data quality. In science generally it is common to express quality in terms such as 'accurate to plus or minus one per cent'. But while such methods are useful for many types of data, they are less so when the data are geographical. The individual items of information in a geographical dataset are typically the result of a long and complex series of processing and interpretation steps. They bear little relationship to the independent measurements of traditional error analysis. Section 2 dealt at length with the data quality issue, and the theme is taken up again in the context of decision-making by Hunter (Chapter 45), and those discussions will not be repeated here. Instead, the following discussion is limited to the particular problems encountered when merging datasets.

While projections and geodetic datums are commonly well-documented for the datasets produced by government agencies, the individual scientist digitising a map may well not be in a position to identify either. The idea that lack of specification could contribute to uncertainty was discussed earlier, and its effects will be immediately apparent if a dataset is merged with one based on another projection or datum. In practice, therefore, users of GIS frequently encounter the need for methods of *conflation*, a topic discussed in detail below.

The individual items of information in a geographical dataset often share lineage, in the sense that more than one item is affected by the same error. This happens, for example, when a map or photograph is registered poorly – all of the data derived from it will have the same error. One indicator of shared lineage, then, is the persistence of error – all points derived from or dependent on the same misregistration will be displaced by the same or a similar amount. Because neighbouring points are more likely to share lineage than distant points, errors tend to show strong positive spatial autocorrelation (Goodchild and Gopal 1989).

Rubber-sheeting is the term used to describe methods for removing such errors on the assumption that strong spatial autocorrelations exist. If errors tend to be spatially autocorrelated up to a distance of x, say, then rubber-sheeting will be successful at removing them, at least partially, provided control points can be found that are spaced less than x apart. For the same reason, the shapes of features that are less than x across will tend to have little distortion, while very large shapes may be badly distorted. The results of calculating areas, or other geometric operations that rely only on relative position, will be accurate as long as the areas are small, but will grow rapidly with feature size. Thus it is important for the user of a GIS to know which operations depend on *relative* position, and over what distance; and where *absolute* position is important (of course the term absolute simply means relative to the Earth frame, defined by the Equator and the Greenwich meridian, or relative over a very long distance).

When two datasets are merged that share no common lineage (for example, they have not been subject to the same misregistration), then the relative positions of objects inherit the absolute positional errors of both, even over the shortest distances. While the shapes of objects in each dataset may be accurate, the relative locations of pairs of neighbouring objects may be wildly inaccurate when drawn from different datasets. The anecdotal history of GIS is full of such examples – datasets which were perfectly adequate for one application, but failed completely when an application required that they be merged with some new dataset that had no common lineage. For example, merging GPS measurements of point positions with streets derived from the US Bureau of the Census TIGER (Topologically Integrated Geographic Encoding and Referencing) files may

lead to surprises where points appear on the wrong sides of streets. If the absolute positional accuracy of a dataset is 50 metres, as it is with parts of TIGER, then such surprises will be common for points located less than 50 metres from the nearest street. In a similar vein but in the context of the fragmented data holdings of UK local authorities, Martin et al (1994) describe the problems and mismatches inherent in matching individual and household information with property gazetteers.

4.2 Semantic integration

Some of the most challenging problems in GIS practice occur in the area of semantic integration, where integration relies on an understanding of meaning. Such problems can occur between geographical jurisdictions, if definitions of feature types, or classifications, or methods of measurement vary between them. It is common, for example, for schemes of vegetation classification to vary from one country to another, making it difficult to produce horizontally merged data (Mounsey 1991). 'Vertical' integration can also be problematic, as for example in merging the information on land classification maps produced by different agencies, or different individuals (Edwards and Lowell 1996).

While some of these problems may disappear with more enlightened standards, others merely reflect positions that are eminently reasonable. The problems of management of ecosystems in Florida are clearly different from those of Montana, and it is reasonable that standards adopted by the two states should be different (see Fisher, Chapter 13). Even if it were possible to standardise for the entire US, one would be no further ahead in standardising between the US and other countries. Instead, it seems a more reasonable approach is to achieve interoperability without standardisation, by more intelligent approaches to system design.

4.3 Conflation

Conflation appears to be the term of choice in the GIS community for functions that attempt to overcome differences between datasets, or to merge their contents. Conflation attempts to replace two or more versions of the same information with a single version that reflects the pooling of the sources; it may help to think of it as a process of weighted averaging. The complementary term *'concatenation'* refers to the integration of the sources, so that the contents of both are accessible in the product. The polygon overlay operation familiar to many GIS users is thus a form of concatenation.

Two distinct forms of conflation can be identified, depending on the context:

1 conflation of feature geometry and topology, and concatenation of feature attributes;
2 conflation of geometry, topology, and attributes.

As an example of the first case, suppose information is available on the railroad network at two scales, 1:100 000 and 1:2 million. The set of attributes available is richer at the 1:2 million scale, but the geometry and topology are more accurate at 1:100 000. Thus it would be desirable to combine the two, thereby discarding the coarser geometry and topology.

As an example of the second case, consider a situation in which soils have been mapped for two adjacent counties, by two different teams of scientists. At the common border there is an obvious problem, because although the county boundary was defined by a process that was in no way dependent on soils, the border nevertheless appears in the combined map. Thus it would be desirable to 'average' the data at and near the boundary by combining the information from both maps in compatible fashion. As these two examples illustrate, the need for conflation occurs both horizontally, in the form of edge matching, and 'vertically'. A further example of the second case is provided by spatially extensive property valuation exercises, such as that which accompanied the introduction of the UK Council Tax (Longley et al 1994): surveyors were each individually responsible for allotted areas and conflation of estimates around the area boundaries was used to enhance consistency.

4.4 Perfect positioning

Some of the problems of conflation, and of relative and absolute positional accuracy, might be expected to dissipate as measurement of position becomes more and more accurate, leading eventually to 'perfect' positioning. Unfortunately, there are good reasons to anticipate that this happy state will never be reached. Although the positions of the Greenwich meridian and various geodetic control points have been established by fixing monuments, fundamental uncertainty will continue to be created

by seismic motions, continental drift, and the wobbling of the Earth's axis. Any mathematical representation of the Earth's shape must be an approximation, and different approximations have been adopted for different purposes. Moreover, there will always be a legacy of earlier, less accurate measurements to deal with. Thus it seems GIS will always have to deal with uncertainty of position, and with the distinctions between relative and absolute accuracy, and their complex implications for analysis.

Instead, strategies must be found for overcoming the inevitable differences between databases, either prior to analysis or in some cases 'on the fly'. Consider, for example, the problems caused by use of different map databases for vehicle routing. Systems are already available on an experimental basis that broadcast information on street congestion and road maintenance to vehicles, which are equipped with map databases and systems to display such information for the driver. In a world of many competing vendors, such systems will have to overcome problems of mismatch between different databases, in terms both of position and of attributes. For example, two databases may disagree over the exact location of 100 Main Street, or whether there *is* a 100 Main Street, with potentially disastrous consequences for emergency vehicles, and expensive consequences for deliveries (see Cova, Chapter 60). Recent trends suggest that the prospects for central standardisation of street naming by a single authority are diminishing, rather than growing.

5 CONCLUSION

The prospects for spatial analysis have never been better. Data are available in unprecedented volume, and are easily accessed over today's communication networks. More methods of spatial analysis are implemented in today's GIS than ever before, and GIS has made methods of analysis that were previously locked in obscure journals easy and straightforward to use. Nevertheless, today's environment for spatial analysis raises many issues, not the least of which is the ability of users to understand and to interpret correctly. Questions are being raised about the deeper implications of spatial analysis, and the development of databases that verge on invasion of individual privacy (Curry, Chapter 55). Our expectations may be unreasonable given the inevitable problems of spatial data quality.

Postmodern scientific discourse has fragmented, and with regard to GIS there is diversity not just in the sources of digital geographical information, but also increasingly (and especially with respect to human systems) in its interpretive meaning. There is a need to communicate clear interpretive conceptions of the rich but widely-distributed and piecemeal data holdings of networked GIS. We need now to think about spatial analysis not just in terms of outcomes, but also in terms of inputs. Metadata will come to fill a crucial role in the comparative assessment *between* different datasets just as, in a previous era, exploratory data analysis allowed *within* dataset assessment to take place. Such assessment and interpretation will become essential in an era in which the relative importance of conventional, governmental data providers is set to diminish. It seems clear that tomorrow's science will be increasingly driven by complex interactions, as data become increasingly commodified, technology increasingly indispensible to science, and conclusions increasingly consensual. New philosophies of science that reflect today's realities are already overdue.

These changes are profound and far-reaching, but they provide grounds for cautious optimism about the future of GIS-based spatial analysis. The established self-perception of rigour among spatial analysts has hitherto been to some extent misplaced, in that data quality, resolution, and richness have not always been commensurate with the sophistication of spatial analytic methods. However nostalgically we may at times now view it, the linear project design was by no means a panacea in practice.

If science and problem-solving are to be constrained by these new realities, then what kinds of spatial analysis are most likely to dominate in the coming years? The points raised in this chapter's discussion suggest that the future environment will favour the following:

1 data whose meanings are clearly understood, making it easier for multidisciplinary teams to collaborate;
2 data which are routinely collected in the day-to-day functioning of society and the everyday interactions between humans and computers;
3 data with widespread use, generating demands that can justify the costs of creation and maintenance;
4 data with commercial as well as scientific and problem-solving value, allowing costs to be shared across many sectors;
5 methods of analysis with commercial applications, making it more likely that such methods will be implemented in widely available form;
6 methods implemented using general standards, allowing them to be linked to other methods using common standards and protocols.

References

Bailey T C, Gatrell A C 1995 *Interactive spatial data analysis*. New York, John Wiley & Sons Inc.

Berry B J L, Marble D F (eds) 1968 *Spatial analysis: a reader in statistical geography*. Englewood Cliffs, Prentice-Hall

Birkin M 1995 Customer targeting, geodemographics and lifestyle approaches. In Longley P, Clarke G (eds) *GIS for business and service planning*. Cambridge (UK), GeoInformation International

Buehler K, McKee L (eds) 1996 *The Open GIS guide*. Wayland, The Open GIS Consortium Inc.

Chorley R J, Haggett P 1965 Trend surface models in geographical research. *Transactions of the Institute of British Geographers* 37: 47–67

Chrisman N R 1997 *Exploring geographic information systems*. New York, John Wiley & Sons Inc.

Edwards G, Lowell K E 1996 Modeling uncertainty in photointerpreted boundaries. *Photogrammetric Engineering and Remote Sensing* 62: 377–91

FGDC 1994 *Content standards for digital geospatial metadata*. Washington DC, Federal Geographic Data Committee, Department of the Interior. *http://www.fgdc.gov*

Fotheringham A S, Rogerson P A (eds) 1994 *Spatial analysis and GIS*. London, Taylor and Francis

Goodchild M F 1988 A spatial analytic perspective on geographical information systems. *International Journal of Geographical Information Systems* 1: 327–34

Goodchild M F, Gopal S 1989 *Accuracy of spatial databases*. London, Taylor and Francis

Goodchild M F, Haining R P, Wise S 1992 Integrating GIS and spatial analysis: problems and possibilities. *International Journal of Geographical Information Systems* 6: 407–23

Haining R P 1990 *Spatial data analysis in the social and environmental sciences*. Cambridge (UK), Cambridge University Press

Harman H H 1976 *Modern factor analysis*. Chicago, University of Chicago Press

Hartshorne R 1959 *Perspective on the nature of geography*. Chicago, Rand-McNally/London, John Murray

Levine G 1981 *Introductory statistics for psychology: the logic and the methods*. New York, Academic Press

Longley P, Clarke G 1995 Applied geographical information systems: developments and prospects. In Longley P, Clarke G (eds) *GIS for business and service planning*. Cambridge (UK), GeoInformation International: 3–9

Longley P, Higgs G, Martin D 1994 The predictive use of GIS to model property valuations. *International Journal of Geographical Information Systems* 8: 217–35

Marascuilo L A, Levin J R 1983 *Multivariate statistics in the social sciences: a researcher's guide*. Monterey, Brooks/Cole

Martin D 1996 *Geographic information systems: socioeconomic applications*. London, Routledge

Martin D, Longley P, Higgs G 1994 The use of GIS in the analysis of diverse urban databases. *Computers, Environment, and Urban Systems* 18: 55–66

Mounsey H 1991 Multisource, multinational environmental GIS: lessons learned from CORINE. In Maguire D J, Goodchild M F, Rhind D W (eds) *Geographical information systems: principles and applications*. Harlow, Longman/New York, John Wiley & Sons Inc. Vol. 2: 185–200

Siegel S 1956 *Nonparametric statistics for the behavioral sciences*. New York, McGraw-Hill

Smith T R, Andresen D, Carver L, Dolin R, and others 1996 A digital library for geographically referenced materials. *Computer* 29(5): 54, 29(7): 14

Taylor P J 1977 *Quantitative methods in geography: an introduction to spatial analysis*. Boston, Houghton Mifflin

Tobler W R 1970 A computer movie simulating urban growth in the Detroit region. *Economic Geography* supplement 46: 234–40

Tukey J W 1970 *Exploratory data analysis*. Reading (USA), Addison-Wesley

Unwin D J 1981a *Introductory spatial analysis*. London, Methuen

Wrigley N 1985 *Categorical data analysis for geographers and environmental scientists*. Harlow and New York, Longman

Introduction

THE EDITORS

Management is what differentiates success from failure wherever GIS is used in any kind of operational context. It can also make a crucial difference in a research environment. Appropriately, a whole new Section has been added to this Second Edition. Given the nature of GIS, it is however impossible to cover all aspects of GIS-related management under one header.

The very term 'management' is redolent of many different interpretations – not least because famous writers like Hamel and Prahalad, Handy, Mintzberg, Porter, or Peters take somewhat different views (and, in some cases, have changed them in successive books). For some people, management still implies a 'top-down' control culture akin to that of the former Soviet Union. For others still employed in traditional government bodies, where the chill wind of the New Public management (Foster and Plowden 1996) has not yet penetrated, it means monitoring of the administration which is carried out within explicit and unchanging rules. For some management gurus who argue for continuous reform, downsizing, and massive outsourcing – sometimes apparently on the basis of sensationalism and idiosyncratic case studies – it is all about 'up front' leadership to secure the future of the firm and its profitability for shareholder benefit. For others, management is all about changing workplace culture and ensuring staff become partners in organisational improvement and lifelong learning.

In this book, a catholic view of management is taken. This is illustrated in Figure 1 which shows some of the main concerns of any able, well-informed, and concerned manager who has anything whatsoever to do with GIS.

MAKING THE GIS EFFICIENT, EFFECTIVE, AND SAFE TO USE

In the first instance, the manager will have a proper concern with the definition of the GIS and the specification of its mandatory and desirable characteristics, including any necessary assessment of the predicted costs and benefits and hence with an investment appraisal. The initial or replacement procurement process is one which needs to be managed very carefully, especially in a public sector environment where transparency of process and fairness is a high priority – yet which must not be allowed to lead to undue delays. These issues are addressed in the first part of this section in the contributions by Tor Bernhardsen, Nancy Obermeyer, and Larry Sugarbaker. The very way in which an organisation perceives and treats the GIS will also influence the likelihood of its success; Heather Campbell summarises three different ways of thinking about the nature of a GIS in an institutional context. Finally, the operations of any system may seem routine but these can cause catastrophic loss of business, data, or reputation if not carried out effectively and with due regard for the errors certain to occur in data (see the contributions to the Data Quality Section of the Principles Part of Volume 1) or the software and for the local legal framework. Gary Hunter and Harlan Onsrud deal with the management of uncertainty and handling of legal liability, respectively.

DATA AS A MANAGEMENT ISSUE

As technology becomes ever more ubiquitous – more and more GIS have the same functionality and the

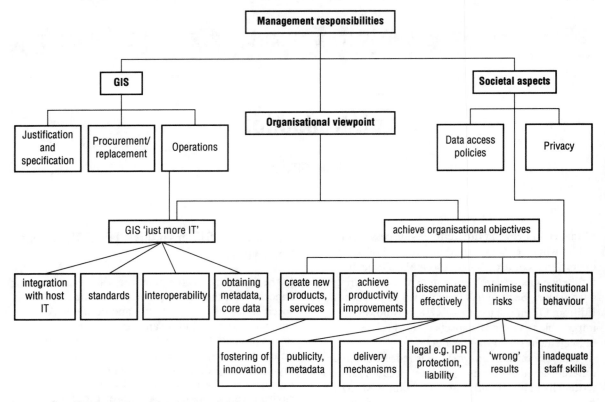

Fig 1. Issues which may face or impact upon all GIS managers.

same software is seen in offices from Albania to Zimbabwe – a key determinant of the ability to use GIS is the availability of data. Whilst any organisation will probably generate some data from its own operational activities, it is almost certain that some 'core data' will best be obtained from other, specialist organisations. Neil Smith and David Rhind, and John Estes and Thomas Loveland describe the sources of 'geographical framework' data and remote sensing imagery. Stephen Guptill summarises recent developments in metadata – data about data and where to find it – and Francois Salgé summarises the confusing multiple attempts to agree international standards relating to geographical information which, once implemented, should facilitate interoperability (i.e. the ability for different systems to work together, especially in a distributed and networked environment, and for data to be transferred safely and easily between systems: see Sondheim et al, Chapter 24).

GIS AS A MANAGEMENT TOOL

To any senior employee, however, the contribution of GIS towards sustaining the whole organisation has to be the major, perhaps overwhelming, concern. Here the relevant question is 'what role can GIS play in achieving my organisation's strategic objectives, helping us to meet our targets, and remaining sustainable and ever more prosperous?' To answer this question sensibly, it needs to be subdivided into others such as:

- How can we do the things we are already doing more effectively and efficiently?
- How can we define and make new products or services which will be valued by our customers or users?
- How best do we ensure that these customers and users know and understand what is available?
- How best can the products and services be delivered?
- How can the organisation be protected against the myriad risks it faces, such as: theft of physical

goods or of intellectual property rights; the risk of liability for nonconforming products; the need to obtain, motivate, and retain staff with appropriate skills; and the need to recognise, minimise, or avoid errors in results produced by the system?

These questions are addressed through the development of three case studies of how GIS have been employed to meet the strategic objectives of different organisations – some real corporations and some virtual entities. Martin Birkin, Graham Clarke, and Martin Clarke describe how GIS tools have been used to assess market potential for a variety of different retail organisations. In contrast, Michael Shiffer sets out the GIS approaches he has employed in a planning context where highly interactive consultation and interaction with the public is a crucial factor. Finally, Jane Smith Patterson and Karen Siderelis translate management issues upwards from the shaping and sustaining of a single organisation to the same concerns with an entire political body, the State of North Carolina, USA.

THE IMPACT OF BROAD SOCIETAL ISSUES ON GIS

Beyond all these organisation-specific issues there are broad societal ones which impact, often indirectly, on management at almost all levels (see Figure 1). Education is one of the most fundamental of contemporary human issues; in GIS its importance goes far beyond technical training. Pip Forer and David Unwin begin by asking 'education for what'? Predicated upon their view of the future utility and use of GIS and the impact of the new technologies described in earlier chapters, they summarise likely or desirable educational developments at all levels. Another fundamental contemporary issue is the right of an individual to privacy. What this means in practice and how geographical information or GIS relate to privacy form questions currently being debated in many countries. In contrast to the very pragmatic approach of Branscomb (1994), Michael Curry here provides an elegant philosophical chapter on GIS and privacy. The last chapter in the Section deals with data policy as espoused by different governments. These differ considerably both in principle and practice and this has many consequences for GI users, software vendors, and the state as a whole. David Rhind describes how the

key issues differ and what might well happen in the future. To a substantial degree – as in the case of new technological developments and intellectual property rights – what happens in the GIS field is at the very least strongly influenced by other actions, and so the wider public policy context is also described.

ACADEMIA, COMMERCE, AND GOVERNMENT AS 'BUSINESS'

In all of the above, the reader could be excused for believing that the focus has been solely on commercial enterprises. Use of words such as 'business' and even 'management' could have fostered such a mistaken view. In practice, however, there has been a substantial convergence between the natures of government, academia, and commercial enterprises in many countries in recent years.

University staff have become increasingly dependent on competitive bidding for research and consultancy projects; heads of the same organisations or departments within them are running what are in effect multi-million dollar enterprises. On the other hand, universities have pioneered the reality of 'networking' and forging of international partnerships which are now common in commerce and becoming appreciated even in government. In national and local government, the New Public Management has become manifest in countries as far apart as Canada, Finland, New Zealand, Sweden, and the United Kingdom (see Foster and Plowden 1996; Rhind 1997). Driven by a common desire to cut back on public expenditure, the result has been to conceptualise government of all types being made up of policy formulation and service delivery functions (the latter of which do not need to be inside government, simply controlled by it). Associated with this is the view that users of the services are in effect customers, paying for service through their taxes, and that it is proper to make public assessments of performance on this basis. Even where this view (of the citizen and organisations as customers) has not been widely adopted – as in the USA – there has been much downsizing of public-sector organisations and encouragement to contract out functions to the private sector and otherwise collaborate with firms in that sector. Hence there are many commonalties between the 'new government' and commercial enterprises – both are in business although some of the ground rules differ. Equally, however, governments and firms are coming to appreciate the ability of the best universities to innovate and to manage highly talented groups of creative individuals who are essential for success in a rapidly changing environment.

THE INADEQUACY OF CLASSIFICATIONS

No classification is ideal for all purposes. Each of the chapters in this Section makes contributions to many of the themes identified above. For example, the nature, provision, and success of education and training is relevant to the individual, to the organisation, and to society as a whole – as is reflected in Figure 1. As a more specific example, the common theme of the law links the chapters by Tor Bernhardsen, Harlan Onsrud, and David Rhind. There is then significant commonalty of

interest in management at all levels in all types of organisations and at most levels within them. GIS can only prosper if it can produce what management requires; equally and increasingly, senior management cannot achieve what they desire without effective (and hence well-managed) contributions from the ubiquitous GIS. That all of this takes place within a wider and ever changing societal context – over which the individual has little control – is what makes management a constantly challenging task.

REFERENCES

Branscomb A W 1994 *Who owns information?* New York, Basic Books

Foster C D, Plowden F J 1996 *The state under stress.* Buckingham, Open University Press

Rhind D W 1997 *Framework for the world.* Cambridge (UK), GeoInformation International

Introduction

THE EDITORS

There is no point whatsoever having a GIS if it does not work efficiently, effectively, and reliably. No organisation functioning in a competitive environment – in commerce, academia, or even government (where benchmarking of performance is becoming more common) – can afford to carry such burdens for long. For this reason, the decisions of whether to have a GIS, which one to have and how best to operate it are central management issues.

Tor Bernhardsen, a senior consultant in the Norwegian firm of Asplan Viak and an author of a successful book on GIS, initiates this Section (Chapter 41). He describes the strategic and tactical issues involved in selection and procurement of a GIS. As the availability of commercial off-the-shelf (COTS) systems has improved, as costs for given functionality have fallen, and as the average size of the enterprises obtaining GIS has decreased, so the approaches to purchasing of systems have evolved. Yet, despite COTS, tailoring of software is normally employed and the 'final' form of the system is rarely easy to define accurately at the outset. As a consequence, much effort is now devoted to selecting a supplier in whom the purchaser has confidence and with whom a constructive and honest relationship, bringing mutual benefits, is readily possible.

Nancy Obermeyer's contribution (Chapter 42) deals with predicting the costs and benefits of use of a GIS. The relationship with Bernhardsen's concerns is obvious. There are some differences in emphasis between the public and private sectors in the weights given to different accounting approaches. In general, however, the formal treatment of tangible benefits is similar and involves assigning a price to the benefits. Yet assigning costs and benefits to the more intangible consequences of implementing GIS is more difficult, especially insofar as general societal benefits are rarely easy to value. Traditional government and private sector organisations tend to take different attitudes to unmeasurable but potentially society-wide benefits!

Assume that senior managers have been convinced by the business case for procurement of a GIS and that selection and installation of the system has proceeded well. How then is the system best managed to ensure it is efficient and effective? Larry Sugarbaker – a veteran of running successful GIS installations – describes the management processes and procedures which underlie reliable and efficient operations (Chapter 43). Given both the number of things that can detract from good performance and that no GIS is unchanged for ever, high order project management skills are required. The need for good relationships between the GIS management – if it is separately managed – and that in the rest of the organisation is also stressed. Finally, 'future proofing' is considered from a management perspective.

If Bernhardsen, Obermeyer, and Sugarbaker deal with the everyday (yet vital) considerations of managers, institutional success requires more than a concern with these issues alone. Heather Campbell relates the findings of much research into the influence of how organisations perceive their information systems (Chapter 44). Those who regard GIS and other information technology as capable of being run simply on 'cookbook' instructions are doomed to failure. She outlines three different managerial perspectives – technological determinism, managerial rationalism, and social interactionism – and argues that adopting the last of these materially enhances the chances of GIS success within an organisation.

As GIS use increases, 'second generation' problems are starting to emerge. Data producers, system vendors, and integrators – and even users – are increasingly concerned with uncertainty in the systems and results. Gary Hunter discusses management approaches to cope with data, software, or human errors or from their interactions with the objective of minimising their consequences and reducing risk (Chapter 45). In so doing, he draws upon other chapters in the Principles and Technical Issues Parts (e.g. Beard and Buttenfield, Chapter 15; Egenhofer and Kuhn, Chapter 28).

For any operational system which produces results which impact on individuals or groups outside the organisation, a significant (and growing) concern is the legal liability for these results. Harlan Onsrud shows how liability is a creation of the law to cope with injurious behaviour, encourage the fulfilment of contractual obligations, and assign losses to those responsible for them (Chapter 46). He outlines some of the key areas in which liability is likely to be incurred in GIS, using scenarios to illustrate the law of tort and contract. Even outside the American context, resort to law is likely to become more commonplace in GIS-related activities.

42

Measuring the benefits and costs of GIS

N J OBERMEYER

The purchase and implementation of a GIS, like that of any other relatively expensive decision-support technology, is not a trivial matter for most organisations, whether they are public or private. Successful implementation of GIS requires a rather substantial commitment in organisational money, staff, and effort. Most organisations will make such a commitment only if the expected payoff justifies it. Justification usually begins with an effort to identify and then assign a price to the benefits and costs of adopting a GIS. It often ends by comparing the benefits with the costs, in what has become known as a benefit–cost (or cost–benefit) analysis. This chapter describes the techniques for measuring the benefits and costs of GIS. The chapter begins by describing the basics of benefit–cost analysis, including a discussion of the traditional and quantifiable costs and benefits in the GIS context. It continues with a detailed discussion of discounting. The chapter concludes with a discussion of the intangible costs and benefits of GIS implementation.

1 BENEFIT–COST BASICS

The terms benefit–cost and cost–benefit are used more or less interchangeably in the literature describing this technique. Using 'benefit–cost' has its advantages, however; the most obvious of these is that this word order implies that benefits outweigh costs in justified projects. Moreover, some scholars (e.g. Zerbe and Dively 1994) have argued that the term 'benefit–cost' implies a richer analysis than the alternative phrase. 'Benefit–cost', therefore, is the phrase used in this chapter.

The end of the twentieth century has brought with it a growing emphasis on economic efficiency within both public and private organisations. 'Doing more with less', 'downsizing', and 'rightsizing' have all become part of everyday language as euphemisms for budget cutting and layoffs (Foster and Plowden 1996; Rhind 1997). In response to calls for increased efficiency, organisations of all types must now provide more reliable and defensible justification for every purchase or new initiative they undertake. Benefit–cost analysis is often the first line of defence

in assuring bosses that organisational GIS initiatives are justifiable; indeed, some organisations require it (Huxhold and Levinsohn 1995).

While private organisations may use benefit–cost analysis, it is most commonly used in the public sector; the private sector may more easily rely on market prices and basic principles of cost recovery to assess the economic validity of implementing a GIS (see Birkin et al, Chapter 51). The thorniest issues related to benefit–cost analysis are those arising from peculiarities in the public sector that make it difficult (and in some cases, impossible) to establish accurate market prices for their products and to externalise many of their costs, as organisations in the private sector can do.

GIS (along with other information technologies) have never been a better buy than they are today. Declining prices have accompanied an explosion in the computing power of GIS and systems that have become remarkably user-friendly in recent years (see Longley et al, Chapter 1). The potential of GIS to increase overall efficiency and productivity in organisations that rely on geographically-referenced data has never been greater. Ironically, the current economic environment also means that the need to justify the purchase and implementation of GIS also has probably never been greater.

The use of benefit–cost analysis as a justification for adopting GIS is well established in the GIS literature (e.g. Aronoff 1989; Dickinson and Calkins 1988, 1990; Grimshaw 1994; Huxhold 1991; Huxhold and Levinsohn 1995; Smith and Tomlinson 1992). Traditional benefit–cost analysis as an economic exercise begins with an organisation identifying the costs associated with implementing a GIS (e.g. costs of hardware, software, transformation of maps and data into digital format, and adding or training of staff), along with the expected benefits of using the technology (greater efficiency and effectiveness, for example). The next step in the analysis requires that the organisation assign economic value (by price) to both the costs and the benefits, sum each of them, then compare the results arithmetically. If the value of the benefits exceeds the value of the costs, there is justification for making the purchase. Benefit–cost analysis typically covers a multiple-year period. This is particularly true for the organisation attempting to justify the implementation of a GIS, because of the high early costs and (potentially) enduring benefits of the technology.

Benefit–cost analysis has, however, received criticism as a 'dogmatic approach that knows the price of everything and the value of nothing' (Zerbe and Dively 1994). Not surprisingly, more sophisticated justifications approach the benefit–cost analysis as an art that recognises the importance of organisational ethics and values as well as the need to consider more interesting and complex questions of economic theory (Zerbe and Dively 1994; and see Campbell, Chapter 44). Huxhold and Levinsohn (1995) recommended an examination of the financial, technical, and institutional feasibility as an alternative to benefit–cost analysis; Grimshaw (1994) suggested a value-added approach to justify a GIS.

1.1 Elements of a benefit–cost analysis

A typical benefit–cost analysis contains several elements. The most rudimentary element is the identification and assignment of a numerical economic value to the benefits and costs associated with an initiative. Costs should include any expense incurred as a result of implementing the project: purchase of any hardware, software, or supplies, the costs of hiring any additional staff or the training of existing staff, along with the cost of transforming maps and data into digital format (see Bernhardsen, Chapter 41). Costs of these types are classified as 'tangible' costs. Tangible costs are readily quantifiable, primarily because they represent costs of products that are bought and sold in the free market. Other products are not so readily quantifiable, creating thorny problems in calculating the benefit–cost ratio.

Along with costs, there are some benefits that the organisation can also classify as tangible. For example, if the organisation expects to be able to reduce its workforce because of the increased efficiencies that the implementation of the technology promises, the numerical value of the salary or wages and benefits of staff members will be both available and quantifiable. Similarly, if the organisation will be able to produce more detailed or more diverse information and information products as a result of implementing a GIS, it may also be able to improve its overall effectiveness. Thus the 'first cut' of benefit–cost analysis is the easiest – quantifying the tangible costs and benefits. As one might expect, the analysis usually becomes much more complicated thereafter.

Huxhold (1991) claimed that there are three major categories of benefits of GIS that should be examined: cost reduction; cost avoidance; and increased revenue. Aronoff (1989) identified five categories: increased efficiency; new non-marketable services; new marketable services; better decisions; and intangible benefits. Aronoff's ideas of increased efficiency and new marketable services broadly correspond to Huxhold's notions of cost reduction and increased revenue, respectively. It is important to recognise, however, that price reductions made possible by the lower costs associated with GIS implementation may actually stimulate demand for some geographical information products. This can result in increased revenues overall because of increased volumes of sales.

Huxhold defined cost reduction as 'the decrease in operating expenses of the organisation, primarily caused by a savings in time by operating personnel performing their tasks more efficiently' (Huxhold 1991). Cost reductions generally accrue because of the improved productivity of staff members responsible for the tasks performed using the GIS.

Cost avoidance is the 'prevention of rising costs in the future caused by projected increases in workload' per staff member (Huxhold 1991). This benefit is consistent with, and more or less an extension of, the first benefit – suggesting that, once a GIS becomes part of an organisation's equipment, it may help to optimise the performance of a variety of both current and future tasks. The improvement in performance may make it unnecessary to hire new employees or at least to postpone such appointments by making the best use of existing employees.

Finally, Huxhold suggested that 'a GIS can increase revenues . . . by selling data and maps, increasing property tax collections, and improving the quality of data used to apply for state and federal grants'. The rationalisation of tasks that the GIS makes possible does indeed bode well for the increase in tax collections and the improvement in data quality. However, Dansby (1991) has pointed out that there may be legal impediments to the sale of such products in the public sector, depending on national, state, and local regulations on copyrights and freedom of information (see Rhind, Chapter 56).

New non-marketable services are 'useful products and services that were previously unavailable' and will be used within the organisation (Aronoff 1989). Aronoff pointed out that the organisation could reasonably anticipate some of these benefits of GIS. Other benefits, however, will not typically become apparent until after the GIS is up and running. Therefore, it will normally be difficult to assess accurately the value of non-market services and include it in the benefit–cost analysis to justify adopting a GIS.

As noted, implementation of GIS will make possible the sale of new geographical information products. These new products are the result of the inherent ability of GIS to extract and combine data in a variety of combinations and permutations, essentially enabling its implementers to deliver customised geographical information products on demand. For example, a city government with a comprehensive, large-scale GIS with current, accurate information can quickly produce a map of vacant downtown retail space for an individual wishing to open a bookstore, along with a table identifying the property owners.

Aronoff (1989) also suggested that the adoption of a GIS would produce 'better decisions'. This will occur, he argued, because 'more accurate information and faster and more flexible analysis capabilities can improve the decision-making process itself'. Again,

determining the economic value of 'better decisions' resulting from GIS adoption is problematic. The large body of literature on organisational decision-making takes a more realistic view, essentially conceding that most decisions are made on the basis of incomplete information (see, for example, Cyert and March 1963; Douglas 1986; Downs 1967; Simon 1945/1976). In many cases, organisations deliberately limit their searches for information because of time and/or financial constraints. In other instances, organisations may be unaware of additional relevant information (seeking and using information incur costs). A GIS cannot eliminate these institutional factors (see Campbell, Chapter 44).

1.2 Variations on basic benefit–cost analysis

There are several variations on benefit–cost analysis: one is cost-effectiveness analysis (Layard and Glaister 1994). Cost-effectiveness analysis provides a comparison of the costs of providing a specific outcome, or performing a specific task, using different means. In adding this step to the benefit–cost analysis, the organisation would compare alternative means of performing the same task; for example, the cost of providing information on property ownership both with and without a GIS. Implicitly, adding this step forces the organisation to demonstrate not just that the benefits of its initiative outweigh the costs, but that a specific strategy for performing a specific task is more cost-effective than other strategies (Layard and Glaister 1994).

Another variation is the calculation of the 'payback period' (Huxhold 1991). This is derived by dividing the total cost of implementing a GIS by the estimated annual benefits of using the system. The resulting figure reveals how many years it would take to accumulate enough benefits to pay for the cost of the system (Huxhold 1991). The benefits may include any or all of the benefits described earlier in this chapter. Not surprisingly, this calculation is fraught with the same difficulties apparent in typical benefit–cost analyses.

Grimshaw (1994) endorsed a third variation, the value-added approach. This approach emphasises the new things technology enables the organisation to do and what it adds to the capacity or worth of the organisation, echoing and extending Aronoff's non-marketable services.

257

1.3 Refinements of benefit–cost analysis

Several other problems arise in performing benefit–cost analysis, some of which apply across the board, others of which are unique to the public sector. There are several refinements of the process to address these difficulties.

1.3.1 Stakeholders

The first of these is the problem of stakeholders (Layard and Glaister 1994; Sen 1994; Zerbe and Dively 1994). Within the context of any organisation's mission, there is a variety of individuals and/or groups who have an interest (or a 'stake') in what the organisation does and the strategies it employs. The most obvious example is that an organisation's customers or clients form a crucial component in its survival (Obermeyer 1990; Weber 1946). The costs and benefits of the actions of an organisation may not be identical for all individuals or groups with a stake in the organisation's actions.

For example, a company whose mission is to produce road maps will include among its stakeholders individuals and groups with varying needs for map detail. The average user who has found the company's maps to be excellent navigation aides is unlikely to be impressed by the company making a decision to provide more detailed maps if that additional detail comes at a higher price. If the company has a competitor which produces a map comparable to the original map at a price lower than the 'new and improved' (and more expensive) version, the company may lose market share and perhaps suffer declining revenues overall as a result of the decision to offer greater detail at a higher price.

The stakeholder problem is even more complex in the public sector where levels of income among end-users vary greatly (Layard and Glaister 1994). For example, a professional nature photographer who can afford to hire a native guide to lead him or her to the lair of an endangered animal (and may also be able to deduct the cost of the guide as a legitimate business expense for tax purposes) has no real need for a detailed, large-scale map of the area. On the other hand, a PhD student trying to study that same animal would probably find such a map to be essential. Thus, trying to anticipate the costs and benefits of all stakeholders can become a complicated, if not impossible, task.

Certainly, a manager cannot afford to ignore the organisation's various stakeholders. In so doing, he or she risks alienating existing and potential customers and clients. Moreover, the organisation may miss an opportunity to report higher benefits arising from its ability to enhance the satisfaction level of existing stakeholders, or by increasing the actual number of stakeholders reported in its benefit–cost analysis. The flexibility of GIS may make it possible for both private and public organisations to increase their product lines and fill new market niches at relatively small additional costs and, as a result, increase their customer and client bases by appealing to a wider audience.

It is up to the manager to estimate the expected value of these potential benefits and include them in the analysis. For example, the director of a local planning agency can build a case for a GIS by first identifying, then estimating, the value of the GIS to the local government itself following, for example, Huxhold's categories. However, the availability of a large-scale, comprehensive GIS will also benefit local utilities, developers, and private businesses by making accessible high-quality 'official' geographical information products that these groups can then use to inform their own decisions and to help in their day-to-day operations.

Not surprisingly, some local governments have exploited the relationship with their stakeholders by working cooperatively with groups such as local utilities and business leaders to build and implement their GIS. For example, the Cincinnati Area GIS (Cincinnati, Ohio, USA) is a joint venture of the city and county governments, the telephone company, the local power and water companies, and local industry (which includes Proctor & Gamble) (Obermeyer 1995). Working with stakeholders has the added advantage of sharing costs among the participants and improving the level of benefits as a result of the specific functional expertise – and data – that each participant brings to the project.

1.3.2 Time and discounting

A second problem that arises in performing benefit–cost analysis is caused by the effects of time and economic inflation (Field 1994; Layard and Glaister 1994; Little and Mirrless 1994; Smith and Tomlinson 1992; Stiglitz 1994). Even when the rate of inflation is low, over time the cumulative effects of inflation erode the economic value of the costs and benefits of any activity. Moreover, people perceive immediate benefits as having greater value than benefits far off in the future. As Zerbe and Dively (1994) put it, 'a benefit received today is worth more than one in the future'. Similarly, a cost that occurs far in the future has less significance than a similar cost today (Field 1994). In order to provide a realistic assessment of costs and benefits, organisations must take this into account and adjust their benefit–cost analysis calculations accordingly.

A refinement designed to address this problem is *discounting* (Field 1994; Smith and Tomlinson 1992). The idea behind discounting is to deflate the costs and benefits in order to remove the effects of inflation. Discounting is needed to provide an accurate assessment of the value of implementing a GIS because of the multi-year life expectancy of a GIS and the result that GIS costs and benefits are also spread over multiple years.

Discounting is not a simple matter, particularly with GIS which typically have their largest outlays early in the life of the project then experience declining costs, but whose benefits can last long into the future. Front-end costs include the purchase of hardware and software and either hiring new staff or paying to educate existing staff (see Sugarbaker, Chapter 43). In addition, organisations in some countries can expect high start-up costs arising from the need to convert analogue (paper) maps into digital form. These start-up costs are likely to seem insurmountable for many small and medium-sized organisations. The perception of insurmountable costs may be compounded in local government by the recognition that they will not begin to realise the benefits of a GIS for several years.

Discounting applies to both costs *and* benefits. Its primary purpose is to aggregate a series of costs and/or benefits which occur over the life of a project. The formula for discounting includes three elements: present (or future) value, the length of time appropriate for the project and an appropriate discount rate (Field 1994).

$$\text{Present value} = \frac{\text{Future value}}{(1 + \text{discount rate})} \text{ years}$$

As a worked example, consider that we need to calculate the present value of US$1000 ten years in the future with bank interest rates at 5 per cent. The formula is applied as follows (Field 1994):

$$\text{Present value} = \frac{\$1000}{(1 + .05)^{10}} = \$613.90$$

Multi-year projects are handled as in the following example, a hypothetical GIS implementation. Assume that the costs and benefits for the seven years of the lifetime of the project are as shown in Table 1, and that the discount rate is 6 per cent.

Table 1 The hypothetical costs and benefits of a GIS.

| | Year | | | | | | |
	1	2	3	4	5	6	7
Costs	100 000	70 000	50 000	25 000	25 000	25 000	25 000
Benefits	0	25 000	50 000	70 000	70 000	70 000	70 000

Using the figures in Table 1, the present values of costs can be calculated using the following formula (Field 1994):

$$PV_{cost} = \$100\,000 + \frac{70\,000}{1+.06} + \frac{50\,000}{(1+.06)^2} + \frac{25\,000}{(1+.06)^3} + \frac{25\,000}{(1+.06)^4} + \frac{25\,000}{(1+.06)^5} + \frac{25\,000}{(1+.06)^6}$$

The present values of benefits may be calculated using the following formula (Field 1994):

$$PV_{benefit} = \$0 + \frac{25\,000}{1+.06} + \frac{50\,000}{(1+.06)^2} + \frac{70\,000}{(1+.06)^3} + \frac{70\,000}{(1+.06)^4} + \frac{70\,000}{(1+.06)^5} + \frac{70\,000}{(1+.06)^6}$$

Choosing an appropriate discount rate is itself not a simple matter. First, there is the issue of real versus nominal interest rates. Nominal interest rates are the actual interest rates available in the market. In order to know the real interest rates, it is necessary to adjust these nominal figures for inflation. For example, if the nominal interest rate is 8 per cent, but the average rate of inflation over the period in question is 3 per cent, then the real interest rate is 5 per cent (Field 1994). In all instances, managers must always consistently use either real costs and real discount rates, or nominal costs and rates (Field 1994).

The plethora of interest rates in use in the world of modern finances complicates the process of discounting. A review of the business/finance section of any reputable newspaper shows a large variety of interest rates from which to choose: rates on normal savings accounts, certificates of deposit, bank loans, and government bonds, to name just a few. There are two views on this issue. The first view suggests that the discount rate should reflect the way people think about time and money. Economists refer to this as the rate of time preference. For example, most people would prefer receiving $1 today, rather than waiting ten years to receive that same amount. This is a positive rate of time preference. Those who support this view would use the average interest rate on a bank savings account as their discount rate (Field 1994).

The second approach to choosing a discount rate is based on the notion of investment productivity. In this view, people anticipate that the value of future returns will offset the cost of investment today. In the public sector, this means that expenditures used for long-term projects should yield rates of return to society that are similar to what the same expenditures could have earned in the private sector (Field 1994). Using this reasoning, an organisation should use a discount rate that reflects the rate which banks charge their investment borrowers; these rates are typically higher than savings account rates (Field 1994).

The nature of this debate on discount rates ultimately ensures that it is up to the manager to choose – and justify – an appropriate discount rate unless the organisation as a whole has well-established rules on how to proceed. One resolution is to perform a sensitivity analysis by repeating the discounting of benefits and costs using two or more different interest rates.

It is not difficult to grasp the impediment that discounting imposes on a benefit–cost analysis for GIS. The high start-up costs of GIS will seem even higher than they are in light of the positive rate of time preference. On the other hand, the benefits of GIS will typically seem smaller after discounting. If one carries out the calculations on the hypothetical seven-year GIS implementation example provided above, it will take the entire period for benefits to begin to outweigh costs. A real-life GIS may take even longer to reach the break-even point.

It is, however, important to remember that the benefits of GIS are often enduring. Once an organisation has paid the high front-end costs, particularly those associated with higher staffing costs and digitising, it should reap the benefits of the technology year in and year out unless the operational needs change dramatically (and that may even provide a large benefit from the use of GIS if the new needs can also be met) – but see Bernhardsen (Chapter 41) and Sugarbaker (Chapter 43) for discussions of how GIS investment should be amortised for accounting purposes. Emphasising the enduring nature of the benefits of GIS can be accomplished by carrying out the analysis for as many years as are required to achieve a favourable benefit–cost ratio. In addition, however, the manager should also make it clear that digitising is a one-time-only expense. Finally, the manager should point out that the investment in GIS is likely to endure for generations to come. Whether this is accepted by management may depend on the level of risk involved.

1.3.3 Uncertainty and risk

Time also influences the level of risk and uncertainty among the benefits and costs of an organisation's initiatives. Humans do not possess perfect knowledge about the present, and it is even more unrealistic to expect them to foresee the future with complete accuracy. Zerbe and Dively (1994) identified two types of uncertainty: uncertainty caused by the unpredictability of future events; and uncertainty caused by limitations on the precision of data (see Fisher, Chapter 13). Both types of uncertainty are relevant to GIS benefit–cost analysis, particularly in the past. Throughout much of the time since GIS has become commercialised, there has been a great deal of uncertainty about both the costs and the benefits of using the technology. For example, lack of experience in the early days of commercial marketing of GIS meant that many organisations underestimated the long-term costs of the implementation such as digitisation costs, consultation fees, and training expenses which often far surpassed initial estimates. Today's turn-key GIS products (such as Maptitude and ArcView) enable GIS adopters to know with greater certainty the cost of the basic package. However, there remains a great deal of uncertainty associated with other critical elements of GIS start-up, namely hiring and/or training staff, digitising maps, and gathering and entering data to customise the GIS (see Bernhardsen, Chapter 41).

In evaluating the wisdom of purchasing a GIS, both the benefits and costs of implementation may be difficult to assess because of the uncertainty surrounding them. It is well known and generally accepted that the costs of implementing a GIS extend beyond the purchase of hardware and software. For example, assembling and maintaining data – along with the training of staff – are two areas that require expenditures after the initial purchase of the GIS. The exact dollar amount of these additional costs is usually difficult to know ahead of time. However, as Smith and Tomlinson (1992) optimistically noted, 'the costs [associated with implementing a GIS] are loaded heavily in the early period whereas the benefits increase . . . and then remain constant'. This assumes a stable organisation and external environment. The wise manager will prepare for unexpected contingencies throughout the life of the system.

These uncertainties surrounding the calculation of benefits and costs of implementing a GIS have been the subject of discussion by several authors (e.g. Aronoff 1989; Huxhold 1991). There are several approaches to handling uncertainty in benefit–cost analyses. The first is to ignore it, which is appropriate if the uncertainty is likely to be minor or where the analysis is intended only to be a rough estimate. It may also be possible to reduce uncertainty by gathering additional information and the organisation should make every reasonable effort so to do. The project manager should also talk with other, similar organisations which have implemented GIS in order to add to their knowledge base. Finally, the organisation can recognise uncertainty and include it in the benefit–cost analysis explicitly (Zerbe and Dively 1994).

1.3.4 Selling data

The sale of geographical information products is often suggested as a benefit to be included in the benefit–cost analysis. Properly managed, these benefits can indeed be significant. For example Rhind (1997) reported that Great Britain's Ordnance Survey generates US $100 million in annual revenues through the sale of geographical information products. Ownership of the copyright to datasets is a prerequisite to having the right to make such sales (Rhind, Chapter 56). In most countries, the national government holds the copyright to all the datasets they develop; the US Federal Government is an exception to this rule (although US cities and states may copyright data).

Difficulties in establishing prices for geographical information products can complicate the assessment of likely benefits, but an organisation can compare its geographical information products with similar products offered for sale by the private sector in order to establish a basic price list. Once products are officially offered for sale, the organisation can adjust the price to try to achieve its desired sales and revenue goals. The sale of data and other geographical information products may expose an organisation to liability risks arising from negative outcomes associated with unintended uses or deliberate abuse of the products (see Onsrud, Chapter 46). The wise manager will consult with the organisation's legal department to resolve these issues in advance of making any commitments.

In short, organisations contemplating the sale of data as a benefit of their GIS should be aware of the pitfalls as well as the benefits. The potential rewards certainly warrant the sale of geographical information products if the organisation is permitted to do so. Rhind (Chapter 56) and Smith and Rhind (Chapter 47) discuss policy aspects regarding the sale of data in more detail.

1.3.5 Externalities and spillovers

Externalities and spillover effects are mirror-image problems that may arise in developing benefit–cost analyses. Externalities arise when a company shifts its costs outside the organisation, usually by ignoring a problem (Papageorgiou 1978). Externalities are particularly troublesome for public institutions which are limited in their ability to externalise. Yet frequently these organisations are involved in cleaning up problems created when private organisations externalise their costs. For example, in the USA the Federal Government has assumed responsibility for cleaning up toxic waste dumps created by the private industry. It is true that the government could ignore the problem but this strategy could lead to problems cropping up elsewhere, for instance in the overall health of people living near the sites. These are ramifications that private companies can – and often do – ignore.

Spillover effects, or positive externalities, are the benefits that an organisation enjoys because the activities of another organisation extend beyond its jurisdictional boundary (Faulhaber 1975). Private companies often enjoy the spillover effects created by public expenditure (e.g. transportation networks, sewer, and water projects), just as some public agencies may benefit from the activities of private companies or other jurisdictions. For example, a GIS firm that includes government census data with its software is able to add value to its product and thus receives tangible economic (spillover) benefits from the data gathering and dissemination activities of the government.

Handling externalities and spillover benefits in the benefit–cost analysis is a matter that merits attention. In the case of governments which are performing a benefit–cost analysis as a prelude to their implementation of a GIS, Smith and Tomlinson (1992) recommended incorporating 'all benefits . . . in the analysis whether or not they accrue to the potential GIS purchaser or the departments that will use the information products'. Among the non-government groups that may realistically expect to benefit from the implementation of a government GIS are taxpayers, private companies, and special service districts.

How does an organisation handle these externalities and spillovers? First, it is necessary to identify them. Perhaps the most significant externality of a GIS is the potential loss of privacy associated with the ability of GIS to disaggregate data (see Curry, Chapter 55). Large public datasets based on national censuses are most likely to raise privacy concerns; however, some private firms have collected large databases that may also threaten the privacy of individuals. It is extremely difficult to place a value on this potential loss of privacy to an individual. Is it $1 per person? £10? More? In this instance, managers are left to make their own assessment.

Spillover effects of a GIS, as Smith and Tomlinson (1992) noted, may accrue to taxpayers and others as they reap the benefits of readily accessible maps, data, and other geographical information products made possible because of the implementation of a GIS. Spillovers, while still problematic, are somewhat easier – and obviously more pleasant – to handle than disbenefits engendered by the export of problems by others. For example, the county assessor might anticipate shorter transaction times for fulfilling requests for basic information, such as a property registration map. In order to assess the value of these time-savings to customers and clients, one should multiply the average number of annual transactions by the economic value of the anticipated time-savings per transaction, which in turn is based on the average hourly wage figure for those involved. Given the range of beneficiaries of spillover effects, there is great value in paying careful attention to assigning benefits to spillovers. Governments in particular, since they have a broad (and in some cases nearly universal) set of stakeholders, can bolster their anticipated benefits by considering spillovers. Whether this is a relevant consideration varies according to government policy (see Rhind, Chapter 56).

1.3.6 Intangible benefits and costs

Many of the benefits and costs that contribute to the development of a benefit–cost analysis are intangible. For example, how can one place a numerical economic value on increased reliability or diminished institutional confusion? Smith and Tomlinson (1992) defined intangibles as '. . . not as much a separate category of benefits as they are a class of benefits that is more difficult to quantify'. The benefits might include such things as better internal communication in the organisation, improved morale, and a better public image. Obviously, placing a specific dollar (or Deutschmark or franc or pound) value on these intangible benefits is not possible. It is, however, still necessary to give an estimate. Organisations may begin by describing these potential benefits and costs in text accompanying the benefit–cost analysis.

Assigning an economic value to intangible benefits is part of the art of the benefit–cost analysis. Assigning such value may be accomplished by using surrogates. For example, improved morale may result in reduced staff turnover, which in turn results in lower costs for personnel searches and training. These items are easier to value than is morale.

Organisations may experience negative changes as they implement GIS (Grimshaw 1994; Huxhold and Levinsohn 1995), resulting in additional intangible benefits and costs. For example, an organisation may find that, as it introduces GIS, those who are most knowledgeable become more important to the organisation; conversely, those who are slow to adapt to the technology may find themselves losing ground and, eventually, their jobs. Some tasks may become deskilled, leading to staff unhappiness. The overall result may be institutional confusion which may in turn temporarily cause a drop in productivity. While a manager might find it impossible to place a precise economic value on institutional disarray, assigning an economic value to time lost to the disruption of the social order of the organisation is easier to do.

Conversely, organisations may find that their foray into the world of GIS may give them increased visibility and an enhanced reputation. For example, the US Department of Housing and Urban Development is attempting to solidify its client base by collaborating with the Caliper Corporation to develop and sell its Consolidated Planning Software GIS program. Similarly, the US Geological Survey noted the value to society of improved decisions made possible by its many mapping products (Bernknopf et al 1993). In the light of the zeal of the calls for downsizing the public sector, solidifying this relationship makes sound organisational sense, given the importance of the relationship between organisations and their client groups (Obermeyer 1990; Weber 1946).

Again, in assigning an economic value to these intangible benefits, organisations need to take a wide perspective. For example, public organisations that make available low-cost or even free geographical information products to citizens might place a value on the goodwill they generate through these actions by calculating the aggregate cost savings that their customers or clients received by using the organisation's products rather than more expensive commercial alternatives. This is fraught with some dangers and must only be carried out with the advice of professional accountants – for instance, such action may actually damage the local private sector, creating externalities of another kind.

Assigning values to intangible benefits and costs can thus be difficult. In the case of benefits, it is extremely important to do so in order to accumulate benefits to offset costs as part of the analysis. In the case of costs, it is necessary to do so in order to achieve fair and honest results. As noted, this part of the analysis is as much art as it is science. Nevertheless, through careful thought, an organisation can usually assign plausible and defensible values to these intangibles, as suggested by Table 2.

Table 2 Summary of the costs and benefits potentially gained through use of a GIS.

Category	Costs	Benefits
Economic (tangible)	Hardware Software Training New staff Additional space	Cost reduction Cost avoidance Increased revenues New market services
Institutional (intangible)	Internal personnel shifts Layoffs	Improved client relationships Better decisions Improved morale

2 CONCLUSION

Benefit–cost analysis is the preferred method to justify the implementation of a GIS, particularly in public organisations. The GIS must be assessed in comparison with existing practices and technology already used in the organisation. In addition, the organisation must consider other alternative means of performing the same tasks. Making these comparisons requires the organisation to perform a separate benefit–cost analysis for each alternative under consideration. The alternative with the highest ratio of benefits to costs is the most efficient one although other factors like cash flow may influence which option is finally chosen.

All of the above should have made it clear that benefit–cost analysis plays an important role in providing an economic rationale for an organisation's decision to adopt a GIS. Performing such analyses is not always easy but it is necessity for some organisations and it is advisable for all others. Taking the time and effort to perform a thoughtful, careful benefit–cost analysis is a first step in building a secure foundation for a successful GIS implementation.

References

Aronoff S 1989 *Geographic information systems: a management perspective*. Ottawa, WDL Publications: 260–1

Bernknopf R L, Brookshire D L, Soller D R, McKee M J, Sutter J F, Matti J C, Campbell R H 1993 *Societal value of geological maps*. Washington DC, US Geological Survey

Cyert R M, March J G 1963 *A behavioral theory of the firm*. Englewood Cliffs, Prentice-Hall

Dansby B 1991 Recovering GIS development costs by copyright use. *GIS World* 4(2): 100–1

Dickinson H J, Calkins H W 1988 The economic evaluation of implementing a GIS. *International Journal of Geographical Information Systems* 2: 307–27

Dickinson H J, Calkins H W 1990 Concerning the economic evaluation of implementing a GIS. *International Journal of Geographical Information Systems* 4: 211–12

Douglas M 1986 *How institutions think*. Syracuse, Syracuse University Press

Downs A 1967 *Inside bureaucracy*. Boston, Little Brown & Co. Inc.

Faulhaber G R 1967 Cross-subsidisation: pricing in public enterprises. *American Economic Review* 65: 966–77

Field B C 1994 *Environmental economics: an introduction*. New York, McGraw-Hill: 119–23

Foster C D, Plowden F J 1996 *The state under stress*. Buckingham, Open University Press

Grimshaw D J 1994 (reprint 1997) *Bringing geographic information systems into business*. Harlow, Longman/ Cambridge (UK), GeoInformation International: 121

Huxhold W E 1991 *An introduction to urban geographic information systems*. New York, Oxford University Press

Huxhold W E, Levinsohn A G 1995 *Managing geographic information system projects*. New York, Oxford University Press

Layard R, Glaister S (eds) 1994 *Cost–benefit analysis*, 2nd edition. New York, Cambridge University Press: 21

Little R, Mirless J 1994 The costs and benefits of analysis: project appraisal and planning 20 years on. In Layard R, Glaister S (eds) *Cost–benefit analysis*, 2nd edition. New York, Cambridge University Press: 199–234

Obermeyer N J 1990 *Bureaucrats, clients and geography*. Chicago, University of Chicago Geography Research Papers, #216

Obermeyer N J 1995 Ameliorating inter-organisational conflict in order to share geographic information. In Rushton G, Goodchild M (eds) *Organisations sharing geographic information*. New Brunswick, Center for Urban Planning Research

Papageorgiou G J 1978 Spatial externalities: parts I and II. *Annals of the Association of American Geographers* 68: 465–92

Rhind D W 1997 Facing the challenges: redesigning and rebuilding Ordnance Survey. In Rhind D (ed.) *Framework for the world.* Cambridge (UK), GeoInformation International: 275–304

Sen A K 1994 Shadow prices and markets: feasibility constraints. In Layard R, Glaister S (eds) *Cost–benefit analysis,* 2nd edition. New York, Cambridge University Press: 59–99

Simon H A 1945/1976 *Administrative behavior: a study of decision-making processes in administrative organisations.* New York, Free Press

Smith D A, Tomlinson R F 1992 Assessing the costs and benefits of geographical information systems: methodological and implementation issues. *International Journal of Geographical Information Systems* 6: 247–56

Stiglitz D 1994 Discount rates: the rate of discount for benefit–cost analysis and the theory of the second best. In Layard R, Glaister S (eds) *Cost–benefit analysis,* 2nd edition. New York, Cambridge University Press: 116–59

Weber M 1946 Bureaucracy. In Gerth H H, Wright Mills C (ed. and trans.) *From Max Weber: essays in sociology.* New York, Oxford University Press: 196–244

Zerbe R O Jr, Dively D D 1994 *Benefit–cost analysis in theory and practice.* New York, HarperCollin

43

Managing an operational GIS

L J SUGARBAKER

The importance of GIS is ultimately judged by how successful they are for the operational purposes for which they have been installed. Good management practices are essential to ensure success. This chapter describes the basic elements of managing and progressively replacing a GIS. It describes the crucial need for customer support, the need to provide support for operations, data management, and applications development. The need for sound project management is stressed. Workload planning and budgeting – the latter varying with stages in the life-cycle of the GIS – are both described, as is the desirable relationship between GIS management and that for the organisation as a whole. Finally, the management implications of likely future changes in technology and societal norms are anticipated.

1 INTRODUCTION

The organisational structure and management direction will determine the type of GIS organisation to be created (see also Campbell, Chapter 44). The three types of installations include corporate (or enterprise), departmental, and project GIS.

The corporate GIS is used by multiple programmes to support mission-critical functions across an organisation. Management responsibility normally lies with a central information technology-support organisation. The corporate GIS requires extensive planning since all of the control and operations functions are needed. Complex business requirements that span multiple departments contribute significantly to the system complexity. Business requirements are satisfied by creating software applications that utilise a corporate database management system linked to the GIS. The benefits of this corporate management approach are often measured in terms of overall productivity or profitability improvements. With the corporate approach, there will also be benefits that are accrued at the department level. While the management overhead cost may exceed other implementation strategies, it may be the only way to achieve these corporate benefits.

The departmental GIS supports one critical business area of an organisation. It is managed within the department that it supports. Often, the departmental GIS receives network, computing, and database services from a central information technology organisation. This management approach is most appropriate when GIS functions within the organisation are only needed by one department. Consequently, all of the benefits and costs attributed to this technology are allocated to the department where it is managed.

The GIS project is a third type of installation. A GIS project has a well-defined deliverable which is a product and not an operational system. There are beginning and ending dates and, upon completion, staff are assigned to other activities. The benefits of this type of installation are determined from the value of the product which is produced. The entire project cost must be considered in the cost–benefit analysis. Residual (depreciated) value of equipment can be recovered upon its sale or transfer to other projects. The GIS project is often managed without oversight or guidance from a central information technology organisation. Consequently, there is little chance of a project GIS evolving into a departmental or enterprise GIS.

2 MANAGING AN OPERATIONAL SYSTEM

Every operational information system has a predictable life-cycle. An organisation has requirements to automate certain business functions. A project is planned and the business application is developed. Following development, the system is put into operation – see Bernhardsen (Chapter 21) and Maguire (Chapter 25) for details of alternative modes of system implementation. Over time, business needs change and technology advances offer new approaches for meeting these needs. Typically, old systems are then replaced and the cycle starts all over again, as shown in Figure 1.

The particular management approach which is adopted can have dramatic impacts upon the characteristics of the GIS life-cycle. From a business perspective, it is highly desirable to modify systems as business requirements change. Further, the cycle of development and operation can create major swings in budget and overall system performance. A disciplined management approach should involve a continuous process of enhancing systems as business needs and technology change. This is easier said than done, although two key factors affect management ability to achieve this objective. First, today's technology is modular – so it should not be necessary to change an entire infrastructure in order to achieve incremental benefits. The second factor is more a function of the management approach. The disciplined manager will be in tune with changing business strategies (see Birkin et al, Chapter 51). Organisational workload planning will facilitate the

change process. Proposed projects are prioritised by their ability to support business needs. It is incumbent upon the manager to propose technology infrastructure projects which meet these demanding business expectations.

GIS management functions include customer support, operations, data management, and application development and support. GIS organisations may also have staff dedicated to project management. Every GIS requires staff support for these functions. The large 'GIS shop' may have multiple staff positions supporting each of these functions. It may organise along these functional lines as well. Staff at smaller installations may perform two or more of the support functions. Typically, when there are fewer than seven or eight employees in a GIS organisation there is little need for multiple supervisors or a multiple level organisational structure. It is important, however, to have clear definitions of support roles and responsibilities for all staff. It is quite common to organise along functional lines. Staff are then assigned to projects that cross all of these lines in order to create a matrix organisation. A different approach is to organise by project teams. Staff are brought together to support all of the functions for a particular project. Upon completion of a project, staff are assigned to new teams.

2.1 Customer support

The customer support function will involve many activities in a typical GIS organisation. Critical to the success of any GIS operation is a customer help

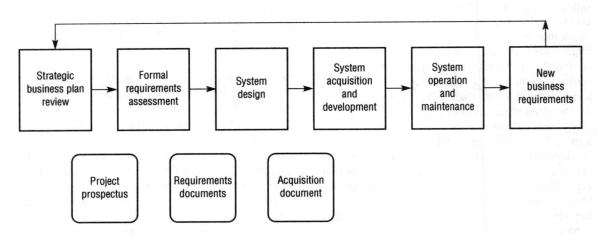

Fig 1. The information system life-cycle.

desk that logs inquiries and responds to customer requests for information. This is usually the first customer contact point and is where impressions of the organisation are formulated. Geographical data requests are usually handled as a support function in this unit. A customer support unit may either arrange for training or provide training on an organisation's internal procedures and GIS software applications. The nature of almost every GIS installation is that a certain amount of ad hoc reporting and special projects are inevitable. This is highly desirable because a GIS should be designed to handle questions which deal with strategic issues. The GIS is then viewed as a tool that supports strategic change, rather than one that imposes rigid institutional structure. A customer support function within any organisation should therefore include GIS analysts who know the data and the technology.

The help desk function can be a very simple call logging and follow-up activity or it can be a problem resolution centre with high-level analysts solving problems as they become known. A smaller organisation will often rotate the help desk function among the staff. Assistance for any given problem is then provided by the individual most capable of resolving an issue. Regardless of how this service is provided, it is useful to log incoming calls and formally close out a call when an issue is resolved. This serves two important purposes. First, the chances that a customer's problem will go unresolved increases as more people become involved. A formal call logging and close-out process insures that customers receive feedback. Second, if problem resolutions are logged as well, the source of reoccurring problems can be identified and actions taken to reduce the level of customer problems. This can take a number of forms, such as increased training, correction of software bugs, or technology infrastructure improvements.

A GIS manager who recognises the value of highly trained staff and users will be in an excellent position to provide quality service. Information technology continues to be very complex and the rate of change is rapid. GIS is no exception. Typically, system administrators and software development engineers will receive training from the system and software vendors. There is, however, an increasing number of firms that provide this specialised training. Customer training to use applications and to perform rudimentary GIS

analysis is often provided by GIS support staff. Training courses can be acquired from the vendors or developed in house.

Applications analysts are called upon to perform complex GIS analysis and create products from the GIS. These specialists need to have a comprehensive understanding of the GIS capabilities and the data resources necessary to perform the most complex analysis. It is always helpful if these individuals also know the business operation well enough to assist in finding creative solutions to real business problems.

2.2 Operations support

System operations include system administration, maintenance, security, backups, technology acquisitions, and a host of other support functions. System administration forms the cornerstone of this important operations function. The systems administrator is often the operations manager or, in a very large operation, the administrator will have key technology responsibilities. Clear service delivery requirements must be understood and written in an operations service document or plan. The operations manager will be challenged on a daily basis to maintain a high up-time operation. GIS installations have peripheral equipment and requirements which are unique to the GIS technology. For example, the network bandwidth requirements used to support spatial data are high. Peripherals such as large scanners and digitisers have their own unique requirements. Operation services such as network support often come from multiple parts of an information organisation. For this reason, it is important that these requirements are well-documented in the operations plan (see also Bernhardsen, Chapter 41).

The operations service document should include service hours and scheduled down-time for system maintenance. Target up-time as well as procedures for reporting system problems form an important part of the document. Without this, the customer's expectation will be for 24 hours per day, seven days a week service with no system failures. Closer examination shows why this is important. Very high up-time system performance is possible but it comes at a high price. Thus the customers need to be a part of the planning process to determine service hours and up-time requirements. It is best to address these issues in the project planning phases rather than when system failures occur. The service document should accurately describe the full range of services that is offered, such as which vendor software products are supported, how new services are requested, and how off-hours support can be obtained if needed.

System security comprises a large number of activities and it is important for a variety of reasons. The common notion is that data security is most compromised by individuals with destructive intentions. But the reality is that a very high percentage of data corruption and losses are caused by inadvertent actions or system failures. So system security involves locks and passwords, but it also includes thought-through procedures to protect data and other system components against inadvertent destruction. Regular system data backups, for example, must be completed according to a plan. How often a backup is taken is a function of the database and the requirements for currency. For example, a banking institution will require a 'point in time' recovery capability since losing *any* customer transaction would be disastrous. GIS almost always have a recovery requirement that is somewhat less demanding than a banking institution. System backups are normally cycled to an off-site location when they become two or three generations old. The first generation backup is usually kept on site so quick recoveries can be completed when a user inadvertently destroys a data file. The second generation backup will be available from the off-site location in the event that a major disaster results in total destruction of the database. It is essential that multiple generations are kept since, in the extreme, a system virus may be planted in a file with a timed release. More typical, however, is that corrupted files resulting from user or application errors go undiscovered for more than one generation of backup. A secure system will also have software environments established for production software as well as development and testing. A programmer check-in and check-out procedure will help secure the production environment.

Other system operation duties are more mundane. Staff routinely install new releases of vendor software, investigate problems, and incorporate patches into the administered software environment. Network address and user identifications are assigned. Systems are monitored for performance and loading. Disk space is managed and projections for when new storage space is needed are maintained.

New hardware and software installations are completed. Maintenance contracts are managed. Peripheral equipment is cleaned and supplies are ordered. All of this may sensibly be carried out within procedures defined within a quality system, perhaps based on ISO 9000 standards.

2.3 Data management support

Base geographical data within a GIS are often managed as part of the system infrastructure. Many of the other chapters in this book entail discussions about data, but the emphasis here will be upon an organisational management perspective. Most information technology applications create data through on-line user transactions that build or modify a database one record at a time. In other cases, the application processes a transaction through a system and upon completion it is discarded. Summary level data may be all that are retained. It is true that many GIS also serve these kinds of function. Examples include permit processing, delivery dispatching, and a host of other applications. GIS, however, have a special requirement for base data upon which customer applications depend. For example, a base data layer of transportation is needed for a delivery company to calculate service routes. The same base transportation data are needed to produce road maps, process building permits, complete environmental assessments, plan road maintenance, and meet many other needs. For this reason, it makes sense to manage the base data in a GIS as part of the basic infrastructure. Indeed software vendors are catching onto this as they begin to bundle their products with base geographical data (see Batty, Chapter 21; Elshaw Thrall and Thrall, Chapter 23).

The concept that base data are a part of the infrastructure is not new. In fact, mapping organisations around the world have produced base cartographic products in hard copy form for years. Many organisations have continued this support function by converting to automated cartography or GIS during the past 20 years. The conversion process is nearing completion in many parts of the world. Now the attention has shifted to maintaining base data and creating ways to move this valuable product throughout the information infrastructure without compromising its integrity. Other terms, such as framework, are evolving to describe the base data (see Guptill, Chapter 49). The concepts of framework are different today in that they also address the issue of maintenance and data sharing (see Smith and Rhind, Chapter 47; Rhind, Chapter 56). Conversely, some of the original methods of moving paper products between and through organisations today are no longer relevant.

If base data are part of the GIS infrastructure, then it follows that data management strategies and procedures are an important component of GIS operations. The GIS base data layers may include the geodetic control, digital terrain, orthoimagery, transportation, boundaries, hydrology, cadastral, and natural resources data (Mapping Science Committee 1995). Historically, organisations have acquired these data from mapping organisations. They have managed the data as part of their internal cartographic support function. This made sense when the final product was a hard copy map. Today, the desired product is the database and the map is a derivative. Many organisations are restructuring to reflect this shift which has been created by advances in GIS and database technologies. The respective roles of the mapping and information technology functions within an organisation thus becomes a strategic direction issue. The primary business functions served will ultimately influence the management solution within any given organisation.

Geographical data collection and support functions are changing in the GIS industry. Today, new base data are being collected by virtually every GIS organisation. This is no longer a task reserved for the map-making organisations. New technologies are also emerging that will further proliferate the potential sources of usable digital data (see Batty, Chapter 21; Lange and Gilbert, Chapter 33). Organisations are still needed to support the base data requirements but their role in the future will be quite different (Morrison 1997). Operations support may mean that coordination, developing standards, and facilitating partnerships becomes a primary activity (Salgé, Chapter 50). Meanwhile, the GIS manager is caught up in this exciting time of reinventing the way data are created, managed, and used (see Goodchild and Longley, Chapter 40).

There are many organisational approaches to the creation and management of base geographical data. Historically – at least in the USA – base data have been digitised from existing maps by multiple organisations as their internal needs dictated. This has resulted in duplicated efforts as datasets are created to meet different business requirements for content and accuracy. While this practice will continue to some degree, the prevailing sentiment will be to economise and work with other organisations to create a common data infrastructure. This opens opportunities to privatise certain functions or to develop consortia to manage base geographical data. Many public organisations seem likely to let contracts to the private sector to collect base data. Upon completion, the public organisation may manage the data and provide information services or copies of the data to other organisations. This may be done for a fee intended to recover costs of development and management. Private sector companies that provide base data services for profit are emerging in this market place as well.

Data administration and database administration are important support functions within any GIS organisation. Depending on the size of the GIS organisation, this function can be part of the GIS data production support activity, the applications support, or a stand-alone function. A smaller organisation will often combine both of these functions into a single position. A database administrator is responsible for ensuring that all stored data meet standards of accuracy, integrity, and compatibility established by the organisation. The database administrator is normally responsible for designing physical storage from logical data models and consulting with developers on accessing data from the database management systems. The data administrator is also responsible for the planning of the organisation's data and metadata (data about data) resources. This individual will consult with business experts in determining information technology project requirements. The data administrator often uses Computer Aided Systems Engineering (CASE) tools to construct logical data models. Data dictionary maintenance typically falls under the responsibility of this position.

2.4 Application development and support

Software application support staff have the responsibility to create new applications to support business needs. The application development process requires that developers continuously learn new software tools. They must be able to adapt to the ever-changing software infrastructure that supports them. A larger organisation may have specialists in spatial and tabular data as well as software engineering. Other organisations may limit the complexity of application they develop or hire contractors to build applications. Regardless of the detail, the application development environment will continue to change very rapidly. This change catalyst means application and development requirements will add complexity to the GIS management environment.

Software documentation standards and procedures are critical to the long-term support of the GIS. A quality project plan for development should allow time to follow these conventions. It is generally too easy to drop documentation requirements or squeeze development time in order to complete a project on time. The long-term impact of these management decisions will affect the ability of a product to be supported over time. From another perspective, over-zealous developers should not insist on exacting procedures and complex development methodologies for an application that has a short life expectancy. This only adds unnecessary costs to a project.

From a management perspective, it is desirable to assign developers full time to a new project. When it is complete, staff reassignment would then be made to the next project. Yet the manager rarely has the luxury of doing this and it is rare that a developer has just one project going at a time. In addition, applications developers usually have software maintenance responsibilities for existing applications. A formal change control process becomes an important management tool for scheduling this type of work. When customers require changes to their applications, it is normal to assume that these take priority over other activities. A closer examination may reveal several things. Changes which are enhancement requests should be scrutinised for priority in the same way as other, completely new, projects. When a system has a large number of change requests on file, this is an indication that either the maintenance support function is understaffed and/or the application is obsolete. A global look at the change request log may reveal that a new application would be more cost-effective.

2.5 Project management

GIS-user organisations will almost always have multiple projects active at any given time. In a smaller organisation, the GIS manager may serve as the project manager and assign tasks to staff as needed to complete work. The project management function becomes increasingly important as projects become larger or more complex. Within organisations where there are multiple large projects, project management specialists may be assigned to oversee staff, to manage budgets, and to direct the completion of all project elements.

3 WORKLOAD PLANNING

The GIS manager has an ongoing responsibility to assign staff to projects and to provide the support necessary to maintain a viable operation. There will always be more work than there are staff resources to accomplish it. Without workload management, there will be little organisational discipline. Staff may tend to work on the things they enjoy doing. Likewise, the GIS manager will likely be responding to operational problems rather than leading the organisation towards its strategic objectives. While workload planning never really ends, there should be a periodic planning activity that coincides with budget development. Customers are brought into the process and a survey of anticipated work is completed. Workload estimates are completed and all proposed projects are prioritised relative to the organisation's strategic business plan. Since it is unlikely that there will be enough funds or institutional capacity to do all projects, it is critical that each project prospectus is evaluated. The prospectus should contain all of the information necessary to complete this evaluation. If prospective customers do not have a project prospectus, it is likely that they have not thought the proposal through. There will also be other projects that are in later stages of development which will require prioritisation and continued funding.

The preliminary workload plan will be organised according to the type of activity to be supported. For example, the plan may include system operations, customer support, application maintenance, and new projects. The workload plan is then linked to budget elements. An example of a workload plan is shown in Table 1. When budgets are set, the plan is adjusted to reflect organisational priorities. A review of the completed plan will reveal a great deal about a mature GIS and its role in an organisation. The plan should include a good mix of activities necessary to maintain a viable organisation. For example, there should be planned projects to

Table 1 GIS organisation workplan, courtesy of State of Washington, Department of Natural Resources.

Geographic Information Section PROJECT NAME/DESCRIPTION	DIV.	FY96 Staff Months	FY97 Staff Months	IT Board	Agency Priority	Status	Project Phase	Strategic Goal, Objective
SUPPORT ACTIVITIES								
Management/Support/IT Coord.	ITD	12.00	12.00			Ongoing		1.1.01
Computing Service	ITD	45.00	45.00			Ongoing		
Training Support	ITD	12.00	15.00			Ongoing		2.1.01
Consulting Support	ITD	24.00	24.00			Ongoing		2.1.01
Standard Data Products	ITD	12.00	12.00			Ongoing		1.1.01
Special Product Requests	ITD	24.00	24.00			Ongoing		1.1.01
Spatial Data Maintenance	ITD	24.00	24.00			Ongoing		1.1.01
Software Configuration Management	ITD	0.00	3.00			New		
Region GIS Coordination	ITD	42.00	42.00			Ongoing		4.1.01
WORKLOAD – Support Activities		195.00	201.00					
SOFTWARE MAINTENANCE								
System Utilities	ITD	6.00	4.00			Ongoing		
TRANS, HYDRO Update Applications	FPD	1.00	1.00			Ongoing		1.1.01
Cadastre Update Application	FRD	1.00	1.00			Ongoing		1.1.01
Land Use Cover Application	FRD	1.00	1.00			Ongoing		
Resource Management Applications	ITD	24.00	18.00			Ongoing		
DISPLAY Application	ITD	4.00	4.00			Ongoing		
Permit System Application	FPD	6.00	12.00			Ongoing		
Planning & Tracking Application	FRD	0.00	8.00			New		
Forest Inventory Application	FRD	12.00	12.00			Ongoing		
WORKLOAD – Maintenance Activities		55.00	61.00					
OPERATIONS PROJECTS								
Conversion to SOLARIS	ITD	6.00	6.00	New		Inactive	Complete	
DBMS Pilot/Transition Plan	ITD	15.00	15.00	Awareness		Active	Development	
Software Configuration Management	ITD		3.00	New		Active	Requirements	
DBMS Transition	ITD	8.00	8.00	Awareness		Active	Pilot	
Spatial Data Framework Pilot	ITD	10.00	10.00	New		Active	Prospectus	1.1.01
Map Display/Query	ITD	4.00	6.00	New		Inactive	Prospectus Needed	
Desktop Applications for Unix Users	ITD	1.00	1.00	New		Active	Pilot	
NAD83 Database Modification	ITD	0.00	3.00	New		Inactive	Prospectus	1.101
Data Dictionary – User Interface	ITD	0.00	3.00	New		Inactive	Prospectus Needed	
Windows NT – Investigation	ITD	3.00	1.00	New		Inactive	Prospectus Needed	
GPS Interface with Update Applications	ITD	1.00	3.00	New		Inactive	Prospectus Needed	
Spatial Metadata – FGDC Grant	ITD-DIS	2.00	2.00	New		Active	Implementation	1.1.01
WORKLOAD – Shared Data Production		50.00	61.00					
FRAMEWORK DATA PROJECTS								
DATA96	ITD	18.00	1.00	Awareness		Inactive	Complete	1.1.01, 4.1.01
Cadastre Framework	FRD	0.00	16.00	New		Active	Scoping	1.1.01, 4.1.01
Transportation Framework	FRD	0.00	3.00	New		Proposed	Prospectus Needed	1.1.01, 4.1.01
Hydrology Framework	FPD	0.00	0.00	New		Proposed	Prospectus Needed	1.1.01, 4.1.01
DEM Framework	ITD	3.00	0.00	New		Active	Development	1.1.01, 4.1.01
Orthophoto Framework	ED	0.00	6.00	New		Proposed	Prospectus Needed	1.1.01, 4.1.01
WORKLOAD – Shared Data Production		21.00	26.00					
BUSINESS APPLICATIONS								
Forest Harvest Permitting – Release 1.0	FPD	24.00	12.00	Mandated	2	Active	Implementation	4.1.01
Forest Harvest Permitting – Release 2.0	FPD	0.00	6.00	Mandated	2	Active	Construction	4.1.01
Planning and Tracking – Release 1.0	FRD	24.00	48.00	3 yr Pri.	5	Active	Construction	4.1.01
Forest Inventory – Release 2.0	FRD	0.00	0.00	3 yr Pri.	6	Active	Construction	4.1.01
Aquatic Ownership Data	ALD	0.25	0.25	3 yr Pri.	7	Active	Requirements	1.1.01, 4.1.01
Harvest Planning – Release 2.0	RPAM	9.00	9.00	Reviewed	8	Active	Development	4.1.01
Geology Data Layer	G&ERD	1.00	2.00	Awareness		Active	Construction	1.1.01, 4.1.01
Natural Heritage Data Layer	FRD	2.00	0.00	Awareness		Active	Requirements	1.1.01, 4.1.01
Reforestation – Nursery Automation	FRD	0.00	0.00	New		Active	Scoping	4.1.01
Resource Planning Data System	RPAM	0.00	0.00	New		Delayed	Prospectus	4.1.01
Landscape Planning	RPAM	0.00	0.00	New		Delayed	Prospectus	4.1.01
Agricultural Data Layer	AgRD	0.00	0.00	New		Active	Prospectus	4.1.01
WORKLOAD – Development Activities		60.25	77.25					
WORKLOAD – All Activities		**360.25**	**400.25**					

Note: All staff month allocations are initial assignments from planning in early 1995. Actual allocations have changed.
Project list includes updated information from staff planning exercises in February, 1996. Several are new proposals.
The FRAMEWORK98 project has been subdivided into a PILOT and several Data Layer related projects.

*Permanent positions in Geographic Information Section and Data Administration Section allocated to GIS support, operations and proposals.

modernise the technology infrastructure on a continuous basis. Nearly every piece of equipment in the GIS world has a useful life of three to five years. Assuming a straight line depreciation, about 20 per cent of the infrastructure should be replaced on an annual basis. This will only be accomplished if replacement is built into the workload plan. Further, an organisation's strategic business goals and objectives will almost certainly be continually defining new directions. There should be technology projects to support these strategic business plans. Finally, ongoing support functions cannot be ignored. An expanding customer base will require additional user and application support. The work planning process is not a simple task: the GIS manager should strive to have a balanced workload that supports each of the important system functions.

4 BUDGETING

Methods of budgeting for any service are always a topic of discussion. Likewise, there are probably as many variations of approaches as there are people in an organisation. Regardless of the budget method utilised, there is usually more work than money or time to do it. It is important that GIS managers understand and articulate the importance of various work activities to their supervisors. Many people in an organisation will be competing for resources in order to complete their work. The GIS manager needs a source of funds to support computing infrastructure. These funds will be the same funds needed to create GIS applications. If a balanced approach to funding infrastructure and application projects does not occur, then a fragmented GIS environment will result. It is important to bear in mind that – unlike some other information technology applications – GIS has a data component that is part of the infrastructure. The GIS workload planning must account for the ongoing costs associated with maintaining the database.

An operational budget will change over time as a system matures. The three primary components of an operational budget include staff, goods and services, and capital investments. Each category may be broken down into more refined tracking units that suit the needs of an organisation. The distribution of costs between these three elements

will shift over time as a system operation matures. In the early years, when initial investments are being made in establishing an infrastructure, capital costs are almost always higher. Likewise, initial goods and services costs are likely to be lower because first year maintenance costs are often included as part of the equipment purchase. Other goods and services costs are also lower since there is usually a smaller user base to support during the start-up period. Table 2 was developed from the author's own budget experience of managing an operational budget for more than 15 years. The per cent breakdown of the major budget element clearly shows a shift in budget over time. As indicated earlier, one important factor to recognise is that infrastructure improvements and replacements are an essential part of any budget. Too many managers assume that capital costs will be very low after the initial acquisitions are complete. The author's own experience and other case studies (see Dickinson and Calkins 1988) clearly suggest that this is not the case.

When contract services are a significant part of the budget, the costs elements will differ significantly from the example provided. Contract services may be acquired for operations support, equipment and software maintenance, application development, and database development and support. Most organisations use at least some level of contract services to support their GIS. It is important to recognise that there is a significant administrative cost for overseeing contracts and services provided by contractors. On the other hand, it is often easier to get access to high quality services through contracting than to hire and train new staff. Another benefit of contracting for some services is that highly specialised work can be accomplished on a schedule without detracting from already planned work.

There are two basic ways in which an organisation can manage a budget and account for

Table 2 Distribution GIS operational budget elements over time based on the author's experience.
(All figures are percentages of the total expenditure.)

Budget object	Year 1–2	Year 1–6	Year 12
Staff and benefits	30	46	51
Goods and services	26	30	27
Equipment & software	44	24	22
TOTAL	100	100	100

Courtesy of the State of Washington, Department of Natural Resources.

costs. The first way is that the GIS unit can receive a direct allocation of funds, along with a clear set of performance and output expectations. Second, there may be no allocation of funds but a pay-for-service (cost recovery) arrangement with other operating programmes. Each way has its advantages and disadvantages. There are many variations of these two approaches that may be used successfully. The cost recovery model is good in that costs become associated directly with the benefiting departments and customers have more control of how their budgets are allocated. The disadvantage of this approach is that corporate infrastructure, data administration, and management support costs are often viewed as unnecessary costs which are built into service fees. By contrast, the centralised GIS budget approach allows for key corporate level business needs to be planned for and met. These objectives and accountability can be more directly associated with budget. However, the GIS organisation that is given an operational budget runs the risk of being viewed as a cost centre. It could become a target every time business costs need to be reduced. A mixture of these two approaches may be the best approach for most GIS organisations. Budget elements which are associated directly to key corporate level business strategies can be allocated to the GIS organisation. Departmental level objectives can receive their own allocation. Departments then pay the GIS service organisation for costs which can be directly linked to services received. These costs would include usage fees for computing services, staff, and materials for maintaining or developing new applications.

5 MANAGEMENT OVERSIGHT

We often hear the GIS technician saying 'if management would only let me get my job done'. Likewise, managers may wish that GIS staff would work on the 'important' things. Regardless of these individual complaints, there is an appropriate level of management interaction with any organisation which is critical to its success. Many of these problems can be avoided when there is an ongoing dialogue between corporate management and the GIS organisation. There should be at least two fora to facilitate this dialogue. First, information technology strategic directions and key project decisions should be made in the context of a corporate technology vision. Steering committees or technology boards are often used to formulate these strategies and recommendations for corporate level decisions. A technology board should have membership representation from all levels of the organisation including customers, users, and corporate management, as shown in Figure 2. The technology board should understand the corporate business goals and it should have knowledge of information technology management principles. This is the organisation that makes final recommendations for corporate budget planning and funding of information technology activities. If this management structure is actively supported, the GIS manager should never be concerned about being in conflict with a corporate vision.

A second forum for communication with the management structure is an ongoing reporting and meeting function. Regular reports which give a clear picture of progress are essential. Progress reports should always include a chart of key project milestones and completion status relative to the project targets. Honest assessments should be made by the project manager about how identified risk factors are being managed. The progress reports become a tool which facilitates useful dialogue with management. It will be possible to address problems before they become too large to solve. Likewise, successes will also receive the attention they deserve (this is less common but is a characteristic of good management).

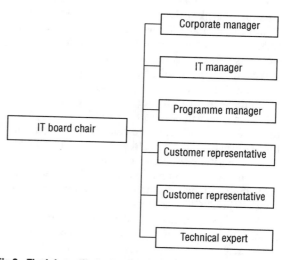

Fig 2. The information technology board: an example of membership and organisation.

6 INFORMATION TECHNOLOGY TRENDS AND THEIR MANAGEMENT IMPLICATIONS

Changes are occurring in and around the GIS industry which will fundamentally change the way work is accomplished. The last 20 years have seen this industry evolve to the point at which the technology is present in most natural resource management and larger government organisations within developed countries. Today, the technology is rapidly moving into the transport and retail industries, as well as small and rural government organisations (see Birkin et al, Chapter 51; Elshaw Thrall and Thrall, Chapter 23; Waters, Chapter 59). Developing nations are seeking ways to introduce the technology effectively into mainstream decision-making. As a result of all this, every GIS manager will be challenged continuously to stay current with technology and remain a leader within their own organisations.

Future trends which are likely to affect the spatial data community were discussed at a workshop in April 1996 (Mapping Science Committee 1997). One of the workshop activities was to identify the technology developments and societal changes that will likely impact on the spatial data community in the year 2010. Taken together, these predictions presented a very different prognosis for GIS technology. It is widely assumed that computing capacity will continue to grow and to get cheaper. A change that we may not be prepared for is that the time lag between data collection and its availability for use will approach zero as real-time data collection systems are realised. Intelligent instrumentation will guide us through real-life situations like driving to work. Or maybe we will all work from home. We will always know where our children are because they will be transmitting their locations in real time. On-board systems will be making decisions based on environmental factors and locations. But modern day systems will provide more than intelligent instrumentation and real-time data. GIS managers will need to work out how to cope with the massive flow of real-time data. Partnerships will involve everyone, not just a few major data producers. From a societal perspective, citizen involvement in government will be in real time. We will also still likely be dealing with privatisation and 'data for free or fee' issues (Rhind, Chapter 56).

As a consequence of all of the above, GIS management will inevitably change. When spatial data become an integral part of every information infrastructure, every information manager will be a GIS manager. Spatial literacy and awareness will open up educational opportunities to which the GIS manager must respond (Forer and Unwin, Chapter 54). The trends in information processing and utilisation in our business systems will continue to place demands on current thinking. Some managers will respond by developing strategies to lead this change. Others will try to slow it down. Thus managing an operational GIS will not be a static or routine endeavour.

This chapter was designed to spur a management realisation that operations continuously change. Projects will come and go. As a backcloth to this there will be an evolving information infrastructure. Strategic planning gives us the chance to examine these change agents. Workload planning allows us to add structure to the operations management process. Other readings have been identified in the reference section (see Aronoff 1989; Korte 1994; Huxhold and Levinsohn 1995; Martin 1991; Obermeyer and Pinto 1994) so that the GIS manager can explore these topics in greater detail.

References

Aronoff S 1989 *Geographic information systems: a management perspective*. Ottawa, WDL Publications

Dickinson H J, Calkins H W 1988 The economic evaluation of implementing a GIS. *International Journal of Geographical Information Systems* 2: 307–27

Korte G B 1994 *The GIS book: the smart manager's guide to purchasing, implementing and running a geographic information system*. Santa Fe, Onword Press

Huxhold W E, Levinsohn A G 1995 *Managing geographic information system projects*. New York, Oxford University Press

Mapping Science Committee 1995 *A data foundation for the national spatial data infrastructure*. Washington DC, National Academy Press

Mapping Science Committee 1997 *The future of spatial data and society: summary of a workshop*. Washington DC, National Academy Press

Martin J 1991 *Rapid application development*. New York, Macmillan Publishing Company

Morrison J L 1997 Topographic mapping in the twenty-first century. In Rhind D (ed.) *Framework for the world*. Cambridge (UK), GeoInformation International: 14–27

Obermeyer N J, Pinto J K 1994 *Managing geographic information systems*. New York, The Guilford

Introduction

THE EDITORS

Acquiring data is usually recognised as being one – perhaps the largest – of the cost elements of setting up a GIS. This Section therefore concentrates on several different aspects of this important management issue.

The first requirement of many users is for 'framework data' (Rhind 1997) – that which define the location of topographic and other key features in the natural, built, or cultural landscape in relation to each other. On this is 'draped' all other 'thematic' data. Neil Smith and David Rhind (Chapter 47) describe the nature and desirable characteristics of these framework data as well as their sources. They consider the role of high resolution remote sensing data as an alternative to existing digital maps and show that both have certain advantages. Finally, they point out that most frameworks to date have been constructed within a national context and that this inevitably leads to problems for those dealing with multinational areas. Few such trans-border problems afflict imagery obtained from satellites (though they suffer other problems); Jack Estes and Tom Loveland describe the potential and sources of such remotely-sensed data for monitoring and management of areas at minimum cost and maximal effectiveness (Chapter 48).

There are many data already in existence and held by governments and other bodies worldwide, although most sources will doubtless have imperfections (e.g. being more out-of-date than is desirable). The problem is often knowing what datasets do exist, who is responsible for them, and how to obtain them. Thus the creation of online metadata – data about data – services has come to be a high priority in many countries of late, not least because the existence of metadata should minimise the duplication of data collection and hence minimise the waste of scarce resources. Stephen Guptill summarises the important

early work done in the USA on metadata and the creation of data catalogues, leading to online 'clearinghouses' (Chapter 49). He points out, however, that we are still at a rather early stage in the evolution of metadata services, not least because we do not yet have tools for establishing whether particular datasets are 'fit for (a particular) purpose'.

The recognition of the need for standards relating to GIS data, let alone those for computer systems more generally, originated at least as early as 1969. Since then, massive resources have been devoted to the definition of standards worldwide, some of this work has been rendered obsolete through the growth of online working. Nevertheless, the importance of standards is universally acknowledged in order to facilitate easy, reliable, and cheap use of data compiled elsewhere and to allow systems created by different vendors to work in tandem. François Salgé (Chapter 50) describes some of the complexities arising from having many different players engaged in the 'standards business', ranging from national data providers, through national and international standards bodies, to the private sector. The result is inevitably complexity and some uncertainty – will Microsoft simply produce *de facto* standards used by everyone and will all the years of effort by many public sector bodies simply be wasted? He argues that the work of the European standards organisation CEN and that of the ISO will lay the foundations for many *de jure* standards for years to come.

Reference

Rhind D W 1997 *Framework for the world*. Cambridge (UK), GeoInformation International

Introduction

THE EDITORS

What can managers use GIS to achieve? Any attempt to answer this question would generate a list of infinite length. Here it is tackled through three case studies, chosen to illustrate the development of business and service applications of GIS which, as Maguire (Chapter 25) also comments in the introduction to the Applications Part of this volume, represents one of the major growth areas of GIS usage. Geographical information represents both a strategic and operational resource for a wide range of business and service organisations, and in many countries (notably the USA) a vast industry of 'value added resellers' (VARs) has grown up to service the needs of advanced information economies. Business and service organisations utilise such data products within GIS for a range of strategic and operational requirements, such as site and market area analysis, sales planning, market research, direct mailing, door-to-door leaflet campaigns, and planning for public services (see Shiffer, Chapter 52).

The term 'geodemographics' has come to be used as a label for small-area typologies derived from population census data that have been shown to exhibit an identifiable correspondence with geographical patterning in consumer behaviour. They remain the most commonly used data products in business information management, having developed out of pre-1960s academic research into geographical techniques for producing summary measures of small area variability in social, economic, and demographic conditions.

Yet recurrent problems with use of census information in such applications have included the outdatedness of infrequent (usually decennial) information, the strictures of data aggregation into (possibly heterogeneous) areas, and the need to rely upon imperfect surrogate data in the absence of variables which bear the most direct relationship with observed consumer behaviour – notably (in most countries) income. Add to these the data pricing policies of some national census agencies and it is possible to understand the recent impetus towards supplementation and/or replacement of census and conventional geodemographic classifications with so-called 'lifestyles' classifications. These are based upon customer surveys, guarantee card returns, and store loyalty programmes. One problem with such sources is that it is not possible precisely to quantify the degree to which they are representative of general populations, and this inevitably restricts their usefulness in many business and service applications. Goodchild and Longley (Chapter 40) discuss this problem in the context of the established scientific basis to geographical data collection.

The first case study in this Section is by Mark Birkin, Graham Clarke, and Martin Clarke (Chapter 51). This is based on their extensive experience within the GMAP firm in Leeds where they have exploited commercially the modelling and spatial analytical techniques they and their colleagues have devised over two decades of academic research (see also Getis, Chapter 16). They show how it is possible to predict important aspects of consumer behaviour. This has great attractions for car manufacturers and other multiple outlet retailers which are continually faced with closing down shops, opening new ones, and investing in improvements to others, taking into account the interactions with other outlets as well as the likely effect on revenues in any one shop. They show that the techniques used in many existing GIS are crude and unreliable but that considerable improvements can be wrought. It is, however, fair to observe that their approach to geospatial data management lies at the most sophisticated and expensive end of the spectrum of business and service planning applications – that is, they are strong advocates of a consultancy-based approach to problem-solving, in which specialised skills are purchased in order to solve strategic as well as operational management problems. Other, usually more mundane but more numerous, business and service applications are developing at the low-cost end of the market, based upon the break-up of GIS software into task-specific desktop systems (Elshaw Thrall and Thrall, Chapter 23) and the movement of GIS vendors into provision of shrink-wrapped limited functionality systems complete with applications data (Batty, Chapter 21).

In Chapter 52 Michael Shiffer's perspective upon decision-making in urban planning sees it as a key process in democratic societies: he argues that there is an important interface between democracy and operational management and that this, as such, reflects the operations of the state in microcosm. It is also becoming increasingly important as local communities come to assert their interests and values against those of officialdom or big business. Central to this interplay of interests is discourse between different parties. Management of planning is therefore a non-linear operation without line managers in any conventional business sense. His chapter describes how combinations of GIS and multimedia can facilitate the active involvement of many different groups in the discussion and management of urban change through planning. It deals with different components of public discourse, notably debate and consensus-building, and with communications between parties sometimes separated by space and time. Shiffer's own experiments clearly demonstrate the practical possibilities and difficulties in the use of these technologies in the planning context.

Governments operate at multiple levels and necessarily embody many of the characteristics of both business management and the democratic, consensual decision-making described in the two earlier chapters. There has been a number of descriptions in the past of how GIS has successfully (or, more rarely, unsuccessfully) supported individual functions in local government. Applications where it operates across the entire range of governmental functions – especially at national or regional level – are, however, very rare. This is not surprising for the complexities of multiple interacting policies, politicised debate, and numerous public and private sector players with sometimes conflicting agendas typify such environments. There is rarely a 'best' answer – and, even if there is, it is unlikely to be thrown up by simplistic spatial analysis generated by a mechanical system. Yet Jane Smith Patterson and Karen Siderelis (Chapter 53) show that GIS can be involved successfully throughout many of the activities of a major governmental organisation, threading through and being influenced by other policies and operations of that government. The State of North Carolina has had strong information technology-based development policies for over 20 years. The North Carolina Information Infrastructure policy is intertwined closely with the State's economic development policies, and both involve public/private sector partnerships. Much of this policy development has a geographical dimension. GIS has played an important role in the development of strategies, the assessment of development options, and the implementation of the strategies: it is a widely used part of the central information infrastructure. The success of these strategies is demonstrated by the substantial inflow of foreign investment into North Carolina, its very low levels of unemployment, and the transformation of its economy from one based largely on primary industries (notably tobacco) to a much more diversified, thriving, and entrepreneurial one.

51

GIS for business and service planning

M BIRKIN, G P CLARKE, AND M CLARKE

Within the realm of business and service planning, traditional GIS are weak so far as predictive modelling is concerned, notably because they have no means of coping with competition. This chapter outlines a framework within which GIS can be linked with other analytical tools in such a way as to offer commercial organisations support in constructing strategic development plans. We have consistently argued that the integration of modelling tools within a GIS environment creates a whole that is substantially more than the sum of the parts. In fact, it leads to the creation of 'decision support systems' for managers that can address a much wider range of operational and strategic issues than 'desktop mapping'. The approach is illustrated by examples drawn from the authors' work for major retailing corporations in both Europe and the USA in terms of optimal siting of outlets and improved distribution mechanisms.

1 INTRODUCTION

Since the first edition of this volume was published in 1991, a significant amount of change has occurred in the area of 'business geographics'. It can be argued (with some important exceptions) that GIS in business has moved from a peripheral, technical role to a mainstream management role. With GIS products such as MapInfo being incorporated into Microsoft Office, the ubiquity of GIS in business is impressive (see Elshaw Thrall and Thrall, Chapter 23). Most, if not all, large commercial and government organisations have access to GIS technology in one form or another. This maturity, following a common life-cycle pattern in IT applications, has naturally led to observers asking 'where do we go from here?' Has GIS become another widely accepted technology that can exist alongside spreadsheets, data warehouses, and other business IT applications – or is there something distinctive about GIS and its relationship with business application that remains untapped and offers significant potential for exploitation? In this chapter we offer, in a pragmatic manner, an approach and an argument that supports the latter rather than the former view.

Viewed from within the GIS industry, business geographics is seen as a tremendous success and we applaud the efforts of the 'GIS in business' community to promote the application of GIS in the commercial area. The number of conferences and the range of participants demonstrates an effective collaboration of developers and appliers. However, viewed from a different perspective (that of contributing to the strategic planning of business organisations), the position is less clear cut. It often appears to be the case that GIS in business is used simply to map out or represent spatially referenced data. At one level this is reasonable and should be supported. At another level, it falls somewhat short of what we have termed 'intelligent GIS' (Birkin et al 1996). We have consistently argued that the integration of modelling tools within a GIS environment creates a whole that is substantially more than the sum of the parts. In fact, it leads to the creation of decision support systems that can address a much wider range of operational and strategic issues than desktop mapping. We would thus like to see GIS move from the mainstream to the strategic (see also Gatrell and Senior, Chapter 66; Sugarbaker, Chapter 43).

Given this background, the aim of this chapter is to outline a framework within which GIS can be linked with other analytical tools in such a way as to offer commercial organisations support in constructing strategic development plans. Such a framework is likely to be built upon the following sequence: data $\Rightarrow$ information $\Rightarrow$ intelligence $\Rightarrow$ action. The chapter will review the role of GIS in business and marketing, and outline its contribution in each of the above categories. Examples will be drawn from the fields of high street retailing, financial services, and automotive retailing. An argument will be made that GIS is excellent for examining 'what is' issues, but poorer at making 'what if' forecasts and predictions. The advantages of GIS for data storage and handling, mapping, and the provision of basic levels of information will be discussed first. Then we shall review the kinds of spatial analysis tools that are currently available or recommended to retail analysts. These include buffer and overlay techniques and network analysis for store location research and customer targeting. A key argument here will be that many of these techniques are now somewhat dated in terms of methods available outside GIS (both in theoretical terms as purveyed through the literature and in applied terms through specialist software and consultancy operations).

In terms of forecasting and predicting the impacts of changes in distribution patterns we then argue that current proprietary GIS are less useful. This has now been recognised by many GIS vendors and 'solved' by introducing suites of location models, particularly location–allocation models and spatial interaction models (see Church, Chapter 20; Getis, Chapter 16). This recognises the usefulness of these tools for forecasting and 'what if' style predictions. However it is our belief that making simple (often one parameter) models available for *applied* problems can be as dangerous as not having such modelling power in the first place, mainly because it is unlikely that such simple models will be able to reproduce accurately the complex consumer behaviour patterns seen in modern retail markets. The alternative is to customise GIS so that the modelling tools become the main analysis tools, supported by good graphics and data processing. We will demonstrate how such (disaggregated) models can help add intelligence in the schema described above through examples of a variety of business situations that require action plans.

2 THE CORPORATE GROWTH MODEL

The geographical development of most retail and business organisations has taken place in a haphazard way. Few organisations can point to a systematic plan for network development which has involved prioritising potential target locations. That is not to say that there has not been an overall 'grand plan'. Indeed, many companies will have decided historically on either contagious diffusion strategies (expansion from a headquarters location: see the example of the UK Kwiksave grocery chain in Sparks 1990) or hierarchical diffusion strategies (involving location in the largest cities first). Some may also have gone as far as publishing lists of target sites (Kwiksave again is a good example – see Langston et al 1997). However, it would seem fair to say that such lists are often merely rankings of shopping centre by size, rather than the result of any proactive spatial analysis of competitive retail markets. This lack of investment in locational analysis does not surprise those who have dealings with senior management in retail organisations. The majority of these managers (if not all) will be accountants, lawyers, or business managers not accustomed or trained to acknowledge the importance of the branch location network in other than a superficial fashion (hence the importance of, and interest in, the arguments in section 3 below). In a fascinating recent appraisal of the importance of dealer networks in the Caterpillar organisation (Fites 1996) there is discussion on the value of dealers in terms of customer care and help in shaping new technical developments. Discussion of the relationship between sales and dealer location is less prominent although Fites does recognise the value of local marketing and knowledge of local customer needs. The only reference to the location of dealers in respect to sales in his publication comes with reference to Mexico where sales dropped during a recession and many competitor dealers were closed:

'When the good times returned, we were the only ones with a viable dealer organisation in Mexico, and we got the vast majority of the business.' (Fites 1996: 92)

Such recognition of national dealer locations needs to be mirrored at the regional and local levels. Part of the problem lies with the fact that many organisations believe few new opportunities exist.

Their networks probably developed in piecemeal fashion over the years, especially where mergers and acquisitions have also added to the uneven nature of spatial corporate growth. In addition, many organisations believe their network is complete in the sense that they no longer concentrate serious resources on location research, often believing home markets to be saturated. When we began to work with W H Smith (the largest bookshop and stationery chain store in the UK; see Birkin and Foulger 1992) they expressed surprise at the opportunities which were revealed through the appraisal of local markets simply because they had got used to the belief that their network of branches was complete and UK shopping opportunities were saturated. This attitude is common amongst retailers despite the fact that, in most retail sectors, company market shares vary significantly from region to region and substantially within regions themselves and that there is widespread regional variation in provision indicators such as floorspace per head of population (Langston et al 1997: see Figure 1). The lack of investment in location research is even more surprising in relation to opportunities provided by overseas markets (Wrigley 1993).

It is good news for those involved with spatial analysis that the situation is slowly changing. This must partly be attributed to the success of GIS vendors faithfully spreading the message at an increasing number of business conferences, through site visits, and through magazine articles. It is also a reflection of an increasing number of academics getting involved with consultancy (see Clarke et al 1995). Before we assess the contribution of GIS to business applications, it is useful briefly to rehearse the arguments concerning the importance of geography in the distribution and targeting of goods and services.

3 THE GEOGRAPHY OF SERVICE PROVISION

Ask any UK retail company about their current market share for different products in the UK and the vast majority will provide the correct answer. There is good market information at this coarse level of spatial resolution. However, ask the same companies for a regional breakdown and they may begin to struggle. This is often more interesting since national figures usually mask very large regional variations. Figures 2 and 3 show the variations in regional market shares for LADA and BMW cars

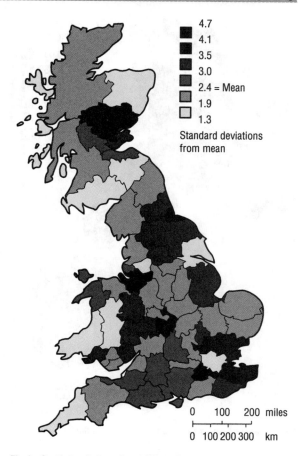

Fig 1. Spatial variations in retail provision: multiple grocer floorspace per household in Britain.
Source: Langston et al (1997)

respectively in the early 1990s. The variations in regional performance are clear to see. Such variations even occur in the business-to-business sectors where historically many companies monopolise markets in traditional strongholds but are very weak in other geographical areas. For example, Shell Chemicals has very strong market shares in northwest England and around Greater London. Referring back to the Fites (1996) article cited above, Caterpillar are equally likely to have uneven international *and* national market shares given the location of its global network of dealers.

Although these regional variations are often striking, the actual variations within regions are normally even greater. The vast majority of service organisations will struggle to provide market share estimates for major cities within regions, let alone for

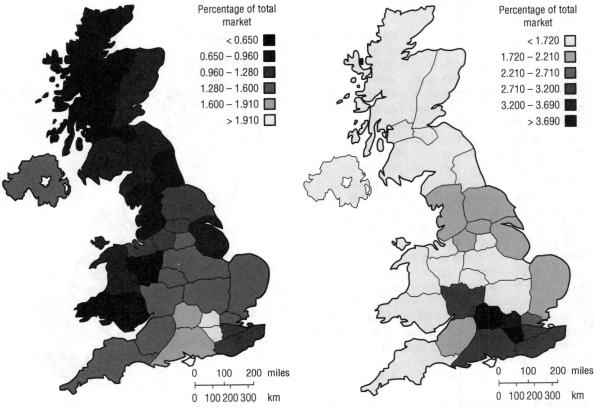

Fig 2. Regional market shares for Lada cars in the UK.

Fig 3. Regional market shares for BMW cars in the UK.

smaller geographical areas such as postal districts. As we mentioned above, this is partly because senior managers are not geographically aware and do not typically think in this way. Figure 4 shows the actual market penetrations for a major car retailer in the two UK counties of Hampshire and Dorset by postal district. This particular manufacturer has a national share of roughly 2.5 per cent. As Figure 4 shows, the postcode district averages within Hampshire and Dorset range between less than half and more than twice this average. The distance decay effect around each dealer is also evident. If geography (or dealer location) were unimportant, then local market penetrations for individual manufacturers would be much flatter and nearer to national figures (even allowing for random variations). This phenomenon is not peculiar to the UK. Figure 5 shows the same relationship between market share and dealer location for a leading motor company in Puerto Rico.

A second obvious factor which influences local geographical performance is the location of

customer 'types'. Kwiksave have a target market of UK social groups D and E, traditionally low- or un-skilled workers more interested in cheaper prices than store layout or store choice ('no frills retailing'). Similarly, Toyota are more likely to sell 'top of the range' cars in areas of high affluence and their two-seater MR2 sports car in areas containing both wealthy and typically young professional workers. Hence customer targeting is very important in strategic planning, and the geodemographics industry aims to supply information to companies on geographical variations in such customer types. Clearly the geographical balancing act is to combine distribution and sales points with the greatest concentration of potential customers.

We have long argued (Birkin et al 1987) that, to understand the geography of service provision, we need to understand the relationships between customers, sales, and markets. For some service sectors these data are readily available but currently remain greatly underused. Financial institutions, for

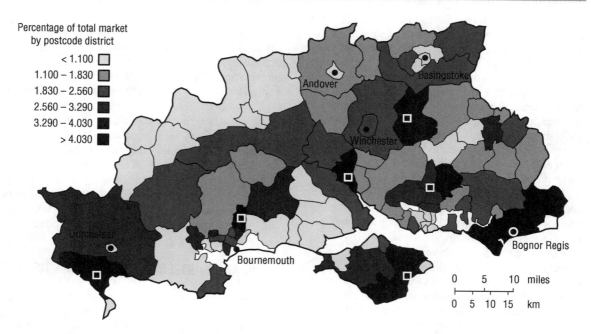

Percentage of total market
by postcode district

< 1.100
1.100 – 1.830
1.830 – 2.560
2.560 – 3.290
3.290 – 4.030
> 4.030

Andover
Basingstoke
Winchester
Dorchester
Bournemouth
Bognor Regis

0   5   10   miles
0   5   10   15   km

Fig 4. Market penetration for a leading motor distributor by postcode district, with dealer locations shown as squares.

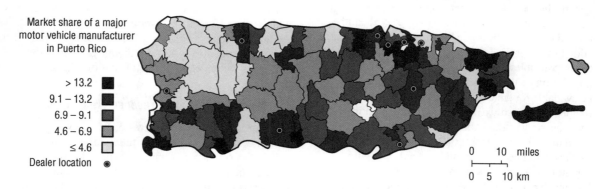

Market share of a major
motor vehicle manufacturer
in Puerto Rico

> 13.2
9.1 – 13.2
6.9 – 9.1
4.6 – 6.9
≤ 4.6
Dealer location ●

0   10   miles
0   5   10 km

Fig 5. Market share of a major motor manufacturer in Puerto Rico. Dealer locations are shown by dots.

example, have superb data on their customers and their place of residence. In Britain, the motor industry can purchase sales information for their company and all the competition from the Society of Motor Manufacturers and Traders (SMMT), including customer postcode addresses by manufacturer and model type. In these two instances, companies are likely to be in the commonly stated position of being 'data rich but information poor' in that there is so much raw data that processing and analysis without computer-based technology is virtually impossible (or at best

extremely time-consuming). Even if data are provided on computer (or by computer tape), it is our experience that it is often simply printed out on reams of computer paper and analysed by traditional eyeballing methods.

Other service organisations (e.g. high street retailers) may not be as fortunate as the ones mentioned above: for them, data may be poorer and less easily accessible. However, it is likely that through a mixture of internal data (invoices, credit card information, transaction data), public domain information (in the UK the Census of Population,

Census of Employment, Family Expenditure Survey, etc.) and agency information (market reports from Euromonitor, the Economic Intelligence Unit, Audits of Great Britain, and more specialist sources such as SMMT for the motor industry and AMI for the chemical industry) it is possible to piece together a large volume of market information; similar market information can be assembled in other countries. Indeed, the first major benefit from developing a computer-based information strategy is often the 'data audit' associated with piecing together relevant material on markets and customers.

The most immediate benefit of GIS itself lies in the ability to store large volumes of such data within a framework which allows quick and efficient retrieval (see Goodchild and Longley, Chapter 40). In geographical terms, this obviously relates to information pertaining to different spatial areas (and possibly spatial scales) as well as the medium of display through modern computer graphics. We have seen in the examples shown above that spatial data and information can be 'brought to life' and hence more easily understood if displayed in map as well as tabular form. It is at this stage of the information system that simple arithmetic operations can greatly add to the information base. Data retrieved can be filtered (or sorted by some key category), ranked, and added together. For example, if the datasets include census variables on characteristics of either the population or the household, then these are likely to be available for small geographical regions. They can then be combined with household information sets available only at the national level to produce new estimates for small geographical regions. Marrying the Census variables on household social class with expenditure on products (in the UK from the annual Family Expenditure Survey) can produce demand estimates by product groups not normally published (see an example of this approach for the estimation of demand for books in Germany shown in Figure 6).

GIS can also provide useful methods of data linkage through polygon overlay where different layers of data can be superimposed (see Martin, Chapter 6). The ability to link data together turns data into information. In the financial services industry, data on different types of account holders could be linked to provide customer profiles of the sorts of persons most likely to have a basic current account, a mortgage, or to be financially active holding many accounts. Targeting new deposit accounts could thus be made more sophisticated

through this data overlay procedure – but see Curry, Chapter 55, for an appraisal of the privacy issues that this sort of operation raises. Whilst many organisations are getting better at producing information, there is still a shortfall in good spatial analysis: we argue that it is the ability to analyse this type of information which provides market intelligence and the ingredients for action. Thus, to continue the financial services example, we could go on to assess the role that different types of customers play in adding to profit levels and trying to work out a strategy for optimising profits. Once we understand the profit implications of new clients, then the search is on to find other cases of these and similar customer types for added value products.

4 GIS AND MODELS IN RETAIL ANALYSIS

We will now illustrate the advantages of combining GIS and models ('intelligent GIS') using four examples of issues or problems which can arise in the retail industry. The first explores the day-to-day activities of marketing and branch network planning. The last two are examples of activities which take place more infrequently but which present obvious geographical problems to solve. We will demonstrate that intelligent GIS can offer solutions which lead to effective action plans.

4.1 Marketing and store revenue predictions

Most organisations are interested in analysing regional or local markets to understand both existing performance ('what is?' analysis) and to predict the impacts of changing the distribution network in some way ('what if?' analysis). Having used simple overlay procedures to identify target sites, the GIS literature often suggests a combination of buffer and overlay analysis to calculate store revenues for existing and new (potential) stores (see Elshaw Thrall and Thrall, Chapter 23; Waters, Chapter 59; Beaumont 1991a, 1991b; Elliott 1991; Howe 1991; Ireland 1994; Reid 1993; Reynolds 1991). This works by first estimating how far consumers are willing to travel to a store (existing or potential). The result of this exercise will be either a travel time (say 20 minutes) or a distance (say no more than three miles). The second stage is then to delimit an area around that store (a buffer) that marks the limit of that time or distance factor in

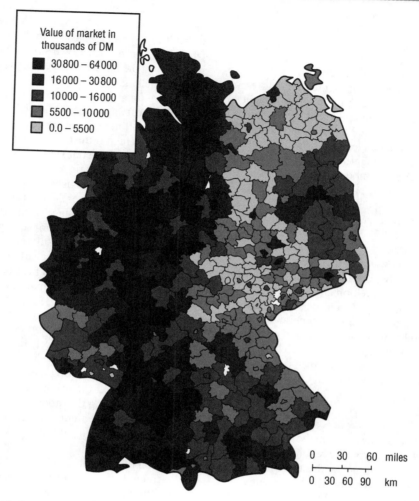

Fig 6. Estimated demand for books in former West Germany.

each direction outwards from the store. The revenue accruing to that store can then be estimated by overlaying the consumer spending power that resides within the postal sectors (or parts of sectors) or zipcodes that lie within that buffer. Unfortunately, the assumption is often made that customers close to the store are as likely to travel to the store as are those at the edge of the buffer. When there are competitor stores within the catchment area buffer, then revenue prediction becomes even more problematic. Normally, the so-called 'fair share' method is used (see Beaumont 1991a, 1991b) where the total amount of revenue generated by all stores in the buffer is divided simply by that number of stores. This could be made more sophisticated by

basing 'fair share' on store size and perhaps store ownership. However, even with these refinements, the revenue predictions made are likely to be fraught with danger. When the supermarket catchments overlap one another, the GIS software merges them into one polygon along with all the spatial attributes associated with that buffer. Clearly, the problem is greater when there are two or more stores which are located close together and thus effectively have very similar catchment areas. Although some of these issues have been addressed in relation to catchment area analysis (see Davies and Rogers 1984), the solutions do not yet appear in many GIS packages.

On the other hand, spatial interaction modelling is ideally suited to the problems of defining realistic catchment areas and estimating store revenues. A study area is divided into a set of residential zones, say postal sectors or zipcodes, and all centres and outlets are identified. The models thus attempt to quantify exactly how many customers in each demand zone will patronise each and every store in the region. Since customers to different stores will come from a variety of demand zones, overlapping catchment areas are expected and explicitly handled. The model is calibrated on existing flow data, after which a new outlet can be introduced and its impact assessed. Although these models have been around for 25 years, only recently has their full potential been realised as better spatial data and computer hardware/software became available (Birkin 1996; Clarke and Wilson 1987). Examples of these models and their use is provided by Birkin et al (1996).

It should be acknowledged that these types of model are increasingly to be found in proprietary GIS (see Maguire 1995 for developments in ARC/INFO). Although this should provide a more sophisticated level of analysis, there remains the general problem of the lack of flexibility in model design and outputs as well as a number of practical problems. The argument concerning the lack of flexibility which is often needed to handle a variety of often unique business questions has been provided by Birkin (1996). The practical problems have been explored by Benoit and Clarke (1997). They found model selection a major problem for the uninitiated. An analysis for a local health care facility will typically require a different model to that for an automotive dealer. A reasonable question from the new investor who has recently purchased the GIS package might be: 'Which model should I use?' Second, there is typically not a useful or related calibration method imparted to the potential

user, so it becomes necessary to devise one's own method. This particular problem makes it eminently difficult for a retail manager initially to comprehend what to do for his or her own store location analysis unless he or she has access to extensive knowledge on retail modelling.

The arguments for integrating specialist modelling software (where highly disaggregated models can often be constructed) and GIS have been made elsewhere and the term 'intelligent GIS' (Birkin 1996; Birkin et al 1996) is broadly synonymous with 'spatial decision support systems' (Shiffer, Chapter 52; Densham 1991; Densham and Rushton 1988). However, it is important here to emphasise the benefits. Since the models deal with spatial interactions, we can not only predict accurate revenues but also analyse resultant flow patterns. This, in turn, facilitates the calculation of a wide range of residence-based and facility-based performance indicators as a matter of routine (see Bertuglia et al 1994; Birkin 1994; Clarke and Wilson 1994) and also allows the analyst to construct area or regional 'typologies'. Examples of these typologies are given in Table 1. In this case, basic data on market performance have been turned into useful information. Having identified appropriate regional strategies, the resulting action plans can then be tested using the models in a 'what if?' fashion. The most common of these is the testing of new store openings. Again, the power of the models over traditional GIS comes with the accurate prediction of new catchment areas which are always likely to be skewed in some way (again because of the location of the competition) rather than the simple, common assumption of circular travel times or distance buffers.

Table 1 Four typologies of regional performance and management actions.

	Low ——— Market share ——→ High	
High ↑ Sales per branch ↓ **Low**	**Type 2** Low market share High sales per branch *Extend branch network*	**Type 1** High market share High sales per branch *Maintain status quo*
	Type 3 Low market share Low sales per branch *Reconfigure branch network*	**Type 4** High market share Low sales per branch *Rationalise branch network*

4.2 Launch of a new product

A special type of 'what if?' analysis concerns the distribution of a new market product. If this is a new type of chocolate bar, then an organisation would probably be safe retailing this from all newsagents and confectionery stores. However, what about a new £30 000+ ($48 000+) motor vehicle? GIS could highlight areas containing the most affluent consumer groups as a useful starting point. Existing dealer locations could then be overlaid to see which dealers might be in the most appropriate locations to sell cars valued at over £30 000. How many cars each dealer would actually sell is a much harder question to address with proprietary GIS. If the company has over 1000 dealers and hopes to sell 10 000 vehicles then it works out at ten per dealer. Realistically, however, some dealers are likely to sell 50, others fewer than five. This suggests we need a selective distribution policy which involves accurately estimating how many each dealer is likely to sell and hence which dealers should take priority. Figure 7 shows the results of a model-based appraisal of the likely sales of a new luxury car for company X from its dealers in the southeast of England and hence the dealers most likely to be able to sell 50 or more.

4.3 Mergers and acquisitions

Historically, mergers and acquisitions have been a key method of generating corporate growth. Most retail organisations have developed their networks and their market shares through acquiring regional players in this way (see Guy 1994; Kay 1987). There are several explicitly geographical issues involved with such action. The first is relevant when deciding on which organisations may be ripe for merger or take over. Different organisations have different regional and local market shares. By considering these variations prior to action, it may be possible to grow rapidly by achieving high market shares in regions where previously market share was very low. It was largely for these reasons that Kwiksave bought out Shoprite and Tesco bought out William Low in Scotland (see Sparks 1996a, 1996b). Once such action is complete, a second set of issues emerges. Does the combined network of stores now give that organisation the best or at least a good distribution network? In the case of the two examples mentioned above, the answer is likely to be 'yes'. However, with many mergers or buy-outs the combined organisation is left with many competing outlets in the same locations (a good recent UK example is the Leeds and Halifax building society

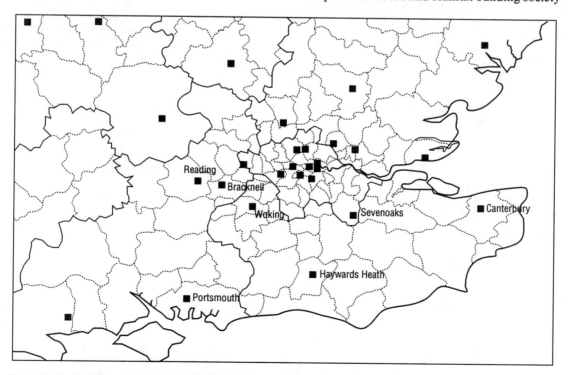

Fig 7. 'Optimal' dealers to sell new luxury cars in southeast England.

(Savings and Loan) merger where both had a high presence in the Yorkshire region as well as similar networks nationally).

If spatial duplication is the case, then the organisation is likely to want to rationalise its network through branch closure. The key questions are then which branches to close and where should transferred accounts be located, whilst at the same time considering the impacts of such closures on the desire to maintain market share. Although the intelligent GIS could be used to evaluate such store closures, a more effective action plan would also include refurbishments, change of fascias, and new branch openings – in other words, a systematic appraisal of all opportunities for the combined organisation. Tables 2 and 3 show an example of recent work we have undertaken for a new financial organisation in the UK following merger. Its main problem is to combine the former branch networks of the two organisations into one network for the new organisation. Table 2 shows a nationwide action plan which has resulted from a set of detailed modelling scenarios based on closures, reconfigurations (changing fascia name from one of the poorer performing branches of one of the old organisations to the name of the new organisation), refurbishments, and new branch openings. The action plan is clearly very different for each major UK geographical region. If we consider the southeast of the UK, we can see the more detailed

action plan in Table 3. Plate 41 shows the combined impacts of these changes on market share. The pressure on the new organisation is to rationalise by closing branches (in order to reduce costs). However, it is also important to minimise the impacts of closure programmes. From our action plan, the closures inevitably result in a loss of market share in the south and east of the region but we believe the impact is minimised with closures in these areas and more than offset by new openings in far more lucrative markets to the north and west. Refurbishment of those branches which are left in the south and the east of the region may well bring back market shares to their previous levels. Swann (Chapter 63) describes analogous problems in the realm of rationalising military operations.

4.4 Optimising retail networks

We saw in section 3 that one of the uses commonly made of a GIS is to overlay different data layers to produce new information. Often this procedure takes place in conjunction with a set of rules governing site locations, such as finding sites where several criteria are met – for example, flat land, proximity to a motorway access point, and areas with a catchment population greater than 100 000. Whilst this is useful in dealing with many environmental problems, it is less useful in business. Again this is mainly because of the problems of dealing adequately with the competition (i.e.

Table 2 National action plans for the amalgamation of two building societies: number of stores affected.

Action	East Anglia	Inner London	Midlands	Yorks and northwest	North
New entry	18	29	25	1	3
Removal	1	4	3	10	5
Rationalise	1	8	0	2	1
Reconfigure	4	22	1	3	0
	Scotland	southeast	southwest	National totals	
New entry	8	16	10	110	
Removal	6	6	3	38	
Rationalise	1	8	0	21	
Reconfigure	5	10	1	56	

where
New entry   =   High potential, no presence: add new stores
Removal   =   Poor performance, low potential: close branches of one former company
Rationalise   =   Poor performance, low potential: close both former company branches
Reconfigure   =   High potential, low market share: invest to upgrade branches

Table 3 Detailed strategy for development in southeast region.

New entry	Hempstead Valley	Ashford (Middlesex)
	Woodley	Romsey
	Frimley	Fleet
	Ryde (Isle of Wight)	Weybridge
	Rainham (Kent)	Ringwood
	Silverhill	Oxted
	Addlestone	Egham
	Hythe (Kent)	Broadstairs
Removal	Newhaven	Midhurst
	New Milton	Rye
	Polegate	Hampden Park
Rationalise	Farnham	Haywards Heath
	Seaford	Bexhill
	Dorking	Southsea
	Littlehampton	Burgess Hill
Reconfigure	Harley	Uckfield
	Steyning	Esher
	Shoreham	Haslemere
	Rustington	Godalming
	Lymington	Tenterden
Expansion[1]	Crawley	Winchester
	Cosham	Sevenoaks
	Newbury	Reigate
	North End	Windsor
	Banstead	Haslemere
	Rustington	

[1] where

Expansion = High potential, low market share: open additional branch

allocating population between stores). Thus, for business applications, other types of optimisation procedure may be more useful. Optimisation methods such as linear programming attempt to find the best solution to a stated problem subject to a number of constraints being satisfied (see Church, Chapter 20). They first appeared in the geographical literature to handle difficult operational research issues such as the 'travelling salesman problem'. Nowadays a much larger class of methods exists, particularly in the non-linear programming area. The great advantage of these methods over traditional GIS approaches is the way in which they can handle spatial interaction. For example, in site location problems they can consider a particular solution (say, 50 particular sites chosen from 500 possible ones) and calculate catchment populations and potential revenues very easily. Where existing GIS *can* help is possibly in identifying local sites. Certainly, a marriage of methods could well prove invaluable.

One methodology which has already been imported into GIS is the location–allocation model which finds the optimal locations for supply points given a spatially non-uniform pattern of demand (see Ghosh and Craig 1984; Ghosh and Harche 1993). The optimisation criterion is usually distance minimisation. The model can be run for any number of supply points. This kind of model is limited in its applied value since existing supply points are likely to be sub-optimally located and extremely difficult or expensive to relocate. However it does give some indication of an idealised pattern of business activity and it is particularly valuable in looking at the best locations for sales representatives or area managers. Given that the primary problem for such 'reps' or area managers is accessibility, then these models are very useful for assigning better new locations from which to serve either an existing pattern of clients to salespersons (or area managers to stores) or a new set (which themselves can be worked out in an optimal sense). Figures 8 and 9 show two examples – the optimal locations for new car dealer locations in the Seattle/Tacoma region of the USA and the Czech Republic, respectively. The optimisation objective in Seattle/Tacoma was to find sites which maximised sales for two new locations for Toyota whilst at the same time minimising the number of deflections from existing dealers. Clarke and Clarke (1995) estimated the potential financial benefits of such an optimisation procedure if this could be replicated throughout the USA: the predicted profits alone ran into many hundreds of millions of dollars! Greater details on the mechanics of optimisation appear in Birkin et al (1995) and Wilson et al (1981).

5 CONCLUSIONS

We have argued that existing GIS for business and service planning do not provide sufficient analytical power to solve many important geographical problems which arise in areas such as sales, marketing, and advertising. We believe highly flexible information systems are required which combine database manipulation and high-quality map and graphical output with spatial modelling techniques. Although some of this can be achieved through existing proprietary GIS packages, these are too often restrictive and unable to offer the right sorts of analyses to business organisations. It is

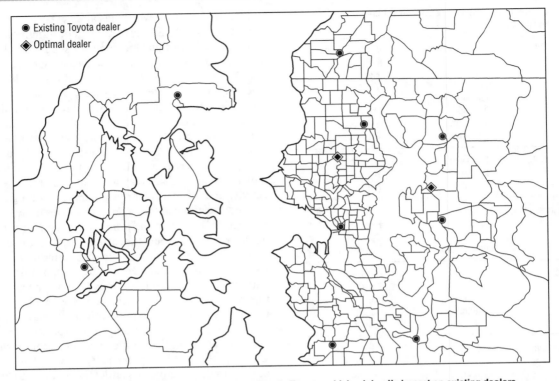

Fig 8. Two new optimal locations for Toyota dealers in Seattle/Tacoma which minimally impact on existing dealers. The land area is divided into census tract units.

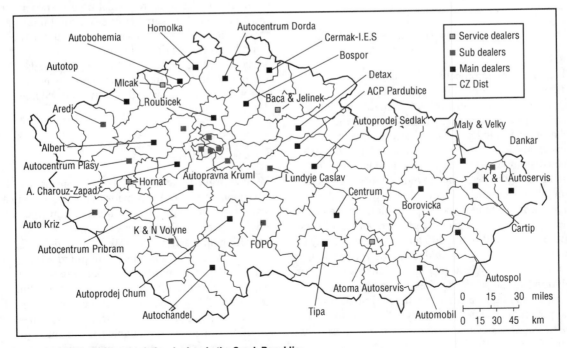

Fig 9. Optimal locations of existing dealers in the Czech Republic.

crucially important that GIS solutions can address the types of key strategic planning issue we have described above, rather than force organisations to 'shoe-horn' their problems into the analytical toolboxes available in current GIS packages. If this is not done, then we believe GIS will remain a low-order planning tool, assisting in the process of data manipulation and mapping but offering few long-term solutions or flexible action plans for distribution planning. Fortunately, the opportunities are still growing. We are convinced there are many new markets for GIS to explore. The area of telecommunications is one (see Fry, Chapter 58), offering enormous potential for good spatial analysis in distribution planning and target marketing as opposed to traditional GIS interests in physical network planning. If we can convince strategic planners in these new areas of the flexibility which is possible with GIS and spatial analysis then exciting times still lie ahead.

References

Beaumont J R 1991a GIS and market analysis. In Maguire D J, Goodchild M F, Rhind, D W (eds) *Geographical information systems: principles and applications*. Harlow, Longman/New York, John Wiley & Sons Inc. Vol 2: 139–51

Beaumont J R 1991b *An introduction to market analysis*. Concepts and Techniques in Modern Geography 53. Norwich, Geo-Abstract Books

Benoit D, Clarke G P 1997 Assessing retail location planning using GIS. *Journal of Retailing and Consumer Services* 4: 239–58

Bertuglia C S, Clarke G P, Wilson A G 1994 (eds) *Modelling the city: planning, performance, and policy*. London, Routledge

Birkin M 1994 Understanding retail interaction patterns: the case of the missing performance indicators. In Bertuglia C S, Clarke G P, Wilson A G (eds) *Modelling the city*. London, Routledge: 105–20

Birkin M 1996 Retail location modelling in GIS. In Longley P, Batty M (eds) *Spatial analysis: modelling in a GIS environment*. Cambridge (UK), GeoInformation International: 207–26

Birkin M, Clarke G P, Clarke M, Wilson A G 1987 GIS and spatial modelling: ships in the night, or the beginnings of a relationship? *Working Paper 498*. School of Geography, University of Leeds

Birkin M, Clarke G P, Clarke M, Wilson A G 1996 *Intelligent GIS*. Cambridge (UK), GeoInformation International

Birkin M, Clarke M, George F 1995 The use of parallel computers to solve non-linear spatial optimisation problems: an application to network planning. *Environment and Planning A* 27: 1049–68

Birkin M, Foulger F 1992 Sales performance and sales forecasting using spatial interaction modelling: the W H Smith approach. *Working Paper 92/21*. School of Geography, University of Leeds

Clarke G P, Clarke M 1995 The development and benefits of customised spatial decision support systems. In Longley P, Clarke G P (eds) *GIS for business and service planning* Cambridge (UK), GeoInformation International: 227–45

Clarke G P, Longley P, Masser I 1995 Business, geography and academia in the UK. In Longley P, Clarke G P (eds) *GIS for business and service planning*. Cambridge (UK), GeoInformation International: 272–81

Clarke G P, Wilson A G 1994 A new geography of performance indicators. In Bertuglia C S, Clarke G P, Wilson A G (eds) *Modelling the city*. London, Routledge: 55–81

Clarke M, Wilson A G 1987 Towards an applicable human geography: some developments and observations. *Environment and Planning A* 19: 1525–42

Davies R L, Rogers D S 1984 *Store location and store assessment research*. Chichester, John Wiley & Sons

Densham P J 1991 Spatial decision support systems. In Maguire D J, Goodchild M F, Rhind, D W (eds) 1991 *Geographical information systems: principles and applications*. Harlow, Longman/New York, John Wiley & Sons Inc. Vol 1: 403–12

Densham P J, Rushton G 1988 Decision support systems for locational planning. In Golledge R, Timmermans H (eds) *Behavioural modelling in geography and planning*. London, Croom Helm: 56–90

Elliott C 1991 Store planning with GIS. In Cadoux-Hudson J, Heywood D I (eds) *Geographic Information 1991*. The Yearbook of the Association for Geographic Information. London, Taylor and Francis: 169–72

Fites D V 1996 Making your dealers your partners. *Harvard Business Review* March–April: 84–95

Ghosh A, Craig C S 1984 A location–allocation model for facility planning in competitive environments. *Geographical Analysis* 16: 39–56

Ghosh A, Harche F 1993 Location–allocation models in the private sector: progress, problems and prospects. *Locational Science* 1: 81–106

Guy C 1994 *The retail development process: location, property and planning*. London, Routledge

Howe A 1991 Assessing potential of branch outlets using GIS. In Cadoux-Hudson J, Heywood D I (eds) *Geographic Information 1991*. The Yearbook of the Association for Geographic Information. London, Taylor and Francis: 173–5

Ireland P 1994 GIS: another sword for St Michael. *Mapping Awareness* 8(3): 26–9

Kay W 1987 *Battle for the high street*. London, Piatkus

Langston P, Clarke G P, Clarke D 1997 Retail saturation, retail location and retail competition: an analysis of British grocery retailing. *Environment and Planning A* 29: 77–104

Maguire D J 1995 Implementing spatial analysis and GIS applications for business and service planning. In Longley P, Clarke G P (eds) *GIS for business and service planning*. Cambridge (UK), GeoInformation International: 171–91

Reid H G 1993 Retail trade. In Castle G (ed.) *Profiting from a geographical information system*. Fort Collins, GIS World Inc.

Reynolds J 1991 GIS for competitive advantage: the UK retail sector. *Mapping Awareness* 5(1): 33–6

Sparks L 1990 Spatial–structural relationships in retail corporate growth: a case study of Kwiksave Group PLC. *Service Industries Journal* 10: 25–84

Sparks L 1996a Space wars: Wm Low and the 'auld enemy'. *Environment and Planning A* 28: 1465–84

Sparks L 1996b Challenge and change: Shoprite and the restructuring of grocery retailing in Scotland. *Environment and Planning A* 28: 261–84

Wilson A G, Coelho J D, MacGill S M, Williams H C W L 1981 *Optimisation in location and transport analysis*. Chichester, John Wiley & Sons

Wrigley N 1993 Retail concentration and the internationalisation of British grocery retailing. In Bromley R D F, Thomas C J (eds) *Retail change: contemporary issues*. London, UCL Press: 41–68

52

Managing public discourse: towards the augmentation of GIS with multimedia

M J SHIFFER

'Urban planning' is a key process in democratic societies: it is the interface between democracy and operational management and, as such, reflects the operations of the state in microcosm. It is becoming increasingly important as local communities come to assert their interests and values against those of officialdom or big business. Central to this planning is discourse between different parties.

Management of planning is therefore a non-linear operation without line managers in any conventional business sense. This chapter describes how combinations of GIS and multimedia can facilitate the active involvement of many different groups in the discussion and management of urban change through planning. It deals with different components of public discourse, notably debate and consensus-building, and with communications between parties sometimes separated by space and time. Recent experiments have already demonstrated the practical possibilities and difficulties in use of these technologies in the planning context. Evolving technology will diminish some of the difficulties in the years to come.

1 INTRODUCTION

There are many ways in which information technology can support the various stages in a city planning process. Much of this has been documented by Armstrong and Densham (1990), Batty and Xie (1994), Couclelis (1991), Harris (1989), and Klosterman (1992) among many others. This chapter concentrates on the subset of planning that involves information exchange for the purposes of managing public discourse. In many of these contexts the information consists of plans, proposals, and alternatives which have already been generated from raw data (frequently with the help of a GIS). Here, the focus is on innovative ways in which the results of GIS-based analyses have been augmented and conveyed effectively to support public discourse. We will also explore how such results can be manipulated in a limited fashion to allow for some degree of additional analysis by a broader group of people than those who undertook

the initial work (see also Cova, Chapter 60; Yeh, Chapter 62).

Public discourse in city planning contexts frequently consists of positioning, consensus-building, argument, and deal-making (among other things) in a variety of settings such as planning meetings, individual review, letter writing, and news reporting. If one were to observe, at a more basic level, what is actually communicated, it would be likely to include some degree of recollection, description, or speculation.

Furthermore, the settings for this discourse can differ by time and place. For instance, community stakeholders might communicate in the same place at the same time (through community meetings), at different places at different times (through the news media), and so on. This chapter describes how recent developments in information technologies (particularly spatial multimedia technologies) provide the capacity to augment traditional GIS to support public discourse. We begin by discussing various types of planning-related discourse in terms of recollection, description, and speculation, and go on to explore how GIS can be augmented with spatial multimedia to support these forms of discourse. The various settings in which planning-related information is accessed and how the augmentations to GIS identified above might be applied are also described. Finally, by drawing from some early experiences with these technologies, some of the issues that have arisen involving their construction, implementation, and use are identified.

2 FORMS OF DISCOURSE

The forms of discourse which occur in the context of planning tend to include:

- the recognition of problems and issues faced by a community;
- understanding of the range of alternative scenarios which can begin to address these problems and issues;
- acknowledgement of the actors and institutional mechanisms available to support action;
- some appreciation of the implications of action based on an understanding of present conditions and past trends.

Knowledge in the planning context is frequently shared through recollections of the past, descriptions of the present, and speculation about the future. This section discusses recollection, description, and speculation. It explains how key applications of spatial multimedia technologies can support these forms of discourse.

2.1 Recollection

In recollecting the past history of a given site or planning issue, conversations may include (among other things) what was said, what was done, or what a place was like. For example, members of a group may try to recall the impact of past interventions on an urban landscape. The purpose of this is to understand and anticipate what may lie ahead, given similar circumstances.

Where the recollection is about fairly structured recent activities (such as past planning meetings), the conversations can be supported with records and systematic documentation of past interactions. However, access to this information can be dependent on a specialised information recording and retrieval 'system' such as a meeting secretary or stenographer. Furthermore, such methods of recollection rarely incorporate spatial referencing with such things as historical maps. Where systematic documentation is lacking, as with recollection of the past environmental conditions of an area, the high degree of dependence on human memory can lead to problems based on the inconsistency of individual memory. For example, one person may recall traffic on a particular street to have been heavy whilst another may think of the same stretch of roadway as lightly travelled.

Where there is a lack of documentation or data to support these recollections, arguments related to inconsistent memories are likely to persist. These arguments can dominate a discussion and shift the focus of a meeting from the matters at hand. Nevertheless, it should be noted that, while various participants in a planning process have their own individual perceptions of a problem, this informal means of knowledge retrieval can be tremendously useful – especially for eliciting informal institutional knowledge that would not have been conveyed in other, more formal, ways.

GIS support of recollection, especially in the contexts of public discourse, has traditionally been somewhat limited. This has been due to a number of factors, including fairly weak historical references in much of the available spatial data. While GIS in combination with historical data can indeed be used to facilitate recollection about characteristics such as demographic trends, property values, and other generalisable information, it is less adept at conveying the past character of an area. Furthermore, using this technology to assist recollection depends on systematic archival of spatial 'snapshots'.

2.2 Description

Description of present conditions generally involves familiarising participants in a collaborative situation with an area being discussed, so that everyone can work from a common base of knowledge. These descriptions frequently include some type of spatial referencing. While many of these references are verbal (i.e. 'over by the river', 'on the site of the former factory') and cognitive (see Mark, Chapter 7), such references may be increasingly inappropriate where familiarity of the participants in the meeting with a particular site is lacking. For example, the term 'on the site of the former factory' is completely meaningless to meeting participants who are unfamiliar with the area being discussed.

The lack of familiarity with a given site can be rectified through an up-to-date map that is used as a central reference point. Individuals describe present conditions verbally and augment this by gesturing at a map spread on a table or tacked to a wall. Such descriptions may be further augmented using thematic data and visual imagery. The thematic data may be provided in the form of land-use maps or demographic conditions of an area. Visual imagery may include photos or video tape of selected sites. The juxtaposition of the above media can strengthen a collective understanding of the various characteristics of a given site.

Until recently, GIS has been unable to support in an effective manner real-time descriptions of existing conditions in collaborative situations. This has been because of issues of speed, the human interface, and a lack of integration with other forms of media (see also Egenhofer and Kuhn, Chapter 28). While many of these issues are being addressed using modern GIS tools that take advantage of interoperability and component software (see Sondheim et al, Chapter 24), the techniques for the effective juxtaposition of this information for retrieval in collaborative contexts still need to be developed further.

2.3 Speculation

Speculation about the future of an area generally involves extrapolating measurable phenomena from past experience and applying the results to the future using informal mental models. For example, a participant in a planning meeting might recall the impact on property values in one area of a previously developed transit station and simply speculate about such an impact at another comparable location under similar conditions. Speculation can also involve hand calculations that can be quite sophisticated. While such speculation is likely to become less accurate as the complexity of the scenario increases, a much more formalised mechanism for speculation in more complex situations has been made available through prediction and modelling using computer-based analysis tools (see, for example, Batty and Xie 1994; Harris 1989; Harris and Batty 1993).

Augmentation of discourse using analytic tools has traditionally been handicapped both by a lack of immediate response and by the production of abstract output that tends to exclude from such conversations those who are not technologically sophisticated. For example, it has been traditionally difficult to interact with an analytic tool, such as a GIS, in the context of a meeting room and expect the immediate response necessary to support a conversation. Furthermore, technical output such as a predicted automobile traffic level of 13 000 cars a day can have little meaning for people who are not transportation specialists.

The issues of speed and responsiveness can be addressed to a limited extent with advances in processing speed of computers, as well as through the employment of direct manipulation computer interfaces that allow users to 'push buttons' and 'slide levers' on a display to elicit an immediate response from the computer. The issue of relatively abstract output can be addressed using multimedia representational aids such as images and sound which can portray the output of analytic tools more descriptively. However, more still needs to be understood about how these techniques can be applied in actual collaborative settings.

3 KEY APPLICATIONS OF MULTIMEDIA

Many of the difficulties described above can be addressed through creative application of multimedia information technologies (see, for example, Câmara et al 1991; Fonseca et al 1993; Jones et al 1994; Laurini and Milleret-Raffort 1990; Polyorides 1993; Shiffer 1992). Recently, the Department of Urban Studies and Planning at Massachusetts Institute of Technology (MIT) has been working with several government agencies and private firms to identify effective combinations of tools and techniques for the support of city planning meetings. The overall goal of this research is to improve the communication of planning-related information with a specific focus on the environmental effects of proposed urban interventions.

3.1 Augmenting recollection: annotation mechanisms

Spatial annotation mechanisms have the capacity to assist in recollection of past annotations by providing an archive that can be accessed based on geographical relevance, chronological relevance, and associative relevance. Geographical relevance allows

295

users of an information system to search for annotations which are related to a specific region or sub-region using typical GIS operations such as buffers and overlays. Chronological relevance allows a user to add the capacity to search for annotations made before, after, or between two stipulated dates. Finally, associative relevance allows searching by keywords or related concepts that could be hyperlinked together in a World Wide Web (WWW)-like associative structure.

An annotation mechanism was implemented as part of a prototype collaborative planning system tested at the National Capital Planning Commission in Washington DC among other locations (Lang 1992; Shiffer 1993, 1995b). The annotation capabilities included the capacity to add text, sound, and video to spatial representations. Initial implementations of such an annotation mechanism in a collaborative setting were not very successful for several reasons. In the first instance, it was awkward to stop meetings for annotations, as such interruptions would tend to disturb the flow of conversation. Second, the issue of which party had the right to annotate a particular representation surfaced. The value of such annotations was not readily apparent due to the fact that there was no precedent for accessing previously annotated maps; and finally, many meeting participants were wary of 'going on the record' with informal comments pertaining to specific regions.

With the maturating of the WWW, it is now possible to incorporate spatial annotation into a distributed planning review process. This practice can help to support scenarios where stakeholders are unable to meet together at the same time. For example, in Washington DC the National Capital Planning Commission has recently experimented with the capacity to annotate a WWW-based virtual streetscape. In Lisbon, two MIT students have separately explored the use of the WWW to gather public comments surrounding various development proposals (see Ferraz de Abreu 1995; Gouveia 1996). While the technical capacity to incorporate a spatial annotation mechanism with GIS now exists, several issues such as the 'legitimacy' of comments and the 'life' of an annotation remain to be resolved.

It is certainly conceivable that GIS-based archival mechanisms can be set up to aid future recollective efforts. But this requires that a substantial spatial data infrastructure be already in place. Since we are only beginning to realise the development of substantial spatial data infrastructures around the world (see

Rhind, Chapter 56), we will need to continue to rely on the (frequently paper-based) libraries of local historical societies for more specific spatial descriptions which can effectively convey the character of a local area. Even in this case, the issue becomes a question of what material is worth maintaining – which has profound implications for the scalability of such a system. For instance, is it reasonable to expect a planning council to archive a spatial representation of every proposal made, along with the corresponding minutes of every planning meeting? If so, what is a reasonable time frame for keeping the record in the archive? Five years? Fifty years? Forever? If not every proposal is archived, then how is the choice of 'what is relevant' to be made?

3.2 Augmenting description: navigational aids

Navigational aids have had the capacity for several years to support descriptions of existing conditions by offering a link between oblique imagery and orthographic maps (see, for example, Batty, Chapter 21; Brand 1987; Kindleberger 1990; Rhind et al 1988; Wiggins and Shiffer 1990). While this technology has traditionally relied upon specialised hardware configurations such as randomly accessible videodisk technology, recent developments such as component GIS and the capacity of digital video clips to incorporate objects now makes it possible to link images (and objects within images) to spatial databases.

For instance, a user can now select an area of a map and see a corresponding motion video of that location playing on the same screen. Users can also 'browse' a region using a navigational video, 'stop' upon identifying an item of interest and see its location highlighted on a corresponding map. More recent developments, such as 'nodal digital video', make it possible to pan, tilt, and zoom from a specific location. Furthermore, these nodes can contain objects that allow users to connect to other video nodes, spatial objects, or other descriptive materials such as those found on the WWW. Much of this is made possible by the standardisation of software-based digital video display tools that allow the use of digital video navigational aids independent of specialised hardware configurations. Finally, the proliferation of component-based WWW browsers and GIS WWW 'plug-ins' makes it possible to access such navigational aids without specialised software (Coleman, Chapter 22).

3.3 Augmenting speculation: multimedia representational aids

The capacity to do urban and regional development forecasting has existed for some time. Yet such forecasts can go unappreciated in city planning meetings where there is a significant mix of participants with various skills (Forester 1982; Schon and Rein 1994). Such a mismatch can be characterised as a 'gap of understanding'. This gap not only exists between (and among) technical specialists (planners, geographers, engineers, architects, etc.) and the public: it also frequently exists between such specialists and key decision-makers and/or other stakeholders. 'Closing the gap' can be accomplished through the augmentation of typically abstract environmental representations using direct manipulation interfaces and multimedia representational aids (Norman 1986).

Direct manipulation interfaces translate human desires into commands that the computer can understand. Multimedia representational aids support information flow in the other direction by augmenting numeric values with graphical representation and associated imagery to transform abstract data into concepts that the human can understand. This results in the use of images and sound to portray the output of analytic tools more descriptively than traditional symbolic representation. The intended result is to make analytic tools and their outputs more manipulable, understandable, and appealing so that information which would normally be meaningless and intimidating to the lay person can be comprehended effectively.

Both direct manipulation and multimedia representational aids have been made available to planning settings through recent increases in computing power available to the masses. While it is important to apply such power to the undertaking of previously unattainable analyses, it is also important to improve the comprehension of existing analytical tools, especially in collaborative contexts with varied participants.

Virtual environments (see Neves and Câmara, Chapter 39) have the capacity to tie all of the above together. Whether future multimedia spatial analysis tools are likely to be manifest as GIS components constructed on top of a WWW architecture, or multimedia components constructed along a GIS backbone, depends to a large extent on the implementation environment.

4 IMPLEMENTATION ENVIRONMENTS

There seem to be four types of environments in which people are likely to access planning-related information. These are:

- meeting at the *same place and same time*, such as in attending a traditional meeting;
- incorporating information supplied by various participants located in *different places at the same time*, such as through the use of human proxies;
- accessing information in the *same place at different times*, such as with individual review of plans that are accessible to the public at a central location;
- bringing together information from *different places at different times*, such as through letter writing and newspaper articles.

This section will describe how recent developments in information technology have augmented discourse in each of these four areas as illustrated in Figure 1, adapted from Armstrong (1993).

4.1 Same place/same time

This is the traditional 'community meeting' environment in which planners are likely to find themselves. It allows for a substantially broad 'bandwidth' of communication because of the capacity of humans to augment verbal communication with non-verbal signals such as gestures. Depending on the institutional mechanisms in place, this environment can lead to a significant degree of participation and discourse with direct verbal interaction.

While this form of communication can be highly efficient, it may be constrained by a lack of access to information or (more recently) a lack of filtering of relevant media. Furthermore, such meeting environments may suffer from a lack of access to analytical tools such as GIS, because of the individual design of many of these tools and a lack of representational aids that would otherwise allow a group of people with diverse backgrounds to participate in a meaningful conversation.

Recent advances have made it possible to augment this kind of environment with a collaborative planning system (CPS). Such tools have been implemented on an experimental basis as described in Shiffer (1993, 1995a, 1995c). Similar approaches have been implemented in several

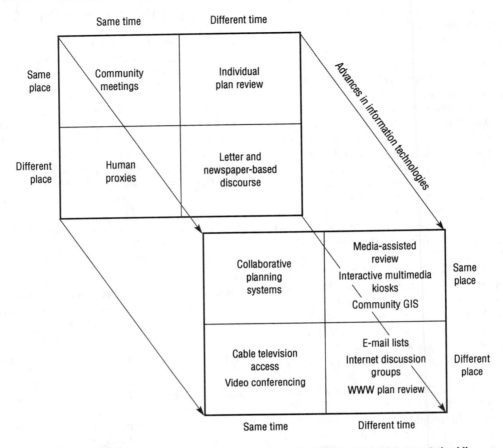

Fig 1. A diagram that outlines how advances in relevant information technologies have augmented public discourse in various settings.

contexts including those described in Dubbink (1989) and Faber et al (1995). A CPS makes significant use of the 'annotation mechanisms', 'navigational aids', and 'multimedia representational aids' described above and projected onto the wall of a meeting room. Participants interact with the system using cordless pointing devices or a technical facilitator to elicit information about selected geographical areas through a direct manipulation graphical interface. The intended result of many of these implementations is to make analytic tools and their outputs more manipulable, understandable, and appealing, so that information that would normally be meaningless and intimidating to the lay person can be more effectively comprehended.

In one case, a CPS was tested in an exploration of alternative uses for a former Air Force base, as described in Shiffer (1995a). In this situation, individuals followed the agenda of the meeting and used the system to access information about the various locations within the Air Force base for display to the group, using a portable computer and projector as illustrated in Plate 42. Such information included descriptive text and panoramic views of the base from key locations identified on a digital map. The environmental noise implications of proposed aircraft operations at selected locations around the former Air Force base could also be heard.

Another example involved the employment of a CPS to support a planning review process by the National Capital Planning Commission in Washington DC (as described by Shiffer 1993, 1995b). Several of the techniques that were originally proposed as part of the prototype CPS have been incorporated into the Commission's monthly review activities. These include the access and use of digitally stored aerial images, site plans, and video clips of physical models.

Several implementation and use issues arose out of early experiences with the CPS. For instance, early tests illustrated a reluctance on the part of users to pick up a pointing device and interact directly with the system during a meeting. Most preferred to interact with the CPS through an 'information expert' – a person who is very familiar with the system's content. Such an individual could anticipate group needs and display relevant information when called upon. Thus, while it would be necessary for the information expert to be familiar with a meeting's agenda, this individual could also track a random conversation and display maps and images as they were discussed.

The development of the prototype CPS has led to the identification of a broad set of issues ranging from institutionalisation to technical infrastructure (some of which are discussed in Shiffer 1995b). The benefits of this approach are that it is fast and self-contained. However, key participants may not be able to attend meetings because of place and time constraints. Other scenarios are therefore described below.

4.2 Different place/same time

While it is difficult to imagine how a geographically dispersed set of individuals could participate in a planning meeting at the same time, such interaction has been traditionally made possible through the use of human proxies (such as spouses, friends, or more formal representatives). If a stakeholder is unable to attend a meeting that is of particular interest, a proxy might attend in his or her absence. This type of situation can allow for limited interaction through conveniently accessible telephones which allow the proxy to report specific events to the stakeholder and give the stakeholder a chance to respond.

This model of communication has recently been augmented by local-access television in many communities. These 'televised meetings' frequently take the form of a fixed camera in a room where a planning-related meeting takes place in an otherwise normal manner. The television broadcast allows individuals not present to see and hear what is going on (at least in front of the camera). While television does indeed add a visual and audible characteristic

to the interaction, frequently this kind of interaction is one-way (flowing from the meeting to the stakeholder). In some advanced cases of local access television, a mechanism might exist for 'phone-in' interaction, where remote participants have the capacity to pose questions. Nevertheless, the remote participant in these situations is frequently at a disadvantage because of a lack of access to (and interactivity with) the media. Even worse, the remote participant or 'caller' feels (and quite often is) completely at the mercy of the meeting's facilitator (or the person accepting the call).

A technical augmentation of this is the capacity to employ teleconferencing mechanisms. While this approach has been instituted successfully in the past to support design interactions that involve a limited number of participants (see, for example, Mitchell 1995), teleconferencing has not taken root in the planning arena in a substantial way. This is largely because of the need for relevant parties to have an adequate infrastructure in place to support such discourse. While Internet video conferencing has recently been popularised with programs such as 'CU-See Me' (as illustrated in Plate 43); a truly high quality connection often still requires a specialised connection and compatible hardware at both ends – something we are likely to see in the not-too-distant future.

Another less technically sophisticated approach is to broadcast the visual display of a spatial multimedia system or GIS during the course of a televised meeting. This method precludes direct interaction with the system on the part of the person who is viewing from 'home'. But if such persons 'had the floor' through a telephone connection, they could interact with the system using an information expert on the other end. Furthermore, skilful switching between images of meeting participants and televised maps (augmented by video) can convey spatial relationships more effectively than a simple broadcast of a meeting.

4.3 Same place/different time

Where it is not possible for a stakeholder to participate in a planning meeting (either in person or remotely), it is still possible to use a centralised location as a base of information sharing and communication. This model of communication represents a situation where physical media (such as paper-based reports, plans, drawings, physical models, etc.) are left in a specific location for the review of a broad group of individuals over a period of time. Interaction in this case is usually limited and has traditionally taken the form of written comments that could be entered into a notebook or posted on a (non-electronic) bulletin board.

In some cases, this means of discourse has been augmented through the provision of video tapes made available for individual review. Like the more traditional media, the tapes might contain proposals, plans, and drawings. These are usually put forth in the context of persuasive arguments for or against a particular development.

More recent implementations of this model of communication have included some limited access to GIS or other types of public service system through the employment of electronic information kiosks (as illustrated in Plate 44). The more sophisticated of these actually have the capacity to collect comments electronically through digitised video or audio facilities. Kiosks make it somewhat easier to control access to the information because they are frequently strategically placed in the communities that they are designed to serve, typically located in libraries and municipal buildings. They also have the capacity to deliver large amounts of information (such as video) much more rapidly than the WWW. Furthermore, software licensing and data costs can be more easily controlled with this model. Some of the more sophisticated kiosks have actually been designed to allow public interaction with GIS through the employment of graphical user interface modifications that are made possible using some of the more recent GIS tool kits (see Egenhofer and Kuhn, Chapter 28).

Individuals with a broad variety of skill levels are likely to interact with public access kiosks. Therefore, specific attention needs to be paid to ergonomics issues related to the design of the software interface (see Batty, Chapter 21; Coleman, Chapter 22; and Elshaw Thrall and Thrall, Chapter 23). The drawbacks are that they cannot be everywhere and a deployment infrastructure is needed; this can be costly.

4.4 Different place / different time

This model of communication involves people who are separate both temporally and geographically. For instance, a person may be unable to attend the meeting of the planning review board in their home community because they may be at work in another

community during relevant meetings. In this case, their capacity to participate is typically limited to traditional forms of correspondence, such as letter writing and reading relevant newspaper articles. A significant difficulty of this is that local newspaper editors often become biased filters through which the information must flow. Furthermore, managing (archiving, categorising, etc.) public comments in such situations can prove to be challenging due to the limitations of print media.

This type of discourse has rarely been augmented with GIS because of the complexities of delivering the same information base to different parties while retaining the capacity for meaningful augmentation. The complexities of delivery tend to be concentrated more in the capacity of the different parties to interact meaningfully with the information base rather than with the technological limitations of delivering the same information base to multiple parties.

Voice mail, e-mail, and now the WWW make it possible to broaden the audience and these permit a more deliberative review. For example, Plate 45 illustrates a WWW-based application of a 'virtual streetscape' for Washington DC that allows the user to explore an area using either an orthographic photo and map overlay, or by using a nodal digital video. A text annotation capability allows the user to attach comments to specific nodes identified on the map. Limited Internet bandwidth and the problems of ensuring access to relevant parties continue to be drawbacks of this approach to planning-related discourse at present.

5 PROSPECTS FOR THE FUTURE

Several years ago we envisaged the capacity to link maps to highly descriptive representations of the past, present, and future by integrating GIS and multimedia with predictive and descriptive models. In many areas, this has become reality. While the exploration of *what* is delivered will continue to advance, the most profound changes in the near future are likely to concentrate on *how* it is delivered to different groups.

The WWW is just beginning to support direct manipulation of decision-aiding algorithms in a manner similar to GIS and stand-alone hypermedia collaborative planning tools. Thus software developers are faced with the challenge of creating directly manipulable components which can be accessed in a distributed manner through a networked infrastructure such as the WWW. The creation of such components is highly dependent on modular approaches to application development that involve the incorporation of relevant objects which are spatially aware. Once such components are created, it is likely that we will see an increasing number of distributed GIS with multimedia components which are organised around a spatial data infrastructure and delivered through wide-area networks such as the WWW.

Object-based development tools will make it possible to move away from the traditional GIS towards specialised applications that can be constructed through combining modular components. As advances are made in compression and network communication, future GIS components will continue to be delivered using this technology without regard to the client architecture. Open architectures will make it possible to deliver these applications to groups using a broad range of devices ranging from conventional televisions in the home to intelligent 'conference tables' (with horizontal displays) in the office and hand-held personal digital assistants in the field. The key to the success of these new architectures will be their capacity to facilitate communication across machines, among users and between the two. Though the WWW already offers unprecedented access to geographically-oriented multimedia information resources on the Internet, the capabilities of forthcoming systems seem likely to dwarf those of the present.

Even now, when interactive control of WWW-based information is still too unpredictable to support the dynamic discourse of a planning meeting, the World Wide Web can effectively support individual browsing or 'hunting and gathering' of multimedia information and planning-related data. In turn, this can be used to feed a locally stored planning support system. Planners and related groups will also benefit in future from the increased use of customisable intelligent agents that will aid problem-solving by identifying and gathering relevant information such as approaches used by others to address problems in comparable situations. In short, discourse in traditional planning settings will be augmented (rather than replaced) by these emerging information technologies as a broader range of individuals acquire the capacity to become active participants in planning problems and cases.

References

Armstrong M P 1993 Perspectives on the development of group decision support systems for locational problem-solving. *Geographical Systems* 1: 69–81

Armstrong M P, Densham P 1990 Database organisation strategies for spatial decision support systems. *International Journal of Geographical Information Systems* 4: 3–20

Batty M, Xie Y 1994 Modelling inside GIS: Part 2. selecting and calibrating urban models using ARC/INFO. *International Journal of Geographical Information Systems* 8: 451–70

Brand S 1987 *The media lab*. New York, Viking

Câmara A, Gomes A L, Fonseca A, Lucena e Vale M J 1991 Hypersnige – a navigation system for geographic information. *Proceedings, European GIS Conference, Brussels* April 1991: 175–9

Couclelis H 1991 Requirements for a planning-relevant GIS: a spatial perspective. *Papers in Regional Science* 70: 9–19

Dubbink D 1989 Use of acoustic examples in airport noise planning and decision-making. *Airport Landside Planning Techniques: Transportation Research Record 1199*. Washington DC, National Research Council

Faber B G, Wallace W, Cuthbertson J 1995 Advances in collaborative GIS for land-resource negotiation. *Proceedings, GIS 95 Ninth Annual Symposium on Geographic Information Systems, Vancouver*. GIS World Inc. 1: 183–9

Ferraz de Abreu P 1995 'Intelligent multimedia in public participation'. Unpublished PhD thesis proposal. Cambridge (USA), MIT

Fonseca A, Gouveia C, Ferreira F C, Raper J, Câmara A 1993 Adding video and sound into GIS. *Proceedings, EGIS 93 Conference Proceedings Genoa, Italy*, April: 176–87

Forester J 1982 Planning in the face of power. *Journal of the American Planning Association* 48: 67–80

Gouveia M C 1996 'Augmenting public participation with information technology in a Portuguese environmental assessment context'. Unpublished master's thesis. Cambridge (USA), Department of Urban Studies and Planning

Harris B 1989 Beyond geographic information systems: computers and the planning professional. *Journal of the American Planning Association* 55: 85–92

Harris B, Batty M 1993 Locational models, geographic information systems and planning support systems. *Journal of Planning Education and Research* 12: 184–98

Jones R M, Edmonds E A, Branki N E 1994 An analysis of media integration for spatial planning environments. *Environment and Planning B: Planning and Design* 21: 121–33

Kindleberger C 1990 Hypermedia for state and local government. *Journal of the Urban and Regional Information Systems Association* 2: 52–5

Klosterman R E 1992 Evolving views of computer-aided planning. *Journal of Planning Literature* 6: 249–60

Lang L 1992 GIS comes to life. *Computer Graphics World* October: 27–36

Laurini R, Milleret-Raffort F 1990 Principles of geomatic hypermaps *Proceedings, Fourth International Symposium on Spatial Data Handling Zurich*: 642–651

Mitchell W 1995 *City of bits: space, place and the Infobahn.* Cambridge (USA), MIT Press

Norman D A 1986 Cognitive engineering. In Norman D A, Draper S W (eds) *User centered system design: new perspectives on human-computer interaction.* Hillsdale, Lawrence Erlbaum: 31–61

Polyorides N 1993 An experiment in multimedia GIS. *Proceedings, European GIS Conference, Genoa, Italy* April: 203–12

Rhind D W, Armstrong P, Openshaw S 1988 The Domesday machine: a nation-wide geographical information system. *The Geographical Journal* 154: 56–8

Schon D A, Rein M 1994 *Frame Reflection.* New York, Basic Books

Shiffer M J 1992 Towards a collaborative planning system. *Environment and Planning B: Planning and Design* 19: 709–22

Shiffer M J 1993 Augmenting geographic information with collaborative multimedia technologies. In McMaster R B, Armstrong M P (eds) *Proceedings AutoCarto 11.* American Society for Photogrammetry and Remote Sensing/American Congress on Surveying and Mapping: 367–76

Shiffer M J 1995a Environmental review with hypermedia systems. *Environment and Planning B: Planning and Design* 22: 359–72

Shiffer M J 1995c Interactive multimedia planning support: moving from stand-alone systems to the World Wide Web. *Environment and Planning B: Planning and Design* 22: 649–64

Wiggins L L, Shiffer M J 1990 Planning with hypermedia: combining text, graphics, sound, and video. *Journal of the American Planning Association* Spring: 226–35

Introduction

THE EDITORS

At any one moment in time, the general environment within which management operates is set by forces and factors far outside the control of any one individual. Thus laws may be set not only by national governments but through the compromise agreements they have struck. European Directives which apply to all members of the European Union form one such supranational agreement. Subscription to the World Trade Organisation Treaty is another, albeit more voluntary, example. Economic, legal, social, educational, employment, environmental, and other policies all impact on the geography of any one country and on the reasons for use of GIS. In some cases, societal values set the operating framework rather than explicit laws or policies. Even though the manager may be able to do little about these, he or she should be aware of their implications since they can be a source of both problems and opportunities for a rapidly globalising business like GIS. As one example, David Rhind points out in his chapter in this Section that the legal protection of databases accorded by the European Database Directive may ensure that US firms would find it advantageous to build some of these in Europe rather than in the USA.

It is a great fallacy to believe that GIS can be reduced to a 'now press button B' process. The complexities of these systems – amply demonstrated elsewhere in this book – render this mass production, deskilled labour viewpoint far from reality. It may well be that the introduction of GIS in operational management can lead to the loss of great swathes of employment (as in the British utility industry where the labour force was halved through privatisation and the introduction of new technology – including GIS). But those who are left have to operate and manage much more complex systems, perform a wider span of activities, and carry greater responsibilities. Moreover, the advent of new industries is primarily based on the trading or use of information to seize competitive advantage: it follows that 'knowledge workers' will be increasingly in demand.

For all these reasons, the importance of education and training has never been greater, both generally and in GIS. To that end, Pip Forer and David Unwin (Chapter 54) examine two cross-cutting aspects of education and training – education 'about what' and 'for whom'. They argue that most education and training thus far has been about systems, ranging from 'which button to press first' approaches to more wide-ranging pedagogy about the nature of GI Systems and how, for what, and when they should be used. In general, they find this to have been quite successful. But they argue that it needs to be superseded by education and training about GI Science (the scientific principles underlying processes operating across space and time) plus GI Studies (the social, legal, and ethical issues associated with use of GIS). Cutting across this 'what should be studied theme', they also examine how and what should be taught at tertiary (higher education) level, in schools, and for continuing professional development: they conclude that the greatest areas for expansion are the last two. Underpinning much of what they say is the view that the advent of the Web and easy-to-use ubiquitous tools changes everything in education and elsewhere in GIS.

The concept of privacy is an elusive one to define in any widely agreed way. Nevertheless it is a matter of great concern and debate in many countries at present, its manifestations ranging from the unwelcome attention of paparazzi to 'profiling' of individuals by examination of their bank accounts and other personal records. Michael Curry (Chapter 55) shows how the concept of privacy has changed over time under the influence of technological developments and how the increasing use of the law to define the nature of privacy has eroded traditional safeguards. It has been argued by some – including Curry – that GIS is playing a role in undermining privacy. Yet it can also be argued that its area aggregation capabilities, allied to statistical threshold safeguards commonplace in population census data, actually render GIS a privacy protection device. Curry's central point, however, is more subtle: he argues that the creation of inferred characteristics of 'virtual people' from aggregate or detailed data becomes an invasion of privacy when these virtual people or fictitious personae are named after real individuals. These digital fictitious personae become reality so far as many commercial or even government processes are concerned. He argues that the trend to geocoding everything via address look-up tables, and such like, will lead to a world without privacy. Is this really so? Even if true, is it unavoidable and does it matter? Or is it simply a minority academic concern expressed by those isolated from business except through having their salaries ultimately paid by the fruits of commercial labours of others?

The final chapter by David Rhind (Chapter 56) discusses national and international geospatial data policies. The existence of such data, their currency, accuracy and consistency, their availability and price, and the terms and conditions of their use are now all major factors in determining the utility, cost, and effectiveness of GIS. The chapter examines the economic, legal, and other policies of the relevant parties, notably data suppliers – and especially those in government since it is government that has been the source of most geospatial data until now. The relevant policies extend far beyond data pricing to include procurement, intellectual property rights, encouragement of markets, monopoly trading, and privacy. Rhind shows that there are considerable variations in these policies between countries; indeed in some countries there is either no information policy at all or multiple, sectorally-based ones. One manifestation – but certainly not the only one – of these international policy variations is the stark difference between US 'cost of copying' provision of data by the Federal Government in contrast to the 'market pricing' policies operated in Canada, New Zealand, the Nordic countries, and the UK. From a technical standpoint, the growth of a commercial data supply sector producing high-resolution satellite imagery and road centreline and other data will impact both on national policies and on government data suppliers in future.

Recent national and international policy initiatives have been set up by governments worldwide to foster wider and safer use of geographical information. An early version of these was the US National Spatial Data Infrastructure. The key features of this, and the similarities and differences between it and two equivalents, are also outlined by Rhind. The variation in national policies, both in principle and in their detailed implementation, has consequences for the nature of GIS developments in different countries and has particular implications for those organisations operating across national frontiers. However – Rhind argues – the situation is unlikely to change dramatically in the near future. Is homogeneity of policy impossible and even perhaps undesirable?

55

Rethinking privacy in a geocoded world

M R CURRY

GIS are playing a role in undermining privacy. This chapter summarises the concept of privacy, how it has changed over time, and how it has been historically influenced by the introduction of new technologies. It is shown that the ways in which technological developments are regarded as autonomous and the increasing use of the law to define the nature of privacy have eroded traditional safeguards. The growth in use of geographically detailed data has led to the growth of geodemographics and the construction of fictitious personae which are now widely used for direct marketing. These digital personae are reality as far as many commercial and even government processes are concerned. Specifically, the author argues that the trend towards geocoding more and more information will lead to a world without privacy – or with a radically different form of privacy – and urges the GIS community to act responsibly to avert such an out-turn.

1 PRIVACY AS AN ISSUE

Today privacy is a difficult and contentious issue. In the USA the very belief in a constitutionally mandated right to privacy is controversial, seen by some as a litmus test to be applied to potential justices and politicians. And if some wonder whether people have in fact been guaranteed a right to privacy, others claim simply that privacy is dead, killed by a computer technology that has created a world in which, increasingly, everything is open to view.

Given the angry and even apocalyptic tenor of this discourse, it is sometimes difficult to see clearly what is happening. Should there be a right to privacy? Can there be? Yet confusing as it is, this is an issue that ought to be faced, and especially by those who develop and use GIS. These systems and their associated technologies – global positioning systems, remote surveillance systems, and computer cartography – are tied to the very practices and institutions of data collection and surveillance that have figured so prominently in the jeremiads for privacy.

In what follows it will be suggested that the advocates of a right to privacy have made an important point. Indeed, the evidence shows that privacy, far from being a luxury, is a necessity for many of those elements of everyday life – including democracy and science – that we take as valuable. At the same time, the development first of information systems and now of GIS, in concert with other geographical and social changes, has not destroyed that right, but rather resulted in its reconfiguration. If in the past the right to privacy could be seen as a right to be left alone, today it needs to be seen, instead, as a matter of the right of control over one's identity. Those who develop and use GIS, at least to the extent that they believe in things like democracy and science, need to be careful about what they do. And indeed, simply being careful is not likely to be enough.

First, development of the explicit right to privacy over the last 100 years will be described briefly, and in doing so the ways in which the right has been redefined in response to technological changes will be shown. Second, more recent developments will be examined. The computer, in particular, has led to a rethinking of the right to privacy; where before to have a right to privacy was to have a right to be left alone, the development of computerised information systems means that privacy is now a matter of the protection of data about oneself. Before, one worried about people peeping in one's window; after the development of the computer, one worries about someone peeping into one's past.

Finally, it will be shown that recent developments, largely fuelled by the easy availability of geographical data, have led to a new way of thinking about privacy. With geographically-coded data it is possible to develop profiles of individuals and neighbourhoods, profiles that suppose what people might be like (see Birkin et al, Chapter 51). Thus it is now easily possible to construct around the name of a person, group, or place a new image – an identity that is in one sense fictional, yet that at the same time appears as real as any other. And this geographically-driven development of data profiles has led to a very real need to rethink the right to privacy. Today the desire for privacy is no longer driven by a desire to be left alone. Rather, it is driven by a desire to have control over those multiple identities – identities lodged in direct marketing agencies, credit reports, geodemographic profiles, and the like.

2 THE GENESIS OF THE RIGHT TO PRIVACY

Although privacy itself is old indeed, the formal codification of a right to privacy is rather new. Granted, there has always – at least in the West – existed a strong expectation of something that we would call privacy. But it was codified only in pieces; in the United States Constitution, for example, the closest to an explicit right is the Fourth Amendment, concerning searches and seizures, which states that:

'The right of the people to be secure in their persons, houses, papers, and effects, against unreasonable searches and seizures, shall not be violated, and no Warrants shall issue, but upon probable cause, supported by Oath or affirmation, and particularly describing the place to be searched, and the persons or things to be seized'.

Through the 19th century, the right to privacy was woven into the pattern of everyday life. In the smallest village there were places where one did not go, things that one did not do or say – at least in public. The strictures against saying the unsayable were seldom stated as formal rules, but rather were enforced through admonition and example. In a small-scale society in which patterns of authority

were well established, it was possible for individuals to maintain something that most of us would identify as privacy, even in the absence of a formal codification of its nature.

But in the same century the development both of an increasingly urbanised society and of technological means for the dissemination of information undercut those patterns of authority. As Georg Simmel bemoaned in his famous 'The metropolis and mental life' ([1903] 1971) and Louis Wirth later formalised for American urban sociology (Wirth 1938, 1969), the urban became a place in which an individual could choose to remain isolated and anonymous.

And so, toward the end of the 19th century a formal right to privacy was first described in the USA, in a famous law-review article by Warren and Brandeis (1890). There they claimed that technological changes, such as the newspaper, allowed an anonymous writer to make claims about individuals, without having to face those individuals. We needed a right to privacy to ensure our ability 'to be left alone'. So the problem for privacy created in this new landscape, where the actions of the individual seemed much less constrained by custom, was solved by the development of a formalised set of privacy guarantees.

In the end – and this should be no surprise, given that privacy has evolved as a set of everyday practices and institutions – this turned out not to be a simple task. Indeed, the Flaherty (1989) catalogue of the elements of the right to privacy shows it to be extraordinarily complex:

- the right to individual autonomy;
- the right to be left alone;
- the right to a private life;
- the right to control information about oneself;
- the right to limit accessibility;
- the right of exclusive control of access to private realms;
- the right to minimise intrusiveness;
- the right to expect confidentiality;
- the right to enjoy solitude;
- the right to enjoy intimacy;
- the right to enjoy anonymity;
- the right to enjoy reserve;
- the right to secrecy.

The right to privacy developed within larger communities, but central to that right has long been the view that the home is the central locus of private activities, the place where one really can be left alone. Or to be more exact, in keeping with common law, what is important has been less that area bounded by four walls than the somewhat larger area within which the intimate activities of everyday life were taking place; this 'curtilage' consisted of (as the Oxford English Dictionary puts it): 'A small court, yard, garth, or piece of ground attached to a dwelling-house, and forming one enclosure with it, or so regarded by the law; the area attached to and containing a dwelling-house and its out-buildings.'

Indeed, in the USA the courts long recognised that the distinction between the dwelling-house and curtilage, and the 'open fields' beyond is 'as old as the common law' (Hester v. United States 1924: 59).

Yet by the 1920s the telephone had begun to undercut the distinction between the home and curtilage and that which lay beyond. This change was quickly reflected in court cases. In 1928, in Olmstead v. United States, the Supreme Court held that:

'By the invention of the telephone, fifty years ago, and its application for the purpose of extending communications, one can talk with another at a far distant place. The language of the [Fourth] Amendment can not be extended and expanded to include telephone wires reaching to the whole world from the defendant's house or office . . . . The reasonable view is that one who installs in his house a telephone instrument with connecting wires intends to project his voice to those quite outside, and that the wires beyond his house and messages while passing over them are not within the protection of the Fourth Amendment . . .' (Olmstead v. United States 1928: 465–6).

In their view, a wiretap involved no invasion of privacy, since it did not involve: 'an official search and seizure of his person, or such a seizure of his papers or his tangible material effects, or an actual physical invasion of his house or "curtilage" for the purpose of making a seizure. We think therefore that the wire tapping here disclosed did not amount to a search or seizure within the meaning of the Fourth Amendment' (Olmstead v. United States, 1928: 4666).

In the USA this was for many years the guiding legal view; the possibility that technological change required a rethinking of the nature of privacy was simply rejected. But by the 1960s the weight of technological change had increased to the point at which it became much easier to engage in just such a rethinking. Indeed, it seems fair to believe that since that time the nature of privacy has been constantly at issue.

For example, in the late 1960s, in Katz v. United States (1967), the Supreme Court for the first time asserted that modern urban life calls for a rethinking of the 'age-old' distinction between the home and the public beyond. Recognising that the telephone is a medium for private conversations, and that people often engage in telephone conversations while away from home, the Court concluded that even within a public telephone booth a person's right to privacy could be violated, simply because by closing the door the individual feels justifiably isolated from the public world outside. In effect the Court recognised that the modern, technologically connected city is a new sort of landscape.

This reforming of the modern urban-industrial landscape, prefigured in Warren and Brandeis but not really acknowledged until much later, was central to the rethinking of privacy through the 1960s. But a second factor has been equally important, and that is the development of the computer.

If the computer has only really become ubiquitous in the last 15 years, there have nonetheless been two moments in which some believed it possible to see the direction in which information processing equipment was pointing. The first was in the era between the 1910s and the 1930s. Then, as Beniger (1986) has pointed out, information processing equipment previously used for purely technical purposes began to be used to 'strengthen the control maintained by the entire bureaucratic structure' (Beniger 1986: 408). Symbolically, this era was captured by the production of America's first Social Security retirement payouts, in 1937, on a punched card. And indeed, as Beniger notes, the arrival of the punched card was greeted in some quarters by a fear that the new system depersonalised people, turning them into mere numbers.

More familiarly, the second moment occurred in the 1960s. At that point, the arrival in government offices of mainframe computers and magnetic tape storage seemed to many to signal an escalation in government's ability to maintain stores of information about its citizens. And for a number of reasons – certainly including the political climate surrounding the Vietnam War and the civil rights movement – there was in this era a public outcry against the possibility of governments creating large databanks of information on individuals and groups (Columbia Human Rights Law Review 1973; Mowshowitz 1976; Rule 1973; Westin 1972).

Central to privacy concerns was the belief that by using computers, governments would be able to combine individual databases into systems that contained comprehensive, cradle-to-grave dossiers on every citizen and resident. If totalitarian countries had previously managed this task using paper records, the ability to create computerised dossiers that could easily be communicated seemed to many to be far more troubling.

In response to political pressure, it was during this era that a number of countries developed systems for the control of these data. In the USA, the issue was hotly debated. Relying in part on the recommendations of a Senate committee (US Senate 1974), Congress passed the Privacy Act (US 1974). At the same time, and relying to varying degrees on proposals developed by the Organisation for Economic Cooperation and Development (OECD 1976), the Commission of the European Communities (1980), and the Council of Europe (1981a), most of the countries of Western Europe passed similar legislation (Bennett 1992; Flaherty 1989).

One feature that the laws enacted in the USA and Europe had in common was a belief that the primary source of threats to individual privacy was government. And the focus of concern, by and large, was the possibility that standard identifiers, such as the US Social Security Number, might be used in a process called 'data matching' to combine dossiers across agencies.

There were, it needs to be granted, privacy concerns directed at non-public bodies. In Europe there has been widespread concern for over 20 years, evinced in documents emerging from the Council of Europe (Council of Europe 1973, 1981a, 1981b, 1984, 1986, 1989, 1990), the European Commission (Commission of the European Communities 1980, 1991), and the Organisation for Economic Cooperation and Development (OECD 1976, 1992). But in the USA, and to a lesser degree in Europe, the fact that such non-public bodies were seen as constituting less of a threat is indicated by the piecemeal approach taken to them. At the extreme, in the USA privacy issues in the private sector have been taken up as they moved into public consciousness, as when the inquiry by reporters into the video-rental records of a nominee to the Supreme Court motivated members of Congress to pass the Video Privacy Act (US 1988).

As Bennett (1992) has noted, one of the striking things about privacy regulations in Europe and America has been their similarity. Although the means for protection have varied, the regulations have in common an appeal to what have come to be known as the fair information principles. These require that any body which maintains a system containing information about individuals must:

- make public the existence of the system;
- give individuals access to data about them;
- give individuals the right to correct erroneous information.

Furthermore,

- personal data should be collected only where necessary;
- personal data should be used only for the purposes for which they were collected;
- personal data should not to be disclosed to another group or agency without some sort of consent;
- personal data should be securely stored.

And so, if through the 19th century privacy was conceived as a matter of the establishment of an appropriate arena within which people could be left alone, the joint development of an industrial, urban society and of the computer had, by 100 years later, led to a refiguration of the concept. The central concern by the 1970s was to protect the data that had been collected by government organisations. The fair information principles laid out what seemed to be appropriate means for the achievement of that goal.

3 THE COMING OF GIS

During the period of the first real flowering of the computer, in the 1960s, concerns about privacy typically centred around the belief that by using a computer some centralised authority – typically a government – would be able to gather large amounts of information about individuals. Indeed, almost every government database relied in the end on the attaching of an identifying number (such as the American Social Security Number or the Canadian Social Insurance Number) to each individual.

But whatever the problems with such systems, they have one merit: the data are only collected about individuals – which was the focus of the fair information principles. With the right set of rules and procedures it would be possible for each individual to check the accuracy of a dossier, making corrections where needed.

It is here that the coming of the geocoded world dramatically recasts the problem of privacy regulation, and then the nature of privacy itself. There are, in fact, two rather separate moments to this refiguration. First, there are changes in the nature and availability of data about individuals, households, and groups. These changes are having a direct effect on the nature of everyday life, both by reshaping those elements of life that have in the past been called 'private' and by creating new elements from which will be constructed a new way of thinking about the private realm. And second, there are certain features of these technological developments that make them seem natural and normal. As a consequence, the accompanying changes in the nature of the right to privacy are, to a degree, seen as equally natural and normal. Putting the matter simply, there are two issues related to the advent of widespread use of GIS: the ways in which privacy is changing and the ways in which those changes are being greeted.

3.1 GIS and the changing nature of privacy

It seems likely that those involved in the development of the GBF-DIME files in the US Bureau of the Census and in the development of the zone improvement plan (ZIP) code in the US Postal Service had no real conception of the future impact of those developments. The GBF-DIME files, after all, were developed primarily as an aid to the streamlining of the decennial census, just as the ZIP code was a way of streamlining the sorting of mail. Yet in the end, the two combined into a branch of GIS that has been deeply connected with the refiguration of private life, and hence with a need to rethink the nature of the right to privacy. This branch is geodemographics.

The development of a computerised system to map the areas occupied by most Americans was a major step toward a refiguration of the right to privacy. This was even more the case because of the ways in which that mapping occurred; in the end it became possible to associate street names and address ranges with geographical coordinates, and so through interpolation geographically to locate the address of any household in urban (and later all) America. Subsequently, much more precise geographical coordinates have been introduced in other nations, notably in Britain.

The general introduction in the late 1960s of the postal ZIP code in America preceded slightly the Census Bureau's computer mapping system. Nonetheless, the ZIP code – the second arm of the geodemographic revolution – was immediately seized upon by the marketing industry. For example, in a 1967 article 'Zip Code – New Tool for Marketers', Baier hailed it as 'a "built-in" and universal means of geographical identification' (Baier 1967: 140). As he put it: '[The] ZIP Code System offers a new, unique opportunity; the way it has been put together (although devised for quite a different reason, namely postal efficiency) just happens to fit many marketing needs' (Baier 1967: 136).

And with the development of computer mapping systems, marketers, and others, were now able to combine data from the Census Bureau's mapping project with data from the Postal Services Carrier Route Information System (which consisted of a comprehensive listing of mailing addresses) and therefore to create lists that provided a geographical location for every address in the nation. Thus was born the geocoded society.

From the point of view of the issue of privacy, two aspects of this technological development were of immediate importance. First, it meant that the locus of information moved from the individual to the household. And second, marketers immediately noted that the new system allowed them a much more powerful way of applying an insight that they had long had, that 'people with like interests tend to cluster' (Baier 1967: 136).

By the early 1970s the first of a growing group of corporations were established with the aim of applying this insight – that people tend to cluster with others of like characteristics – in the context of increasingly powerful and affordable computer technology. These companies, engaging in what they termed 'geodemographics', moved rapidly away from the mere collection of information about individuals (Curry 1992, 1997; Goss 1995; Larson 1992; Weiss 1988). Faced with restrictions (such as the American Fair Credit Reporting Act of 1970: US 1970) on the collection of individual information, they began to take the household as the primary unit. At the same time, relying on the theory that people cluster with others like themselves, they began to develop methods of 'data profiling'.

In data profiling the aim is not to say what an individual or household is like. Rather, it is to say what that individual or household is *probably* like.

Drawing upon a wide array of government data, market surveys, and purchasing records, producers of geodemographic profiles divide an area into a small number of regional types and then attribute the characteristics of that type to each household (see Birkin et al, Chapter 51). If household or individual data are available, those data may be added to enhance the discrimination capability and guide automated decisions made about which household is to receive which piece of advertising, which version of a news magazine, which offer of credit.

It is this profiling which undercuts traditional methods for the protection of the right to privacy and at the same time results in a refiguration of that right. For if it seemed possible, at least in theory, to apply the fair information principles to individual data, just how that might be done in the case of geodemographic data is not at all clear. Geodemographics creates a whole new range of individuals – what has been called 'digital individuals' (Agre 1994) or the 'digital personae' (Clarke 1994) – and households. If in the past everyone had a reputation, these reputations are now codified, stored in computers, and bought and sold. Through geocoding we now have a world of virtual households and virtual selves, taken by many people to be more real than old-fashioned physical ones.

All this indicates that the nature of privacy itself needs to be rethought as it was 100 years ago. For if the traditional functions of privacy are to persist, that is to allow people to control what aspects of themselves can be seen, to allow them to develop their personalities, to allow the testing of ideas that is essential to democracy, to science, and to economic innovation, what counts as privacy needs to change. People need to have control over the virtual individuals and virtual selves that others create.

4 IMPEDIMENTS TO A NEW RIGHT TO PRIVACY

There are two impediments to the sort of rethinking of privacy suggested above. Both, as it happens, are directly concerned with issues raised not simply by geodemographics, but rather by GIS more generally.

The first of these impediments is a very general way of thinking about technology and social change. It will be useful here to turn away from GIS to one of the predecessors of these systems, the paper map. The paper map has two striking features. Just to the

extent that it includes identifiable elements it appears to contain the possibility of being linked with other maps, and indeed, with all other maps. It seems possible, in principle, to create one very large map that is a compendium of all other maps. Here, in fact, the map – just because of the way in which it points to the face of the Earth – appears to provide just the model of the organisation of knowledge that has eluded those who have attempted to organise discursive knowledge.

At the same time, when we see an object or event located on a map it seems perfectly possible to locate that event with increasing accuracy. If we had a map constructed at a one-to-one scale, we could quite literally show exactly where I am sitting as I write this, down to the smallest fraction of a millimetre.

In these two ways, by seeming intrinsically to allow for the concatenation of maps and by seeming intrinsically to involve a concept of absolute accuracy, the map appeals to and seems to support the idea that the development of the map is preordained. That is, the map seems to support the view that left unfettered, maps would come to be more and more accurate, and more and more interconnected. The map seems to support the notion that the development of technologies is, as Langdon Winner has termed it, 'autonomous' (Winner 1977).

As appealing as this idea is, and it has a long history of support, there is little to justify it. Indeed, Mackenzie (1990) has shown in the similar case of missile guidance systems just how little this view of accuracy can be justified. But justified or not, the view of the map – and now the GIS – as developing autonomously has worked its way both into popular culture and into common sense. And there, true or not, it has an effect on the understanding of privacy. To the extent that people believe the development of privacy-related technologies to be an inexorable process with a preordained goal, they are likely to see that process as natural and normal, and are likely to see the right to privacy as something that must undergo a regular process of diminution. Indeed, this is the very understanding of the right to privacy that has over the last 20 years been enunciated in the US courts (Curry 1997a). So even ignoring geodemographics, the very existence of the family of technologies that make up GIS – and especially remote surveillance systems and global positioning systems – has an impact on the right to privacy.

There is a second impediment to the reconceptualisation of privacy as an associated right. It was suggested earlier that in the USA the right to privacy as a formal right was articulated about 100 years ago. A critical feature of that articulation, and one that was not mentioned, was that in the famous Warren and Brandeis law review article they not only defined a right to privacy, but defined that right as a legal right. Before that time, privacy had been something that people had by virtue of being in particular places and social situations; the violation of privacy was therefore a social violation. But after Warren and Brandeis, and more so after Griswold v. Connecticut (1965), privacy became a legal right.

In fact, a fundamental impediment to the rethinking of privacy is the extent to which the discourse around privacy has come to be seen in the first instance as a legal discourse. And here too, though in a less obvious way, the development of GIS can be seen as supportive of the increased hegemony of legal discourse (see Onsrud, Chapter 46), and the accompanying decline of other forms of moral and social discourse, and control. As noted elsewhere (Curry 1994, 1995, 1996), an important feature of GIS has been their role in recasting traditional systems of ownership and authority. The creator of works in science could at one time expect to obtain ownership rights in those works, where those rights were in a fundamental way outside the legal system. Newton, to take an obvious example, did not need to hire a lawyer in order to get credit for his laws. The credit was given within the context of a system that defined that credit in social and moral terms.

But GIS, like the rest of big science (Pickering 1989; Price 1963), move us away from the traditional model of the scientist as heroic loner, to the new model of the institutionally and technologically GIS-supported research manager, allocating rights and responsibilities in detailed and formalised ways. And to the extent that a scientist is using hardware, software, and data of the complexity of those involved in the typical GIS, he or she is embracing a move from social means of guaranteeing appropriate behaviour to legal means. The upshot of the use of the systems in geography, as elsewhere in science, is a social system in which rights and responsibilities are allocated through that legal system; it is a system that sees the problems of human interaction as definable in strictly legal terms.

5 RETHINKING THE NATURE OF PRIVACY

It may now seem as though I have painted a bleak picture. We are moving toward a geocoded society, a society in which for every physical self there are a dozen or more 'virtual selves', applying for credit, buying magazines, renting apartments, looking for the perfect spouse. These selves, constructed from bits and pieces of data, much of it associated with me only because of where I live, work, or shop, are taken by many people to be more real than the real one. And in an important sense these selves are not only beyond my control, they are beyond the control of anyone – at least if we rely on the fair information principles. For those principles were developed for a world without geocoding.

Yet as far as the rethinking of the nature of privacy is concerned, one has to say on the evidence that it *is* occurring. One need not look just at academic articles in order to find a wide-ranging dissatisfaction with current ways of thinking about the issue. Certainly this was expressed several years ago in the Lotus MarketPlace dispute (Bain 1991; Culnan 1991; Gurak 1995; Seymour 1991). When Lotus announced a software/data product that would work on a desktop computer and would provide to small business the same detailed geodemographic and personal information about most of the households in the country that had long been available to large data users, there was a widespread and vigorous response. In the end, having received more than 30 000 requests to have individual information deleted, Lotus decided not to introduce the product.

A similar response greeted the 1996 rumour – that Lexis-Nexis would introduce a system called 'P-Trak', that would consist of records containing individuals' social security numbers and the maiden names of their mothers (a common security identifier). Although Lexis-Nexis denied that there was ever a plan to contain highly personal data in the system, they did back down, and removed social security numbers (Aguilar 1996; Flynn 1996a, 1996b). In both cases the public complained not about the existence in these databases of confidential or inaccurate data, but rather about the release of those data in ways that constituted uncontrolled, digital identities.

So whatever the impediments to the rethinking of the nature of privacy, it appears that that rethinking is in fact occurring, although not always in the name of privacy. But this leaves the second question: if people are now rethinking the ways in which they wish information about themselves to be available, if they are rejecting the wholesale proliferation of virtual selves, how ought these changes be reflected in the work of the everyday user of a GIS?

The first and most obvious answer – that the creator and the user of a GIS ought to be careful and ought not to produce works that will patently be damaging – is both an inadequate one and a truism. It needs to be said, though, simply because in the current frenzy of data buying and selling and in the current political climate there are many who would take the position of Lexis-Nexis, who responded to the P-Trak controversy by saying the company could not be held responsible for what is done with the service's information (Aguilar 1996). Yet to be careful is not enough, as the history of the right to privacy that I have sketched above shows; surely none of the inventors or early promoters of the telephone or the computer could have predicted its impacts.

A second step, and a step in the right direction, would be to recognise that new technologies can make old laws obsolete. Nowhere is this more true than in the case of public records. For public records, such as property assessments and court documents, were made public in a time in which getting those records was recognised to be not easy. It required going to the appropriate office in the appropriate city at the appropriate time, waiting in line, laboriously copying materials by hand and sometimes giving a well-placed gift.

Once placed on computers, those same records can easily be accessed by a person with a modem and a little free time. Where before few would have bothered to look at their neighbours' property assessments, now many more can and with much greater ease. Rules for access that under one set of technological conditions meant one thing now mean something very different.

Even this recognition – and an accompanying action – may well leave the user of GIS in the position of unwittingly supporting a means of undercutting the right to privacy. One solution here, and one that has some support in Europe, would simply outlaw the mixing of data pertaining to aggregations with individual data; it would assume that, in the mixing, the data become personal and hence it would make illegal much of what is now the direct marketing business.

This tack has the advantage that it appeals to an existing, and even common, model of a privacy-friendly way of thinking about individuals and households. That is the model that we use in thinking about our relations with our own work and the relations of other people to theirs. It is a model that abjures the traditional American and British conception of ownership, and instead operates in terms of a European conception. This European conception, often termed the 'personality' or 'moral-right' model, has long been used, and is in fact incorporated into national and international copyright agreements (see, for instance, Grelot 1997). It specifies that individuals have the right to control the ways in which that which they make can be used and modified, just because the products that individuals create are intrinsic to their identities. As scientists we routinely appeal to this very system; it is, after all, the one that we are taught in graduate school. And it is one that in important ways serves us well.

We would do well, in our work, however, to recognise this fact and, where possible, take it to heart, yet it seems extraordinarily unlikely that such an approach will be formally recognised in the USA. Indeed, and as I have suggested elsewhere, the USA is involved not in adopting such a view but rather in promoting its abolition in those places, primarily in western Europe, where it is now held (Curry 1996). And recent initiatives, such as the World Intellectual Property Organisation's (1996) 'Draft Treaty on Intellectual Property in Respect of Databases', appear in the law to signal the end of the moral-right view and with it the demise of the traditional means of regulating the exchange of scientific data.

There are ways though in which one can create a GIS that does not fit so easily into systems that undercut personal privacy. One major way in which the systems contribute to a diminution in personal privacy arises from their use of geocoding. But paradoxical as it sounds, one can in fact create GIS that do not use geocoding; and such systems, to the extent that they are much more difficult to combine with others, offer a degree of protection. So geographers ought – if they are serious about privacy – also to take seriously the need to question the ways in which they carry out their work. In particular, they need to question the belief that, where geocoding is possible, it ought to be done. In the end, a world in which everything is geocoded will be a world in which virtual individuals and households are everywhere and in which they cannot be controlled. It will be a world without privacy.

References

Agre P E 1994 Understanding the digital individual. *The Information Society* 10: 73–6

Aguilar R 1996 Service pulls Social Security numbers. *C/Net* June 12 1996

Baier M 1967 Zip code – new tool for marketers. *Harvard Business Review* 45: 136–40

Bain G D 1991 Lotus primes MarketPlace for desktop marketing. *MacWEEK* 5: 31ff

Beniger J R 1986 *The control revolution: technological and economic origins of the information society.* Cambridge (USA), Harvard University Press

Bennett C J 1992 *Regulating privacy: data protection and public policy in Europe and the United States.* Ithaca, Cornell University Press

Clarke R 1994 The digital persona and its application to data surveillance. *The Information Society* 10: 77–94

Columbia Human Rights Law Review 1973 *Surveillance, dataveillance and personal freedoms.* Fair Lawn, R E Burdick

Commission of the European Communities 1980 *Commission communication on the protection of individuals in relation to the processing of personal data in the Community and information security.* Brussels, European Community

Commission of the European Communities 1991 *Opinion on the proposal for a Council Directive concerning the protection of individuals in relation to the processing of personal data.* Brussels, European Community

Council of Europe 1973 *Protection of the privacy of individuals vis-á-vis electronic data banks in the private sector.* Strasbourg, Council of Europe

Council of Europe 1981a *Convention for the protection of individuals with regard to automatic processing of personal data.* Strasbourg, Council of Europe

Council of Europe 1981b *Regulations for automated medical data banks.* Strasbourg, Council of Europe

Council of Europe 1984 *Protection of personal data used for scientific research and statistics.* Strasbourg, Council of Europe

Council of Europe 1986 *Protection of personal data used for the purposes of direct marketing.* Strasbourg, Council of Europe

Council of Europe 1989 *Protection of personal data used for employment purposes.* Strasbourg, Council of Europe

Council of Europe 1990 *On the protection of personal data used for payment and other related operations.* Strasbourg, Council of Europe

Culnan M J 1991 The lessons of the Lotus MarketPlace: implications for consumer privacy in the 1990s. Paper read at The First Conference on Computers, Freedom and Privacy

Curry D J 1992 *The new marketing research systems: how to use strategic database information for better marketing decisions.* New York, John Wiley & Sons Inc.

Curry M R 1994 Image practice and the hidden impacts of geographic information systems. *Progress in Human Geography* 18: 441–59

Curry M R 1995b Rethinking rights and responsibilities in geographic information systems: beyond the power of the image. *Cartography and Geographic Information Systems* 22: 58–69

Curry M R 1996 Data protection and intellectual property: information systems and the Americanisation of the new Europe. *Environment and Planning A* 28: 891–908

Curry M R 1997a Geodemographics and the end of the private realm. *Annals, Association of American Geographers* 87: 681–99

Flaherty D H 1989 *Protecting privacy in surveillance societies: the Federal Republic of Germany, Sweden, France, Canada, and the United States.* Chapel Hill, The University of North Carolina Press

Flynn L J 1996a Company stops providing access to social security numbers. *New York Times* June 13 1996

Flynn L J 1996b Lexis-Nexis flap prompts push for privacy rights. *New York Times* October 13 Cyber Times Extra

Goss J 1995 'We know who you are and we know where you live' the instrumental rationality of geodemographic information systems. *Economic Geography* 71: 171–98

Grelot J-P 1997 The French approach. In Rhind D W (ed.) *Framework for the world.* Cambridge (UK), GeoInformation International: 226–34

Griswold v. Connecticut. 1965 381 *US 479 (1965)*

Gurak L J 1995 Rhetorical dynamics of corporate communication in cyberspace. The protest over Lotus MarketPlace. *IEEE Transactions on Professional Communication* 38: 2–10

Hester v. United States. 1924 265 *US 57 (1924)*

Katz v. United States. 1967 389 *US 347 (1967)*

Larson E 1992 *The naked consumer: how our private lives become public commodities.* New York, Penguin

Mackenzie D 1990 *Inventing accuracy: an historical sociology of nuclear missile guidance.* Cambridge, MIT Press

Mowshowitz A 1976 *The conquest of will: information processing in human affairs.* Reading (USA), Addison-Wesley

OECD 1976 *Policy issues in data protection and privacy.* Paris, Organisation for Economic Cooperation and Development

OECD 1992 *Privacy and data protection – issues and challenges.* Paris, Organisation for Economic Cooperation and Development

Olmstead v. United States. 1928 277 *US 438 (1928)*

Pickering A 1989 Big science as a form of life. Paper read at 'The restructuring of the physical sciences in Europe and the United States 1945–1960' 19–23 September 1988, Singapore

Price D J de S 1963 *Little science big science.* New York, Columbia University Press

Rule J B 1973 *Private lives and public surveillance.* London, Allen Lane

Seymour J 1991 Lotus' MarketPlace succumbs to media hysteria. *PC Week* 8: 57

Simmel G [1903] 1971 The metropolis and mental life. In *Georg Simmel on individuality and social forms*. Chicago, University of Chicago Press: 324–39

US 1970 Fair Credit Reporting Act of 1970. 15 *USC. Sec. 1681*

US 1974 The Privacy Act of 1974. *PL 93-579 15 USC 552a Sec. 3 (e) (4)*

US 1988 Video Privacy Act of 1988. 18 *USC Sec. 2901*

US Senate Subcommittee on Constitutional Rights of the Committee on the Judiciary 93d Congress 2d session, 1974 *Federal data banks and constitutional rights*. Washington DC, US Government Printing Office

Warren S, Brandeis L D 1890 The right of privacy. *Harvard Law Review* 4: 193–220

Weiss M J 1988 *The clustering of America*. New York, Harper and Row

Weiss M J 1994 *Latitudes and attitudes: an atlas of American tastes, trends, politics, and passions*. Boston, Little, Brown and Co.

Westin A F 1972 *Databanks in a free society: computers record-keeping and privacy*. New York, Quadrangle Books

Winner L 1977 *Autonomous technology: technics-out-of-control as a theme in political thought*. Cambridge (USA), MIT Press

Wirth L 1938 Urbanism as a way of life. *American Journal of Sociology* 44: 1–24

Wirth L 1969 Rural-urban differences. In Sennett R (ed.) *Classic essays in the culture of cities*. New York, Appleton-Century-Crofts: 165–9

World Intellectual Property Organisation 1996 *Draft treaty on intellectual property in respect of databases*. Geneva

56
National and international geospatial data policies

D W RHIND

As the technology of GIS becomes ever more ubiquitous and apparently ever easier to use, other factors condition its development. Perhaps the key factor is the availability of 'content' – the data and information which act as the 'fuel' for geographical information and other computer systems. The existence of such data, their currency, accuracy, and consistency, their availability and price and the terms and conditions of their use, are now all major factors in determining the utility, cost, and effectiveness of GIS. This chapter examines the economic, legal, and other policies of the relevant parties, notably data suppliers – and especially those in government since that is the source of most geospatial data until now. It also examines the consequences of different policies operating within and between different nations. The relevant policies extend far beyond data pricing to include procurement, intellectual property rights, encouragement of markets, monopoly trading, and privacy.

Finally, recent national and international policy initiatives have been set up by governments worldwide to foster wider and safer use of geographical information. An early version of these was the United States National Spatial Data Infrastructure (US NSDI). The key features of this and the similarities and differences between it and two equivalents are outlined. It is concluded that national policies vary enormously both in principle and in their detailed implementation. This has some consequences for the nature of GIS developments in different countries and has particular implications for those organisations operating across national frontiers. However the situation is unlikely to change dramatically in the near future.

1 INTRODUCTION

It is now well established that the use of GIS and geographical information (GI) can play a key role in human activities. This role ranges from the affairs of individual businesses (e.g. Birkin et al, Chapter 51) through the governance and economic development of large areas or countries (e.g. Smith Patterson and Siderelis, Chapter 53; MacDevette et al, Chapter 65) to international development programmes (Htun 1997).

In the early phases of the GIS revolution, the primary concerns of users were hardware and software which functioned reliably and speedily (Coppock and Rhind 1991). In the last half decade

or so, however, the primary concern has shifted to the information required to use the GIS. Since acquisition of certain types of geographical information or geospatial data is difficult and/or costly to collect for many individuals, this concern has translated into how to get the data from other parties – where to find them; how to obtain their costs, currency, and reliability; and the terms under which they can be used and the liabilities incurred. For historical reasons, the great bulk of data commonly used in GIS have long been collected by governments at various levels throughout the world. The policies of these governments are therefore primary influences on the current and future use of GIS and provide the main focus of this chapter. Here the main concern is not what should or could be, but rather what is, the situation in terms of the current availability and utility of geographical information: the chapter therefore takes a very different approach to the Barr and Masser (1997) paper.

The gathering of GI or geospatial information has hitherto been localised. Rarely have attempts been made to collect consistent data over multiple countries though there have been previous attempts to coordinate summaries of it. Of the latter, the earliest and best known is the International Map of the World project initiated in 1891 which petered out in the 1980s (Thrower 1996). Even within individual countries, however, such GI has typically been collected on a basis defined largely by individual government departments either for their own purposes or for use by a limited number of the organs of the state. The only consistency has derived from two sources – historical legacies and inherited frameworks. Thus many government bodies have collected data on a periodic basis in a manner akin to that in which they have previously collected it (e.g. through population censuses) and they have used the 'topographic template' or geographical framework (Smith and Rhind, Chapter 47) as the basis on which to do so. In short, diversity in the manifestations of phenomena to be described, in the classification of features, in spatial resolution and accuracy, in the periodicity of measurement, in the manner in which the data are stored, and in the policies and practices of data dissemination have been the norm at the national and international levels. The results of this are becoming clearer as we gain the technological capacity to assemble and exploit data on a pan-national basis: Mounsey (1991) described the CORINE project of the European Commission in which the greatest rate of change in some commonly used physical parameters were found to occur at national boundaries! A similar situation may be found within large federally organised countries where much data collection occurs at the state or provincial level.

Efforts to foster greater coherence have begun in a significant number of countries. Typically they have been manifested by idiosyncratic national initiatives, often denoted by the title of 'spatial infrastructures'. Most reflect national priorities: thus Mooney and Grant's (1997) description of the Australian National Spatial Data Infrastructure differs considerably from Tosta's (1997) characterisation of the US NSDI.

Concern with data and information policy matters is not new but has expanded greatly in recent years. Any single chapter can only summarise the key issues involved. Highly relevant contributions to this topic have recently been made by Branscomb (1994), Harris (1997), Masser (1998), Onsrud (1995), Onsrud and Rushton (1995), and Rhind (1992, 1996). The reader is, however, strongly advised not only to use these more detailed sources but to seek professional advice in order to understand how his/her organisation is liable to be affected by the matters which are described below.

1.1 The changing context in which policies are set

The creation of geospatial data policies is driven by a number of factors which interact differently in different regional or national domains, some of which are in the process of rapid change. These factors are primarily:

1 the impact of new technologies on data collecting and providing organisations, changing what they do, how they do it, and the consequences for their customers or users. One example of this is a concern to safeguard by legal means the Intellectual Property Rights for information distributed in digital form – and especially for that made available on the World Wide Web (WWW);

2 rapid change in the expectations of users. Few are now content to be told what they can have and, as a result of this and financial changes, the power of the customer or user is now much greater. As a consequence, fewer geospatial data providers are now production led;

3 changes in society values, such as the greater concern for privacy, a diminution of trust in government, and the shift of responsibilities to lower level (e.g. communities) from higher level governments;

4 the effects of reform in government, such as major reductions in staffing, new approaches to financing and management, and public exposure of successes and failures;

5 the advent of significant commercial sector GI providers;

6 the effects of regionalisation and globalisation of business and even government. In Europe, directives made centrally within the European Union force change in national laws on intellectual property, trading practices, and much else. At the

global scale, the work of the World Trade Organisation may well impact on information trading. All of these business-related developments should be contrasted with the need for global data for scientific purposes where little funding is usually available to pay for the information (see also Goodchild and Longley, Chapter 40; Collins and Rhind 1997; Draeger et al 1997).

The most notable areas in which these attentions have focused are on the 'core' or, more specifically, the 'framework' data (Smith and Rhind, Chapter 47). The reasons for this are self-evident: these data are the most widely used and are central to the use of other datasets. Without them, the use of GIS is severely constrained. That said, the principles and practices of many other data collecting and using parties also have a feedback effect on the collectors and providers of these core data. There is therefore much merit in considering them as a whole.

1.2 Who sets data and information policy?

Governments – or amalgamations of them (see section 1.3) – set policies for nations or for the subsets of the nation state for which their jurisdiction extends. Unlike the policies of commercial organisations which are manifested through strategic objectives and targets, major governmental policies are typically made operative through promulgation of statute or law. It should be noted that policy-making takes many different forms worldwide, related notably to the nature of the nation state. A substantial difference used to exist, for instance, between the centrally directed model as represented in the pre-1989 Soviet Union and that in some democratically organised societies. However, similarities as well as differences occur – President Clinton's 1994 Executive Order on the US National Spatial Data Infrastructure is a classic example of 'top down', central direction for agencies of federal government. It has something in common therefore with article 71, item 'p' of the Russian Constitution which defines the state's responsibility for geodesy, cartography, and the naming of geographical features (Zhdanov 1997).

A significant difference in policy-making exists between those nations presently organised as a unified state (e.g. Britain or France) and those organised explicitly as federations (e.g. Australia, Canada, Germany, and the USA). Within federal countries there may well be important differences in

policy by geographical area whereas this is much less likely in unitary states. There are numerous reasons for this. One reason is that there are large variations in the powers of the subsidiary bodies in these federal entities: for example, the powers of the US States remain significantly less than those of their Australian equivalents. Responsibility for different functions varies between the federal and state governments even within the same country: thus the population census is a federal responsibility in Australia but topographic mapping of the country is primarily achieved through the states. Parenthetically, one exception to this heterogeneity in data policy and practice occurs when the federal government has a monopoly in data collection of a given type, the best example being the collection of civilian remotely-sensed data in the USA in the past by the National Aeronautics and Space Administration (NASA) and its availability through the US Geological Survey (see section 4.4).

Although geographical homogeneity of policy is much more common in unitary states, differences between policies involving particular data themes may occur under either system of government. Thus, even within any one government, there are usually many different policies which impact upon the organisations responsible for collecting and disseminating geospatial data. Coopers and Lybrand (1996) reviewed the situation in relation to data providers within British government and found considerable differences in policy and practice. Rhind (1996) itemised no fewer than ten sets of laws, treaties, or agreements which influence policies within a single government department in relation to the collection and provision of data. The end result is that national, subnational, and sectoral patterns of data policy are surprisingly varied; the most significant differences to the end-user may also lie in the detail rather than in the fundamentals of the policy.

Considerable variation also exists in the relationships between public and private sectors. The concept of state-owned trading enterprises selling data or services which is common in France and Nordic countries is largely absent from the USA. This is another manifestation of different views of the state and its role: Chamoux and Ronai (1996), Grelot (1997), and Lummaux (1997), have discussed the role of the state in geospatial data provision from a French perspective, while Sandgren (1997) has summarised some important developments in Sweden (see also Dale and McLaren, Chapter 61).

All of the above (and much of what follows) concentrates on governments. Yet other policies and policy-makers are increasingly relevant to the geospatial data world. Dominant suppliers, notably Microsoft, are strongly influential, especially in setting *de facto* standards and in setting price levels. The advent of commercial 'players' in the car guidance information market, in high resolution satellite imagery, and in added value products could change the nature of the policy debate, especially since they often have very different agendas to government (Rhind 1996).

1.3 Regional or global versus national approaches to policy-making

As indicated above, the creation of binding policy is most frequently initiated at national levels in national parliaments or legislatures or through official devices such as Executive Orders. There are, however, at least two other formal ways in which policy may be initiated. These both occur through multinational governmental agreements. In one case, these bind all of the individual states into a regional agreement (e.g. through Directives prepared by the European Commission and approved within the governmental framework of the European Union). In the other, some multinational thematic agreement is forged and accepted (at least in principle) by a self-defined set of national governments. Examples of the latter include Agenda 21 (Htun 1997) and the World Trade Agreement.

In addition to these intergovernmental agreements, however, there are many other less formal attempts to set international policy which, in time, may lead to formal *de jure* or to *de facto* policies. Some national policies also take on an international role, notably those of the US military and of commercial data suppliers. Activities known in mid-1997 which may well lead directly or indirectly to geospatially relevant policy are as follows:

1 'bottom up' assembly of map-derived databases by groups of national mapping organisations, such as MEGRIN (see Smith and Rhind, Chapter 47);
2 the Japanese-led plans for the development of a global digital map database at 1:1 million scale (Warita and Nonomura 1997);
3 'bottom up' assembly of map-derived databases by the US military and its partners (Lenczowski 1997);

4 assembly of road and related feature databases for car guidance purposes by multinational private sector consortia (see Waters, Chapter 59);
5 the advent of new satellite systems which provide digital imagery data of resolutions comparable to some maps yet provide global consistency of datum and sensing tools. Most notably these are the forthcoming high resolution commercial satellites (Calvert et al 1997), but also include science missions by NASA, the European Space Agency and others (see Barnsley, Chapter 32; Estes and Loveland, Chapter 48);
6 continuing efforts to harmonise data-related standards on a global basis through official bodies like Comité Europeén de Normalisation (CEN) or International Organisation for Standardisation (ISO) and through the activities of industry-led bodies like the Open GIS Consortium (OGC) (see Salgé, Chapter 50);
7 official policy initiatives like the European Union's extended attempts to define pan-European GI polices (EC 1997) plus 'unofficial', collaborative attempts to formulate and foster governmental acceptance of policy such as that culminating in the Santa Barbara Declaration (Htun 1997) and in the Global Spatial Data Infrastructure meetings (Coleman and McLoughlin 1997; Rhind 1997).

It is again evident that there are huge variations in practice in how and why policy is formed in regard to geospatial data. One fundamental factor, however, strongly influences all these variations and this is now considered.

2 THE BIG ISSUE: THE ROLE OF GOVERNMENT

Many governments worldwide are reviewing their roles, responsibilities, and taxation policies and reforming their public services as a consequence. In many cases this is driven by a changed view of the role of the state (Foster and Plowden 1996). Fundamental reviews and subsequent reforms have taken place in New Zealand (Douglas 1994; Scott 1994), Australia, the Nordic countries, the UK (Rhind 1996), the USA (Gore 1993; Osborne and Gaebler 1992), and elsewhere, leading to some dramatic changes in what government does and how it does it. In some other countries, the process seems to be at a relatively early stage: in Japan, for instance, the creation of government agencies with public targets was being discussed in mid 1997. Moreover, in some cases the driving factor has been relatively short term: for instance, in Germany stringent financial pressures have arisen from the need to meet the convergence criteria for a single European currency laid out under the Maastricht Treaty: one manifestation of this is that the mapping agency of Lower Saxony has become a type of executive agency and has been instructed to charge customers and thus generate revenues.

In other cases, the imperative has been ideological: a strongly held view of some politicians is that the private sector is necessarily better than the public sector at any production and distribution function (though this view typically takes no account of the equity or accountability issues which form a significant part of government operations). As a result, some functions previously carried out by government have been outsourced though this is not an invariant out-turn of fundamental reviews: the review of the possible commercialisation of some of the US Geological Survey's roles carried out under the Reagan Presidency led to no major policy changes.

Despite these exceptions, the bulk of these reviews has directly impacted on almost all activities where the government is a producer of goods which are widely used by the populace and where taxation has previously funded the activities. Thus they have had significant impacts on the collection and dissemination of geospatial data.

Such reviews have usually been based on at least two general considerations: (1) what does government actually need to ensure occurs in a particular policy area? and (2) how can government's objectives or obligations be met whilst minimising the overall cost and maximising the overall benefit to the taxpayer? With respect to the first consideration, the US Congress has described the inalienable, core role of federal government but its list is couched in rather general terms. Elsewhere the first question has often only been answered through a study of what has happened (i.e. the 'policy' is manifested through the accumulation of its individual actions). One didactic response to the second consideration is provided by ideology – turn over all data collecting and dissemination functions and responsibilities to the private sector. A more sophisticated variant on this is found in New Zealand, where the national mapping organisation has been partitioned into two parts: these are an ongoing regulatory and purchasing body and a production and marketing operation, founded as a state enterprise but which may well move at some stage into the private sector and whose government work is protected only for a limited period (Robinson and Gartner 1997). Other data-collecting bodies in the New Zealand government have been treated in equivalent ways.

There are, however, more fundamental aspects to this question, especially since it must subsume the cultural, legal, and other public policy elements of the first question. In the USA, for instance, it is widely held that the 'near free' cost of federal data provision has led to the creation of an industry which, in paying taxes and in improving decision-making through use of its products, has led to a substantial national benefit (see Elshaw Thrall and Thrall, Chapter 23). Other countries have taken a very different view, based on the contention that charging the market rate for data has various advantages.

This dichotomy of approach to maximising benefit and minimising costs has not been restricted to nation states. Various international bodies or treaties have urged or mandated the merits of one or other of these approaches. Thus the Organisation for Economic Cooperation and Development (OECD), recognising that 'member countries are increasingly financing government services with user charges', has defined the best practice basis for proper charges for public services to reflect their full cost (OECD 1997); on the other hand, the Bangemann Report of the European Union has argued the case for public sector information to be made available cheaply and readily as part of the necessary infrastructure of the 'information society'. This view is replicated in drafts of a Green Paper on Public Sector Information produced by the European Commission (the public servants of the European Union).

To some degree these statements are simply advocacy of principles, but one situation where the tensions between cost recovery and 'free availability' have already arisen is in the operations of the national meteorological organisations. These bodies need to exchange data to maximise the quality of their weather predictions. The data provided by other nations to the US are immediately available over computer networks under various statutes (see section 4 below) to US private sector bodies who can then sell their data-based services in the originating nations. Since these US-based private firms – unlike the original data suppliers – pay little or nothing for the data, they are at a cost advantage compared to public or private sector service suppliers in the other countries. This situation led to fierce discussions in the World Meteorological Organisation in 1995–96 which were partially resolved by recognising both the need for the free exchange of 'raw' data and the Europeans' wish to charge for (and hence restrict access to) 'value-added' data (e.g. predictions).

3 ELEMENTS OF GEOSPATIAL INFORMATION POLICY

Although they interact, the key elements of geospatial data policy and their 'drivers' are best considered separately in the first instance. Individual policies may, however, be supported by reference to multiple, different benefits. Thus there is some support (and the contrary) on the grounds of economic theory for dissemination of government geospatial data at marginal cost (see below) but Perritt (1994) has argued the same case on US constitutional and technical grounds and urged extension of the same policy to state and local governments in that country.

3.1 The economics of geospatial data

3.1.1 The theory

Coopers and Lybrand (1996) and Didier (1990) have published important contributions in this area. It is usually argued that information is in general a public good in that consumption by one person does not affect its availability to others. It is an optional public good in that – unlike defence – it is possible to opt to take it or not. Love (1995), however, has pointed out that the accessibility and cost of the systems which permit use of information influence whether, in practice, it is a public good (see also Curry, Chapter 55). He argued that information is best defined as a quasi-public good since it may be non-rival but its consumption can also be excluded and controlled. He also points out that the pecuniary value of information may well depend on restricting its availability whilst its social value may be enhanced by precisely the opposite approach. The approach to economics which is adopted partially determines the end result.

Taking geospatial information as a simple subset of information generally (Arrow 1986), it seems to have two possible uses:

1 For consumption: individuals will decide how much value they assign to it based upon their valuation of pleasure, time saving or some other metric, and their awareness of the uses of the potential benefits of the information.

2 As a factor of production where the information is used as part of a good or service: the end-user of that good or service will make his/her decision on the uses to which it can be put, its availability to others, and the ease of substitutability. One obvious example is the use of geospatial information in making planning decisions in government and in commerce. It follows that some decisions will place a high economic value on the information whilst others will give it a high social value. Measuring these values is rather difficult so discussion thus far has typically been at a very high conceptual level.

Economic theory suggests that marginal cost copying is the most efficient allocative procedure. However this is only true for data already collected and if information is regarded in isolation from the rest of the economic system (Coopers and Lybrand 1996). For 'framework data' at least, there is ongoing need for regular updates to be collected and hence for ongoing spending in perpetuity. Moreover, where there is any pressure on budgets (e.g. opposition to continuing high levels of taxation), such considerations will lead to allocative distortions elsewhere in the economy. Indeed, in the longer term, the absence of any contribution from users to the costs of data collection will ensure that there is no information about the appropriate level of resources that should be dedicated to this task. Since governments are in practice increasingly interested in controlling their expenditure, there will be ongoing if uninformed pressure to reduce data collection costs under the marginal cost model, especially since measurement of externalities arising from government actions is difficult and may be impossible.

3.1.2 The actual market for geospatial data

Markets for geospatial data vary greatly between countries, for different types of information and by its spatial and temporal resolution. Most of the market sizing studies done to date provide only limited indications of price elasticity and other important characteristics of these markets. Nevertheless, one general characteristic of the market for geospatial information is its immaturity compared to that of other types of information (e.g. financial) – and certainly compared to that for other commodities. The reasons for this include:

1 the expectation of many customers is that geospatial information is a free good as well as a public good since it has been provided historically by governments to taxpayers;
2 certain detailed geospatial data have some of the characteristics of a natural monopoly – for instance, once a complete set of information is available for every house in the country and the majority of the costs are sunk ones, it is unlikely that duplicate information sources can be made competitive or are in the national interest;
3 the relative lack of transience of much traditional geospatial data which is analogous to 'reference books';
4 the skills of the user and the available software determine how much can be done with the data and both of these are still developing within national markets;
5 the value varies by many orders of magnitude to different users and knowledge of this on the part of the information producers is limited at present;
6 the extent of linkage with other data (and extraction of added value) which is possible;
7 the poor level of quality specification of most existing products and the relatively untested legal liability issues involved in their use (see Onsrud, Chapter 46).

All of this is largely of theoretical interest to those commercial data suppliers working mainly in the short term. They can (and often must) treat the provision of information as much more akin to that of any other commodity than can government. The information provider in the latter sector who seeks to generate revenue on a significant scale may have some freedoms (e.g. a limited scope for pricing to foster markets in the UK) but is constrained by rules outlawing exclusive deals, differential pricing on a major scale, etc., and s/he is expected to behave to the highest standards of equity, probity, propriety, transparency, and consistency. There are awesome penalties for failing in any one of these – such as public cross-examination on television by the House of Commons Public Accounts Committee in the UK and the associated costs of senior management time. In the USA, some of these considerations do not loom large for the Federal Government where recovery of costs is restricted to the costs of preparing the material for dissemination and for the dissemination itself (see section 4 below).

A final, practical consideration relates to how benefits are measured in government. There is an apparent paradox between the 'atomisation' of government into units whose performance is individually assessed on financial and other criteria (see Foster and Plowden 1996) and, on the other hand, the 'Gestalt' economic benefits which are widely if uncritically anticipated from ubiquitous use of the information superhighway (see, for instance, POST 1995). It has been argued that the latter can only flourish if individual government organisations make available their information in a readily accessed, consistently priced (or free), and frequently updated way. Yet the costs to the 'atomised units' inherent in producing nebulous overall benefits seem almost certain to complicate the meeting of performance targets individual to each separate organisation. The likely out-turn in these circumstances is self-evident! The generalisation of this paradox is that the benefits and costs of making government information widely available usually fall on different organisations – a point of real significance in establishing National Spatial Data Infrastructures (see section 5 below).

3.1.3 Policies on creating added value
The least heralded yet most fundamental advantage of GIS may well turn out to be its capacity to add value. Various ways of adding value to geospatial data have been identified. These include linkage of multiple datasets to facilitate the range of applications which may be envisaged and improving the quality of data through logical intersection of independently compiled databases to identify inconsistencies. Yet the options are much wider than that, as Perritt (1994) has suggested. There is considerable variation in national policies on whether value adding is permitted by the public sector. Indeed, what constitutes 'added value' is difficult to define in regard to geospatial data (see Coopers and Lybrand 1996).

3.1.4 The merits and demerits of cost recovery
All of what has been described above indicates that different legitimations exist for charging a market price for government's geospatial data and, alternatively, for making it available freely or at the cost of copying. The arguments used by proponents of the different views are summarised below, with contrary arguments set out in *italics*. It should be noted that there are legal preconditions for some economically-related policies to be feasible (notably some form of intellectual property rights protection of the information to be exploited; see section 3.2 below).

Arguments for cost recovery

- charging which reflects the cost of collecting, checking, and packaging data actually measures 'real need' and forces organisations to establish their real priorities *(but equal charges to all do not necessarily force setting of priorities since not all organisations have equivalent purchasing power e.g. utility companies can pay more than charities)*;
- users exert more pressure where they are paying for data and, as a consequence, data quality is usually higher and the products are more 'fit for purpose';
- it is equitable since the number of data users is presently small compared to the number of taxpayers. Hence cost recovery minimises the problem of subsidy of some individuals at the expense of the populace as a whole *(some users are acting on behalf of the populace, such as local governments)*;
- empirical evidence shows that governments are more prepared to part-fund data collection where users are prepared to contribute meaningful parts of the cost. Hence full data coverage and update is achieved more rapidly where cost recovery of some significant level is achieved *(but once the principle is conceded, government usually seeks to raise the proportion recovered inexorably)*;
- it minimises frivolous or trivial requests which may well require much resource and detract from the specified functional objectives of the government body concerned *(what is the real role of government? See section 2 above)*;
- it enables governments to reduce taxes in comparison to what would otherwise have been the case.

Arguments for dissemination of data at zero or copying cost

- the data are already paid for hence any new charge is a second charge on the taxpayer for the same goods *(this is unlikely to be true if the data are constantly being updated; moreover, there are many more taxpayers than users)*;
- the cost of collecting revenue may be large in relation to the total gains *(since the latter – including benefits of prioritisation – are unmeasurable, this is unprovable even though it may be true. It does not seem to be true in the private sector)*;
- maximum value to the citizenry comes from widespread use of the data through intangible benefits or through taxes paid by private sector added-value organisations *(the first of these is unmeasurable and the second may fall in the same category, though possibly capable of being modelled; as a result, the contention is unprovable. In any case, this point is irrelevant where costs and benefits are reckoned by government on an organisation-by-organisation basis rather than as integrated national accounts)*;
- the citizen should have unfettered access to any information held by his/her government *(this is a matter of political philosophy rather than economics so is not considered further in this section)*.

Irrespective of the merits of the different arguments, the fact is that different nations adhere to different policies (see section 4). These have had dramatic effects on the data-providing organisations and had some effects on the nature of use of the data. The first of these is illustrated in Figure 1 which compares the sources of financing of two National Mapping Organisations: (Ordnance Survey (OS) and the National Mapping Division of the US Geological Survey (NMD/USGS) by appropriations from Parliament and Congress. This is shown indirectly through the 'cost recovery level'. In the case of Ordnance Survey, this is based on standard commercial accounting practices, including interest on capital: the statistical values shown relate trading revenues from all sources to total cost of operations. In simple terms, the gap is funded by Parliament. In the NMD/USGS case, the philosophy is that the scientific information provided is a public good with benefits that accrue to the nation at large. The type of information provided is multipurpose in nature, and is available to a wide variety of users. As a result, 'cost recovery' is shown as total reimbursements (including programmes joint-funded by other federal, state, and other bodies plus product sales), as a fraction of total expenditures. It should be noted that the NMD/USGS is restricted by statute to recovering from sales only the costs associated with reproducing and distributing natural science information products and the costs of time and materials for scientific services, whilst the OS attempts to recover total costs of operation to include research and development. NMD/USGS is forbidden by statute in establishing copyrights on public domain information products. In the case of joint-funded programmes, the NMD/USGS is restricted to collecting a 50 per cent cost share. Again, it can be considered that the gap is funded by Congress, although the interpretation is less straightforward. In 1996–7, the total cash expenditure of Ordnance Survey was approximately

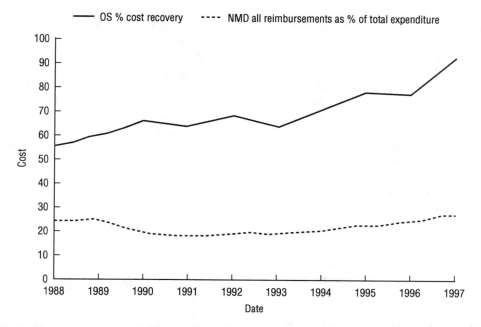

Legend: OS % cost recovery ---- NMD all reimbursements as % of total expenditure

Fig 1. The levels of funding of two National Mapping Organisations (Ordnance Survey and the National Mapping Division of the US Geological Survey) by appropriations from Parliament and Congress, respectively.

US$143m whilst that of NMD/USGS was US$183m. This comparison is made solely to show the different financing models, political philosophies, and laws associated with agency funding in the UK and US governments. Because of the different policy objectives involved, it should be regarded as indicative rather than definitive.

3.2 The legal policy framework for geospatial data

The law impinges on most aspects of the collection, description, conditions of sale, protection, and liability aspects of geospatial data. Defined in the widest sense, at least eight types of law influence the provision of information worldwide (see, for instance, Perritt 1994). Some of these of course do not apply in any one country; some of them apply only to the public sector. They are:

1 statutory or Ministerial authority for public agencies to involve in trading activities;
2 public access laws such as Freedom Of Information Acts;
3 human rights laws;
4 fair trading (or anti-trust) laws;
5 copyright and other Intellectual Property Right (IPR) laws;

6 data protection laws;
7 public procurement laws or regulations;
8 legal liability laws.

An extensive discussion of some of the issues involved, at least in a US context, is in Onsrud (1995). The same author has described one key element of the legal framework – liability for use of such data – elsewhere in this book (Onsrud, Chapter 46) so no further consideration of it will be made here. Since Curry (Chapter 55) has discussed related privacy issues in some detail, these will not be considered in this chapter.

3.2.1 Copyright basics

For the present purposes, only IPR laws embracing patents, copyright, and trade secrets will be considered, with particular emphasis on copyright law. The greatest advocates of strengthening legal protection of their information and high technology industries have been the developed countries, notably the USA and some countries in Europe. Many developing countries have favoured an easier flow of information and technology so as to facilitate their own economic development (Echoud and Hugenholtz 1997). Particular emphasis will be laid in this section on European law since that has been subject to much discussion, debate, and some change in the mid 1990s and it may well serve as a model for what is adopted elsewhere in future.

In general, there are great international variations in contemporary IPR laws: according to Echoud and Hugenholtz (1997), the greatest differences often occur between civil law countries (e.g. those on mainland Europe) and the common law countries (e.g. the UK, Republic of Ireland). To take one example, the 1992 French IPR legislation makes no distinction between the status (private or public sector) of an organisation holding ownership of these rights. Protection is given against reproduction, representation, adaptation, and transformation by all means and on all media though facts, ideas, and raw data are not covered. The protection covers two areas (Grelot 1997): moral protection and material protection. Under the material protection provision, the author can control who uses the information, the uses to which it can be put, and the duration of the licence.

Finally, it is clear in general terms that online access and distribution do not fit very well with many existing national copyright laws.

3.2.2 Originality

So far as copyright law in general is concerned, a key feature in many nations is the requirement for and definition of originality in the protected work. But defining originality is not simple: Echoud and Hugenholtz (1996) cite a number of court judgements which suggest contrary interpretations being made in different countries. In the UK and in Australia (Masser 1998), databases and maps are protected under copyright provided that 'judgement, skill and labour' have been employed. In the USA, however, the Feist Publications, Inc. v. Rural Telephone Service Company, Inc. case (Karjala 1995) has indicated that US courts under present law regard many databases as comprised of 'facts'. In the absence of strong evidence of intellectual creativity and originality, they will not support copyright protection on the basis of 'sweat of the brow' efforts of the compiler of such databases. As Karjala (1995) has also pointed out, this is likely to introduce disincentives to investment in the creation of databases – a point which the European Union's (EU) Database Directive is designed to combat (European Parliament and Council Directive 96/9/EC).

Whatever the formal legal position, it seems inconceivable that all maps and geospatial databases can be regarded as lacking originality. In practice,

the most appropriate interpretation is often not obvious – especially to those inexpert in geospatial data. For example, all small scale maps contain an element of artistry because of the generalisation which is inherent in them and in any digitally encoded version (see Weibel and Dutton, Chapter 10; Kraak, Chapter 11). In the extreme case, the maps may be used as propaganda, involving substantial originality (Monmonier 1996). The same is true for certain classes of large scale maps where these have been compiled by methods which involve interpretation of evidence. On the other hand, maps created through automated interpretation of automatically collected imagery seem very unlikely to be protected under copyright law in the USA or in various parts of Europe (except possibly under the EU's Database Directive).

3.2.3 Restricted acts and limitations

National laws appear to be extremely heterogeneous (Hugenholtz 1994) both as to the scope of the protected (or restricted) acts and to the scope and content of the limitations. In some respects the 1988 UK law is at one extreme: whilst all countries have limitations on reproduction of electronic information stored in 'permanent' form (e.g. on magnetic or optical disks), only the UK law defines transient electronic reproduction as a restricted act e.g. display on a screen.

As in other aspects of IPR law, the 'fair use' conditions of copyrighted material vary widely from country to country. Whilst the UK has strong copyright protection for creators and/or publishers, that country's law also has a 'breathtaking set of library privileges' (Hugenholtz 1994: 36) so far as limitations to statutory restrictions are concerned. Samuelson (1994) states that four factors generally determine whether courts find for 'fair use' in the USA: whether the copying is for personal, non-commercial use; the nature of the copyrighted work (whether entertainment or factual works); the 'substantiality of the taking' (judged qualitatively as well as quantitatively); and the harm or potential harm to the market for the copyrighted work arising from the non-approved copying. These issues are not academic: a specific example where the use of fair trading provisions formed the basis for a long-running court case was the action between the Belgian national mapping agency (IGN) and the commercial organisation TeleAtlas.

3.2.4 Moves to provide and harmonise database protection

Though the EC has issued a Green Paper on copyright, there is presently no standard approach in Europe other than a recent harmonisation of the period of protection. However, many of the aspects of protection of intellectual property have long been under consideration (Echoud and Hugenholtz 1997). Within the European Union, a Directive on the protection of databases which took eight years to agree came into general operation in January 1998. It may well enforce changes in the method of protection available for some databases and this could have major ramifications: under the Database Directive, the period of protection is much shorter than under copyright protection (though databases in existence prior to the date of publication of the Directive in 1986 and previously regarded as covered by copyright protection will continue to be so considered). The period of protection is 15 years though a further period of protection may be obtained if substantial investment has been applied to updating the database. In some countries in the European Union, however, the Directive will provide the first legal protection of certain kinds of databases. This may encourage some non-European data providers to create databases inside the Union and thereby achieve some level of protection; but to do this they have to be operating normally and for an extended period in the EU.

Attempts to harmonise the protection of intellectual property rights on a global basis have met with mixed success (Echoud and Hugenholtz 1997). The most notable recent achievements have been the 1993 global agreement on the trade related aspects of intellectual property rights (TRIPS), which formed part of the seventh (Uruguay) Round of the General Agreement on Tariffs and Trade (GATT), and the 1996 Copyright Treaty of the World Intellectual Property Organisation, an agency of the United Nations (see *http://www.wipo.int*). The latter treaty will come into effect when 30 nations have become signatories to it. Its effects on European IPR protection seem unlikely to be great but this may not be true in many other countries.

The complexity of IPR protection is such that professional advice should always be taken before creating any database which it is intended to exploit commercially. Equally, such advice should be sought before using data from a third party unless the provenance and terms of use of that data are very clear and the data provider has a long track record.

3.2.5 Tools for protecting databases

Some geospatial data are, for most purposes, much more ephemeral than others – thus the bulk of the value of meteorological data is much more transient than that of geology. The moral for information providers – if they are only information providers – is that they should build transience into their data wherever they can persuade customers to accept it and impose a leasing arrangement, rather than one-off sales. Parenthetically, it is clear that both users and data vendors presently understand little of the ephemeral transience of information and the 'half life' of its value, though organisations such as Microsoft are betting heavily on the value of historical material in purchasing libraries like the Bettmann Archive of photographs.

Periodic refurbishment of a stock item, such as the 'localisation' of the information in Microsoft's Encarta CD-ROM, is a good secondary strategy to re-energise the information market. A more comprehensive approach where professional markets are concerned is to build services and supply 'solutions' on top of the supply of information.

Despite the above advice, the problem of protection of investment remains, since much geospatial data is slow to change – though the data specification is a primary influence, typically 1 per cent or less of many 'natural environment databases' change annually. Some solutions have been proposed but many have been found wanting. For instance, encryption of data may be valuable as a form of protection, provided it does more than protect only the 'first time read' of the data rather than any subsequent onward transmission. Perritt (1994), in acknowledging the problem of illegal copying, argued that the solution is for a pricing regime which charges modest amounts via credit card use for small amounts of text retrieved from a database using search keys (e.g. as in the legal database Lexus), with downloading of the whole database being impossibly expensive. It is not clear how this can be translated into the geospatial domain: there is often huge value to be obtained by a pass through the database and finding a null or minimal response (e.g. of competing businesses in a defined area). To permit legal action against data theft, data publishers will have to devise an armoury of techniques to investigate and 'reverse engineer' images and digital maps to assert and prove provenance. 'Finger printing' of geospatial data with distinctive but invisible features thus seems certain to become more widespread.

325

4 SOME CASE STUDIES

4.1 Australia

Masser (1998) has summarised much information related to GI policies at both Commonwealth (national) and state level in Australia. Essentially, land related matters are dealt with at state and Territory level whilst the Commonwealth is concerned with collection of statistics, the creation of small scale national mapping and the coordination of national policy.

The states' systems for describing land and their survey and mapping systems vary somewhat but have many similarities. The policy aspects do, however, vary somewhat. In the state of Victoria, following a major study by Tomlinson Associates, it was decided that a multi-agency GIS body was a government priority; an Office of Geographic Data Coordination (OGDC) was accordingly set up, housed in the Department of Transport and Finance. It was charged with contributing to increased economic growth for the state, creating a flourishing information industry, supporting greater (Australian) competitiveness in global markets, creating efficiencies across government, and generating a major capital asset for government. The OGDC has identified 61 high priority information products, assessed the cost/benefit of developing these products, and begun work on creating some of them. In addition, OGDC has set up a separate agency in conjunction with Melbourne Water to maintain and upgrade the State Digital Map Database; this operates as a commercial agency to package and market cadastral, topographic, and road centreline datasets. The operational aspects of running this database were contracted to Intergraph Inc. and no copy of the database was retained in government (Mike Smith 1996, personal communication).

The national level coordination of land information is achieved through the Australian and New Zealand Land Information Council (ANZLIC), serviced by the Australian Surveying and Land Information Group (AUSLIG) – a Commonwealth body responsible for the small scale national mapping and known until the 1980s as NATMAP. Members of ANZLIC represent a coordinating body within their own jurisdiction. A great deal of work has already been carried out by ANZLIC in defining and agreeing technical standards relating to the geodetic datum (including acceptance of use of WGS-84 for all mapping in future; see Seeger, Chapter 30), to data models, dictionaries, and transfer standards, and to the national metadata system.

The inherent nature of Australia's government structure is reflected in ANZLIC's statement on the national GI strategy:

'The primary objective of a national data infrastructure is to ensure that users of land and geographic data who require a national coverage, will be able to acquire complete and consistent datasets meeting their requirements, even though data is collected and maintained by different jurisdictions' (Smith and Thomas 1996, quoted in Masser 1998).

The completion of the 'first pass' national digital map produced by the Public Sector Mapping Agencies for use by the Australian Bureau of Statistics in running the 1996 Census (Mooney and Grant 1997) is an example of such a 'middle up' operation in practice.

Perhaps the area of least agreement between the various component bodies has been in regard to data dissemination policies. A draft statement prepared in 1993 for ANZLIC has not (at the time of writing) been formally agreed. It states:

'Government spatial data, already collected and funded in the public interest, could be generally made available at the average cost of transfer, subject to certain conditions [However] any collection, upgrade or further processing of public interest data to meet client needs may be subject to additional charges over and beyond recovering the full cost of distribution' (CSDC 1994).

This policy was intended to apply only to public interest uses of the data, with commercial charging being applied at higher levels. In practice, charging levels to all users vary considerably throughout the country. Licensing data for particular purposes is common.

As elsewhere, pressures to reduce the call on the taxpayer and ensure that users pay for data are evident in the Australian system, including for certain types of data provided by the Australian Bureau of Census. What marks Australia out is the extended, serious, and informed debates on the issues of public good, monopoly trading, public/private sector relationships, and technical issues, all within the context of a highly diverse and large country – aided, of course, by the fact that there are only eight main players in the states and Territories (see *http://www.anzlic.org.au*).

4.2 The USA

The US Federal Government sees distribution of data it holds at the cost of dissemination (or less) as a matter of principle. Perritt (1994) has pointed out the mutually reinforcing roles of the Freedom of Information Act (FOIA) and the Office of Management and Budget's Circular A-130 (the latter now having been codified as section 3506(d) of The Paperwork Reduction Act 1995) in realising this principle.

Circular 130 (OMB 1993) states that 'the free flow of information between the government and the public is essential in a democratic society'. It also states that charges should be set at 'a level sufficient to recover the cost of dissemination but no higher'. Section 105 of the US 1976 Copyright Act specifies that copyright protection is not available for any work of the US government that is prepared by an employee or officer of the government as part of that person's specific duties.

Thus Federal Government is not generally able to assert copyrights and any other party is free to copy, disseminate, license, or sell data held by the Federal Government at will and without seeking permission or notifying the originators of the uses to which they will be put. There are, however, some potentially important exceptions to this rule. For example, where material has been supplied by foreign governments under specific agreements permitting internal business use (e.g. of data for military purposes), this information may not be passed on and copyright remains with the supplier.

The consequences of this situation are multiple and, in some cases, quite subtle. It has facilitated extensive use of the Internet for data dissemination which in turn has both increased access to the federal geospatial data (Calvert et al 1997; see also Coleman, Chapter 22) and shifted some costs of data acquisition to the end-user. However, the great bulk of financing for data collection in the US Federal Government necessarily arises from appropriations in Congress (see Figure 1) and the level and use of these line budget funds are subject to short term political expediency. In addition, the limited ability to protect information made available by others (especially commercial and some state governments), has made some intersectoral data

sharing cooperations impossible or highly tortuous.

The situation in state and local governments varies considerably. Though many states have statutes enabling public free or low cost access to certain types of records, Archer (1995) summarised twenty cases where states had amended or were in the process of amending these statutes to permit cost recovery through higher levels of charging. Masser (1998) provides a summary of various other studies on state and local government approaches. Given the strong US pressure to codify and enforce international copyright laws (see section 3.2.1), it is no surprise that the US commercial sector strongly asserts its copyright on most occasions.

Although the US federal policies on data dissemination are well established, they are not without their critics (e.g. Dando 1996) and occasional re-examinations are carried out. For instance, the National Academy of Public Administration (NAPA) was asked by the American Congress on Survey and Mapping and by four federal agencies in 1997 to study current geographical information functions in the USA and how these could be best structured and performed. NAPA has studied the role of GI in US competitiveness, the most appropriate roles of federal, state, and local governments, and the private sector in the GI 'industry', whether some government functions can be commercialised or transferred to other bodies, and whether there are other means to effect economies in GI in the federal government. The final report reaffirms the merits of existing federal government information policies but argues the need for wider active participation in the National Spatial Data Infrastucture (see section 5).

4.3 The UK

As indicated in various parts of this chapter, the information policy situation in the UK varies not by geography (as in some other countries) but rather by government department (as in the Netherlands: Masser 1998). Coopers and Lybrand (1996) summarised the variability in policy and practice in the UK and Table 1 illustrates the financial part of it. This variability arises from the different remits and targets given to government bodies – especially the Executive Agencies (which have significant freedoms to decide how to achieve these ends) – by Ministers. Information policy is thus generally a second order matter driven by the need to meet explicit public targets for quality of service, efficiency, and financial performance. All government-generated information is regarded as Crown copyright and its use is licensed though the level of charging may be zero. Some Executive Agencies have delegation of responsibility for administering Crown copyright whilst others follow central direction.

As in any other system, this has advantages and disadvantages to different organisations or individuals. Certainly the level of use of digital geospatial data in the UK has expanded very greatly in the last five years; much of this, however, has been in major organisations and a slower takeup has occurred in smaller organisations (see the general introduction to the Applications Part of this volume). The extent to which this reflects data pricing or the nature of society and availability of computing resources is unclear (and it is in any case changing quite rapidly).

4.4 Remote sensing

Harris (1997) has reviewed national and international policies in regard to collection, access, charging, and storage of remote sensing data as a whole, while Barnsley (Chapter 32) and Estes and Loveland (Chapter 48) provide reviews of the technology and data sources. Draeger et al (1997) and Williamson (1997) have similarly reviewed policies but focused very much on the lessons learned through 25 years of operation of the LANDSAT programme, including the effects of the

Reagan era commercialisation of the data supply. All three are particularly helpful studies since civilian satellite remote sensing data have hitherto differed considerably from some other types of geospatial data. They have largely been collected under the 'open skies' policy proposed by the US government in the 1950s and only governments or government-licensed bodies have been able to collect them. The open skies policy was internationally accepted as a United Nations treaty in 1967, with national governments being given implicit responsibility for licensing any national satellite operators. This has ensured that the effects of official policy have been applied much more homogeneously than in other sectors. In the case of the USA, clear policy guidance on the distribution of data from civilian satellite land remote sensing has been available through the 1967 UN Outer Space Treaty, the 1984 US Land Remote Sensing Commercialization Act (Public Law 98-365), the 1986 UN Principles on Remote Sensing, the 1987 US Department of Commerce Private Remote Sensing Licensing Regulations, the 1992 Land Remote Sensing Policy Act (Public Law 102-555), and the Executive Branch's National Space Policy of 1996.

In essence, the LANDSAT programme was run as a US government operation until 1979 when its management was transferred from NASA to the National Oceanographic and Atmospheric Administration (NOAA) and from NOAA to the private sector (EOSAT Corporation) in 1985. The rationale underlying the latter transfer was that

Table 1 Expenditure, revenue, and percentage cost recovery for various UK government information providers in 1994/95.

	Expenditure (£m)	Revenue (£m)	% cost recovery
Central Statistical Office*	49.5	1.9	4
Office for Population Censuses and Surveys*	70.0	38.0	54
Meteorological Office (includes research revenue)	141.0	57.0	40
British Geological Survey (includes research revenue)	40.0	24.0	60
Hydrographic Office	37.9	22.0	70
Ordnance Survey	74.8	58.6	78
Registers of Scotland (cadastral organisation)	29.6	31.5	106
Her Majesty's Land Registry (cadastral organisation)	197.4	235.6	119

* CSO and OPCS merged in April 1996 to form the Office for National Statistics

revenues from product sales and ground station fees would exceed costs, government subsidies would be eliminated and a profitable commercial enterprise would result. In the event, this did not occur and led to heated debates on pricing strategies and the wisdom of the transfer. Table 2 indicates how the volume of LANDSAT imagery disseminated changed over the period from 1979 to 1989. Despite the market's familiarity with the LANDSAT products and the commercialisation policy, the revenue generated was soon overtaken by that generated from imagery from the French SPOT satellite system (Table 3). Since then, other data providers (e.g. India, the European Space Agency, and the Canadian government) have also all sought to generate revenues from satellite imagery with some greater success.

There are several reasons why the initial commercialisation of LANDSAT failed. One of these is that it may have been premature, given the market's then limited capability to use such geospatial data. Another may well be the strong opposition to charging in the USA engendered by free or low cost access to other federal government data (see section 4.2). Yet

another may well be the resolution and timeliness of the data. But another factor could well be the staff culture of those who set up and ran the LANDSAT programme. As scientists and even visionaries, some at least were strongly opposed to the commercialisation. Draeger et al (1997) concluded that:

'It is clear that the management and data distribution policies and practices of the US government and its attempts at commercialisation have prevented the LANDSAT programme from living up to the vision and expectations of its early proponents.'

Despite all these setbacks, a number of commercial firms (see Estes and Loveland, Chapter 48; Barnsley, Chapter 32) have decided to launch high resolution satellites and charge for their imagery on a commercial basis; Williamson (1997) provides a fascinating description of the policy issues faced by the US government in considering the licensing and sale of information and satellite systems to non-nationals. One commercial organisation has bought the firm (EOSAT) involved in the commercialisation of LANDSAT to exploit its

Table 2 LANDSAT data sales and price history. *Source:* Draeger et al 1997

Year	Film items sold (000s)	Average film price (US$)	Computer compatible tape items sold (000s)	Average CCT price (US$)
1979	134.4	15	3.0	200
1980	128.3	15	4.1	200
1981	128.8	15	4.4	200
1982	115.0	20	5.0	250
1983	76.6	30	5.6	500
1984	35.0	60	5.0	500
1985	39.1	60	6.7	500
1986	19.1	125	7.8	1000
1987	12.4	150	8.3	1000
1988	9.1	150	8.5	1000
1989	3.6	150	9.1	1000

Table 3 Sales revenues from LANDSAT and SPOT data.
Source: Williamson 1997

	SPOT data sales (US$m)	EOSAT's LANDSAT data sales (US$m)
1986	5	15
1987	10	17
1988	16	16
1989	22	18
1990	32	22

distribution channels. Meantime, in the government domain, Williamson (1997) has claimed that LANDSAT 7 data will be available from 1998 under the terms of Circular A-130 and thus be made available at the cost of copying and distribution but Draeger et al (1997) are more cautious, saying that the data price will be 'based on the requirement to offset all costs of spacecraft and mission operations as well as ground processing'.

5 THE ADVENT OF NATIONAL SPATIAL DATA INFRASTRUCTURES

5.1 The US National Spatial Data Infrastructure (NSDI)

The bulk of this section is drawn from a summary of the history of the NSDI by Tosta (1997). During the early 1990s, the Mapping Science Committee (MSC) of the US National Research Council began to investigate the research responsibilities and the future of the National Mapping Division of the US Geological Survey. The MSC coined the phrase 'National Spatial Data Infrastructure' and identified it as the comprehensive and coordinated environment for the production, management, dissemination, and use of geospatial data. The NSDI was conceived to be the totality of the policies, technology, institutions, data, and individuals that were producing and using geospatial data within the USA. The MSC (1993) report proposed a number of actions and responsibilities for various agencies and for the Federal Geographic Data Committee (FGDC) which related to their vision of the NSDI whilst another report a year later urged the use of partnerships in creating the NSDI (MSC 1994).

The FGDC adopted the term NSDI to describe a 'national digital spatial information resource' and discussed the concept of the NSDI with the Clinton Administration teams which were exploring means to 'reinvent' the Federal Government in early 1993. The NSDI was recognised as an idea and a means to foster better intergovernmental relations, to empower State and local governments in the development of geospatial datasets, and to improve the performance of the Federal Government. In September 1993, the NSDI was listed as one of the National Performance Review (NPR) initiatives to reinvent Federal Government. Vice-President Gore stated that '[in] partnership with State and local governments and private companies we will create a National Spatial Data Infrastructure' (Gore 1993).

One of the primary means of implementing the initiatives arising from the NPR was through Presidential Executive Orders. In April 1994, Executive Order #12906: 'Coordinating Geographic Data Acquisition and Access: The National Spatial Data Infrastructure' was signed by President Clinton, directing that federal agencies carry out certain tasks to implement the NSDI (see also Lange and Gilbert, Chapter 33). These tasks were similar to those that had been outlined by the FGDC in its Strategic Plan a month earlier and which had since been updated (FGDC 1997). The Executive Order created an environment within which new partnerships were not only encouraged, but required. In the USA, Presidential Executive Orders are only applicable to federal agencies but, in this case, these agencies were directed to find partners (specifically among other levels of government). In practice, state and local governments will often voluntarily cooperate with federal agencies if this makes it likely to result in funding or improve their access to data. In addition, the Executive Order had significant effects in increasing the level of awareness about the value, use, and management of geospatial data among federal agencies specifically. Perhaps more importantly, it raised the political visibility of geospatial data collection, management, and use, both nationally and internationally.

The NSDI is defined in the Presidential Executive Order as 'the technology, policies, standards, and human resources necessary to acquire, process, store, distribute, and improve utilisation of geospatial data' (Clinton 1994). That Order and the FGDC identified three primary areas to promote development of the NSDI. The first activity area is the development of standards, the second improvement of access to and sharing of data by developing the National Geospatial Data Clearinghouse, and the third is the development of the National Digital Geospatial Data Framework. All of these efforts were to be carried out through partnerships among federal, state, and local agencies, the private and academic sectors, and non-profit organisations.

5.1.1 Standards

One component of the Federal Geographic Data Committee is a series of subcommittees based on different themes of geospatial data (e.g. soils, transportation, cadastral), each chaired by a different federal agency (see Tosta 1997). Several working groups have been formed to address issues on which there is a desire among agencies to coordinate and which cross sub-committee interests (e.g. Clearinghouse, Standards, Natural Resource Inventories). Many of these groups are developing standards for data collection and content, classifications, data presentation, and data management to facilitate data sharing. For example, the Standards Working Group developed the metadata standard, which was formally adopted by the FGDC on 8 June 1994 and mandated in the NSDI Executive Order for use by all federal agencies on new geospatial data collected after January 1995. After a review, work began in 1996 to refine the metadata standard in conjunction with the ISO (see Salgé, Chapter 50). As of mid 1997, 11 different thematic standards were in development by FGDC committees with those on cadastral data and classification of wetlands being endorsed. All of the FGDC-developed standards undergo an extensive public review process that includes nationally advertised comment and testing phases plus solicitation of comments from state and local government agencies, private sector firms, and professional societies. The NSDI Executive Order mandated that federal agencies use all FGDC-adopted standards.

5.1.2 National Geospatial Data Clearinghouse

The second activity area is intended to facilitate access to data, with the goal of minimising duplication and assisting partnerships for data production where common needs exist. This is being done by helping to 'advertise' the availability of data through development of a National Geospatial Data Clearinghouse. The strategy is that agencies producing data describe the existence of the data with metadata and serve those metadata on the Internet in such a way that they can be accessed by commonly used Internet search and query tools (see Guptill, Chapter 49). The FGDC-adopted metadata standard describes the content and characteristics of geospatial datasets. The NSDI Executive Order, besides requiring that federal agencies describe their data using the metadata standard, also stipulated that agencies make these metadata accessible electronically. Nearly all federal agencies, as well as most States and numerous local jurisdictions have become active users of the Internet for disseminating geospatial data. This model does not necessarily assume that data will be distributed for free. Obtaining some of these datasets requires the payment of a fee, others are free. The Clearinghouse can also be used to help find partners for database development by advertising interest in or needs for data.

5.1.3 Digital Geospatial Data Framework

The third activity area is the conceptualisation and development of a digital geospatial framework dataset that will form the foundation for the collection of other data to, hopefully, minimise data redundancy and to facilitate the integration and use of geospatial data (see Smith and Rhind, Chapter 47). The Executive Order directed the FGDC to develop a plan for completing initial implementation of the framework by the year 2000, considering the requirements for developing a database useful for the decennial census by 1998. During 1994, a vision and conceptual plan were developed by the FGDC Framework Working Group (consisting of representatives of state, local, and regional government as well as federal agencies) and was published as a Framework Report (FGDC 1995).

Organisations from different levels of government and occasionally the private sector are increasingly forming consortia in their geographical area to build and maintain digital geospatial datasets that meet a diversity of needs. Examples include various cities in the US where regional efforts have developed among major cities and surrounding jurisdictions (e.g. Dallas, Texas), between city and county governments (e.g. San Diego, California), and between state and federal agencies (e.g. in Utah). The characteristics of these partnerships vary depending on the level of technology development within the partner jurisdictions, on institutional relations, on the funding, and on the type of problems being addressed. Because investments in geospatial data development at the local level are significant and often result in higher resolution data than can easily be collected by states or the federal government, the FGDC aims to foster their development in such a way as to comply with minimal national standards and thus be capable of being integrated, aggregated, or generalised to build datasets over ever larger geographical areas up to and including the nation. Various pilot projects investigating different ways of building the framework were launched in 1996 (FGDC 1997).

In its short lifetime, NSDI has generated huge levels of interest in the USA and beyond (see, for instance, Masser 1998). Some considerable successes have been achieved, notably in the formulation of some standards and the creation of the clearinghouse of metadata. Perhaps its greatest success however has been as a catalyst, acting as a policy focus, publicising the importance of geospatial data, and focusing attention on the benefits of collaboration – especially important in a country as large and governmentally complex as the USA. The process continues on several fronts; the MSC, for instance, has attempted to anticipate the most significant GIS developments in the period up to 2010 (MSC 1997).

Inevitably, many problems have arisen in NSDI, notably about incentivising different organisations to work together and in ensuring that benefits arise for all organisations incurring costs. The concept of 'bottom up' aggregation of data to form national datasets now being explored is also an intrinsically complex one since the logistics alone of drawing together data from many thousands of other organisations (e.g. US counties) – which vary greatly in resources and inclinations – is daunting. That said, NSDI has been and remains a considerable achievement (*http://www.fgdc.gov/nsdi2.html*) and has triggered equivalents elsewhere, including multinational collaborations (Majid 1997).

5.2 Some NSDI equivalents

5.2.1 The Netherlands
Masser (1998) has described how Ravi – the Dutch Council for Real Estate Information – was restructured in 1993 as a national consultative body for geographical information. The Board of the new Ravi includes most of the main data providers and users in the public sector. The coordination responsibility for geographical information in the Dutch government lies with the Minister for Housing, Spatial Planning, and the Environment. Core funding is provided by his Ministry but other bodies contribute. Private sector bodies are involved through a business forum.

Ravi has, despite modest resources and an inability to enforce any decisions on its component bodies, been rather successful in a variety of projects, including the forging of agreement on creation of a 1:10 000 scale national core database. It has carried out internationally respected work on copyright and other GIS-relevant matters and acted as the host for the European Umbrella Organisation for Geographic Information (EUROGI). Its definition of a National Geographic Information Infrastructure differentiates between 'core' and 'thematic' data, arguing that the primary task is to improve the cohesion between those data falling into the former category. It has also set up a National Clearinghouse for Geographic Information (see *http://www.euronet.nl/users/ravi/english.html*).

5.2.2 The UK National Geospatial Data Framework (NGDF)
In Britain, observation of the US NSDI led in 1995 to what became the NGDF. It was recognised that the situation in the UK differed from that in the US in many ways: notably the structure and role of government is very different, there is already greater availability of high quality geospatial data in the UK, and government policy is very different with regard to dissemination of information.

The problems to be resolved in the UK are seen as:

1 information about geospatial datasets is difficult to obtain;
2 the information available varies greatly in quality between organisations;
3 valuable datasets are held, especially by government bodies, but are not currently available for many reasons;
4 existing datasets have been collected to different specifications so it is not easy to integrate data safely from multiple sources;
5 data are often not easy to access physically;
6 there are presently few services based on data combinations and extraction of added value.

As a result, the NGDF has been designed as a facilitator with a mission 'to develop an over arching UK framework to facilitate and encourage efficient linking, combining, and widespread use of geospatial data which is fit for purpose'. The objectives of NGDF are to: facilitate and encourage collaboration in the collection, provision, and use of geospatial data; facilitate and encourage the use of standards and best practice in the collection, provision, and use of geospatial data; and facilitate access to geospatial data.

If NGDF is successful, data quality will improve as data owners and custodians see new revenue-earning opportunities from sale of data and new added-value derived datasets will be created by combining data from two or more sources. This should lead to growth of the UK geospatial data market through improved access to data, growth of focused data linkage applications and services (including 'one stop shopping'), reduced data collection costs by reducing data duplication, and better decision-making. To bring all this about, the NGDF Board (comprised of data providers in the public and private sectors), with advice and support from the NGDF Advisory Council (comprising users from all sectors, software vendors, academics, and others), has set up a programme of work for various Task Forces under a NGDF Programme Director. Details of this and progress may be found on the NGDF Web site at *http://www.ngdf.org.uk.*

6 CONCLUSIONS

Some obvious conclusions emerge from all of the above. The first is that there are huge variations in the national policies relating to geospatial data or GI. Many of these are being reconsidered under the impetus of other policy matters (e.g. to force cost reductions in government or to stimulate the private sector information industry). But one trend in recent years has certainly been that many governments are coming to see their geospatial data holdings as assets capable of being financially exploited. Despite the many national initiatives, however, there is little international coherence in the policies underpinning the geospatial data 'business' and this statement is also true in many individual countries. The opinions of key opinion-formers are also far from coherent (Burrough et al 1997). As a consequence, there is no possibility of ensuring complete global harmony in all aspects relating to GI. The range of actors, their range of agendas, and the historical legacies are already much too broad for this to be successful. This presents both opportunities and challenges to those who create geospatial data, who operate multinationally, and who are 'footloose' in their operations: whilst the national variations are troublesome, they also offer advantages in the selection of sites to take maximum advantage of the legal framework as well as low labour costs.

For government and the citizen alike, data policy is a 'no single best solution' field because of the interaction of legal, economic, public policy, and political threads. The interactions between different factors are sometimes highly complex. It is typical – and inevitable – therefore that many of these different policy, elements, or influences produce conflicts which must be resolved by examining the particular tradeoffs involved.

Two particular problems are, however, generic to most national geospatial data policy arenas. The first is how to ensure that the user needs – especially latent ones and those of users with modest resources – are considered in the creation and provision of geospatial data. This is particularly acute when governments are the data providers. The second generic problem is to ensure that those organisations bearing the costs of any new policy also reap tangible benefits; without this, little will happen for economic and political systems are rarely altruistic. The ultimate success of NSDIs and, more generally, of GIS will be very strongly influenced by how well such problems are addressed.

References

Archer H N 1995 Establishing a legal setting and organisational model for affordable access to government-owned information management technology. In Onsrud H (ed.) *Law and information policy for spatial databases.* Orono, NCGIA: 13–24

Arrow K 1986 The value of and demand for information. In McGuire C B, Radner R (eds) *Decision and organisation*, Minneapolis, University of Minneapolis Press

Barr R, Masser I (1997) Geographic information: a resource, a commodity, an asset or an infrastructure? In Kemp Z (ed.) *Innovations in GIS* 4. London, Taylor and Francis: 234–48.

Branscomb A W 1994 *Who owns information?* New York, Basic Books

Burrough P, Craglia M, Masser I, Rhind D 1997 Decision-makers' perspectives on European geographic information policy issues. *Transactions in GIS* 2: 61–71

Calvert C, Murray K, Smith N S 1997 New technology and its impact on the framework for the world. In Rhind D (ed.) *Framework for the world.* Cambridge (UK), GeoInformation International: 133–59

Chamoux J P, Ronai M 1996 *Exploiter les données publiques.* Paris, A Jour

Clinton W J 1994 Executive Order 12906. *Coordinating geographic data acquisition and access: the National Spatial Data Infrastructure.* Washington DC: 2 pp

Coleman D J, McLoughlin J D 1997b Towards a global geospatial data infrastructure. *Proceedings, Second Global Spatial Data Infrastructure Conference, University of North Carolina,* October

Collins M, Rhind D 1997 Developing global environmental databases: lessons learned about framework information. In Rhind D (ed.) *Framework for the world.* Cambridge (UK), GeoInformation International: 120–9

Coopers and Lybrand 1996 *Economic aspects of the collection, dissemination and integration of government's geospatial information.* Southampton, Ordnance Survey

Coppock J T, Rhind D W 1991 The history of GIS. In Maguire D, Goodchild M F, Rhind D W (eds) *Geographical information systems: principles and applications.* Harlow, Longman/New York, John Wiley & Sons Inc. Vol.1: 21–43

CSDC 1994 *Commonwealth Spatial Data Committee Annual Report 1993–94.* Canberra, Australian Government Printing

Dando L P 1995 A case for the commercialisation of public information. In Onsrud H (ed.) *Law and information policy for spatial databases.* Orono, NCGIA:32–6

Didier M 1990 *Utilité et valeur de l'information géographique*. Paris, Economica

Douglas R 1994 *Unfinished business*. London, Penguin

Draeger W C, Holm T M, Lauer D T, Thompson R J 1997 The availability of Landsat data: past, present and future. *Photogrammetric Engineering and Remote Sensing* 63: 869–75

European Commission 1997 Towards a European policy framework for geographic information: a working document. In Rhind D (ed.) *Framework for the world*. Cambridge (UK), GeoInformation International: 202–5

Echoud M van, Hugenholtz P B 1997 *Legal protection of geographical information: copyright and related rights. Bottlenecks and recommendations*. Amersfoort, EUROGI

European Parliament and Council Directive 96/9/EC of 11 March 1996 on the Legal Protection of Databases, 1996. *Official Journal of the European Communities* No. L 77/20–28

FGDC 1995a, b *Development of a National Digital Geospatial Data Framework* and *The national geospatial data clearinghouse*. Washington DC, Federal Geographic Data Committee

FGDC 1997 *A strategy for the National Spatial Data Infrastructure*. Washington DC, Federal Geographic Data Committee

Foster C D, Plowden F J 1996 *The state under stress*. Buckingham, Open University Press

Gore A 1993 *From red tape to results: creating a government that works better and costs less*. Report of the National Performance Review. Washington DC, US Government Printing Office

Grelot J-P 1997 The French approach. In Rhind D (ed.) *Framework for the world*. Cambridge (UK), GeoInformation International: 226–34

Harris R 1997 *Earth observation data policy*. Chichester, John Wiley & Sons

Htun N 1997 The need for basic map information in support of environmental assessment and sustainable development strategies. In Rhind D (ed.) *Framework for the world*. Cambridge (UK), GeoInformation International: 111–19

Hugenholtz P B 1994 Analysis of law relating to copyright and electronic delivery in Europe. *Proceedings, Conference of European Commission's Legal Advisory Board on 'Legal aspects of multimedia and GIS', Lisbon, 27–8 October*: 33–8

Karjala D S 1995 Copyright in electronic maps. *Jurimetrics Journal* 35: 395–415

Lenczowski R 1997 The military as users and producers of global spatial data. In Rhind D (ed.) *Framework for the world*. Cambridge (UK), GeoInformation International: 85–110

Love J 1995 Pricing government information. In *Agenda for access: public access to Federal information for sustainability through the Information Superhighway*. Report prepared by the Bauman Foundation, Washington DC: Chapter 10

Lummaux J C 1997 Partenariat public privé. Paper given to Workshop organised by CNIG/AFIGÉO/EUROGI on geographic information: public/private partnerships, Paris, 2 April. See also *http://www.frw.ruu.nl/eurogi/eurogi.html*

Majid D A 1997 Geographical data infrastructure in Asia and the Pacific. In Rhind D (ed.) *Framework for the world*. Cambridge (UK), GeoInformation International: 206–10

Mapping Science Committee 1993 *Toward a coordinated spatial data infrastructure for the Nation*. Washington DC, National Academy Press: 171 pp

Mapping Science Committee 1994 *Promoting the National Spatial Data Infrastructure through partnerships*. Washington DC, National Academy Press: 113 pp

Mapping Science Committee 1997 *The future of spatial data and society: summary of a workshop*. Washington DC, National Academy Press: 68 pp

Masser I 1998 *Governments and geographic information*. London, Taylor and Francis

Monmonier M 1996 *How to lie with maps*, 2nd edition. Chicago, University of Chicago Press

Mooney D J, Grant D M 1997 The Australian National Spatial Data Infrastructure. In Rhind D (ed.) *Framework for the world*. Cambridge (UK), GeoInformation International: 197–205

Mounsey H M 1991 Multi-source, multi-national environmental GIS: lessons learned from CORINE. In Maguire D J, Goodchild M F, Rhind D W (eds) *Geographical information systems: principles and applications*. Harlow, Longman/New York, John Wiley & Sons Inc. Vol. 2: 185–200

OECD 1997 *Best practice guidelines for user charging for government services*. Paris, Organisation for Economic Cooperation and Development

OMB 1993 *Circular A-130: Management of Federal Information Resources*. June 25. Washington DC, US Office of Management and Budget

Onsrud H (ed.) 1995 *Law and information policy for spatial databases*. Orono, NCGIA

Onsrud H J, Rushton G (eds) 1995 *Sharing geographic information*. New Brunswick, Center for Urban Policy

Osborne D, Gaebler T 1992 *Reinventing government*. Reading (USA), Addison-Wesley

Perritt H H Jr 1994 Commercialisation of Government Information: comparisons between the European Union and the United States. *Internet Research* 4(2): 7–23

POST 1995 *Information superhighways: the UK National information infrastructure*. London, Parliamentary Office for Science and Technology, House of Commons

Rhind D W 1992 Data access, charging and copyright and their implications for GIS. *International Journal of Geographical Information Systems* 6: 13–30

Rhind D W 1996a Economic, legal and public policy issues influencing the creation, accessibility and use of GIS databases. *Transactions in Geographical Information Systems* 1: 3–11

Rhind D W 1997c Implementing a global geospatial data infrastructure. *Proceedings, Second Global Spatial Data Infrastructure Conference, University of North Carolina, 19–22 October*

Robinson W A, Gartner C 1997 The reform of national mapping organisations: the New Zealand experience. In Rhind D (ed.) *Framework for the world*. Cambridge (UK), GeoInformation International: 247–64

Samuelson P 1994 Copyright's Fair Use doctrine and digital data. *Comm. ACM* 32: 21–7

Sandgren U 1997 Merger, government firms and privatisation? The new Swedish approach. In Rhind D (ed.) *Framework for the world*. Cambridge (UK), GeoInformation International: 235–46

Scott G 1994 Civil service reform in New Zealand. In *Reforming government in practice: lessons and prospects*. European Policy Forum/BDO binder. London, Hamlyn: 15–30

Smith M, Thomas E 1996 National spatial data infrastructure: an Australian viewpoint. *Proceedings, Emerging Global Spatial Data Infrastructure Conference*. Bonn: Paper 7: 16 pp

Thrower N 1996 *Maps and civilization*. Chicago, University of Chicago Press

Tosta N 1997 National spatial data infrastructures and the roles of national mapping organisations. In Rhind D (ed.) *Framework for the world*. Cambridge (UK), GeoInformation International: 173–86

Warita Y, Nonomura K 1997 The national and global activities of the Japanese national mapping organisation. In Rhind D (ed.) *Framework for the world*. Cambridge (UK), GeoInformation International: 31–47

Williamson R A 1997 The Landsat legacy: remote sensing policy and the development of commercial remote sensing. *Photogrammetric Engineering and Remote Sensing* 63: 877–85

Zhdanov N 1997 Mapping in Russia: the present state of development. In Rhind D (ed.) *Framework for the world*. Cambridge (UK), GeoInformation International: 71–82

PART 4: APPLICATIONS

Introduction

THE EDITORS

The previous Parts of this Book have considered the principles, technical issues, and management issues of GIS. This Part deals with the applications of GIS. In many respects applications are the most important aspect of GIS since the only real point of working with geographical information systems is to solve substantive real-world problems. This might include performing an existing task better, cheaper, or faster (e.g. automating the process of producing maps, or tracking building permits), or allowing new problems to be tackled (e.g. directing cruise missiles using onboard digital terrain analysis, or looking for clusters of infrequently occurring diseases such as childhood leukaemia). GIS is perhaps best considered a methodology or collection of tools which when *applied* can bring great benefit.

The applications of GIS are legion, and indeed several pages of this Book could be filled by a list of the constellation of areas to which GIS has been applied. From archaeology to zoology GIS can contribute a great deal to our study of patterns and processes on the surface of the Earth. There are many books which are themselves devoted to the description of GIS applications and those who care to peruse the bibliography of this Book will discover an extended list of application areas.

Even though the innovation of GIS is itself quite recent, it is possible to classify GIS applications as traditional, developing, and new. Traditional GIS application fields include military, government, education, and utilities. The developing GIS application fields of the mid 1990s include a whole raft of general business uses (e.g. banking and financial services, transportation logistics, real estate, and market analysis). New application areas, which are probably due for takeoff in the next decade, include small office/home office (SOHO) and personal or consumer applications. This simple classification, although useful in itself, hides a complexity of approaches to applying GIS. To choose but one example, utilities frequently undertake applications which fall into traditional, contemporary, and forward looking application classes identified above. Traditional utility applications include creating asset inventories (e.g. databases of pipes, valves, manholes, customers, pumping stations) and mapping networks. Contemporary developing applications include outage analysis (tracing the cause of faults and predicting shortages during planned maintenance) and work order processing. New, forward looking applications include integrated SCADA (Supervisory Control and Data Access – field-based dataloggers) systems and automated network load balancing. Further details of utility applications are provided by Jeffrey Meyers (Chapter 57) and Carolyn Fry (Chapter 58).

Diverse though the range of GIS applications is, many nevertheless share common themes. A convenient way to group them, and the one used in this Section, is on the one hand those dealing with operational issues and on the other those dealing with more general social and environmental issues. At the risk of overgeneralisation it is possible to say that the former typically focus on very practical issues, such as cost effectiveness, service provision, system performance, competitive advantage, and database creation/access/use; in contrast, the latter are often more concerned with model sophistication, the social and environmental consequences of results, and the precision and accuracy of findings.

In part by design, but also because of the background and interests of the authors, the chapters in this first Section, 'Operational applications', deal with the 'nitty-gritty', often day-to-day, practical issues of applying GIS. Most of the authors are GIS practitioners and would classify themselves first and foremost as direct users or managers of GIS. The chapters in the second Section, 'Social and environmental applications', take a wider look at the issues with which they are concerned. Whilst all of these authors are also GIS users, their backgrounds as academics and researchers have allowed them to write more general chapters from a wider perspective.

The information about GIS applications in this book is not solely contained in this Section. Other important contributions on GIS applications include those by Birkin et al (Chapter 51) which considers distribution and other business planning applications, Elshaw Thrall and Thrall (Chapter 23) which covers desktop and business applications, and Forer and Unwin (Chapter 54) which deals with educational application issues.

COMPARISON WITH THE FIRST EDITION

To see how GIS applications have changed over the past decade it is interesting to compare the current situation with that described in the first version of this Book (Maguire et al 1991). One of the most noticeable differences is in the structure of the applications Part. In 1991 the editors felt it necessary to provide general reviews of the state of GIS applications at the national level. This was mainly because the field was comparatively immature at the time. A further significant factor was that there was comparatively little written about who was using GIS. Close examination of the applications Sections of the first edition also reveals that in the early 1990s there were relatively few areas with sufficiently well-established operational GIS to allow authors to write on the general characteristics and lessons learnt from several years of activity. How things have changed!

In 1991 Maguire, Goodchild, and Rhind called for more to be written on failed GIS and cost–benefit analysis (Maguire et al 1991). More than half a decade later there are still very few descriptions of failed GIS. This is probably for the same reason identified in 1991, namely, that few people are prepared to admit that they 'failed', let alone write about it and then publish it for all to see. It is more pleasing to report that there has been considerable work in the intervening period

on cost–benefit analysis and cost justifying GIS. Much of this work has confirmed what we all suspected: GIS *do* add value beyond their cost. The most important work on cost–benefit analysis is summarised in Obermeyer (Chapter 42).

GIS APPLICATION TRENDS

The earlier part of this discussion classified GIS applications as traditional, developing, and new. A different way of examining trends in GIS applications is to look at the diffusion of GIS use. Figure 1 shows the classic model of GIS diffusion originally developed by Everett Rogers (see Brown 1993). Rogers' model divides the adopters of an innovation into five categories:

- Venturesome Innovators – willing to accept risks and sometimes regarded as oddballs.
- Respectable Early Adopters – regarded as opinion formers or 'role models'.
- Deliberate Early Majority – willing to consider adoption only after peers have adopted.
- Sceptical Late Majority – overwhelming pressure from peers is needed before adoption occurs.
- Traditional Laggards – people oriented to the past.

Applying this to a generalisation of the adoption of GIS applications (Figure 2), it can be seen that GIS conforms fairly well. In the late 1990s GIS seems to be in the transition between the Early Majority and the Late Majority stages. The Innovators who dominated the field in the 1970s were typically based in universities and research organisations. The Early Adopters were the users of the 1980s, many of whom were in government and military establishments. The Early Majority,

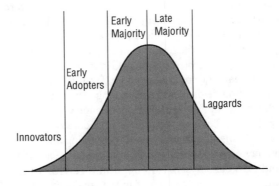

Fig 1. The classic model of GIS diffusion.

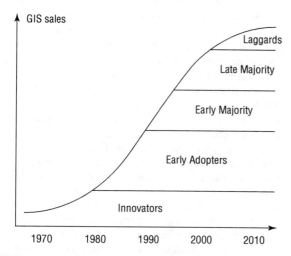

Fig 2. Classic diffusion of innovation applied to GIS.

typically in government and private businesses, came to the fore in the mid 1990s. Whether history will see this analysis as correct only time will tell, but it seems to parallel the introduction of similar technologies such as database management systems and statistical analysis software. The current question for potential users appears to be: do you want to gain competitive advantage by being part of the Majority user base or wait until the technology is completely accepted and join the GIS community as a Laggard?

CURRENT STATUS AND TRENDS

It is interesting to look at the current status of GIS applications and examine some of the current and envisaged trends. The predictions will probably hold at best for the two–three years following publication of this Second Edition, and the only certainty is that because the field is moving so quickly, no one can have any clear idea about what will happen beyond that (but see the Epilogue, Chapter 72, for some general predictions).

1997 was probably the year in which the number of GIS software systems installed passed one million. At a conservative estimate of an average of two users for every system installed, this means a user base of around two million. These crude estimates are only for core GIS and do not include associated systems which also have some GIS capabilities (usually mapping or data management), or the massive number of users who now experience GIS over the Internet using a conventional Web browser (Longley et al, Chapter 1). Examples of

such associated systems include AutoCAD (claimed user base of three million), Datamap in Excel (claimed user base exceeding one million), and the many electronic atlases (Elshaw Thrall and Thrall, Chapter 23). The very rapid rise in the deployment of Internet Map Servers and mapping products has also very significantly increased the number of GIS users. A conservative estimate is that at the end of 1997 over two million maps were being made and viewed over the Internet each day. This would mean that the number of Internet GIS users is several million. If the definition of a GIS were extended to include associated systems and Internet users then the estimate of the size of the GIS community would probably be as large as between eight and ten million. It is interesting to note that in the First Edition of this book, the editors did not forecast a GIS population beyond 580 000 even in the year 2000 (Maguire et al 1991).

As well as there being more GIS users today, there are also more large mature user sites. It is not uncommon for a large government agency, university, or utility to have more than 100 GIS seats. There is also a significant number of the largest sites with site licences for software products and more than 1000 seats.

The reasons for this explosive growth of GIS applications are legion, and for each individual person or organisation they are unique. In general the key factors include:

- *Greater awareness of the potential GIS has to offer*. It is estimated that over 1000 universities now teach degree level courses in GIS and there are many others which touch it in passing. At the school level a number of curricula require GIS to be taught and, in the USA at least, GIS is now working its way into elementary and middle schools. GIS is now diffusing quite rapidly into many organisations as education increases and awareness spreads.

- *Better technology to support applications*. The substantial research and development investment by many GIS software vendors in the 1980s and early 1990s is now beginning to pay dividends in many areas, particularly visualisation, data management, and analysis. Most GIS software packages now have customisation capabilities suitable for use by third-party application developers and technical users within organisations (see Maguire, Chapter 25, for further details). This has led to a proliferation of many domain-specific, highly focused applications which has swelled the size of the user community. A further reason for the success of GIS is their ability to link to other software systems (Goodchild et al

1992), such as corporate databases, statistical and spatial analysis systems (Maguire 1995; Gatrell and Senior, Chapter 66; Yeh, Chapter 62) and dataloggers (Larsen, Chapter 71).

- *More, cheaper data.* Clearly GIS are almost worthless without data. In the past few years a number of projects have implemented well-designed data collection programmes and have delivered databases which are persistent and application independent. There is a significant number of substantial local, national, and global databases. Some of these are in the public domain, others are under the control of national mapping agencies, and some are privately held. With the increasing commodification of spatial information, there is already a substantial and growing market for data. Since it is almost always cheaper to buy data than capture it first hand, this has led to significant cost reductions for several projects (Smith and Rhind, Chapter 47; Rhind, Chapter 56).

- *Improved ease of use.* Most general-purpose GIS software systems have now adopted standard windowing environments: X Windows on UNIX or Microsoft Windows on PCs. The latter is becoming particularly important as the norm for client or desktop GIS user interface. Because of many users' familiarity with Windows, some of the previous barriers to user acceptance and usability have been removed.

- *Reduction in price.* The price reduction of GIS hardware and software and the economies of scale reflecting the increase in the size of the market have both increased the attractiveness of GIS. The significant reduction in the price of GIS hardware and software mirrors the general reduction in the IT industry. In the case of the hardware used for GIS, the well-known hardware law devised by Intel Corporation founder Gordon Moore, holds true to date. Moore's law states that computer processing power will double and its cost will halve every 18 months.

- *Availability of applications.* The GIS market has also been stimulated by the development of end-user applications which are available commercially off-the-shelf (COTS) or 'ready to run out of the box'. Within the field of GIS there is a thriving and growing community of companies which produce end-user oriented applications. In a type of positive feedback loop this further stimulates the GIS market, which in turn encourages more developers to produce applications, and so on.

GIS is now less driven by technological considerations and many decisions are based on firm cost–benefit cases (see Obermeyer, Chapter 42). Today there are strong methodologies for developing cost–benefit models and several case studies of implementations which have positive economic benefits. It is also the case that there are well understood and tried and tested methodologies for GIS implementation, customisation, and ongoing management (Maguire, Chapter 25; Sugarbaker, Chapter 43). The so-called 'institutional issues' associated with implementing a GIS are also better known. Factors such as lack of support from senior management, problems of user acceptance, establishing a strong management team, the legality of using geographical information, and conducting business in the GIS world are now much better understood.

Once GIS applications become established within an organisation it is not uncommon for GIS to spread widely (Figure 3). Much of the early focus of implementations is typically on technical issues and many of the applications are little more than spatial data processing. For some this type of activity is sufficient to warrant adoption, but it was not until GIS started to become integrated with corporate (or enterprise) information systems that it really started to takeoff. During this phase the emphasis moves from a technological to an analysis/modelling focus.

Integration of GIS with corporate information system (IS) policy, planning, and systems is an essential prerequisite for success in many organisations. Until GIS technology can utilise corporate databases, interface with existing corporate information systems, and become accepted as a core part of corporate IS strategies, the full benefits will not be realised. Most of today's large-scale GIS

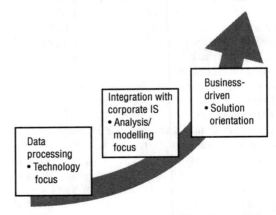

Fig 3. Changing business focus of GIS within organisations.

implementations are in this phase of GIS development. Users make widespread use of the analytical and modelling capabilities of modern GIS.

As implementations mature, GIS can become increasingly business-oriented with a focus on problem-solving and business benefits. This trend is evident within organisations even today. The initial focus of many GIS applications is technological, with data processing the major activity. As an implementation matures, the focus changes to that of integration with existing applications and there is an interest in analysis and modelling. In the most mature sites GIS is driven by business requirements and providing cost-effective solutions to business problems.

GIS can be used by organisations at the Operational, Tactical, and Strategic levels (Grimshaw 1994), as shown in Figure 4. Operational activities are the basic day-to-day activities of many organisations. They include performing site maintenance, deliveries, and scheduling. GIS can be used, for example, to create and manage inventories of facilities (sites, people, deliver routes, etc.). Much of the data relating to operational activities is collected and maintained inhouse. Operational decision-making tends to be highly structured.

Tactical activities are typically the domain of middle managers. The decision-making process is often semi-structured and is based on a combination of internal and external data. Within businesses, for example, external data include geodemographics and industry pricing information. Examples of tactical applications of GIS include locating land for 'new to industry' sites and territory management.

Strategic activities involve senior management. Strategic decision-making is frequently unstructured and intuitive. It involves ad hoc assembly and timely analysis of internal and external data, much of it collected on disparate spatial bases, at different resolutions, and on different projections. Examples of

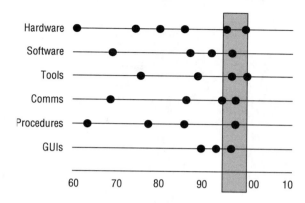

Fig 5. Innovation in GIS. During the five-year window of a typical corporate GIS application project (shaded box) it is highly likely that there will be several major technological GIS innovations (black circles) (after Maguire and Dangermond 1995).

strategic activities are planning sites for a new territory or evaluating an expansion of a product line.

All the talk of technological change sometimes makes people nervous about buying GIS and associated technology. It is a sobering thought that over the five-year lifetime of an average corporate application project (Figure 5) there are likely to be as many as ten significant technological innovations. This suggests that projects should be planned on an incremental rather than a 'big bang' basis and that project managers should not worry overly about the latest technology since much of it will be outdated before the end of a large project!

References

Brown R 1993 *Market focus: achieving and sustaining market effectiveness*. Oxford, Butterworth-Heineman

Goodchild M F, Haining R P, Wise S 1992 Integrating GIS and spatial data analysis: problems and possibilities. *International Journal of Geographical Information Systems*: 407–23

Grimshaw D J 1994 *Bringing geographical information systems into business*. Cambridge (UK), GeoInformation International

Maguire D J 1995 Implementing spatial analysis and GIS applications for business and service planning. In Longley P A, Clarke G P (eds) *GIS for business and service planning*. Cambridge (UK), GeoInformation International: 171–91

Maguire D J, Dangermond J 1995 Future GIS technology. In Green D R, Rix D (eds) *The AGI sourcebook for geographic information systems*. London, Association for Geographic Information: 113–22

Maguire D J, Goodchild M F, Rhind D W 1991 *Geographical information systems: principles and applications*. Harlow, Longman/New York, John Wiley & Sons Inc.

Fig 4. Different levels at which GIS can be used within organisations.

Introduction

THE EDITORS

Out in the real world of public and private commercial implementations the vast majority of today's large GIS are performing operational tasks. These include recording millions and, in some cases, billions of dollars of assets (utility outside plant, transportation infrastructure, and public and private land and property). This world is driven by the accounting metrics of actual benefits realised and costs saved. It is ironic that while almost every significant GIS acquisition requires formulation of a detailed cost–benefit case and many firms of consultants make a tidy living from selling the methodology repeatedly, so little of it is documented and publicly available (but see LGMB 1993; Obermeyer, Chapter 42).

The chapters in this Section describe the ways in which GIS are being used on a daily basis to record the location of features (e.g. land parcels, network devices, and highways signs), to perform update transactions (e.g. the acquisition and disposal of properties and changes to telecommunication devices), and to calculate the delivery routes for vehicles (e.g. for drinks vendors and long-distance haulage companies). Such prosaic applications are the everyday business of many GIS users.

In the first chapter Jeff Meyers (Chapter 57) reviews how GIS are being used in electric, gas, and water utilities to help them stay competitive in a rapidly changing marketplace. The combined external pressures of business re-engineering, privatisation, and open energy marketing are all profoundly affecting utilities. In this context, GIS can be seen as both a facilitator and a consequence of change. Meyers describes how GIS in utilities have evolved technically to become open systems, linked to a commercial database management

system (DBMS) with a Windows interface. The combined demands of large databases, the need to model complex objects (e.g. transformers, pumping stations, and gas networks) and the very large numbers of concurrent users, means that in general utility systems push GIS technology more than any others. A modern utility GIS is likely to have been created using object-oriented modelling and programming techniques. It is also likely to have the capability to support multi-user transactional access on a continuous shared database. Data automation remains a major consideration for utilities, because of the time and cost involved. Collecting the basic data is now only the first part of the process of building an automated mapping/facilities management (AM/FM) GIS application. Many of the current applications require access to an intelligent network model, which is topologically correct and linked to other useful data, such as customer addresses and facilities data. Any assessment of the costs of data in an AM/FM/GIS should not underestimate the costs of maintenance which may exceed initial capture. Key utility GIS applications include: map production in support of many utility operations (e.g. inventories, gas pressure zones, rights-of-way); map (database) editing; facility queries; design/work order processing; trouble call/outage analysis; and utility executive information systems. Also of note in this chapter is Meyers' clear demonstration that the benefits of utility GIS far exceed the costs in many cases.

Although there are many similarities between the use of GIS in telecommunications and other utilities, there are also significant differences. It is generally accepted that telecommunication networks offer the most challenging environments for implementing

GIS. In part this is because of the very large size of telecommunication networks and therefore the tasks are often huge and substantial sums of money are at stake. It is also because telecommunication networks are very difficult to model in current GIS. The complexity of the networks (many links along a single fibre and many cables in a single duct) and the large number of devices also add to the difficulty. Carolyn Fry (Chapter 58) reviews a range of telecommunications applications and highlights the impact of deregulation, the availability of new technologies (such as fibre optic cables, more efficient terrestrial broadcasting, and satellites offering greatly increased bandwidth), and the rise in customer expectations as key drivers for change. Creating and maintaining up-to-date records on network infrastructures, engineering works, buildings, transport routes, and customers, has become vital for the survival of telecommunication networks. More recently, as the telecommunications market has become more competitive, GIS is also being used in marketing (e.g. to locate new customers), in cellular trouble call analysis (e.g. to map areas with poor signal), and in customer care (e.g. to determine the impact of planned maintenance programmes).

From the very start of GIS, transportation has been a fertile area for GIS applications. As Nigel Waters shows (Chapter 59), the overwhelming focus on networks presents many interesting problems for GIS. Among these is the need to manage transportation infrastructure and associated assets, and to provide efficient and effective solutions to network analysis problems. Most local and/or regional governments in the developed world have large departments devoted to managing and maintaining highways. They are concerned with the road and pedestrian pavements as well as associated street furniture (all the fixed features of roads, including rubbish bins, traffic lights, pavement markings, and even roadside verges). A specific purpose data structure, called 'dynamic segmentation' or linear referencing, has been developed to handle such information. The field of transportation logistics has the potential to make massive savings in the operating costs of public and private organisations. From routing pizza delivery vehicles, to rubbish and postal collections, the applications seem endless. Underlying this work are network analytical solutions such as shortest path, vehicle routing, spatial interaction modelling, and urban transport

modelling. The characteristics of these models and their applications are described in this chapter.

In the next chapter Thomas Cova (Chapter 60) shows how GIS can be used in the field of emergency management. Many organisations have a mandate to prepare emergency response plans and to act quickly following an emergency event. Traditionally, plans have been developed manually and many are remarkably unsophisticated. Cova's review illustrates how GIS can be used to create comprehensive emergency management strategies focusing on mitigation, preparedness, response, and recovery. GIS have been used in emergency management because of the capabilities they offer for: integrating disparate data; modelling everything from the possible impact of hazards to the best evacuation routes under rapidly changing circumstances; recording the location of assets and resources (e.g. food and building supplies, emergency services, and specialist personnel such as doctors); and educating people about the possible hazards in an area and any necessary response plans. In spite of the great benefits of applying GIS to emergency management it seems surprising that they are not more widely used. This probably reflects a lack of awareness and, paradoxically, given the massive potential savings of lives and resources, the cost of setting them up.

Peter Dale and Robin McLaren (Chapter 61) take a look at the application of GIS in land administration. They examine the technical, institutional, organisational, and business issues associated with implementing a national system of land administration. One of the features of this chapter is that it discusses the application of many of the technical and management issues discussed at length in the previous Parts of this Book. As the authors point out, many countries have large, sophisticated, mature systems of land administration. These go by various titles, such as (using English translations as necessary) the Austrian Database of Real Estates, the Hungarian National Land Registration System, the Swedish Land Data Bank System, the UK National Land Information Service, and the Australian Land Information System. Some of these have been developed using private funding according to the non-interventionist model (e.g. the US Realtors Information System), others use public money and a centralist model (the Swedish Land Data Bank System), while some emerge as a hybrid of the two (the Dutch National Cadastre). Although the early systems were predominantly text-based, more recent developments have seen a significant move to incorporating spatial information as a core part of such systems. In the simple systems this may mean incorporation of a land parcel centroid, but the more sophisticated include a detailed description of the parcel boundary and its relation to other features. Not only do such facilities improve the accuracy and precision of registration, they also offer the scope for creating maps for use in land transactions, inventory assessments, taxation, and so forth. Land administration systems can assist enormously in the process of data integration and the creation of a truly national land (geographical) information system.

Closely related to land administration and management, and often to be found in the same operational department, is the function of urban planning. As Tony Yeh notes (Chapter 62), GIS is a well-proven operational and affordable information system for urban planning. Urban planners use GIS for general administration, development (building) control, and plan making. The first two of these tasks can be regarded as routine operations usually undertaken on a daily basis. For example, applications for new developments need to be logged and their status tracked through the planning approval process. Plan making, on the other hand, is often undertaken less frequently and is regarded as a strategic planning tool. It requires new local and regional plans to be devised. Mapping has proved to be an efficient and effective way to encode and visualise urban planning information. Today, most advanced government planning departments have computerised their operational activities. Increasingly, this involves using a fully integrated GIS for resource inventory, the analysis of existing situations, modelling and projection, the development of planning options, selection of planning options, plan implementation, and plan evaluation, monitoring, and feedback.

Defence agencies were among the earliest users of GIS. The military advantages the technology offers persuaded many governments to invest heavily in advanced research and development projects. Rather surprisingly, this has not led to the widespread adoption of GIS across wide areas of the military today. David Swann argues in Chapter 63 that GIS uptake has been comparatively slow in some areas of the military because of problems of affordability, the excessive burden it places on training, the limited hardware characteristics (the absence of voice input and high disk and display capacity), the need for improved software (especially better terrain analysis, visualisation, and data management), as well as the lack of data (particularly high-resolution imagery and large-scale vector data for very large areas). These limitations have most impact on battlefield-oriented systems. In stark contrast, base plant, barrack, and general mapping applications are well developed. Here, military applications have much in common with the utility, transportation, emergency management, land management, and urban planning applications featured earlier.

The final chapter in this operational GIS subsection is by Prudence Adler and Mary Larsgaard (Chapter 64). It looks at GIS and libraries: both how libraries can facilitate access to GIS technology, geographical data, and geoprocessing, and how GIS can be used to enhance the traditional work of libraries. Within the last few years, libraries of all types – research, public, academic, and special-purpose – have become active users and providers of GIS resources. The interest in GIS has been sparked by a number of factors. First is the utilisation of new technologies such as GIS in support of education, research, and effective access to information resources. Second, collecting, maintaining, preserving, and providing access to spatial resources is not new to libraries, but the advent of GIS has resulted in libraries exploring new approaches to many, if not all, of these 'traditional' library functions. Third, there has been rapid expansion and utilisation of networked services, particularly within the academic sector, as communication and educational tools present new opportunities for libraries to address the information needs of a diverse clientele. These changes are occurring at the same time as libraries are in a state of transition, experimentation, and transformation. It is clear that the library of the future will be increasingly digital and that GIS will have a significant role to play in the cataloguing and visualisation of information.

Reference

LGMB 1993 *Geographic information systems (GIS) case studies. Experiences in geographic information management.* Luton, Local Government Management Board

Introduction

THE EDITORS

The chapters in the previous Section focused on the operational applications of GIS. This section takes as its theme the social and environmental applications of GIS. These chapters look at a series of application areas from a wider perspective, considering the ways GIS has been used in planning, decision support, and modelling environmental and social problems. The applications range in scale from the global (Wilson, GIS and agriculture), to the national (MacDevette et al, reconstruction and redevelopment of South Africa) and the local scale (Larsen, environmental monitoring and assessment of Ringkøbing Fjord, Denmark). Other chapters provide extensive, contemporary reviews of key GIS application areas: health and health care (Gatrell and Senior), politics (Horn), land cover and land-use (Bibby and Shepherd), landscape conservation (Aspinall), agriculture (Wilson), and environmental monitoring and assessment (Larsen). Collectively, these applications demonstrate a wide range of environmental and social uses of GIS.

In the opening chapter, David MacDevette, Robert Fincham, and Greg Forsyth examine the role of GIS in the reconstruction and development programme of South Africa (Chapter 65). Despite the succession of Nelson Mandela's government in 1994, South Africa remains a land of stark contrasts with classical first- and third-world elements. The chapter focuses on the role for GIS in development planning, management, and policy formulation at the national, provincial, and local levels. The examples highlight the role of GIS both in information dissemination and as an adjunct to the planning process, especially in the context of improved decision support. Two points illustrate clearly the issues associated with applying GIS in a third-world context. First, MacDevette et al argue that decision-makers in the Third World prefer to work from maps, because they are particularly effective for 'hands on' activities. Second, they highlight the need for customised mapping conventions which are sensitive to context – citing, for example, the need to avoid inappropriate Western symbology such as blue for rivers.

The health and health-care applications chapter by Tony Gatrell and Martyn Senior (Chapter 66) provides a wide-ranging and detailed review of the field of health research. In this chapter the authors review spatial databases for health research using the traditional organising framework of points, lines, areas, fields, and interaction data. They briefly review some of the spatial object transformations necessary to improve the usability of standard datasets (e.g. creating hospital catchment areas from point locations). The remainder of their chapter embraces two main areas of health research: spatial epidemiology (the incidence of disease) and the spatial operation of health-care facilities. In many respects the study of epidemiology poses classic geographical problems which can benefit greatly from the use of GIS. Gatrell and Senior discuss many examples of the use of GIS in visualising, exploring, and modelling the geographical incidence of disease. These range from studies involving the use of global positioning systems (GPS) to study malaria in South Africa, to the creation of improved disease maps in Britain, to the postulated effects of low frequency radio towers on leukaemia in Hawaii. The issues of accessibility and utilisation in the location, configuration, and planning of health-care facilities is addressed in the second part of the chapter. Using examples ranging from urban northwest England to rural Goa, India, the authors show how standard GIS tools like 'drive time analysis' and 'polygon overlay' can be used to improve decision support of the planning process.

Mark Horn (Chapter 67) examines the role of GIS in defining and using electoral areas. His chapter is primarily concerned with the theoretical and practical challenges that arise in the delineation of electoral districts. The significance of geographical elements varies under different electoral frameworks. The creation of equitable electoral districts is an inherently geographical task. It involves consideration of several key factors including: economic, social, and regional interests; patterns of population; physical characteristics of areas (use of natural features such as rivers and mountains to delineate regions); and the paths of any existing areal boundaries. There are now several commercially available redistricting extensions to GIS software systems which can be used to create optimal districts. In the wrong hands, however, they can be subject to the abuses of gerrymandering (manipulation of electoral constituencies to pack and split voters with particular voting preferences) and malapportionment (substantial deviation from equal population representation). The chapter concludes with a case study of the use of redistricting techniques in Australia.

In Chapter 68, Peter Bibby and John Shepherd consider how GIS has been used in studies of land-use and land cover for urban and regional planning purposes. The authors describe how 'land-use' implies a social purpose rather than a set of physical qualities. 'Land cover', on the other hand, is the material which clothes the surface of the Earth, including vegetation, water bodies, and urban areas. Land-use information is vital for the development of planning policy rules and the application of planning policy to individual cases. The discussion and case studies draw on the English experience, although much of it is also applicable to other parts of the world. Bibby and Shepherd deal with examples such as the construction of land-use policy areas, the definition of rural settlements, synthesis of shopping centre areas, and projection of urban growth and change. The discussion highlights the importance of the conceptual distinction between land-use and land cover because the boundaries of geographical objects demonstrably expand and contract as purpose shifts. For this reason the notion that land-use applications of GIS are limited merely to vector mapping of land parcels should be rejected.

Closely related to studies of land-use and land cover is the topic of landscape conservation and Richard Aspinall reviews recent and current GIS usage (Chapter 69). The subject of landscape conservation is considered in its broadest sense in the context of integrated ecological systems that include both physical and human components. The chapter concentrates on three themes: how biodiversity and sustainability provide guiding principles in conservation efforts; the role of geography as an integrative science and 'landscape' as an object for analysis; and the role of ecological understanding for planning and managing specific resources in particular locations and contexts. The traditional approach to landscape conservation has been through establishment of designated sites that are considered independently of their surroundings and within which conservation is treated as the major, or sole, land-use. More recently the wider geographical context of designated sites has begun to be considered and approaches based on cooperation have replaced competition between land-uses. A particularly interesting application of GIS in landscape studies is the analysis of scenes based on viewshed analysis.

Agriculture is an inherently geographical phenomenon and it not surprising that this, together with the extremely large sums of money involved, means that it is a natural application for GIS. In Chapter 70 John Wilson reviews agricultural applications of GIS at global, regional, and local scales. Several projects have been initiated during the past decade in order to build spatially distributed databases that cover continents and, in some instances, the entire globe. These have concentrated on building global-scale topographic, climatic, soil, and land cover databases. GIS techniques have been used for farm-related assessments at national and regional scales for many years. These techniques have been combined with GIS and remotely-sensed data to support assessments of land capability, crop condition and yield, range condition, flood and drought, soil erosion, soil compaction, surface and ground water contamination, pest infestations, weed eradication, and climate change impacts. The number and variety of local agricultural GIS applications have increased dramatically during the past five years. Some applications target individual farms. Most of these field- and subfield-scale applications are connected with precision or site-specific farming, which aims to direct the application of seed, fertiliser, pesticide, and water within fields in ways that optimise farm returns and minimise chemical inputs and environmental hazards. Most site-specific farming systems utilise some combination of GPS receivers, continuous yield sensors, remote sensing, geostatistics, and variable rate treatment applicators with GIS.

In the final chapter in this Section, Lars Larsen examines GIS in environmental monitoring and assessment (Chapter 71). The emphasis of monitoring systems is placed on data collection, pre-processing, and quality control. Analysis systems focus on using tools to manipulate and model data. There are many different types of monitoring systems, most of which automate the process of data collection and (pre-) processing – a task often hidden from the ordinary user. By

introducing GIS as this stage it is possible to utilise mapping functions to display objects such as measurement stations. Because environmental monitoring operations are expensive to set up and maintain, a reduced field-based monitoring programme is often combined with mathematical modelling which can be used to estimate parameters. The author describes the application of environmental monitoring and modelling in the Ringkøbing Fjord, a lagoon area in the western part of Denmark. The monitoring system installed has the dual purpose of documenting the state of the environment and functioning as an alarm system should a critical situation arise. Data collected using this process are fed into a GIS database which organises and stores the data.

APPENDIX: GLOSSARIES AND ACRONYMS

GLOSSARIES

Extensive glossaries of GIS terminology can be found at the following sources:

Books

Arlinghaus S L 1994 *Practical handbook of digital mapping terms and concepts*. Boca Raton, CRC Press
Burrough P A, McDonnell R A 1998 *Principles of geographical information systems*, 2nd edition. Oxford, Oxford University Press
Clarke K C 1997 *Getting started with geographic information systems*. Upper Saddle River, Prentice-Hall
DeMers M N 1996 *Fundamentals of geographic information systems*. New York, John Wiley & Sons Inc.
McDonnell R A, Kemp K 1995 *International GIS dictionary*. Cambridge (UK), GeoInformation International
Padmanbhan G, Leipnik M R, Yoon J 1992 *A glossary of GIS terminology*. NCGIA Technical Papers Series 92–13, Santa Barbara, NCGIA
Worboys M F 1995 *GIS: a computing perspective*. London, Taylor and Francis

World Wide Web sites

There is a huge range of online GIS glossaries, including:
http://www.geo.ed.ac.uk/agidict/welcome.html
http://www.esri.com/base/users/glossary.glossary.html
http://www.lib.berkeley.edu/EART/abbrev.html

ACRONYMS

AACR	Anglo-American Cataloguing Rules
AAG	Association of American Cartographers
ABS	Australian Bureau of Statistics
ACIC	Automated Cartographic Information Center
ACID	Atomicity; Consistency; Isolation; Durability
ACM	Association of Computing Machinery
ACMLA	Association of Canadian Map Libraries and Archives
ACSM	American Congress on Surveying and Mapping
ADL	Alexandria Digital Library
ADT	abstract data type
AEC	Australian Electoral Commission
AGI	Association for Geographic Information
AGILE	Association of Geographic Information Laboratories in Europe
AGNPS	agricultural non-point source
AI	artificial intelligence
ALA	American Library Association
ALE	Association Liègeoise d'Electricité
AM/FM	automated mapping/facilities management
AML	Arc Macro Language
ANC	African National Congress
ANN	artificial neural network
ANSI	American National Standards Institute
ANZLIC	Australia and New Zealand Land Information Council
API	application programming interface
AR	augmented-reality/arc routing
ARMGS	autonomous route management and guidance system
ASAP	All-hazard Situation Assessment Programme
ASPRS	American Society of Photogrammetry and Remote Sensing
ATCS	advanced travel conditions system
ATKIS	Amtliches Topographisch-Kartographisches Informations-system
ATM	asynchronous transfer mode
ATOS	advanced travel orientation system
ATPS	advanced trip planning system
ATSR	Along-Track Scanning Radiometer
AUSLIG	Australian Surveying and Land Information Group
AVDS	Automatic Vehicle Dispatch System
AVHRR	Advanced Very High Resolution Radiometer
AVI	automatic vehicle identification
AVIRIS	advanced visible and infrared imaging spectrometer
AVLS	automatic vehicle location system
AVMS	automatic vehicle monitoring system
AVNS	automatic vehicle navigation system

AZP	automated zone design problem
BCS	British Cartographic Society
BEAR	base-wide environmental analysis and restoration
BEV	Bundesant für Eich- und Vermessungswesen (Austria)
BFKG	Bundesamst für Kartographie und Geodäsie
BIH	Bureau International de l'Heure
BLG	binary line generalisation (tree)
BLPU	(British Standard) basic land and property unit
BMDP	BioMedical Data Processor
BMS	Battlefield Management System
BPI	bits per inch
BRDF	bidirectional reflectance distribution function
BSI	British Standards Institution
BSP	binary space partitioning
BT	British Telecommunications
CAD	computer-aided design
CADD	computer-aided design and drafting
CAG	Canadian Association of Cartographers
CAM	computer-aided mapping
CAPDU	Canadian Association of Public Data Users
CARL	Canadian Association of Research Libraries
CASE	computer-assisted software engineering
CBRED	Central Board for Real Estate Data (Sweden)
CCD	census collector district
CD-ROM	compact disk, read-only memory
CDV	cartographic data visualiser
CEM	comprehensive emergency management
CEN	Comité Européen de Normalisation
CEOS	Committee on Earth Observation Systems
CERCO	Comité Européen des Responsables de la Cartographie Officielle
CGA	colour graphics adapter
CGIA	California Geographic Information Association; also Center for Geographic Information and Analysis
CGIS	Canadian Geographic Information System
CHEST	Combined Higher Education Software Team
CHGIS	critical history of GIS

CI	computational intelligence
CIS	Customer Information System
CLI	Canadian Libraries Initiative
CMLS	chemical movement through layered soils
CNES	Centre National d'Etudes Spatiales
CNIG	Comité de l'Information Géographique
COBOL	common business-oriented language
CODASYL	Conference on Data Systems and Language
COGO	coordinate geometry
COM	common object model
CORBA	common object request broker architecture
COTS	commercial-off-the-shelf
CPD	Central Postcode Directory
CPGIS	Chinese Professionals in GIS
CPR	Continuing Property Records
CPS	Collaborative Planning System
CPU	central processing unit
CRC	class, responsibilities, and collaborators
CRESP	crisis and response prototype (NATO)
CRT	cathode ray tube
CSCW	computer-supported cooperative work
CSEP	cumulative seasonal erosion potential
CSIR	Council for Scientific and Industrial Research (S Africa)
CSIRO	Commonwealth Scientific and Industrial Research Organisation
CSR	complete spatial randomness
CTI	compound terrain index
C3I	command control and communication information
CTS	conventional terrestrial reference system
DBTG	database task group
DBMS	database management system
DCDB	digital cadastral database
DCM	digital cartographic model
DCP	distributed computing platform
DCW	digital chart of the world
DED	district electoral division
DEM	digital elevation model
DEMON	digital elevation model networks
DGIWG	Digital Geographic Information Working Group
DGPS	differential global positioning system
DIF	directory interchange format
DIGEST	digital geographic information exchange standard

DIME	dual independent map encoding
DLG	digital line graph
DLI	Data Liberation Initiative
DLL	dynamic link library
DLM	digital landscape model
DN	digital number
DoD	Department of Defense (US)
DOE	Department of Environment (UK)
DOP	dilution of precision
DOS	disk operating system
DOT	Department of Transportation
DPI	dots per inch
DPW	digital photogrammetric workstation
DSP	depository services programme
DSS	decision support system
DTM	digital terrain model
DXF	digital exchange format
EASD	Empowerment for African Sustainable Development
ECDIS	electronic chart display
ECU	European currency unit
ED	enumeration district
EDA	exploratory data analysis
EDAMS	Electoral Distribution and Mapping System
ED-50	European Datum 1950
EDIS	economic development information system
EEA	European economic area
EGA	enhanced graphics adapter
EIA	environmental impact assessment
EIS	executive information system
ENV	(European standards)
EOS	Earth observing system
EPIC	erosion productivity impact calculator
EPZ	emergency planning zone
ERA	entity–relationship–attribute
ERS	European resource satellites (of the European Space Agency and of Japan)
ERTS	Earth resources technology satellite
ESDA	exploratory spatial data analysis
ESRI	Environmental Systems Research Institute Inc.
ESTDM	event-based spatio-temporal data model
ETRS	European terrestrial reference system
EU	European Union
EUROGI	European UmbRella Organisation for Geographic Information
FACC	feature and attribute coding catalogue

FAO	Food and Agriculture Organisation (United Nations)
FDDI	fibre distributed data interface
FDLP	Federal Deposit Library Program
FEMA	Federal Emergency Management Agency (US)
FEPD	Forum for Effective Planning and Development (S Africa)
FGDC	Federal Geographic Data Committee (US)
FHWA	Federal HighWay Administration
FINDAR	facility for interrogating the National Directory of Australian Resources
FIPS	Federal Information Processing Standard
FMV	fuzzy membership value
FOIA	Freedom of Information Act
FORTRAN	formula translation
FRAMME	facilities rule-based application model management environment
GADS	geodata analysis and display system
GAM	geographical analysis machines
GAP-tree	generalised area partitioning
GATT (1994)	General Agreement on Tariffs and Trade, 1994
GBF/DIME	geographic base file/dual independent map encoding
GCEM	geographical correlates exploration machine
GCM	global circulation model
GCP	ground control point
GDF	geographic data files
GDIS	geodemographic information system
GENIE	Global Environmental Network for Information Exchange (UK)
GFS	geo-facilities information system
GGI&S	Global Geospatial Information and Services
GIC	geospatial information community
GICC	Geographic Information Coordinating Council (North Carolina)
GILS	Government Information Locator Service
GISDK	GIS developers kit
GISRUK	GIS Research UK
GISy/GISc	subdivisions of GIS: systems/science
GKS	graphics kernal system
GLEAMS	groundwater loading effects of agricultural management systems
GLIS	global land information system

GNU	Government of National Unity (South Africa)	IERS	International Earth Rotation Service
GOTS	government-off-the-shelf	IETF	Internet Engineering Task Force
GPAC	Government Performance Audit Committee (US, North Carolina)	IFLA	International Federation of Library Associations
GPO	Government Printing Office (US)	IFOV	instantaneous field-of-view
GPS	global positioning system	IGBP	International Geosphere Biosphere Programme
GRASP	greedy randomised adaptive search procedure	IGES	Initial Graphic Exchange Standard
GRASS	geographic resource analysis support system	IGN	Institut Géographique National
GRE	ground resolution element	IHO	International Hydrographic Organisation
GRIA	global regional interchange algorithm	IMap	intelligent mapping, accounting, and provisioning
GRIDS	geographic roadway information display system	IP	Internet Protocol
GRSA	GIS-relevant spatial analysis	IRMC	Information Resources Management Commission (US; formerly the Information technology Council)
GSDI	global spatial data infrastructure		
GUI	graphical user interface	ISBD	international standards for bibliographic description
GVD	generalised Voronoi diagram		
HARN	statewide high accuracy reference network	ISBN	international standard book number
HAS	highway analysis system	ISDN	integrated services digital network
HCI	human computer interaction	ISM	industry structure model
HDT	hierarchical Delaunay triangulation	ISO	International Standardisation Organisation
HIPSS	health information for purchaser planning system	ISPRS	International Society for Photogrammetry and Remote Sensing
HIRIS	High Resolution Imaging Spectrometer	ITA	interactive territory assignment
HMD	head-mounted display	ITS	intelligent transportation system
HMLR	Her Majesty's Land Registry	ITU	International Telecommunications Union
HPC	high performance computing		
HPF	highly parallel FORTRAN	JERS	Japanese Earth Resources Satellite
HRTF	head-related transfer functions	KA	knowledge acquisition
HRV	high resolution visible	KB	kilobyte
HSR	hidden surface removal	KBLIMS	knowledge-based land information management system
HTML	hypertext markup language		
http	hypertext transmission protocol	KHz	kilohertz
ICA	International Cartographic Association	LAI	leaf area index
		LAN	local area network
ICT	information and communication technology	LANDSAT	LAND resources assessment SATellite system (US)
IDGIS	interactive distributed geographical information system	LAPC	Land and Agriculture Policy Centre
		LDBS	land data bank system
IDNDR	International Decade for Natural Disaster Reduction	LEACHM	leaching and chemistry estimation
		LGMB	Local Government Management Board
IDW	inverse distance weighted interpolation		
IEC	International Electrotechnical Commission	LIDARS	light detection and ranging system
		LIS	land information system
IEEE	Institute of Electrical and Electronics Engineers	LISA	local indicators of spatial analysis
		LISS	linear imaging self-scanning system

LOD	level of detail management algorithm
LP	linear programming
LP/IP	linear programming/integer programming
LUNR	land use and natural resource inventory (State of New York)
LWIR	long wavelength infrared
MAGI	Maryland automated geographic information
MAN	metropolitan area network
MAP	map analysis package
MAPS	Montana agricultural potential system
MARBI	machine-readable bibliographic information
MARC	machine-readable cataloguing
MAUP	modifiable areal unit problem
MB	megabyte, or mbyte
MBR	minimum bounding rectangle
MBS	minimal bounding sphere
MC&G	mapping, charting, and geodesy
MCE	multicriteria evaluation
MCNC	Microelectronics Center of North Carolina
MEL	Master Environmental Library
MERIS	medium resolution imaging spectrometer
MESSR	multispectral self-scanning radiometer
MHZ	megahertz
MIMD	multiple instruction, multiple data
MIPS	million instructions per second
MIS	management information system
MISR	multi-angle imaging spectrometer
MIT	Massachusetts Institute of Technology
MLRA	major land resource area
MODIS	moderate resolution imaging spectrometer
MOPTT	Ministry of Post and Telecommunications (Saudi Arabia)
MOS	marine observation satellite
MPI	message passing interface
MPO	multi product operations
MPP	massively parallel processor
MSC	Mapping Science Committee
MSR	Microwave Scanning Radiometer
MSS	multispectral scanner
MST	minimum spanning tree
MVA	multiple view angle
NAD-27	North American Datum 1927
NAD-83	North American Datum 1983
NAFTA	North American Free Trade Agreement

NAPA	National Academy of Public Administration
NRSA	National Remote Sensing Agency
NASA	National Aeronautical and Space Administration (US)
NASDA	National Space Development Agency (Japan)
NATO	North Atlantic Treaty Organisation
NATSGO	national soil geographic database
NAVSTAR	navigation satellite timing and ranging system
NCGA	National Computer Graphics Association
NCGIA	National Center for Geographic Information and Analysis
NCIH	North Carolina Information Highway
NC-REN	North Carolina Research and Education Network
NCSA	National Centre for Supercomputing Applications
NDVI	normalised difference vegetation index
NESPAL	National Environmentally Sound Production Agriculture Laboratory (US)
NFM	network flow model
NGDF	national geospatial data framework
NGS	National Geodetic Survey
NISO	National Information Standards Organisation (US)
NLIS	national land information system
NLS	National Land Survey
NMA	National Mapping Agency
NMAS	national map accuracy standards
NMCA	National Mapping and Cadastre Agency
NMO	National Mapping Organisation
NNI	(Netherlands Standardisation Organisation)
NNSDP	National Nutrition and Social Development Programme (S Africa)
NNWG	Natal Nutrition Working Group
NOAA	National Oceanographic and Atmospheric Administration (US)
NP-Hard	non-deterministic polynomial-hard
NPR	national performance review
NRIS	national resource information system
NRSA	National Remote-Sensing Agency (India)
NSF	National Science Foundation (US)
NSDI	national spatial data infrastructure
NTF	national transfer format
OCLC	online computer library centre

ODMG	Object Database Management Group
OECD	Organisation for Economic Cooperation and Development
OEEPE	(European Organisation for Experimental Photogrammetric Research)
OGC	Open GIS Consortium
OGDC	Office of Geographic Data Coordination
OGIS	open geodata interoperability specification
OGM	open geodata model
OHD	'off the head' display
OID	object identifier
OLE	object linking and embedding
OLE/COM	object linking and embedding/component object model
OMB	Office of Management and Budget (US)
OMG	Object Management Group
OMT	object modelling technique
ONS	Office of National Statistics (UK)
OODB	object-oriented database
OODBMS	object-oriented database management system
OOP	object-oriented programming
OOPL	object-oriented programming language
OO2VD	ordered order-2 Voronoi diagram
OpenGIS	Open Geodata Interoperability Specification
OPS	optical sensor
ORNL	Oak Ridge National Laboratory
OS	operating system
OS (GB)	Ordnance Survey (Great Britain)
OS (NI)	Ordnance Survey (Northern Ireland)
OSM	OGIS Services Model
O2VD	order-2 Voronoi diagram
OVD	ordinary Voronoi diagram
OVP	ordinary Voronoi polygon
PACE	prediction and coverage estimation
PAIS	Prototype Allied Command Europe Intelligence System
PC	personal computer
PCT	personal construct theory
POES	polar orbiting environmental satellite
POLDER	polarisation and directionality of the Earth's reflectances
POST	Parliamentary Office for Science and Technology
PPS	precise positioning service

PR	proportional representation
PRN	pseudo random noise
PROM	programmable read-only memory
PSNP	primary school nutrition programme
PSS	planning support system
PSTN	public standard telephone network
QA/QC	quality assurance/quality control
QTM	quaternary triangular mesh
RAM	random access memory
RDBMS	relational database management system
RDP	Reconstruction and Development Programme (S Africa)
RFP	request for proposals
RICS	Royal Institution of Chartered Surveyors
RIN	realtors information network
RINEX	receiver independent exchange
RMSE	root mean square error
ROM	read-only memory
RS	remote sensing
RSG	regular square grid
RST	regulated spline with tension
RUSLE	revised universal soil loss equation
RVIS	reliability visualisation tool
SA	selective availability
SAIF	spatial archive and interchange format (Canadian)
SAR	synthetic aperture radar
SAS	statistical analysis system
SBC	Santa Barbara climate
SCADA	supervisory control and data acquisition
SCDF	spatial cumulative distribution function
SDA	spatial data analysis
SDE	spatial data engine
SDSS	spatial decision support system
SDTS	spatial data transfer standard
SERDP	strategic environmental research and development program
SFWMD	South Florida Water Management District
SIGMOD	special interest group on the management of data
SIMD	single instruction, multiple data
SINES	spatial information enquiry service
SLC	spatial location code
SLM	structure line model
SLOSH	sea, lake, and overland surge from hurricane

SLR	satellite laser ranging
SMI	Strategic Mapping Incorporated
SMMT	Society of Motor Manufacturers and Traders
SMSA	standard metropolitan statistical area (US)
SNR	signal-to-noise ratio
SOC	soil organic carbon
SOHO	small office/home office
SoLIM	soil land inference model
SPA	shortest path analysis
SPOT	Satellite Pour l'Observation de la Terre
SPS	standard positioning service
SPSS	statistical package for the social sciences
SQL	structured query language
SRAM	static random access memory
SSC	Swiss Society of Cartography
SSM	system status management
STAC	space–time attribute creature
STATSGO	State soil geographic database (US)
STEP	standard for the exchange of product model data
STV	single tranferable voting
SWIR	short wavelength infrared
TAZ	traffic analysis zone
TIGER	topologically integrated geographic encoding and referencing
TIMS	transportation information management system
TIN	triangulated irregular network
TM	thematic mapper
TLP	Telefones de Lisboa e Porto
TPS	thin plate spline
TRIPS	trade-related aspects of intellectual property rights
UCC	uniform commercial code
UCGIS	University Consortium for GIS
UDMS	Urban Data Management Society (Europe)
UGISA	University Geographic Information System Alliance
UKMARC	United Kingdom machine-readable cataloguing
UNCRD	United Nations Centre for Regional Development
UNDRO	Office of the United Nations Disaster Relief Coordinator
UN ECE	United Nations Economic Commission for Europe
UNESCO	United Nations Economic, Social, and Cultural Organisation
UNIDROIT	International Institute for the Unification of Private Law
UNITAR	United Nations Institute for Training and Research
UPRN	unique property reference numbers
URISA	Urban and Regional Information Systems Association
URL	uniform resource locator
USBC	United States Bureau of Census
USGS	United States Geological Survey
USMARC	United States machine-readable cataloguing
UCT	universal time coordinated
UTM	universal transverse mercator
UTMS	urban transportation model system
VAR	value added reseller
VE	virtual environment
VET	virtual environment theatre
VGIS	virtual geographical information system
ViSC	visualisation in scientific computing
VLBI	very long baseline interferometry
VNIR	visible and near-infrared radiometer
VOA	Valuation Office Agency
VR	virtual reality
Vro	vehicle routing
VRT	variable rate treatment
VTIR	visible and thermal infrared radiometer
WAN	wide area network
WCED	World Commission on Environment and Development
WGS-84	World Geometric System 1984
WIM	worlds in miniature
WIMP	windows, icons, mice, and pointers
WLS	weighted least squares
WOFOST	World Food Studies
WTO	World Trade Organisation
XGA	extended graphics adapter
ZDES	zone design system
ZIP	zone improvement plan

CUSTOMER NOTE: **IF THIS BOOK IS ACCOMPANIED BY SOFTWARE, PLEASE READ THE FOLLOWING BEFORE OPENING THE PACKAGE.**

This software contains files to help you utilize the models described in the accompanying book. By opening the package, you are agreeing to be bound by the following agreement:

This software product is protected by copyright and all rights are reserved by the author, John Wiley & Sons, Inc., or their licensors. You are licensed to use this software on a single computer. Copying the software to another medium or format for use on a single computer does not violate the U.S. Copyright Law. Copying the software for any other purpose is a violation of the U.S. Copyright Law.

This software product is sold as is without warranty of any kind, either express or implied, including but not limited to the implied warranty of merchantability and fitness for a particular purpose. Neither Wiley nor its dealers or distributors assumes any liability for any alleged or actual damages arising from the use of or the inability to use this software. (Some states do not allow the exclusion of implied warranties, so the exclusion may not apply to you.)